Advances in Neurobiology

Volume 48

Advances in Neurobiology covers basic research in neurobiology and neurochemistry. It provides in-depth, book-length treatment of some of the most important topics in neuroscience including molecular and pharmacological aspects. The main audiences of the series are basic science researchers and graduate students as well as clinicians including neuroscientists (neurobiologists and neurochemists) and neurologists. *Advances in Neurobiology* is indexed in PubMed, Google Scholar, and the Thompson Reuters Book Citation Index.

Ege T. Kavalali
Editor

Nano-organization of the Synapse

From Structure to Function

Springer

Editor
Ege T. Kavalali
Department of Pharmacology and Vanderbilt Brain Institute
Vanderbilt University
Nashville, TN, USA

ISSN 2190-5215 ISSN 2190-5223 (electronic)
Advances in Neurobiology
ISBN 978-3-032-12593-4 ISBN 978-3-032-12594-1 (eBook)
https://doi.org/10.1007/978-3-032-12594-1

This Springer imprint is published by the registered company Springer Nature Switzerland AG
The registered company address is: Gewerbestrasse 11, 6330 Cham, Switzerland

Preface

Synapses are the fundamental units of information processing in the nervous system. In many ways, synapses in neuroscience are analogous to atoms in physics. Just as atoms were originally proposed by Democritus as the basic building blocks of matter, synapses have long been considered indivisible functional entities. However, while the twentieth century saw nuclear and particle physics reveal the complex subatomic structure of atoms, recent advances in neurobiology have uncovered a functionally diverse subsynaptic organization within just a few hundred nanometers of the synaptic junction.

This deeper insight into synaptic architecture has been made possible by the development of high-resolution electrophysiology and imaging techniques, allowing researchers to investigate synaptic function at the level of individual synapses and to visualize synaptic structures in unprecedented detail. These studies have challenged traditional views, revealing that distinct forms of neurotransmission can be independently regulated and often transmit parallel, non-overlapping signals to the postsynaptic apparatus—serving a range of diverse functions.

The inspiration for this book originated at the Gordon Research Conference on Synaptic Transmission held in 2016. That early vision took shape thanks to Springer Nature starting in 2023, culminating in the current volume. Many of the contributors to this collection were speakers at the original conference, where the initial findings on synaptic nanostructure were shared. Since then, the scope of this work has expanded far beyond the discussions of that meeting.

We are fortunate to include chapters by leading investigators at the forefront of synaptic research. This volume spans a wide spectrum of synapse biology—from atomic-level structural analyses to studies of presynaptic components such as vesicle pools and active zones, transsynaptic adhesion complexes, and postsynaptic structures explored using super-resolution imaging. It also highlights advances in dynamic functional imaging and electrophysiology. Even so, this volume can only serve as a snapshot of a rapidly evolving field, where new insights and discoveries emerge on a monthly basis.

I would like to express my deep gratitude to all my colleagues who have collaborated with me on single-synapse research over the past two decades. I am also sincerely thankful to all the contributors to this volume, who have brought their unique perspectives to bear on fundamental questions about synaptic structure and function. We hope readers will find this collection as inspiring and valuable as we did while assembling it.

Nashville, TN, USA
April 25, 2025

Ege T. Kavalali

Contents

1 **Nano-organization of the Synapse: A Requisite Framework for Synaptic Signaling** 1
Ege T. Kavalali

2 **Recent Structural Insights into the Molecular Architecture of Synapses** 11
Axel T. Brunger, Richard G. Held, Yousuf A. Khan, Jeremy Leitz, Jiahao Liang, Chuchu Wang, and K. Ian White

3 **Formation, Organization, and Maintenance of Synaptic Vesicle Pools: From Molecules to Structure and Dynamics** 39
Natali L. Chanaday

4 **Active Zone: Linking Resolution Levels of Microscopic Modalities** 79
Maksim Galkov, Paulina Nemcova, Dirk Dietrich, and Susanne Schoch

5 **Synaptic Cell Adhesion: A Functional Perspective** 107
Kelly Honkanen and Richard C. Sando

6 **Synaptic Cell Adhesion: A Structural Perspective** 151
Sumit J. Bandekar, Szymon P. Kordon, and Demet Araç

7 **Scales of Postsynaptic Nanostructure: Molecules, Nanoclusters, and Nanodomains** 191
Martina Damenti, Aaron D. Levy, and Thomas A. Blanpied

8 **Functional Nano-segregation of Distinct Forms of Neurotransmission** 217
Natalie J. Guzikowski, Camille S. Wang, and Ege T. Kavalali

9 **Optical Approaches to Dissect the Structure and Dynamics of the Synapse at Nanoscale Resolution** 259
Jongyun Myeong and Vitaly A. Klyachko

10 Visualizing Postsynaptic Density in Excitatory Synapses with Electron Tomography 289
Rong Sun and Qiangjun Zhou

11 Roles of Glia in Synapse Nano-organization 331
Young-Jin Choi and Won-Suk Chung

Correction to: Recent Structural Insights into the Molecular Architecture of Synapses C1
Axel T. Brunger, Richard G. Held, Yousuf A. Khan, Jeremy Leitz, Jiahao Liang, Chuchu Wang, and K. Ian White

Index 355

About the Editor

Ege T. Kavalali was born and raised in Istanbul, Turkey. He studied Electrical Engineering at Boğaziçi University in Istanbul. In 1995, he received his Ph.D. degree in Biomedical Engineering from Rutgers University. In 1999, he completed his postdoctoral studies at Stanford University in the Department of Molecular and Cellular Physiology. He then joined the faculty at the Center for Basic Neuroscience at UT Southwestern Medical Center at Dallas where he rose to the rank of Professor in the Departments of Neuroscience and Physiology. He also held the Rosewood Corporation Chair, in addition to serving as the Chair of the Neuroscience Graduate Program between 2008 and 2016. In 2018, he joined the Department of Pharmacology at Vanderbilt University as a Professor and the William Stokes Chair in Experimental Therapeutics. He currently serves as the Chair of the Department.

Chapter 1
Nano-organization of the Synapse: A Requisite Framework for Synaptic Signaling

Ege T. Kavalali

Abstract Synapses are the fundamental units of communication and information processing in the nervous system. They show remarkable functional autonomy, as well as speed of signaling, bidirectional plasticity, and diversity. Synapses shape complex network functions, such as learning, memory, pattern separation, pattern completion, and many others. Synaptic dysfunction is responsible for a wide range of neurological and psychiatric diseases. Many of these diseases are emerging as synaptopathies, but the precise disease mechanisms are unknown. Given their micron-scale small size and seemingly compact structure, synapses were generally thought to be functionally indivisible structures. However, with the advent of high-resolution electrophysiology and imaging techniques, it is increasingly clear that synapses harbor functionally specialized nanomodules. This expanding nanobiology of the synapse provides a new perspective on synaptic signaling. The novel insight gained from this work is critical to understand disease mechanisms and to guide the development of appropriate therapeutic strategies for major brain diseases ranging from neurodegenerative and neurodevelopmental disorders to neuropsychiatric disorders such as depression and schizophrenia.

Keywords Multidisciplinary research · Fast neurotransmission · Slow neurotransmission · PET imaging · Rapid antidepressant action · Synapse nano-organization

1.1 Evolution of Synapse Research

In a lecture delivered to Cambridge audience in 1959, the physicist-turned novelist Charles Percy (C.P.) Snow lamented the drift between the cultures of humanities and sciences, despite their shared origins embodied in polymaths that dominated

E. T. Kavalali (✉)
Department of Pharmacology, Vanderbilt University, Nashville, TN, USA

Vanderbilt Brain Institute, Vanderbilt University, Nashville, TN, USA
e-mail: ege.kavalali@vanderbilt.edu

E. T. Kavalali (ed.), *Nano-organization of the Synapse*, Advances in Neurobiology 48,
https://doi.org/10.1007/978-3-032-12594-1_1

scientific inquiry from the antiquity until the twentieth century (Snow 1998). In science itself, scientists' tendency toward overspecialization in a narrow domain has concerned scholars such as the renowned physicist Erwin Schrödinger, who also conveyed his frustration in a lecture in Dublin in 1950 (Schrödinger 1996). Even within a subdiscipline such as neuroscience, overspecialization among scientists can create a similar bifurcation of cultures. While focus is a commendable aspect of research, scientists often ignore developments beyond their immediate domain even in neighboring fields at their own peril, as most breakthroughs occur not within a field or a subfield but at junctions where they border other domains of inquiry. It is often forgotten that fields, disciplines, and specializations are human creations, whereas nature functions in a manner that is oblivious to our self-imposed scholarly divisions.

Studies on synaptic transmission require such a broad interdisciplinary framework spanning atomic-level structural biology of synaptic proteins to their organismal level functions underlying cognition and behavior. This multidisciplinary effort is exacerbated by the fact that investigation of mechanisms underlying the secretion and actions of neurotransmitters and neuropeptides spans various neuroscience disciplines. Early studies on chemical neurotransmission largely depended on testing the impact of psychoactive compounds on behavior (Carlsson 2003). This effort led to the identification of several neurotransmitters, where in many instances the neuromuscular junction and the peripheral nervous system were used to determine their direct actions assessed via electrophysiological or mechanical measures. The early success of drugs that target the monoaminergic system in the treatment of neuropsychiatric and neurological conditions, such as selective serotonin reuptake inhibitors for the treatment of depression or antipsychotics for the treatment of psychosis, represented an important advance in neuropsychopharmacology research. These findings largely shifted the field toward a focus in "slow" neurotransmission (Greengard 2001), where "fast" chemical neurotransmission was more typically being viewed as a relatively passive target of neuromodulation. While studies on slow neurotransmitters that primarily activate G-protein-coupled receptors and elicit downstream second messenger signaling have provided important information to the field, there are caveats in some of the assessments regarding their temporal and spatial specificity. For example, studies on slow neurotransmitters often overemphasize bulk electrochemical or chemical measurements, summating microscale signaling events over large volumes and longer time scales, potentially overlooking key properties of individual signaling events. In contrast, neurophysiologists studying synaptic plasticity, particularly those focused on Hebbian forms of plasticity—long-term potentiation and long-term depression—largely focused on fast neurotransmission carried out by the excitatory neurotransmitter glutamate.

The initial work elucidating the molecular mechanisms of fast neurotransmission in the brain took advantage of high-resolution electrophysiological methods to measure this form of synaptic communication. These studies largely focused on glutamatergic and GABAergic transmission as a "Rosetta Stone" to interrogate mechanisms of fast regulated secretion. The precision of these measurements was enabled by their dependence on ionotropic receptors that physiologically respond to released neurotransmitters within a millisecond. These studies elucidated mechanisms of fast neurotransmission in exquisite molecular and biophysical detail (Südhof 2013). To date, the application of similar approaches to study the release of monoaminergic transmitters has been limited to a few accessible preparations. For instance, catecholamine release from chromaffin cells has been extensively studied as a biophysical model, and the underlying principles of monoamine release have been found to mostly overlap with fast glutamatergic synapses in the central nervous system. A key reason that hinders work on monoaminergic or neuropeptide neurotransmission is that it is hard to measure these release events due to their non-ionotropic basis and limited amenability to high-resolution electrophysiology. This constraint compels reliance on techniques with limited temporal and spatial resolution. Moreover, perspectives gained from approaches with low temporal and spatial resolution are then often applied to fast neurotransmitters, where, for instance, glutamate signaling has been studied in the same way as volume transmission, leading to inferences from bulk measurements rather than spatially and temporally precise readouts. This is exemplified by initial work using techniques such as microdialysis that possessed chemical specificity—an early necessity in a field where the chemical identities of the fast amino acid neurotransmitters were in doubt—but lacked the necessary resolution to elucidate spatial and temporal aspects of signaling.

1.2 Recent Findings Requiring a New Framework for Synaptic Signaling

In the last decade, a key discovery emerging from the clinic began to change the landscape of neuropsychopharmacology. Clinical studies have demonstrated that administration of a low dose of ketamine, an ionotropic glutamatergic NMDAR (N-methyl-D-aspartate receptor) antagonist, produces rapidly acting antidepressant responses in patients with major depressive disorder as well as those resistant to classical antidepressants that target monoaminergic signaling. These findings resulted in a resurgence of clinical and preclinical research on fast glutamatergic synaptic signaling and plasticity mechanisms (Krystal et al. 2024). An important implication of ketamine's rapid clinical efficacy is that symptoms of depression can be alleviated in patients with a long history of resistance to conventional treatments within hours, linking temporally and spatially precise synapse-specific action of glutamate to the rapid regulation of mood. However, these initial clinical findings also presented a paradox. Decades of studies on Hebbian forms of synaptic plasticity have unequivocally shown that activity-dependent regulation of synaptic strength largely relies on the activation of NMDA receptors via coincidence of presynaptic glutamate release and postsynaptic activity. Therefore, the rapid behavioral action of an NMDA receptor blocker necessitated revisiting this tenet with a closer look at mechanisms of fast glutamatergic neurotransmission and how its suppression may lead to plasticity that outlasts the presence of ketamine. This effort identified tonic (spontaneous or activity-independent) NMDA receptor-driven signaling pathways that once inhibited give rise to behavioral responses as well as plasticity that outlasts the initial receptor block. This work brought together several concepts that originated from the studies of fast glutamatergic neurotransmission to the forefront of neuropsychopharmacology. Specifically, the biophysics of NMDA receptors, homeostatic synaptic plasticity, signaling driven by spontaneous neurotransmission, and the metaplasticity mechanisms underlying the long-term effects of ketamine are all increasingly discussed to explain the mechanisms of rapid antidepressant action.

The expanding focus and need for rapidly acting antidepressants facilitate direct communication between the fundamental studies of synaptic transmission and clinical studies that present a strong framework guiding basic research. For example, the rapid antidepressant responses can persist for days to more than a week in some patients and outlast the initial administration and systemic presence of ketamine, implicating a role for synaptic plasticity processes. In addition, memantine, an NMDA receptor blocker with similar properties to ketamine, does not produce antidepressant action necessitating a closer look into the biophysics of NMDA receptor block by the two compounds. Finally, repeated administration of ketamine augments its therapeutic efficacy, suggesting a form of synaptic metaplasticity elicited by the initial administration of the drug. These strong clinical observations require in-depth examination of synaptic mechanisms elicited by ketamine action, while they also enable the construction of a mechanistic model that can be tested systematically to make predictions for the future design of rapidly acting psychiatric medications. Along these lines, emerging evidence suggests unraveling the neurotherapeutic action of psychedelics will likely require a similar level of detailed molecular cellular analysis.

In addition to the renewed focus on fast glutamatergic signaling within the context of mood disorders, the development of optical probes is changing prevailing perspectives on monoaminergic transmission and enabling the same analytical scrutiny afforded to fast neurotransmission (Wu et al. 2022). The molecular manipulation of proteins involved in neurotransmitter release as well as synaptic active zone scaffolds in combination with these novel probes fosters a more precise understanding of monoaminergic transmission. Monoaminergic volume transmission which may provide less specific measurements than synaptic transmission, nevertheless, is governed by similar molecular mechanisms and tightly regulated at spatiotemporal scales similar to fast neurotransmission (Özçete et al. 2024). Studies have shown that a single neurotransmitter such as dopamine may convey multiple signals based on its release kinetics (phasic vs. tonic), form of release (e.g., evoked vs. spontaneous) or proximity, and type of target receptors. The same diversity of action may also be applicable to neurotrophic factors such as Brain-Derived Neurotrophic Factor (BDNF), where its signals may differ depending on the location, source, target, and temporal characteristics of transmission. Recent studies increasingly support the fact that neuromodulatory slow neurotransmitters act akin to a fast neurotransmitter such as glutamate, where their signaling function and information content are dictated by their kinetics, location, and targets rather than merely by their identity. There is hope that techniques such as positron emission tomography

(PET) imaging will bring a level of precision to studies of neurotransmitter signaling in humans, as exemplified by the latest arsenal of probes targeting specific synaptic proteins such as the synaptic vesicle protein SV2A (Krystal et al. 2024).

1.3 Insight into the Organization of the Synapse at Nanoscale

Synaptic contacts can exhibit remarkable plasticity and function as autonomous elements as they possess mechanisms for activity-dependent changes in synaptic strength, local protein synthesis and degradation, and compartmentalized signaling pathways. These features allow individual synapses to adapt and independently respond to specific stimuli. The molecular machinery underlying synaptic function encompasses mechanisms for vesicle fusion and recycling, trafficking of organelles and receptors, and nanoscale calcium signaling events. In the last two decades, high-resolution methods for optical imaging of synaptic vesicle trafficking and visualization of synaptic transmission at single synaptic terminals were developed. In this way, we could focus on the examination of synaptic transmission at the level of individual synaptic contacts, single synaptic vesicles, and postsynaptic receptor clusters. When combined with these advanced functional tools, our insight into the molecular organization of synaptic terminals increased. These functional studies are bolstered by the extensive work that molecularly identified the protein components of synaptic vesicle pools, the active zone, postsynaptic densities, and synaptic cell adhesion complexes.

Recent advancements in super-resolution imaging and cryo-electron microscopy have revealed a complex nano-organization within individual synapses. This organization includes the alignment of neurotransmitter release sites with postsynaptic receptor clusters and the spatial segregation of distinct neurotransmitter release systems. Increasing lines of evidence suggest that, within a single synapse, sites of neurotransmitter release and clusters of postsynaptic receptors are aligned via synaptic cell adhesion molecules that cross the synaptic cleft (Chen et al. 2018; Guzikowski and Kavalali 2021). While this nano-organization is thought to render evoked neurotransmission faster and more precise, it also allows for functional specialization across the active zone, where distinct neurotransmitter release modes can function in parallel and convey diverse signals (Guzikowski and Kavalali 2021). Interestingly, such spatial segregation of distinct forms of neurotransmission at the nanoscale has been observed in invertebrates, suggesting evolutionary conservation of this key feature. This complex nano-organization within individual synapses, along with their parallel signaling capabilities and functional specialization, emphasizes the importance of understanding synaptic function at the nanoscale level to comprehend neural computation and behavior. As discussed above, ketamine's rather anomalous action—triggered via blockade rather than activation of NMDA receptors—became clearer once considered within the functional nano-organization of the glutamatergic synapse, where spontaneous "tonic" and evoked "phasic" forms of transmission occur in parallel and ketamine action on spontaneous signaling is necessary and sufficient to deliver the biochemical signaling required to elicit plasticity (Krystal et al. 2024; Guzikowski and Kavalali 2021). The same framework likely applies to dopamine signaling as the molecular manipulation of active zone components and release machinery has resulted in surprising findings indicating distinct behavioral impacts of different forms of dopamine release (Özçete et al. 2024).

1.4 Toward a Unified Analysis of Chemical Neurotransmission at Multiple Scales

Super-resolution imaging techniques have revolutionized our understanding of synaptic structure and function at the nanoscale level. These advanced imaging methods, combined with novel optical reporters for neurotransmitters, are driving a convergence of previously separate subfields in neuroscience. There is also increasing evidence that fast and slow forms of neurotransmission often coexist within individual synapses as the co-release of monoamines and the fast neurotransmitters glutamate and GABA is observed throughout the brain. These observations are consistent with the premise that release machinery components are precisely arranged at individual synapses facilitating the transmission and processing of multiple signals. This nanoscale organization of synapses has significant implications not only for our understanding of neurotransmission but also for the development of novel neurotherapeutics, as it suggests that different forms of neurotransmission within individual synapses can be selectively modulated.

As our knowledge of synaptic nano-organization expands, several key areas emerge for future research. One key direction is the integration of diverse model systems to understand how despite their vast differences in brain architectures and circuit connectivity, they nevertheless rely on similar neurotransmitter systems to yield analogous behavioral outcomes. Additionally, investigating neurotrophic factors like BDNF, which may act with high temporal and spatial precision at the nanoscale level, is crucial for elucidating their roles in neurotransmission.

Taken together, the merging of essential concepts from synaptic neurobiology, neuropsychopharmacology, and clinical neuroscience will lead to a more comprehensive understanding of neurotransmitter signaling across the nervous system. Key clinical observations, as exemplified by the discovery of rapidly acting antidepressants, will yield new insights or present anomalies that will require reassessment of our established views of neurotransmission and neuronal signaling. This integrated approach will pave the way for breakthroughs in addressing intractable neurological and neuropsychiatric disorders, highlighting the importance of interdisciplinary communication in advancing neuroscience research. Moving forward, neuroscientists will increasingly need to merge concepts and perspectives across multiple fields, seeking synergy among diverse approaches regardless of how unrelated they may seem at first.

This volume seeks to follow this guiding principle by bringing together expert voices from a wide range of backgrounds, all focused on studying synapse organization across multiple scales.

The chapters in this collection span atomic-level structural analyses to studies of presynaptic components such as vesicle pools and active zones, trans-synaptic adhesion complexes, and postsynaptic structures explored using super-resolution imaging. The chapters also highlight the advances in dynamic functional imaging and electrophysiology that help us gain exquisite insights into the functional implications of this nano-organization. The chapter by Axel Brunger and colleagues explores the structural biology of the neurotransmitter release machinery (Brunger et al. 2025), while Natali Chanaday examines the composition and physical localization of synaptic vesicle pools (Chanaday 2025). Susanne Schoch and her team discuss the latest developments in the organization and composition of active zone proteins (Galkov et al. 2025). Chapters by Rick Sando and Demet Arac delve into the functional and structural aspects of synaptic cell adhesion complexes that connect the pre- and postsynaptic sides of the synapse (Honkanen and Sando 2025; Bandekar et al. 2025). These are followed by a contribution from Tom Blanpied and colleagues, who use super-resolution microscopy to analyze the nano-organization of postsynaptic receptors and scaffolds (Damenti et al. 2025). Building on this, the chapter by Guzikowski and colleagues emphasizes the relationship between structural nano-organization and the functional diversity of different forms of neurotransmission (Guzikowski et al. 2025). Vitaly Klyachko and his team present research utilizing optical approaches to dynamically resolve synaptic function at the level of single organelles (Myeong and Klyachko 2025). In parallel, Qiangjun Zhou and colleagues discuss the growing capabilities of cryo-electron microscopy in revealing synaptic structures with remarkable detail (Sun and Zhou 2025). Finally, Won-Suk Chung and his collaborators highlight the crucial roles of glial cells in shaping synaptic nanostructure and signaling (Choi and Won-Suk Chung 2025).

This collection of chapters presents the current landscape of studies aimed at uncovering the inherent complexity of synaptic transmission from the perspective of single synaptic contacts. As this field rapidly develops, it fosters the hope that more in-depth understanding of synaptic structure and function will be instrumental in elucidating the mechanisms of drug action and design of better therapeutics for intractable brain disorders. The insights from this work will also be instrumental for the emerging field of neuromorphic circuit design that replicates synaptic function in hardware with higher fidelity, thus enabling more efficient and capable artificial intelligence networks.

References

Bandekar SJ, Kordon SP, Araç D. Synaptic cell adhesion: a structural perspective. In: Nano-organization of the synapse: from structure to function advances in neurobiology. Springer; 2025. In Press.

Brunger AT, Held RG, Khan YA, Leitz J, Liang J, Wang C, White KI. Recent structural insights into the molecular architecture of synapses. In: Nano-organization of the synapse: from structure to function advances in neurobiology. Springer; 2025. In Press.

Carlsson A. A half-century of neurotransmitter research: impact on neurology and psychiatry. In: Jörnvall H, editor. Nobel lectures, physiology or medicine 1996–2000. Singapore: World Scientific Publishing Co.; 2003.

Chanaday NL. The formation, organization and maintenance of synaptic vesicles: from molecules to structure and dynamics. In: Nano-organization of the synapse: from structure to function advances in neurobiology. Springer; 2025. In Press.

Chen H, Tang AH, Blanpied TA. Subsynaptic spatial organization as a regulator of synaptic strength and plasticity. Curr Opin Neurobiol. 2018;51:147–53.

Choi YJ, Won-Suk Chung WS. Roles of glia in synapse nano-organization. In: Nano-organization of the synapse: from structure to function advances in neurobiology. Springer; 2025. In Press.

Damenti M, Levy AD, Thomas A, Blanpied. Scales of postsynaptic nanostructure: molecules, nanoclusters, and nanodomains. In: Nano-organization of the synapse: from structure to function advances in neurobiology. Springer; 2025. In Press.

Galkov M, Nemcova P, Dietrich D, Schoch S. Active zone: linking resolution levels of microscopic modalities. In: Nano-organization of the synapse: from structure to function advances in neurobiology. Springer; 2025. In Press.

Greengard P. The neurobiology of slow synaptic transmission. Science. 2001;294:1024–30.

Guzikowski NJ, Kavalali ET. Nano-Organization at the synapse: segregation of distinct forms of neurotransmission. Front Synaptic Neurosci. 2021;13:650334.

Guzikowski NJ, Wang CS, Kavalali ET. Functional nano-segregation of distinct forms of neurotransmission. In: Nano-organization of the synapse: from structure to function advances in neurobiology. Springer; 2025. In Press.

Honkanen K, Sando RC. Synaptic cell adhesion—a functional perspective. In: Nano-organization of the synapse: from structure to function advances in neurobiology. Springer; 2025. In Press.

Krystal JH, Kavalali ET, Monteggia LM. Ketamine and rapid antidepressant action: new treatments and novel synaptic signaling mechanisms. Neuropsychopharmacol Rev. 2024;49:41–50.

Myeong J, Klyachko VA. Optical approaches to dissect the structure and dynamics of the synapse at nanoscale resolution. In: Nano-organization of the synapse: from structure to function advances in neurobiology. Springer; 2025. In Press.

Özçete ÖD, Banerjee A, Kaeser PS. Mechanisms of neuromodulatory volume transmission. Mol Psychiatry. 2024; https://doi.org/10.1038/s41380-024-02608-3.

Schrödinger, E. (1996). Nature and Greeks and science and humanism., Canto Edition. (Cambridge: Cambridge University Press).

Snow CP. The two cultures., Canto Edition. Cambridge: Cambridge University Press; 1998.

Südhof TC. Neurotransmitter release: the last millisecond in the life of a synaptic vesicle. Neuron. 2013;80:675–90.

Sun R, Zhou Q. Visualizing postsynaptic density with electron tomography. In: Nano-organization of the synapse: from structure to function advances in neurobiology. Springer; 2025. In Press.

Wu Z, Lin D, Li Y. Pushing the frontiers: tools for monitoring neurotransmitters and neuromodulators. Nat Rev Neurosci. 2022;23:257–74.

Chapter 2
Recent Structural Insights into the Molecular Architecture of Synapses

Axel T. Brunger, Richard G. Held, Yousuf A. Khan, Jeremy Leitz, Jiahao Liang, Chuchu Wang, and K. Ian White

Abstract Synaptic transmission between pre- and postsynaptic neurons occurs when the presynaptic neuron terminal is temporarily depolarized upon action potential arrival, opening voltage-gated Ca^{2+} channels at synapses. Ca^{2+} will flow into the presynapse, and it will trigger the fusion of neurotransmitter-filled synaptic vesicles with the presynaptic membrane in less than a millisecond. Neurotransmitter molecules are then released into the synaptic cleft and bind to receptors in the postsynaptic membrane. Understanding the mechanisms that underlie this complex cellular process requires detailed knowledge of the spatial and structural organization of the macromolecular components of the synapse. This chapter focuses on recent structural insights into the presynaptic machinery and the spatial relationship between synaptic vesicles, presynaptic factors, and postsynaptic receptors.

Keywords Synaptic transmission · Synaptic vesicle fusion · SNARE · Synaptotagmin · Complexin · N-ethylmaleimide-sensitive factor · SNAP · Munc13 · Munc18 · AMPA receptor

---

The original version of the chapter has been revised. A correction to this chapter can be found at https://doi.org/10.1007/978-3-032-12594-1_12

A. T. Brunger (✉) · R. G. Held · Y. A. Khan · J. Leitz · J. Liang · C. Wang · K. I. White
Department of Molecular and Cellular Physiology, Stanford University, Stanford, CA, USA

Department of Neurology and Neurological Sciences, Stanford University, Stanford, CA, USA

Department of Structural Biology, Stanford University, Stanford, CA, USA

Department of Photon Science, Stanford University, Stanford, CA, USA

Howard Hughes Medical Institute, Stanford University, Stanford, CA, USA
e-mail: brunger@stanford.edu

E. T. Kavalali (ed.), *Nano-organization of the Synapse*, Advances in Neurobiology 48,
https://doi.org/10.1007/978-3-032-12594-1_2

2.1 Introduction

Neuronal synapses are specialized signaling compartments that control information transfer in the nervous system. Synaptic transmission between pre- and postsynaptic neurons occurs when the presynaptic neuron terminal is temporarily depolarized upon action potential arrival, opening voltage-gated Ca^{2+} channels at synapses. Because the extracellular Ca^{2+} concentration is much higher than the basal cytoplasmic concentration, Ca^{2+} will flow into the neuron. In turn, Ca^{2+} will trigger the fusion of neurotransmitter-filled synaptic vesicles with the presynaptic membrane in less than a millisecond (Rothman 2014; Südhof 2013). Neurotransmitter molecules are then released into the synaptic cleft and bind to receptors in the postsynaptic membrane. Essential components of the synaptic vesicle fusion machinery include the neuronal SNAREs (*S*oluble *N*-ethylmaleimide-sensitive factor *A*ttachment protein *RE*ceptor), the Ca^{2+} sensor synaptotagmin, the activator/regulator complexin, the assembly factors Munc18 (*m*ammalian *unc*oordinated-18) and Munc13 (*m*ammalian *unc*oordinated-13), the AAA+ ATPase N-ethylmaleimide-sensitive factor (NSF, Sec18 in lower eukaryotes), and soluble NSF attachment proteins (SNAPs, Sec17 in lower eukaryotes) (Fig. 2.1) and are summarized in references Brunger et al. (2018) and Rizo (2022). In addition, other active zone proteins (including RIM and RIM-BP) connect the voltage-gated Ca^{2+} channel to the fusion machinery and Rab effector proteins, as summarized in references Held

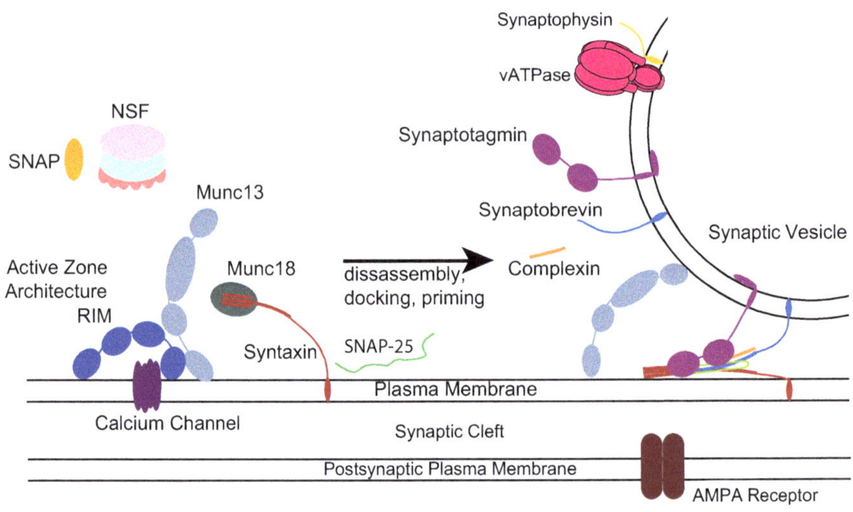

Fig. 2.1 A simplified schema of the presynaptic neurotransmitter release machinery that is discussed in this chapter. Shown are a subset of presynaptic proteins (RIMs, Munc13, Munc18, P/Q- or N-type Ca^{2+}-channels), SNAREs (SNAP-25, syntaxin, and synaptobrevin), complexin, synaptotagmin, NSF, and α-SNAP. Also shown are the synaptic vesicle V-ATPase and synaptophysin, as well as the AMPA receptor in the postsynaptic membrane

and Kaeser (2018) and Südhof (2012). For the proteins in the synaptic cleft and the postsynaptic side, we refer to references Südhof (2012, 2017).

Understanding the mechanisms that underlie this complex cellular process requires detailed knowledge of the spatial and structural organization of the macromolecular components of the synapse. This chapter focuses on recent structural insights into the presynaptic machinery and recent new insights on how these presynaptic factors relate to postsynaptic receptors.

2.2 Technical Developments: Cryo-Electron Tomography and FIB Milling

Recent advances in cryo-electron microscopy (cryo-EM) and tomography (cryo-ET) have been instrumental in the three-dimensional imaging of cellular samples frozen in a near-native state (Turk and Baumeister 2020). These developments have enabled high-resolution in situ structure determination of large protein complexes in cells, including ribosomes, proteasomes, nuclear pore complexes, and muscle sarcomeres (Mosalaganti et al. 2022; Schuller et al. 2021; Tegunov et al. 2021; Wang et al. 2022a). These in situ structures required accurate localization of target proteins within tomographic volumes, relying on their large size, distinctive shape, or ultrastructural arrangement. When successful, subtomogram averaging and in situ structure determination can reveal protein structure and conformation in a native environment, including interactions with known and unknown binding partners. Cryo-ET requires thin enough samples (~100–200 nm thickness) to achieve good image contrast and minimal background. Focused-ion beam (FIB) milling of cryo-frozen samples is often essential to achieve such thin samples. A key advantage of using cryo-fixation is that artifacts due to chemical fixation or freeze substitution techniques can be avoided (Korogod et al. 2015).

In principle, these advances in cryo-ET promise to visualize proteins and uncover the molecular architecture of cells, particularly synapses. However, there are significant challenges to achieving molecular or atomic resolution due to the relatively small size of many synaptic proteins, and their association with membranes can obscure the protein densities. This chapter provides examples illustrating the current state-of-the-art imaging of synapses and synaptic proteins.

2.3 Presynaptic Architecture

Figure 2.2a shows one slice (thickness of ~14 nm) of a 3D cryo-ET reconstruction of a FIB-milled lamella of primary cultures of mouse hippocampal neurons (Held et al. 2024a). On the presynaptic side, many synaptic vesicles are typically observed (Fig. 2.2a) with a bimodal distance distribution to the active zone membrane (Held

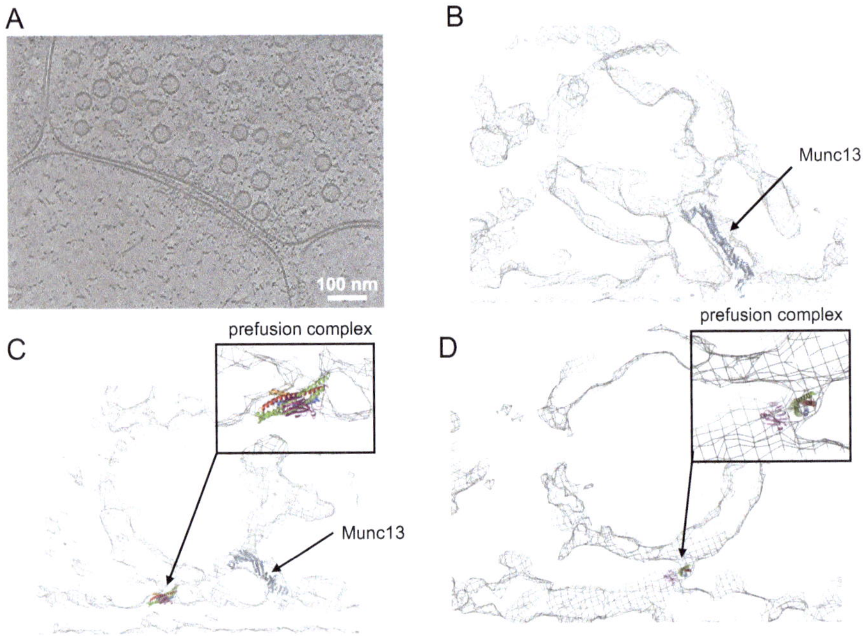

Fig. 2.2 Cryo-ET images of FIB-milled intact synapses (Held et al. 2024a). (**a**) A single XY section through a tomogram of a FIB-milled synapse reconstructed at bin 8 (1.36 nm/pixel) and denoised with cryo-CARE (Buchholz et al. 2018) (TS001 in EMDB EMD-45161) (Held et al. 2024a). (**b**–**d**) Examples of membrane-proximal synaptic vesicles with tether and globular densities. The Wiener-filtered (Tegunov and Cramer 2019) cryo-ET map is shown as a gray mesh at a contour level of 0.5, 0.7, and 1 σ in panels (**b**), (**c**), and (**d**), respectively. The "prefusion complex" (PDB ID 5W5C) (Zhou et al. 2015, 2017), consisting of the SNARE complex (syntaxin: dark red, SNAP-25: green, synaptobrevin: light blue), the C2B domain of synaptotagmin-1 (magenta) interacting with the SNARE complex via the primary interface, and the core domain of complexin-1 (orange), was fitted to globular densities in panels (**c**) and (**d**). The C_1-C_2B-MUN-C_2C fragment of Munc13 (Grushin et al. 2022) (PDB 7T7V, slate blue) was fitted to tether densities in panels (**b**) and (**d**). Munc13 is in an upright position in (**b**) and a lateral position in (**c**). The insets in panels (**c**) and (**d**) show close-up views of the globular densities with the fitted prefusion complex

et al. 2024a). There is a population within less than 10 nm and a broad population beginning at ~10 nm, consistent with previous work; for example, see Imig et al. (2014), which studied the role of priming factors in establishing this distance distribution. The membrane-proximal (<10 nm) synaptic vesicles likely correspond to loosely or tightly primed synaptic vesicles (Lin et al. 2022). In contrast, the more distant synaptic vesicles likely constitute the reserve pool (Denker et al. 2011). However, static cryo-ET reconstructions do not allow one to distinguish between functionally distinct primed/docked synaptic vesicles.

2.4 Intermembrane Densities Between Synaptic Vesicles and the Plasma Membrane

Cryo-ET reconstructions of FIB-milled mouse synapses revealed interfacial densities that connect membrane-proximal synaptic vesicles to the plasma membrane, which can be approximately divided into two broad groups: globular densities and elongated (~20–30 nm long) tethers (Held et al. 2024a) (Fig. 2.2b–d). One to three distinct globular densities were typically observed with no regular or symmetric arrangement. These connecting densities are similar to cryo-ET densities observed in synaptosomes from mouse hippocampal organotypic slice cultures (Orozco-Borunda et al. 2025; Papantoniou et al. 2023) (Fig. 2.3), where, again, no regular arrangements were found. Symmetric arrangements of connecting densities have been reported in subtomogram average maps of cryo-ET reconstructions of cultured hippocampal neurons (Radhakrishnan et al. 2021). However, density maps of individual membrane-proximal synaptic vesicles did not show well-defined regular

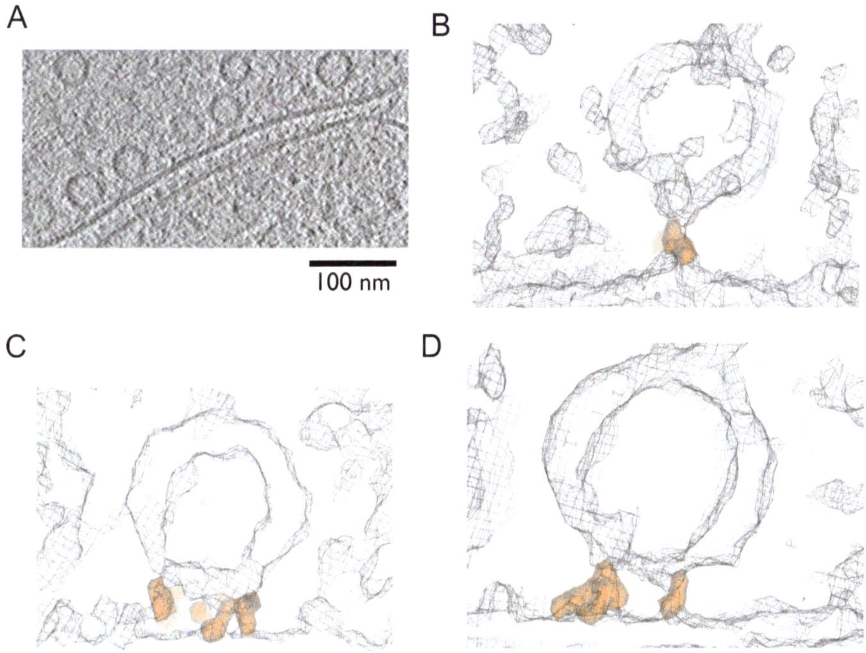

Fig. 2.3 Cryo-ET images of synaptosomes from mouse hippocampal organotypic slice cultures (EMDB EMPIAR-12512) (Orozco-Borunda et al. 2025; Papantoniou et al. 2023). (**a**) A single XY section through a tomogram of a synaptosome at bin 4 (1.756 nm/pixel). (**b**–**d**) Examples of membrane-proximal synaptic vesicles with various densities (orange, detected by Pyto (Lučić et al. 2016)) that connect them to the plasma membrane. The Weiner-filtered (Tegunov and Cramer 2019) cryo-ET map (gray mesh) is shown as a mesh at a contour level of 1, 0.5, and 0.8 σ in panels (**b**), (**c**), and (**d**), respectively

arrangements, which is consistent with the other studies (Held et al. 2024a; Papantoniou et al. 2023).

All these cryo-ET studies support the notion of stable intermembrane densities at resting Ca^{2+} concentrations between synaptic vesicles and plasma membranes. It is tempting to speculate that these complexes occur for primed synaptic vesicles that undergo fast synchronous release, but further experiments are required to investigate this fascinating question. As such, these prefusion complexes are expected to *prevent* fusion since they keep the membranes too far apart for fusion (i.e., more than the critical distance of 0.9 nm) (Aeffner et al. 2012), but they set the stage for fast-triggered fusion upon Ca^{2+} binding to synaptotagmin molecules. As suggested by the recent analyses of neurotransmission of the calyx of Held synapses (Lin et al. 2022), most of these prefusion complexes probably correspond to tightly docked synaptic vesicles.

What are these intermembrane densities? An earlier work revealed similar globular and tether densities using individual cryo-ET reconstructions of synthetic vesicles with reconstituted neuronal SNAREs, synaptotagmin-1, complexin-1, and the C1C2BMUN fragment of Munc13 in the Ca^{2+}-free state (Gipson et al. 2017). The addition of Ca^{2+} resulted in the disappearance of most of these intermembrane densities by vesicle fusion. However, only globular densities were present when Munc13 was omitted. Subsequently, a large cryo-ET dataset of a mixture of isolated glutamatergic synaptic vesicles (ISVs) from a mouse brain homogenate and synthetic vesicles mimicking the plasma membrane was collected in the presence of complexin, but MUN instead of the longer C1C2BMUN fragment (Peters et al. 2022). Subtomogram averages revealed globular, but not tether, densities linking the two types of vesicles (Peters et al. 2022). Again, relatively small numbers of intermembrane densities were observed for a given pair, and there were no apparent regular arrangements between ISVs and acceptor vesicles (Peters et al. 2022). Thus, based on these studies, the globular densities likely contain a "prefusion" complex between neuronal SNAREs, synaptotagmin-1, and complexin-1, especially considering that these were the only proteins used in some conditions in reference Gipson et al. (2017). Recent subtomogram averages of cryo-ET densities observed in synaptosomes from mouse hippocampal organotypic slice cultures (Orozco-Borunda et al. 2025) further support this notion.

What is the molecular composition of the prefusion complex? Before synaptic vesicle fusion, the neuronal SNAREs synaptobrevin-2 (also called VAMP2—*V*esicle *A*ssociated *M*embrane *P*rotein 2) on the synaptic vesicle and syntaxin-1A and SNAP-25A on the plasma membrane initially form a *trans*-SNARE complex (Fig. 2.1), with the transmembrane domains of synaptobrevin-2 and syntaxin-1A in the synaptic vesicle and plasma membranes, respectively. Other synaptic proteins interact with the fully assembled (post-fusion) *cis*-SNARE complex (Chen et al. 2002; White et al. 2018; Zhao et al. 2015; Zhou et al. 2015), and, likely, some of these interactions are also present in the *trans*-SNARE complex (Fig. 2.1).

The crystal structure of the SNARE–synaptotagmin-1–complexin-1 complex (Zhou et al. 2015, 2017) (Fig. 2.4) can fit some of the globular densities (Fig. 2.2c, d) observed in the cryo-ET reconstructions of synapses (Held et al. 2024a). The

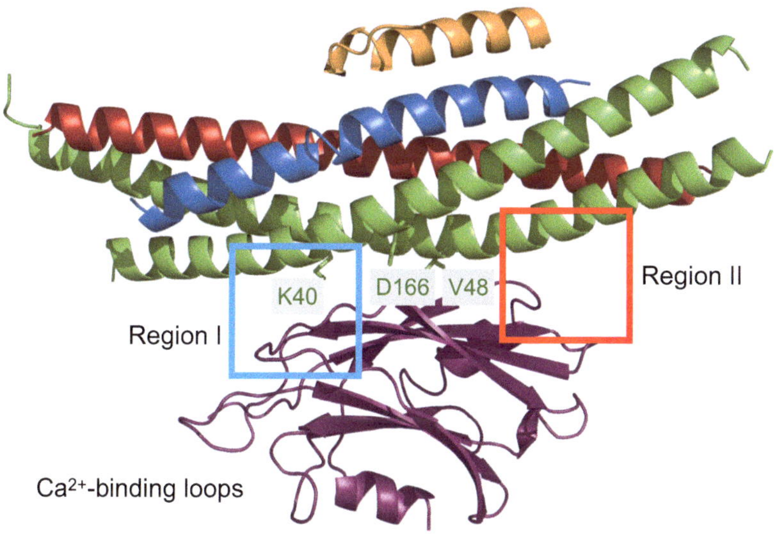

Fig. 2.4 Crystal structure of the SNARE complex (light blue, synaptobrevin-2; dark red, syntaxin-1a; green, SNAP-25) and its interaction with the C2B domain of synaptotagmin-1 (magenta) via the so-called primary interface and its interaction with complexin-1 (orange) (PDB ID 5W5C) (Zhou et al. 2015, 2017). SNAP-25 residues 40, 166, and 48 (sticks) are implicated in DEEs (Alten et al. 2021). Regions I and II of the primary interface (Zhou et al. 2015) are indicated by blue and red rectangles, respectively

structure shown in Fig. 2.4 includes the so-called primary interface between the C2B domain of synaptotagmin-1 and the SNARE complex (Zhou et al. 2015, 2017) and the subcomplex between the SNARE complex and complexin-1 (Chen et al. 2002). Qualitatively similar fits were also obtained using subtomogram average maps obtained from the tomograms of synaptosomes (Orozco-Borunda et al. 2025). The fitted model of the SNARE complex is, of course, an approximation of the expected *trans*-SNARE complex conformation between the synaptic vesicle and plasma membrane. Unfortunately, the tomographic reconstructions and subtomogram averages are not yet of sufficient resolution to determine the details of the *trans*-SNARE complex and its interactions with synaptotagmin and complexin. Nevertheless, these observations are consistent with the notion that the SNARE–synaptotagmin-1–complexin-1 complex is situated between the membrane proximal synaptic vesicles and the plasma membrane in the resting state of the synapse.

As mentioned above, in addition to the globular densities, elongated tether densities were observed in tomographic reconstructions of synapses (Fig. 2.2b, c) (Held et al. 2024a) and in subtomogram average maps of synaptosomes (Orozco-Borunda et al. 2025; Papantoniou et al. 2023). These densities can fit the structures of the C1C2BMUNC2C fragment of Munc13 (Grushin et al. 2022) (Fig. 2.2b, c).

The following sections discuss the prefusion complex and Munc13 in more detail.

2.5 Physiological Relevance of the SNARE—Synaptotagmin-1—Complexin-1 Complex

The primary interface between the SNARE complex and synaptotagmin-1 has an interface area of ~720 Å^2 with a complementary pattern of electrostatic interactions. The primary interface has been observed in several crystallization conditions (Zhou et al. 2015, 2017), in solution by a^{1}H-^{15}N TROSY-HSQC experiment (Voleti et al. 2020), and by single-particle cryo-EM on lipid nanotubes (Grushin et al. 2019). The primary interface is not close to the Ca^{2+} binding loops of synaptotagmin-1; remarkably, crystal structures in both the absence and presence of Ca^{2+} or Mg^{2+} are very similar (Zhou et al. 2015, 2017). The possible models of Ca^{2+} triggering are discussed below.

The primary interface was tested with a reconstituted fusion assay that included neuronal SNAREs, synaptotagmin-1, and complexin-1 (Zhou et al. 2015). Mutations of both synaptotagmin-1 C2B and SNAP-25A were designed to disrupt the primary interface based on the crystal structures. All mutations disrupted Ca^{2+}-triggered fusion and had varying impacts on vesicle association. Additional support for the primary interface came from electrophysiological experiments of mutations in the C2B domain of synaptotagmin-1 (Zhou et al. 2015, 2017). These mutations disrupted evoked release, while mutations in region II (Fig. 2.4) and the polybasic region increased spontaneous release. All mutations resulted in facilitation during high-frequency stimulation compared to depression for the wildtype cultures. More support came from disease-related mutations of SNAP-25A that are at the primary interface (SNAP-25A Lys40Glu, Asp166Tyr, Val48Phe, Fig. 2.4) (Alten et al. 2021). These particular mutants were uncovered by next-generation DNA sequencing techniques from patients suffering from developmental and epileptic encephalopathies (DEEs) (Melland et al. 2021). The three mutations were individually tested in SNAP-25 knockout mice that were rescued with wildtype or mutant SNAP-25B. All the three mutations result in substantially reduced eIPSC amplitudes and increased slopes; in other words, less Ca^{2+}-triggered neurotransmitter release and less synchrony (Alten et al. 2021).

More recent electrophysiological studies of synaptotagmin mutations of the primary interface suggest that the primary interface is essential for Ca^{2+}-evoked release, priming, and clamping of spontaneous release (Toulmé et al. 2024). Binding of synaptotagmin-1 C2B domain arginine residues to SNAP-25 acidic residues at one face of region II (Fig. 2.4) of the primary interface is crucial for Ca^{2+} triggering. In contrast, other SNAP-25 and syntaxin-1 acidic residues at region I (Fig. 2.4) of the primary interface are primarily involved in priming and clamping of spontaneous release. Importantly, the Glu295Ala mutation stabilizes the primary interface, but it reduces evoked release, suggesting that the wildtype primary interface is relatively dynamic and it sets the stage for the transition to a different state upon Ca^{2+} binding (see Sect. 2.9).

A hydrocarbon-stapled peptide, named SP9, was designed to disrupt Ca^{2+}-triggered membrane fusion by interfering with the primary interface (Lai et al.

2022). SP9 strongly suppressed Ca^{2+}-triggered fusion with reconstituted systems with neuronal synaptic proteins or their airway homologs (Brunger and Leitz 2023; Lai et al. 2022). Conjugation with the cell-penetrating peptide (CPP) penetratin (PEN) to the stapled peptide resulted in efficient delivery into cultured human airway epithelial cells and mouse airway epithelium, where it markedly and specifically reduced stimulated mucin secretion in both systems and substantially attenuated mucus occlusion of mouse airways. In contrast, SP9 did not affect baseline secretion in the IL-13 (metaplastic) cultures. Although SP9 is imperfect because it binds relatively weakly or variably to the synaptotagmin-1 C2B domain (Lai et al. 2022), it could serve as a therapeutic starting point as baseline mucin secretion is left unaffected, which is essential for maintaining the healthy state of airways.

A second interface, the so-called tripartite interface, was also observed by X-ray crystallography (Zhou et al. 2017), where the synaptotagmin-1 C2B domain binds to the SNARE–complexin-1 subcomplex via interactions with both the SNARE and complexin-1 components. However, thus far, the strongest support for the tripartite interface comes from electrophysiology experiments (Zhou et al. 2017) and molecular dynamics simulations (Brunger and Leitz 2023), but it has not yet been detected in solution (Jaczynska et al. 2023). Thus, further validation of the tripartite interface will be required. If the interface exists, it should ideally be visualized by cryo-ET reconstructions of docked/primed synaptic vesicles.

2.6 Tethering of Synaptic Vesicles by Munc13

Munc13-1 and its homologs are primarily brain-specific, cytoplasmic proteins in the presynaptic terminal involved in synaptic vesicle priming and short-term synaptic plasticity (Südhof 2012, 2013). In terms of domain structure, Munc13-1 consists of a C2 domain (C2A), a calmodulin-binding domain, a diacylglycerol (DAG)-binding C1 domain (C1), a Ca^{2+}/PIP2-binding C2 domain (C2B), the so-called MUN domain, and another C2 domain (C2C). The crystal structure of the C1C2BMUN fragment of Munc13-1 revealed interdomain interactions between the C1, C2B, and MUN domains and a folded linker region between the C1 and C2B domains (Xu et al. 2017). The C2A domain of Munc13-1 interacts with the Rab3 effector RIM, forming a tripartite complex (Dulubova et al. 2005), which is essential for the priming function of Munc13-1 and for the presynaptic localization in mice (Kaeser et al. 2011). RIM binding releases Munc13 from an autoinhibitory homodimer (Camacho et al. 2017; Deng et al. 2011), and genetic deletion of RIM reduces tethered synaptic vesicles (Fernández-Busnadiego et al. 2013).

Cryo-ET reconstructions of synapses (Fig. 2.2b, c) (Held et al. 2024a) and of synaptosomes (Orozco-Borunda et al. 2025; Papantoniou et al. 2023) suggest that Munc13-1 bridges synaptic vesicles to the plasma membrane via its C1C2B and C2C domains, as previously found (Camacho et al. 2021; Grushin et al. 2022; Quade et al. 2019) based on in vitro experiments. There are two different interaction modes of the C1C2B region with the plasma membrane: (i) one involving a polybasic face that is expected to yield an upright of Munc13-1 and hinder release and (ii) another involving the DAG-Ca^{2+}-PIP2-binding face that is predicted to result in a lateral orientation and facilitate release (Camacho et al. 2021; Grushin et al. 2022). Indeed, the cryo-ET reconstructions of synapses (Held et al. 2024a) revealed roughly two different conformations of the tether densities in the tomograms of synapses (Fig. 2.2b, c) that are likely related to these two conformations of Munc13 when bridging synaptic vesicles and the plasma membrane.

2.7 Structural Biogenesis of *trans*-SNARE Complexes and Synaptic Vesicle Priming

This section focuses on how fusogenic *trans*-SNARE complexes (Fig. 2.1) are established.

2.7.1 Role of NSF

The AAA+ ATPase N-ethylmaleimide-sensitive factor (NSF, Sec18 in lower eukaryotes) and soluble NSF attachment proteins (SNAPs, Sec17 in lower eukaryotes) disassemble various SNARE complexes both before and after fusion (Huang et al. 2019; Khan et al. 2025; White et al. 2018, 2025; Zhao et al. 2015; Yousuf et al. 2021). Combined with functional studies with reconstituted systems (Lai et al. 2017; Ma et al. 2013; Stepien et al. 2019), NSF has also been established as a critical factor for synaptic vesicle priming before fusion. In particular, NSF and SNAP disassemble nonfusogenic SNARE complexes (White et al., 2025; Gao et al., 2023; Lai et al. 2017; Ma et al. 2013). Disassembly of off-pathway SNARE complexes allows neurons to recycle SNARE components locally and may act as a quality control mechanism for SNARE complex assembly. NSF and SNAP, along with an ATP regeneration system, improve the efficiency of subsequent Ca^{2+}-triggered fusion using a system with reconstituted components and isolated synaptic vesicles (Leitz et al. 2024a). Supporting the notion that NSF and SNAP disassemble nonfusogenic *cis*-SNARE complexes, several atomic-resolution cryo-EM structures were determined of NSF and SNAP in complex with the ternary neuronal SNARE complex (Huang et al. 2019; White et al. 2018; Zhao et al. 2015), a syntaxin tetramer, and two different binary complexes of syntaxin–SNAP-25 (White et al. 2025) (Fig. 2.5). All these complexes involve four-helix bundles of SNAREs, with the N-terminal end of one syntaxin or one SNAP-25 SNARE domain bound within the catalytic pore of the D1 ATPase hexamer of NSF.

What is the potential biological relevance of the syntaxin tetramer structure (Fig. 2.5a)? Syntaxin forms dense nanodomains at the active zone, and NSF directly colocalizes to these syntaxin clusters (Sieber et al. 2007; White et al. 2025). Thus, NSF and SNAP likely disassemble syntaxin tetramers and binary SNARE

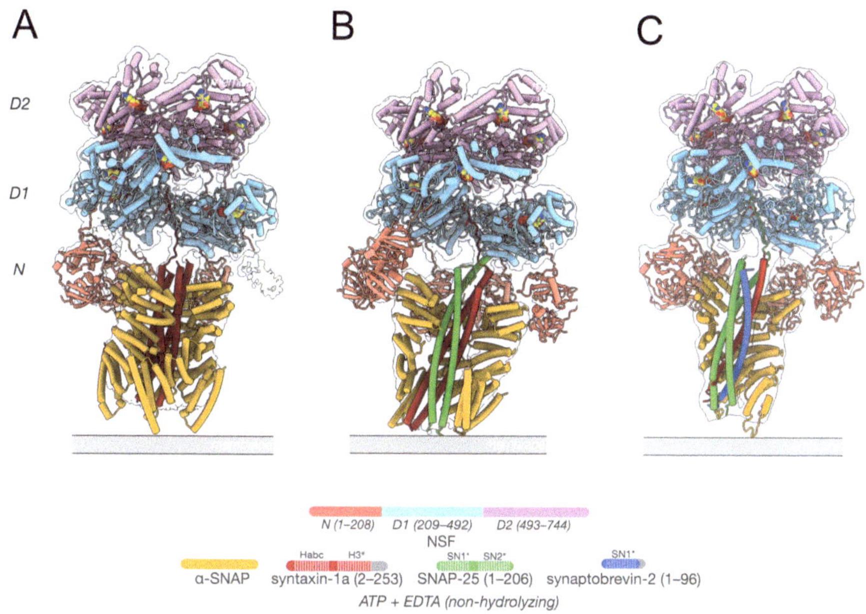

Fig. 2.5 (**a**) Hexameric NSF engages an α-SNAP–syntaxin tetramer subcomplex via its N-domains (salmon), several of which engage four α-SNAP molecules (gold) surrounding the soluble, tetrameric syntaxin H3 complex (dark red); the N-terminal end of one syntaxin is engaged in the D1 pore (light blue) (White et al. 2025). (**b**) Hexameric NSF engages the subcomplex consisting of α-SNAP and the 2:1 syntaxin–SNAP-25 (binary) SNARE complex via multiple N-domains (salmon), which engage three α-SNAP molecules (gold) surrounding the binary complex composed of two syntaxin H3 domains (dark red) and the SN1 and SN2 SNARE domains of a single SNAP-25 molecule (green). Part of the linker N-terminal to one of the syntaxin H3 domains is found within the catalytic D1 ring of NSF (light blue). (**c**) Hexameric NSF engages the subcomplex consisting of α-SNAP and the ternary SNARE complex. Hypothetical membranes are illustrated in gray. For the SNARE proteins, domains visible in the reconstruction of the SNARE complex are indicated with an asterisk

complexes on the perimeter of a given cluster to maintain a population of liberated syntaxin molecules. Munc18 can then capture individual syntaxin molecules liberated from clusters by NSF. Syntaxin nanodomains could thus effectively act as high-density storage sites for syntaxin, and NSF could tap these stores under periods of high-frequency vesicle release. It is also possible that NSF activity is regulated by post-translational modification (Gao et al. 2023; Zhou et al. 2013; Matveeva et al. 2001).

SNARE domains and TM regions introduce topological constraints for the disassembly process. Syntaxins are a particular case in point since they contain a large N-terminal (Habc) domain: a cryo-EM structure of a yeast SNARE complex, Sec18 (the NSF orthologue), and Sec17 (SNAP orthologue) in a nonhydrolyzing condition revealed that the SNARE Sso1 (the syntaxin-1A orthologue) is threaded through the D1 and D2 ATPase rings of Sec18, with its folded, N-terminal Habc domain

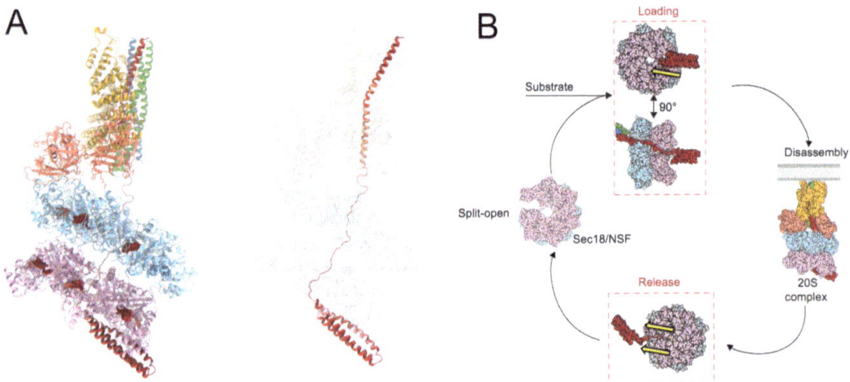

Fig. 2.6 (**a**) Sso1 (syntaxin homolog, dark red) is threaded through the D1 and D2 domains of Sec18 (NSF homolog). The N-terminal (Habc) domain of Sso1 interacts with the D2 ATPase ring of Sec18 (Khan et al. 2025). Left subpanel: cartoon representation of the complex of the yeast SNARE complex (consisting of Sso1, Snc1 (light blue), and Sec9 (green)), Sec18 (salmon, light blue, pink), and Sec17 (α-SNAP homolog, gold). Right subpanel: All parts are shown at high transparency except for Sso1. (**b**) Mechanism of substrate side-loading and side-release for Sec18/NSF

interacting with the D2 ring (Fig. 2.6) (Khan et al. 2025). Moreover, the N-terminal domain of syntaxin does not unfold during Sec18/NSF activity, as shown with single-molecule FRET experiments with NSF. Cryo-EM structures under hydrolyzing conditions revealed substrate-released and substrate-free states of Sec18 with a coordinated opening in the side of the ATPase rings. These findings suggest that Sec18/NSF can operate by substrate side-loading and unloading topologically constrained SNARE substrates. Sec18/NSF can load and release SNARE substrates via a coordinated opening in the D1 and D2 rings. This mechanism does not preclude the more conventional notion of complete substrate threading through the pores of Sec18/NSF if the SNARE substrate is not topologically restricted by membrane domains or anchors or by N-terminal domains or complexes.

How does NSF recognize SNARE substrates? Considering that we observe either syntaxin or SNAP-25 to be bound to the D1 pore (Khan et al. 2025; White et al. 2018, 2025), as well as part of the N-terminal domain in a self-inhibiting state of NSF (Khan et al. 2025), the interaction with D1 is likely highly promiscuous. The specificity of SNARE substrates is, therefore, likely conferred by the interactions between SNAP molecules and SNAREs, and these interactions are primarily driven by charge–charge complementarity (White et al. 2018). Consequently, complexin-1 and/or synaptotagmin-1 binding is expected to interfere with some SNAP–SNARE interactions. Indeed, complexin inhibits NSF-mediated disassembly of neuronal SNARE complexes in a dose-dependent fashion (Choi et al. 2018).

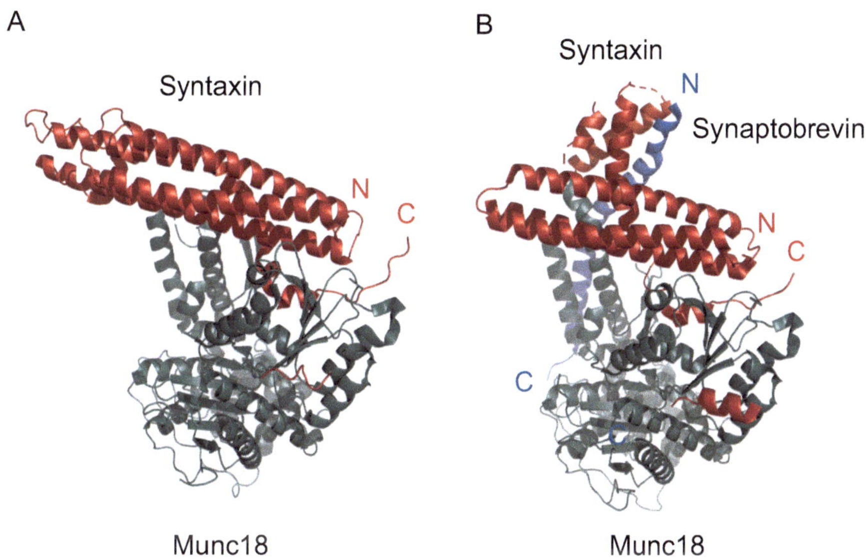

Fig. 2.7 (**a**) Crystal structure of the syntaxin-1a (dark red)–Munc18a (dark teal) complex (Burkhardt et al. 2008; Misura et al. 2000). (**b**) Cryo-EM structure of the synaptobrevin (light blue)–Munc18-1 (dark teal)–syntaxin-1 (dark red) complex (Stepien et al. 2022). The N- and C-terminal ends of the visible fragments of syntaxin and synaptobrevin are labeled

2.7.2 Role of Munc18 and Munc13

The activity of NSF establishes syntaxin–Munc18 complexes (Fig. 2.7a) that can be considered the starting point for synaptic vesicle priming. The MUN domain of Munc13-1 catalyzes the transit of syntaxin-1A from the syntaxin-1A–Munc18-1 complex into the ternary *trans-SNARE* complex (Basu et al. 2005; Ma et al. 2013; Yang et al. 2015) and promotes the proper assembly of the SNARE complex (Lai et al. 2017). Before the action of Munc13, Munc18 forms a scaffolding complex with the cytoplasmic regions of synaptobrevin and syntaxin (Stepien et al. 2022) (Fig. 2.7b). This complex positions the C-terminal ends of the cytoplasmic regions of synaptobrevin and syntaxin, such that they could reach the synaptic vesicle and plasma membranes, respectively. This complex would initiate interactions between synaptobrevin and syntaxin that lead to the *trans*-SNARE complex formation through the action of Munc13 which changes the conformation of syntaxin and allows the SNARE domain of syntaxin to interact with synaptobrevin and SNAP-25. Synaptotagmin-1 then binds to the assembled *trans*-SNARE complex via the primary interface (Zhou et al. 2015), and complexin-1 also binds to the *trans*-SNARE complex (Chen et al. 2002).

When the soluble fragments of SNARE proteins are simply mixed in solution, improper configurations may occur (Choi et al. 2016; Lou et al. 2015; Ryu et al. 2015; Weninger et al. 2003). In particular, nonfunctional antiparallel SNARE complex configurations were observed using single-molecule fluorescence resonance energy transfer (FRET) experiments (Choi et al. 2016; Weninger et al. 2003). Thus, the assembly of a functional SNARE complex is an active process, and the ancillary proteins involved in the assembly (and disassembly) of the SNARE complex are crucial to sustained neuron function.

2.8 Dynamic Character of Docked Synaptic/Primed Synaptic Vesicles

The ability to capture membrane-proximal synaptic vesicles in electron microscopy/cryo-ET reconstructions of resting neurons gives the impression of a relatively static state. However, flash freeze and simulation experiments suggested that vesicle docking and priming are dynamic and may be reversible (Kusick et al. 2022; Murthy and Stevens 1999; Neher and Brose 2018; Silva et al. 2021). In support of this notion, dissociation of previously docked synaptic vesicles was observed in a reconstituted system with isolated synaptic vesicles, plasma membrane-mimicking vesicles, NSF, α-SNAP, Munc13, and Munc18 (Leitz et al. 2024a). It is likely that the membrane-bridging/tethering activity by the C1C2BMUNC2C fragment of Munc13 (Camacho et al. 2021; Grushin et al. 2022) (Fig. 2.2b, c) physically permits the exchange and clearance of α-SNAP from the synaptic vesicle–plasma membrane interface. In addition, the presence of the C1, C2B, and C2C domains effectively increases the local concentration of Munc13 near the synaptic vesicle docking site. The lateral conformation of C1C2BMUNC2C (Camacho et al. 2021; Grushin et al. 2022; Quade et al. 2019) brings the synaptic vesicle closer to the plasma membrane (compare Fig. 2.2b and c). Munc18 and Munc13 act together (Prinslow et al. 2019; Stepien et al. 2019), domain 3a of Munc18 interacts with the full-length assembled SNARE complex (Parisotto et al. 2014), and Munc13 interacts directly with synaptobrevin and SNAP-25 during addition to the SNARE complex to ensure parallel SNARE complex formation (Kalyana Sundaram et al. 2021; Lai et al. 2017; Shu et al. 2020; Wang et al. 2019). Thus, these processes likely promote the formation of the primed SNARE–synaptotagmin-1–complexin-1 prefusion complex (Fig. 2.4) (Zhou et al. 2015, 2017) and protect the *trans*-SNARE complex from α-SNAP binding and disassembly (Choi et al. 2018). In contrast, improperly or incompletely formed SNARE complexes may not efficiently bind complexin-1 and synaptotagmin-1, consequently enabling α-SNAP binding for subsequent disassembly by NSF and effectively inhibiting any fusion activity by these complexes. In support of this notion, there is an increase in spontaneous fusion when the ATP concentration is not maintained in a reconstituted system (Leitz et al. 2024a).

2.9 Models of Ca^{2+} Triggering of Synaptic Vesicles

The SNARE complex provides the energy necessary for membrane fusion (Sutton et al. 1998; Weber et al. 1998). During fusion, the *trans*-SNARE complex completely zippers into the fully assembled *cis*-SNARE complex, where the soluble core consists of a parallel, four-α-helix bundle. At least two synaptobrevin-2 molecules, and presumably two SNARE complexes, are required for fast Ca^{2+}-triggered exocytosis (Sinha et al. 2011). However, binding to synaptotagmin-1 and complexin-1 arrests the *trans*-SNARE complex, so it cannot promote fusion at resting Ca^{2+} concentration (Voleti et al. 2020; Zhou et al. 2015). The synaptotagmin-1 C2B domain contains many charged residues consisting of clusters of basic residues on the surface of the β-sandwich, including the so-called polybasic region, which can interact with anionic membranes (Pérez-Lara et al. 2016; Voleti et al. 2020). These membrane interactions enhance the primary interface at resting Ca^{2+} concentration by simultaneously binding the polybasic region and the SNARE complex to the membrane (Grushin et al. 2019; Voleti et al. 2020; Zhou et al. 2015).

Upon Ca^{2+} binding to the C2 domains of synaptotagmin, the inhibition is released, and fusion is triggered on a fast time scale. Several models have been proposed for the release of this inhibition: (1) The synaptotagmin C2B molecule that forms the primary interface induces membrane puckering in conjunction with the SNARE complex upon Ca^{2+} binding of the membrane near the fusion site (Zhou et al. 2015) (Fig. 2.8a). (2) Ca^{2+} triggering of synaptic vesicle fusion causes a rearrangement of the primary interface involving dissociation of region I, while region II remains bound (Jaczynska et al. 2025; Toulmé et al. 2024). This rearrangement may act as a

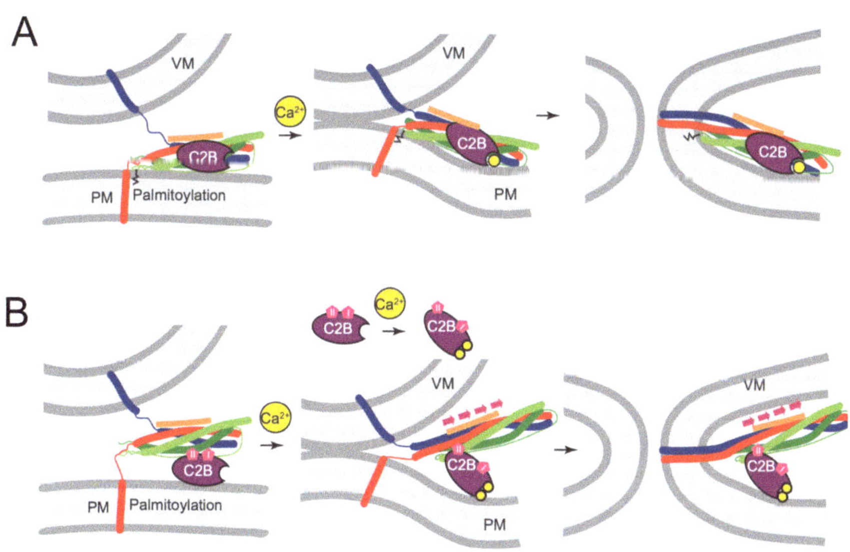

Fig. 2.8 Models of Ca^{2+} triggering. (**a**) Membrane bending model proposed in reference Zhou et al. (2015). (**b**) Pivot model (Toulmé et al. 2024). VM: vesicle membrane, PM: vesicle membrane

lever that pulls the SNARE complex when Ca^{2+} causes reorientation of the C2B domain, facilitating zippering and membrane fusion (Fig. 2.8b). (3) The complexin-1 accessory helix inhibits release through electrostatic (and perhaps steric) repulsion of the synaptic vesicle by its location between the vesicle and plasma membranes (Rizo 2022; Trimbuch et al. 2014). (4) If the tripartite interface between a second synaptotagmin-1 C2B domain and SNARE–complexin subcomplex exists, it may dislodge upon Ca^{2+} binding, further enhancing the effect of the C2B domain that is initially bound to the primary interface (Zhou et al. 2017). Finally, several of these mechanisms may act together. Further studies will be needed to clarify the Ca^{2+}-triggering mechanism.

2.10 New Insights into the Biogenesis of Synaptic Vesicles

Before docking and priming, synaptic vesicles are formed, primarily through clathrin-mediated endocytosis (Heuser and Reese 1973; Maycox et al. 1992; Watanabe et al. 2014). Synaptic vesicles have a remarkably well-defined proteome and copy number distribution (Takamori et al. 2006). Interaction maps of synaptic vesicle proteins have been obtained based on cross-linking mass spectrometry (Wittig et al. 2021), although the structural details are unclear. As a first step to decipher the molecular architecture of synaptic vesicles, glutamatergic ISVs were

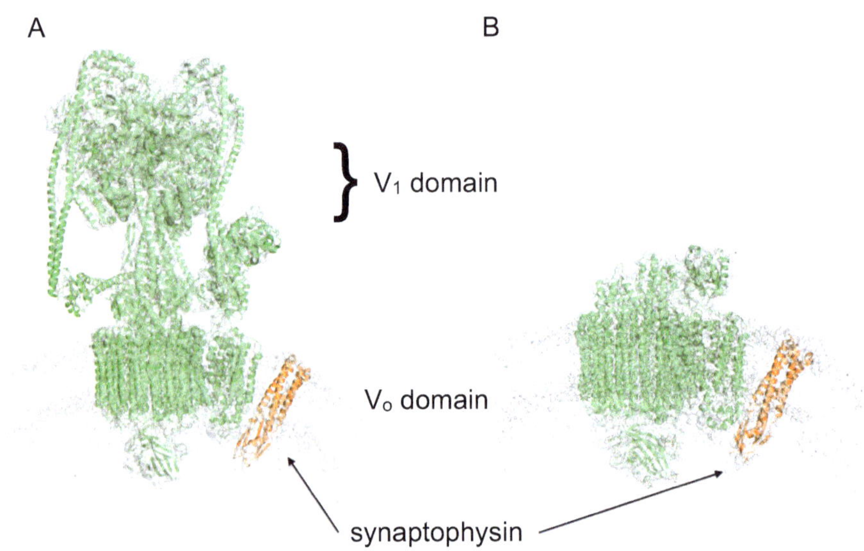

Fig. 2.9 (**a**) In situ structures of the V-ATPase (PDB ID 9BRQ) and (**b**) V0-only V-ATPase (PDB ID 9BRZ) assembly in synaptic vesicle membranes (Wang et al. 2024). V-ATPase is colored green, and synaptophysin is colored orange. The EM density maps (gray) are superimposed. The gray arc represents the ISV membrane density

immuno-isolated and purified from mouse brains and imaged by cryo-EM and cryo-ET (Leitz et al. 2024a, b; Takamori et al. 2006). The ISVs were functional as assessed by a Ca^{2+}-triggered vesicle–vesicle fusion assay (Leitz et al. 2024a).

In situ structures of V-ATPases in functional, glutamatergic mouse ISVs were obtained by subtomogram averaging of cryo-ET data and by reconstructions with single-particle mode cryo-EM data (Wang et al. 2024) (Fig. 2.9). Both intact and V0-only V-ATPase assemblies were observed, consistent with the observation that V-ATPases reversibly dissociate the V1 domain upon synaptic vesicle acidification (Bodzęta et al. 2017). The cryo-ET and cryo-EM structures revealed that synaptophysin is bound to the synaptic vesicle V-ATPase (Wang et al. 2024). Two other research groups independently observed a similar interaction at similar resolutions (Coupland et al. 2024; Kravcenko et al. 2024). Considering the limited resolution of these reconstructions and to further confirm the identity of the binding partner, synaptophysin knockout ($Syp^{-/-}$) mice were used (Wang et al. 2024), and the resulting

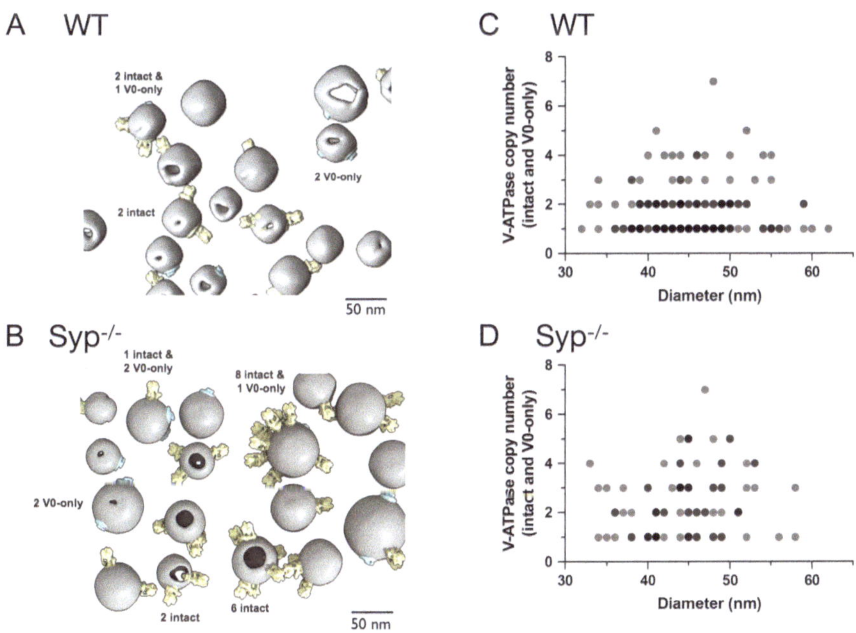

Fig. 2.10 Renderings of representative tomographic reconstructions of isolated synaptic vesicles from (**a**) wildtype (WT) and (**b**) $Syp^{-/-}$ mouse brains. The synaptic vesicle membrane, the intact, and V0-only V-ATPase assemblies are colored in gray, yellow, and cyan, respectively. (**c**, **d**) Scatter plots of the ISV diameters and V-ATPase copy numbers (including both intact and V0-only assemblies) for wildtype (**c**) and $Syp^{-/-}$ ISVs (**d**) from six randomly selected tomograms, respectively. The gray scale of the dots indicates the number of cases (dark: more frequent cases). The simple linear regression *p*-value (analyzed by GraphPad Prism V.10) between the WT and $Syp^{-/-}$ datasets is 0.56 and 0.08, respectively, indicating no statistically significant correlation between ISV diameters and V-ATPase copy numbers. (The source data are associated with Extended Data Fig. 8i in reference Wang et al. (2024))

cryo-EM maps revealed the absence of a binding partner to V-ATPase in the $Syp^{-/-}$ ISVs, thus confirming the identity of the binding partner to wildtype synaptic vesicle V-ATPases. Considering that synaptoporin and synaptogyrin (close homologs of synaptophysin) are present in the $Syp^{-/-}$ mice (Wang et al. 2024), the interaction between V-ATPase and synaptophysin is, therefore, highly specific to synaptophysin.

The copy number of V-ATPases roughly doubled in $Syp^{-/-}$ ISVs compared to wildtype ISVs (Wang et al. 2024) (Fig. 2.10). The size distributions of the ISVs are similar, and there is no correlation between the V-ATPase copy numbers and ISV diameters for both wildtype and $Syp^{-/-}$ ISVs; in other words, the average ISV diameter is similar (Wang et al. 2024) (Fig. 2.10c, d). However, in synaptophysin/synaptoporin/synaptogyrin-1/synaptogryin-3 quadruple knockout mice, the diameter of synaptic vesicles is substantially increased (Park et al. 2024), suggesting that the loss of synaptophysin in $Syp^{-/-}$ mice is partially compensated by the presence of the other members of the synaptophysin family in determining the synaptic vesicle size. A possible explanation for the increased copy number of V-ATPases in $Syp^{-/-}$ ISVs is that the absence of the V-ATPase–synaptophysin interaction effectively decreases the lateral membrane cross section of the combined molecular assembly. This smaller membrane cross section will reduce molecular crowding by decreasing the excluded surface area (Ellis 2001), resulting in a higher V-ATPase copy number in $Syp^{-/-}$ ISVs. However, the effect of a single synaptophysin molecule interacting with the V-ATPase may not be large enough to explain the roughly twofold effect on the V-ATPase copy number (Wang et al. 2024). It is possible, or even likely, that synaptophysin is also involved in interactions with synaptobrevin (Adams et al. 2015), and synaptophysin itself can form oligomers (Arthur and Stowell 2007; Bera et al. 2023). These additional interactions would then effectively increase the cross section of this supramolecular complex. However, these additional interactions with synaptophysin have not yet been observed in cryo-ET/cryo-EM studies of synaptic vesicles, possibly due to heterogeneity. In any case, we speculate that the interactions involving the synaptophysin family of proteins play a fundamental role in synaptic vesicle biogenesis.

At the gross morphological level, $Syp^{-/-}$ mice form normal synapses and synaptic vesicles (Eshkind and Leube 1995), corroborated by the cryo-ET reconstructions of $Syp^{-/-}$ ISVs (Wang et al. 2024). $Syp^{-/-}$ mice exhibit normal EPSC and mEPSC release probabilities, though a small but significant increase in EPSC quantal amplitude was previously observed (McMahon et al. 1996). More profound changes were observed with double genetic deletion of both synaptophysin and synaptogyrin, reducing short-term and long-term plasticity in neurons (Janz et al. 1999). Quadruple knockout studies of the entire brain-enriched synaptophysin family (synaptophysin/synaptoporin/synaptogyrin-1/synaptogryin-3) produced an even more significant increase in quantal amplitude (Raja et al. 2019). Importantly, $Syp^{-/-}$ mice exhibit a severe seizure phenotype under pharmacological treatment with a glutamate agonist (Wang et al. 2024), concordant with the increase in quantal size observed for both

Syp$^{-/-}$ and quadruple knockout (McMahon et al. 1996; Raja et al. 2019). The increased V-ATPase copy number in Syp$^{-/-}$ ISVs offers a possible explanation for the observed seizure phenotype, concordant with the increase in quantal size observed in previous Syp$^{-/-}$ knockout studies (McMahon et al. 1996; Raja et al. 2019). The increased quantal size could be different in excitatory and inhibitory synapses, as suggested by different synaptophysin copy numbers in glutamatergic and GABAergic wildtype ISVs (Grønborg et al. 2010).

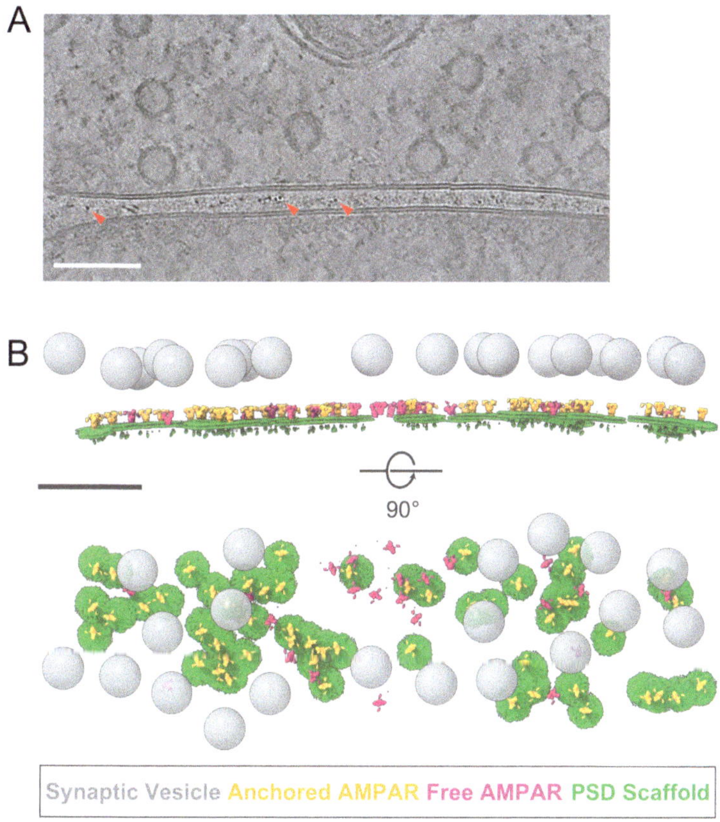

Fig. 2.11 (**a**) Slice (1 nm thick) through a 3D cryo-ET reconstruction of a synapse (1 nm/pixel). Scale bar: 100 nm. Red arrows indicate selected Au-NP positions. (**b**) Representative AMPAR-ECD (yellow and magenta) and PSD (green) sub-tomogram average maps back-plotted to their tomogram positions and displayed along with membrane-proximal synaptic vesicles (gray (Held et al. 2024b)). AMPARs engaged with PSD scaffolds are colored yellow, whereas un-anchored AMPARs are colored magenta (Held et al. 2024b)

2.11 New Insights into the Nanotopography of Synapses

Although docking presynaptic proteins and their complexes into densities is compelling (Fig. 2.2b–d), subtomogram averaging will be required to confirm the identity of the observed densities. Moreover, most observed densities will likely require labeling or genetic knockout approaches for identification. As a first step to establish the utility of labeling, endogenously expressed AMPA receptors were labeled with gold nanoparticles (AuNPs) using SpyCatcher/Tag and mapped the molecular topography of AMPARs mapped with cryo-ET reconstructions, relating receptor localizations to synapse ultrastructure in near-native conditions (Held et al. 2024b) (Fig. 2.11a).

Supramolecular clusters of active zone and postsynaptic density scaffolds were observed on the scale of tens of nanometers and trans-synaptic alignment of said nanoclusters into nanocolumns (Held et al. 2024a, b), consistent with previous studies using super-resolution fluorescence microscopy (MacGillavry et al. 2013; Nair et al. 2013; Sakamoto et al. 2018; Tang et al. 2016) and cryo-ET studies of unlabeled synaptosomes (Martinez-Sanchez et al. 2021). However, in the studies of intact synapses, membrane-proximal synaptic vesicles are mostly offset from AMPAR nanoclusters and trans-synaptic scaffold nanocolumns (Held et al. 2024a, b) (Fig. 2.11b). Thus, synaptic vesicles are not preferentially aligned with nanocolumns, although there are a very small number of membrane-proximal (membrane–membrane distance less than 10 nm) synaptic vesicles aligned with AMPARs.

What could be the mechanism for these offsets? Active zone nanoclusters may act at an upstream step to capture synaptic vesicles and allow the formation of initial membrane tethering interactions by Munc13 (Quade et al. 2019). These synaptic vesicles could then disperse from active zone nanoclusters and proceed with priming processes. While the resulting offset of synaptic vesicles may only amount to shifts of tens of nanometers, such distances can potentially impact downstream signal transduction (Eggermann et al. 2012; Raghavachari and Lisman 2004).

What is the consequence of these spatial offsets between synaptic vesicles and AMPAR nanoclusters? It is tempting to speculate that these offsets are related to different forms of synaptic vesicle fusion. Synaptic vesicles fuse both synchronously and asynchronously in response to action potentials (Südhof 2013), and fusion sites for both types of release are likely spatially segregated (Li et al. 2021; Malagon et al. 2023). Additionally, spontaneous release probably occurs at distinct release sites (Wang et al. 2022b) that could be spatially separated from evoked release sites. Moreover, the lateral offsets between AMPAR nanoclusters and membrane-proximal synaptic vesicles likely determine synaptic strength and plasticity. Monte Carlo simulations based on tomographic reconstructions suggest that offsets between synaptic vesicles and nano-clustered AMPARs introduce larger variability in postsynaptic response amplitude than expected if AMPARs were randomly distributed throughout the synapse (Held et al. 2024a), as had been observed in electrophysiological recordings of hippocampal neurons many years ago (Bekkers et al. 1990). Further studies are needed since the present cryo-ET reconstructions do not allow one to infer the vesicular release probability of membrane-proximal synaptic vesicles.

2.12 Conclusions

The recent results presented in this chapter illustrate the potential of in situ cryo-EM/cryo-ET imaging methods for discovering new interactions or uncovering the molecular details of a known interaction. We foresee rich opportunities to pursue other proteins and their complexes in the synapse and to visualize the predominant information transfer path in the nervous system at a molecular level. Such knowledge of the spatial organization of the synapse will provide a framework for a deeper understanding of the molecular mechanisms that underlie neuronal communication.

Acknowledgments We thank Josep Rizo, Liv Jensen, Paul Lapios, and Vladan Lučić for stimulating discussions and Vladan Lučić for providing the cryo-ET maps shown in Fig. 2.3. Funding was provided by the US National Institutes of Health (NIH; RO1MH63105 to A.T.B.).

References

Adams DJ, Arthur CP, Stowell MHB. Architecture of the synaptophysin/synaptobrevin complex: structural evidence for an entropic clustering function at the synapse. Sci Rep. 2015;5(1):13659.

Aeffner S, Reusch T, Weinhausen B, Salditt T. Energetics of stalk intermediates in membrane fusion are controlled by lipid composition. Proc Natl Acad Sci USA. 2012;109(25):E1609–18.

Alten B, Zhou Q, Shin OH, Esquivies L, Lin PY, White KI, et al. Role of aberrant spontaneous neurotransmission in SNAP25-associated encephalopathies. Neuron. 2021;109(1):59–72.e5.

Arthur CP, Stowell MHB. Structure of synaptophysin: a hexameric MARVEL-domain channel protein. Struct Lond Engl 1993. 2007;15(6):707–14.

Basu J, Shen N, Dulubova I, Lu J, Guan R, Guryev O, et al. A minimal domain responsible for Munc13 activity. Nat Struct Mol Biol. 2005;12(11):1017–8.

Bekkers JM, Richerson GB, Stevens CF. Origin of variability in quantal size in cultured hippocampal neurons and hippocampal slices. Proc Natl Acad Sci USA. 1990;87(14):5359–62.

Bera M, Radhakrishnan A, Coleman JK, Sundaram RV, Ramakrishnan S, Pincet F, et al. Synaptophysin chaperones the assembly of 12 SNAREpins under each ready-release vesicle. Proc Natl Acad Sci USA [Internet]. 2023;120(45):e2311484120. Available from: https://pnas.org/doi/10.1073/pnas.2311484120.

Bodzęta A, Kahms M, Klingauf J. The presynaptic v-ATPase reversibly disassembles and thereby modulates exocytosis but is not part of the fusion machinery. Cell Rep. 2017;20(6):1348–59.

Brunger AT, Leitz J. The core complex of the Ca2+-triggered presynaptic fusion machinery. J Mol Biol. 2023;435(1):167853.

Brunger AT, Choi UB, Lai Y, Leitz J, Zhou Q. Molecular mechanisms of fast neurotransmitter release. Annu Rev Biophys. 2018;47(1):469–97.

Buchholz TO, Jordan M, Pigino G, Jug F. Cryo-CARE: content-aware image restoration for cryo-transmission electron microscopy data. In: 2019 IEEE 16th international symposium on biomedical imaging (ISBI 2019). IEEE; 2018. p. 1–5.

Burkhardt P, Hattendorf DA, Weis WI, Fasshauer D. Munc18a controls SNARE assembly through its interaction with the syntaxin N-peptide. EMBO J. 2008;27(7):923–33.

Camacho M, Basu J, Trimbuch T, Chang S, Pulido-Lozano C, Chang SS, et al. Heterodimerization of Munc13 C2A domain with RIM regulates synaptic vesicle docking and priming. Nat Commun. 2017;8(1):15293.

Camacho M, Quade B, Trimbuch T, Xu J, Sari L, Rizo J, et al. Control of neurotransmitter release by two distinct membrane-binding faces of the Munc13-1 C1C2B region. elife. 2021;10:1–34.

Chen X, Tomchick DR, Kovrigin E, Araç D, Machius M, Südhof TC, et al. Three-dimensional structure of the complexin/SNARE complex. Neuron. 2002;33(3):397–409.

Choi UB, Zhao M, Zhang Y, Lai Y, Brunger AT. Complexin induces a conformational change at the membrane-proximal C-terminal end of the SNARE complex. elife. 2016;5(9):e16886.

Choi UB, Zhao M, White KI, Pfuetzner RA, Esquivies L, Zhou Q, et al. NSF-mediated disassembly of on- and off-pathway SNARE complexes and inhibition by complexin. elife [Internet]. 2018;7:e36497. Available from: https://elifesciences.org/articles/36497.

Coupland CE, Karimi R, Bueler SA, Liang Y, Courbon GM, Di Trani JM, et al. High-resolution electron cryomicroscopy of V-ATPase in native synaptic vesicles. Science. 2024;385(6705):168–74.

Deng L, Kaeser PS, Xu W, Südhof TC. RIM proteins activate vesicle priming by reversing autoinhibitory homodimerization of Munc13. Neuron. 2011;69(2):317–31.

Denker A, Kröhnert K, Bückers J, Neher E, Rizzoli SO. The reserve pool of synaptic vesicles acts as a buffer for proteins involved in synaptic vesicle recycling. Proc Natl Acad Sci USA. 2011;108(41):17183–8.

Dulubova I, Lou X, Lu J, Huryeva I, Alam A, Schneggenburger R, et al. A Munc13/RIM/Rab3 tripartite complex: from priming to plasticity? EMBO J. 2005;24(16):2839–50.

Eggermann E, Bucurenciu I, Goswami SP, Jonas P. Nanodomain coupling between Ca2+ channels and sensors of exocytosis at fast mammalian synapses. Nat Rev Neurosci. 2012;13(1):7–21.

Ellis RJ. Macromolecular crowding: obvious but underappreciated. Trends Biochem Sci. 2001;26(10):597–604.

Eshkind LG, Leube RE. Mice lacking synaptophysin reproduce and form typical synaptic vesicles. Cell Tissue Res. 1995;282(3):423–33.

Fernández-Busnadiego R, Asano S, Oprisoreanu AM, Sakata E, Doengi M, Kochovski Z, et al. Cryo-electron tomography reveals a critical role of RIM1 α in synaptic vesicle tethering. J Cell Biol. 2013;201(5):725–40.

Gao Y, Khan YA, Mo W, White KI, Perkins M, Pfuetzner RA, et al. Sensory deficit screen identifies nsf mutation that differentially affects SNARE recycling and quality control. Cell Rep. 2023;42(4):112345.

Gipson P, Fukuda Y, Danev R, Lai Y, Chen DH, Baumeister W, et al. Morphologies of synaptic protein membrane fusion interfaces. Proc Natl Acad Sci USA. 2017;114(34):9110–5.

Grønborg M, Pavlos NJ, Brunk I, Chua JJE, Münster-Wandowski A, Riedel D, et al. Quantitative comparison of glutamatergic and GABAergic synaptic vesicles unveils selectivity for few proteins including MAL2, a novel synaptic vesicle protein. J Neurosci. 2010;30(1):2–12.

Grushin K, Wang J, Coleman J, Rothman JE, Sindelar CV, Krishnakumar SS. Structural basis for the clamping and Ca2+ activation of SNARE-mediated fusion by synaptotagmin. Nat Commun. 2019;10(1):1–12.

Grushin K, Kalyana Sundaram RV, Sindelar CV, Rothman JE. Munc13 structural transitions and oligomers that may choreograph successive stages in vesicle priming for neurotransmitter release. Proc Natl Acad Sci USA [Internet]. 2022;119(7):e2121259119. Available from: https://pnas.org/doi/full/10.1073/pnas.2121259119.

Held RG, Kaeser PS. ELKS active zone proteins as multitasking scaffolds for secretion. Open Biol. 2018;8(2):170258.

Held RG, Liang J, Brunger AT. Nanoscale architecture of synaptic vesicles and scaffolding complexes revealed by cryo-electron tomography. Proc Natl Acad Sci USA [Internet]. 2024a;121(27):e2403136121. Available from: https://pnas.org/doi/10.1073/pnas.2403136121.

Held RG, Liang J, Esquivies L, Khan YA, Wang C, Azubel M, et al. In-situ structure and topography of AMPA receptor scaffolding complexes visualized by cryoET. bioRxiv. 2024b. https://doi.org/10.1101/2024.10.19.619226.

Heuser JE, Reese TS. Evidence for recycling of synaptic vesicle membrane during transmitter release at the frog neuromuscular junction. J Cell Biol. 1973;57(2):315–44.

Huang X, Sun S, Wang X, Fan F, Zhou Q, Lu S, et al. Mechanistic insights into the SNARE complex disassembly. Sci Adv. 2019;5(4):eaau8164.

Imig C, Min SW, Krinner S, Arancillo M, Rosenmund C, Südhof TC, et al. The morphological and molecular nature of synaptic vesicle priming at presynaptic active zones. Neuron. 2014;84(2):416–31.

Jaczynska K, Esquivies L, Pfuetzner RA, Alten B, Brewer KD, Zhou Q, et al. Analysis of tripartite synaptotagmin-1-SNARE-complexin-1 complexes in solution. FEBS Open Bio. 2023;13:26–50.

Jaczynska K, Esser V, Xu J, Sari L, Lin MM, Rosenmund C, et al. A lever hypothesis for synaptotagmin-1 action in neurotransmitter release. Proc Natl Acad Sci USA. 2025;122(1):e2417941121.

Janz R, Südhof TC, Hammer RE, Unni V, Siegelbaum SA, Bolshakov VY. Essential roles in synaptic plasticity for synaptogyrin I and synaptophysin I. Neuron. 1999;24(3):687–700.

Kaeser PS, Deng L, Wang Y, Dulubova I, Liu X, Rizo J, et al. RIM proteins tether Ca2+ channels to presynaptic active zones via a direct PDZ-domain interaction. Cell. 2011;144(2):282–95.

Kalyana Sundaram RV, Jin H, Li F, Shu T, Coleman J, Yang J, et al. Munc13 binds and recruits SNAP25 to chaperone SNARE complex assembly. FEBS Lett. 2021;595(3):297–309.

Khan YA, White KI, Pfuetzner RA, Singal B, Esquivies L, Mckenzie G, et al. SNARE disassembly requires Sec18/NSF side loading. Nat Struct Mol Biol. 2025;32:1708-1720. Nature Structural & Molecular Biology. 2025; 32:1708–20.

Korogod N, Petersen CC, Knott GW. Ultrastructural analysis of adult mouse neocortex comparing aldehyde perfusion with cryo fixation. elife. 2015;4:e05793.

Kravcenko U, Ruwolt M, Kroll J, Yushkevich A, Zenkner M, Ruta J, et al. Molecular architecture of synaptic vesicles. bioRxiv. 2024. https://doi.org/10.1101/2024.04.11.588828.

Kusick GF, Ogunmowo TH, Watanabe S. Transient docking of synaptic vesicles: implications and mechanisms. Curr Opin Neurobiol. 2022;74:102535.

Lai Y, Choi UB, Leitz J, Rhee HJ, Lee C, Altas B, et al. Molecular mechanisms of synaptic vesicle priming by Munc13 and Munc18. Neuron. 2017;95(3):591–607.e10.

Lai Y, Fois G, Flores JR, Tuvim MJ, Zhou Q, Yang K, et al. Inhibition of calcium-triggered secretion by hydrocarbon-stapled peptides. Nature. 2022;603(7903):949–56.

Leitz J, Wang C, Esquivies L, Pfuetzner RA, Peters JJ, Couoh-Cardel S, et al. Beyond the MUN domain, Munc13 controls priming and depriming of synaptic vesicles. Cell Rep. 2024a;43(5):114026.

Leitz J, Wang C, Esquivies L, Peters JJ, Gopal N, Pfuetzner RA, et al. Observing isolated synaptic vesicle association and fusion ex vivo. Nat Protoc [Internet]. 2024b;19:3139–61. Available from: https://www.nature.com/articles/s41596-024-01014-x.

Li S, Raychaudhuri S, Lee SA, Brockmann MM, Wang J, Kusick G, et al. Asynchronous release sites align with NMDA receptors in mouse hippocampal synapses. Nat Commun. 2021;12(1):677.

Lin KH, Taschenberger H, Neher E. A sequential two-step priming scheme reproduces diversity in synaptic strength and short-term plasticity. Proc Natl Acad Sci USA. 2022;119(34):e2207987119.

Lou X, Shin J, Yang Y, Kim J, Shin YK. Synaptotagmin-1 is an antagonist for Munc18-1 in SNARE zippering. J Biol Chem. 2015;290:10535.

Lučić V, Fernández-Busnadiego R, Laugks U, Baumeister W. Hierarchical detection and analysis of macromolecular complexes in cryo-electron tomograms using Pyto software. J Struct Biol. 2016;196(3):503–14.

Ma C, Su L, Seven AB, Xu Y, Rizo J. Reconstitution of the vital functions of Munc18 and Munc13 in neurotransmitter release. Science. 2013;339(6118):421–5.

MacGillavry HD, Song Y, Raghavachari S, Blanpied TA. Nanoscale scaffolding domains within the postsynaptic density concentrate synaptic AMPA receptors. Neuron. 2013;78(4):615–22.

Malagon G, Myeong J, Klyachko VA. Two forms of asynchronous release with distinctive spatio-temporal dynamics in central synapses. elife. 2023;12:e84041.

Martinez-Sanchez A, Laugks U, Kochovski Z, Papantoniou C, Zinzula L, Baumeister W, et al. Trans-synaptic assemblies link synaptic vesicles and neuroreceptors. Sci Adv. 2021;7:eabe6204.

Maycox PR, Link E, Reetz A, Morris SA, Jahn R. Clathrin-coated vesicles in nervous tissue are involved primarily in synaptic vesicle recycling. J Cell Biol. 1992;118(6):1379–88.

Matveeva EA, Whiteheart SW, Vanaman TC, & Slevin JT, Phosphorylation of the N-ethylmaleimide-sensitive factor isassociated with depolarization-dependent neurotransmitter release from synaptosomes. J. Biol. Chem. 2001; 276:12174–12181.

McMahon HT, Bolshakov VY, Janz R, Hammer RE, Siegelbaum SA, Südhof TC. Synaptophysin, a major synaptic vesicle protein, is not essential for neurotransmitter release. Proc Natl Acad Sci USA. 1996;93(10):4760–4.

Melland H, Carr EM, Gordon SL. Disorders of synaptic vesicle fusion machinery. J Neurochem. 2021;157(2):130–64.

Misura KM, Scheller RH, Weis WI. Three-dimensional structure of the neuronal-Sec1-syntaxin 1a complex. Nature. 2000;404(6776):355–62.

Mosalaganti S, Obarska-Kosinska A, Siggel M, Taniguchi R, Turoňová B, Zimmerli CE, et al. AI-based structure prediction empowers integrative structural analysis of human nuclear pores. Science. 2022;376(6598):eabm9506.

Murthy VN, Stevens CF. Reversal of synaptic vesicle docking at central synapses. Nat Neurosci. 1999;2(6):503–7.

Nair D, Hosy E, Petersen JD, Constals A, Giannone G, Choquet D, et al. Super-resolution imaging reveals that AMPA receptors inside synapses are dynamically organized in nanodomains regulated by PSD95. J Neurosci. 2013;33(32):13204–24.

Neher E, Brose N. Dynamically primed synaptic vesicle states: key to understand synaptic short-term plasticity. Neuron. 2018;100(6):1283–91.

Orozco-Borunda DH, Papantoniou C, Brose N, Lucic V. Direct cryo-ET detection of native SNARE and Munc13 protein bridges using AI classification and preprocessing. bioRxiv. 2025. https://doi.org/10.1101/2024.12.18.629213.

Papantoniou C, Laugks U, Betzin J, Capitanio C, Ferrero JJ, Sánchez-Prieto J, et al. Munc13- and SNAP25-dependent molecular bridges play a key role in synaptic vesicle priming. Sci Adv. 2023;9(25):eadf6222.

Parisotto D, Pfau M, Scheutzow A, Wild K, Mayer MP, Malsam J, et al. An extended helical conformation in domain 3a of Munc18-1 provides a template for SNARE (soluble N-ethylmaleimide-sensitive factor attachment protein receptor) complex assembly. J Biol Chem. 2014;289(14):9639–50.

Park D, Fujise K, Wu Y, Luján R, Del Olmo-Cabrera S, Wesseling JF, et al. Overlapping role of synaptophysin and synaptogyrin family proteins in determining the small size of synaptic vesicles. Proc Natl Acad Sci USA. 2024;121(29):e2409605121.

Pérez-Lara Á, Thapa A, Nyenhuis SBB, Nyenhuis DAA, Halder P, Tietzel M, et al. PtdInsP $_2$ and PtdSer cooperate to trap synaptotagmin-1 to the plasma membrane in the presence of calcium. elife. 2016;5:1–22.

Peters JJ, Leitz J, Guo Q, Beck F, Baumeister W, Brunger AT. A feature-guided, focused 3D signal permutation method for subtomogram averaging. J Struct Biol. 2022;214(2):107851.

Prinslow EA, Stepien KP, Pan YZ, Xu J, Rizo J. Multiple factors maintain assembled trans-SNARE complexes in the presence of NSF and αSNAP. elife [Internet]. 2019;8:e38880. Available from: https://elifesciences.org/articles/38880.

Quade B, Camacho M, Zhao X, Orlando M, Trimbuch T, Xu J, et al. Membrane bridging by Munc13-1 is crucial for neurotransmitter release. elife [Internet]. 2019;8:e42806. Available from: https://elifesciences.org/articles/42806.

Radhakrishnan A, Li X, Grushin K, Krishnakumar SS, Liu J, Rothman JE. Symmetrical arrangement of proteins under release-ready vesicles in presynaptic terminals. Proc Natl Acad Sci USA. 2021;118(5):e2024029118.

Raghavachari S, Lisman JE. Properties of quantal transmission at CA1 synapses. J Neurophysiol. 2004;92(4):2456–67.

Raja MK, Preobraschenski J, Del Olmo-Cabrera S, Martinez-Turrillas R, Jahn R, Perez-Otano I, et al. Elevated synaptic vesicle release probability in synaptophysin/gyrin family quadruple knockouts. elife. 2019;8:1–23.

Rizo J. Molecular mechanisms underlying neurotransmitter release. Annu Rev Biophys. 2022;51(1):377–408.

Rothman JE. The principle of membrane fusion in the cell (Nobel lecture). Angew Chem Int Ed. 2014;53(47):12676–94.

Ryu JK, Min D, Rah SH, Kim SJ, Park Y, Kim H, et al. Spring-loaded unraveling of a single SNARE complex by NSF in one round of ATP turnover. Science. 2015;347(6229):1485–9.

Sakamoto H, Ariyoshi T, Kimpara N, Sugao K, Taiko I, Takikawa K, et al. Synaptic weight set by Munc13-1 supramolecular assemblies. Nat Neurosci. 2018;21(1):41–9.

Schuller AP, Wojtynek M, Mankus D, Tatli M, Kronenberg-Tenga R, Regmi SG, et al. The cellular environment shapes the nuclear pore complex architecture. Nature. 2021;598(7882):667–71.

Shu T, Jin H, Rothman JE, Zhang Y. Munc13-1 MUN domain and Munc18-1 cooperatively chaperone SNARE assembly through a tetrameric complex. Proc Natl Acad Sci USA. 2020;117(2):1036–41.

Sieber JJ, Willig KI, Kutzner C, Gerding-Reimers C, Harke B, Donnert G, et al. Anatomy and dynamics of a supramolecular membrane protein cluster. Science. 2007;317(5841):1072–6.

Silva M, Tran V, Marty A. Calcium-dependent docking of synaptic vesicles. Trends Neurosci. 2021;44(7):579–92.

Sinha R, Ahmed S, Jahn R, Klingauf J. Two synaptobrevin molecules are sufficient for vesicle fusion in central nervous system synapses. Proc Natl Acad Sci USA. 2011;108(34):14318–23.

Stepien KP, Prinslow EA, Rizo J. Munc18-1 is crucial to overcome the inhibition of synaptic vesicle fusion by αSNAP. Nat Commun. 2019;10(1):4326.

Stepien KP, Xu J, Zhang X, Bai XC, Rizo J. SNARE assembly enlightened by cryo-EM structures of a synaptobrevin–Munc18-1–syntaxin-1 complex. Sci Adv. 2022;8:eabo5272.

Südhof TC. The presynaptic active zone. Neuron. 2012;75(1):11–25

Südhof TC. Neurotransmitter release: the last millisecond in the life of a synaptic vesicle. Neuron. 2013;80(3):675–90.

Südhof TC. Synaptic neurexin complexes: a molecular code for the logic of neural circuits. Cell. 2017;171(4):745–69.

Sutton RBB, Fasshauer D, Jahn R, Brunger AT. Crystal structure of a SNARE complex involved in synaptic exocytosis at 2.4 Å resolution. Nature. 1998;395(6700):347–53.

Takamori S, Holt M, Stenius K, Lemke EA, Grønborg M, Riedel D, et al. Molecular anatomy of a trafficking organelle. Cell. 2006;127(4):831–46.

Tang AH, Chen H, Li TP, Metzbower SR, MacGillavry HD, Blanpied TA. A trans-synaptic nanocolumn aligns neurotransmitter release to receptors. Nature [Internet]. 2016;536:210–4. Available from: http://www.nature.com/doifinder/10.1038/nature19058.

Tegunov D, Cramer P. Real-time cryo electron microscopy data preprocessing with Warp. Nat Methods. 2019;16(11):1146–52.

Tegunov D, Xue L, Dienemann C, Cramer P, Mahamid J. Multi-particle cryo-EM refinement with M visualizes ribosome-antibiotic complex at 3.5 Å in cells. Nat Methods. 2021;18(2):186–93.

Toulmé E, Salazar Lázaro A, Trimbuch T, Rizo J, Rosenmund C. Neurotransmitter release is triggered by a calcium-induced rearrangement in the synaptotagmin-1/SNARE complex primary interface. Proc Natl Acad Sci USA. 2024;121(42):e2409636121.

Trimbuch T, Xu J, Flaherty D, Tomchick DR, Rizo J, Rosenmund C. Re-examining how complexin inhibits neurotransmitter release. elife. 2014;3:e02391.

Turk M, Baumeister W. The promise and the challenges of cryo-electron tomography. FEBS Lett. 2020;594(20):3243–61.

Voleti R, Jaczynska K, Rizo J. Ca2+-dependent release of synaptotagmin-1 from the SNARE complex on phosphatidylinositol 4,5-bisphosphate-containing membranes. elife [Internet]. 2020;9:e57154. Available from: https://elifesciences.org/articles/57154.

Wang S, Li Y, Gong J, Ye S, Yang X, Zhang R, et al. Munc18 and Munc13 serve as a functional template to orchestrate neuronal SNARE complex assembly. Nat Commun. 2019;10(1):69.

Wang Z, Grange M, Pospich S, Wagner T, Kho AL, Gautel M, et al. Structures from intact myofibrils reveal mechanism of thin filament regulation through nebulin. Science. 2022a;375(6582):eabn1934.

Wang CS, Chanaday NL, Monteggia LM, Kavalali ET. Probing the segregation of evoked and spontaneous neurotransmission via photobleaching and recovery of a fluorescent glutamate sensor. elife. 2022b;11:e76008.

Wang C, Jiang W, Leitz J, Yang K, Esquivies L, Wang X, et al. Structure and topography of the synaptic V-ATPase–synaptophysin complex. Nature [Internet]. 2024;631(8022):899–904. Available from: https://www.nature.com/articles/s41586-024-07610-x.

Watanabe S, Trimbuch T, Camacho-Pérez M, Rost BR, Brokowski B, Söhl-Kielczynski B, et al. Clathrin regenerates synaptic vesicles from endosomes. Nature [Internet]. 2014;515:228–33. Available from: http://www.nature.com/doifinder/10.1038/nature13846.

Weber T, Zemelman BV, McNew JA, Westermann B, Gmachl M, Parlati F, et al. SNAREpins: minimal machinery for membrane fusion. Cell. 1998;92(6):759–72.

Weninger K, Bowen ME, Chu S, Brunger AT. Single-molecule studies of SNARE complex assembly reveal parallel and antiparallel configurations. Proc Natl Acad Sci USA. 2003;100(25):14800–5.

White KI, Zhao M, Choi UB, Pfuetzner RA, Brunger AT. Structural principles of SNARE complex recognition by the AAA+ protein NSF. elife [Internet]. 2018;7:e38888. Available from: https://elifesciences.org/articles/38888.

White KI, Khan YA, Qiu K, Balaji A, Choi UB, Couoh S, et al. Structural remodeling of target-SNARE protein complexes by NSF enables synaptic transmission. Nat Commun. 2025; 16:8371.

Wittig S, Ganzella M, Barth M, Kostmann S, Riedel D, Pérez-Lara Á, et al. Cross-linking mass spectrometry uncovers protein interactions and functional assemblies in synaptic vesicle membranes. Nat Commun. 2021;12(1):858.

Xu J, Camacho M, Xu Y, Esser V, Liu X, Trimbuch T, et al. Mechanistic insights into neurotransmitter release and presynaptic plasticity from the crystal structure of Munc13-1 C 1 C 2 BMUN. elife [Internet]. 2017;6:e22567. Available from: http://elifesciences.org/lookup/doi/10.7554/eLife.22567.

Yang X, Wang S, ShengY, Zhang M, Zou W, Wu L, et al. Syntaxin opening by the MUN domain underlies the function of Munc13 in synaptic-vesicle priming. Nat Struct Mol Biol. 2015;22(7):547–54.

Yousuf A. Khan, K. Ian White & Axel T. Brunger. The AAA+ superfamily:a review of the structural and mechanistic principles of these molecular machines, Critical Reviews in Biochemistry andMolecular Biology, 2021;57(2):156–187. https://doi.org/10.1080/10409238.2021.1979460.

Zhou, H. et al. Toward a comprehensive characterization of a human cancer cell phosphoproteome. J. Proteome Res. 2013;12, 260–271.
Zhao M, Wu S, Zhou Q, Vivona S, Cipriano DJ, Cheng Y, et al. Mechanistic insights into the recycling machine of the SNARE complex. Nature. 2015;518(7537):61–7.
Zhou Q, Lai Y, Bacaj T, Zhao M, Lyubimov AY, Uervirojnangkoorn M, et al. Architecture of the synaptotagmin–SNARE machinery for neuronal exocytosis. Nature. 2015;525(7567):62–7.
Zhou Q, Zhou P, Wang AL, Wu D, Zhao M, Südhof TC, et al. The primed SNARE–complexin–synaptotagmin complex for neuronal exocytosis. Nature. 2017;548(7668):420–5.

Chapter 3
Formation, Organization, and Maintenance of Synaptic Vesicle Pools: From Molecules to Structure and Dynamics

Natali L. Chanaday

Abstract Synaptic vesicles are small trafficking organelles that store neurotransmitters. Proper communication between neurons depends on the tight temporal and spatial regulation of synaptic vesicle fusion and recycling. Synapses evolved to have different pools of synaptic vesicles, defined both by their functional properties and molecular composition, allowing them to adapt the kinetics and type of neurotransmitter released to the functional demands of the neuron circuit. In this chapter, we will discuss the life cycle of synaptic vesicles, from biogenesis to fusion and regeneration at synapses, integrating our knowledge of their molecular identity, functional properties, and spatial organization at presynaptic axon terminals.

Keywords Synaptic vesicles · Neurotransmitter release · Release site organization · Synaptic vesicle recycling · Synaptic vesicle pools · Presynaptic calcium sources

3.1 Introduction: What Are Synaptic Vesicles?

In the 1950s, Bernard Katz and colleagues observed something interesting: they measured small ("miniature") end plate potentials in the muscle that resulted from the release of the neurotransmitter acetylcholine from the motor neuron axon in the absence of action potentials (Fatt and Katz 1950, 1952). After careful studies, they concluded that those miniature events represented the minimal unit of neurotransmission, or quanta, suggesting that neurotransmitters were not released continuously but in packets of a conserved size (Del Castillo and Katz 1954). How was that possible? Around the same time, Eduardo De Robertis and colleagues found "granular or vesicular" organelles in nerves from frogs and earthworms using electron microscopy, which they named synaptic vesicles (De Robertis and Bennett 1955).

N. L. Chanaday (✉)
Department of Physiology, Perelman School of Medicine, University of Pennsylvania, Philadelphia, PA, USA
e-mail: natali.chanaday@pennmedicine.upenn.edu

E. T. Kavalali (ed.), *Nano-organization of the Synapse*, Advances in Neurobiology 48,
https://doi.org/10.1007/978-3-032-12594-1_3

Thus, one hypothesis was that neurotransmitters were stored inside synaptic vesicles and the action potential would cause the exocytosis of these vesicles releasing a discrete amount of neurotransmitters. In the next decade, the laboratories of De Robertis and Victor Whittaker designed methods to isolate nerve terminals (synaptosomes) and, from them, obtain a purified fraction of synaptic vesicles. They were able to prove that synaptic vesicles contained acetylcholine (De Robertis et al. 1963; Whittaker et al. 1964) and even estimated the number of acetylcholine molecules per vesicle (Whittaker and Sheridan 1965). But something was missing: there was no direct evidence of synaptic vesicles fusing with the plasma membrane in response to action potentials. This last piece of the puzzle came in the 1970s from the work of Thomas Reese, John Heuser, and colleagues. Using electron microscopy and freeze-fracture methods, they caught vesicles "in the act" of fusing, showed the amount of fusion (exocytic load) correlates with stimulation intensity, that exocytosis happens independently at specific locations (now known as release sites), and presented the first evidence of synaptic vesicle endocytosis and reformation from coated pits and intermediate cisternae (possibly clathrin-mediated endocytosis and endosomes, respectively) (Heuser and Reese 1973; Heuser et al. 1974, 1979). How is the fusion of these synaptic vesicles coupled to the arrival of action potentials? Katz and colleagues knew that the amplitude of end plate potentials in the muscle scaled with calcium concentration in the extracellular space, suggesting that acetylcholine release was somehow modulated by calcium. Two contemporaries of Katz, Alan Hodgkin and Andrew Huxley, were trying to understand the molecular basis of electrical signals in neurons and muscle. In a series of publications in the 1950s, using elegantly designed experiments and careful interpretation, Hodgkin and Huxley were able to prove that ions, including sodium and potassium, diffuse across biological membranes by passing through ion channels that are activated by changes in voltage (i.e., membrane potential) explaining how action potentials are generated (Hodgkin and Huxley 1952a, b; Hodgkin et al. 1952). Thus, a link between membrane potential and ion flow was made. The presence of voltage-gated calcium channels in axons, and the fact that the consequent calcium influx facilitates neurotransmitter release, was then experimentally demonstrated by Katz, Ricardo Miledi, Rodolfo Llinás, George Augustine, Wade Regehr, and others using electrophysiological methods (Augustine and Eckert 1984; Augustine et al. 1985; Katz and Miledi 1965, 1967, 1970; Simon and Llinás 1985) as well as optical fluorescence (Sabatini and Regehr 1996; Smith et al. 1993). Then, a new question emerged: how does calcium trigger synaptic vesicle exocytosis? The answer came in the 1990s when a protein capable of binding calcium and phospholipids was discovered in synaptic vesicles: Synaptotagmin (Brose et al. 1992; Perin et al. 1990). Further work by Thomas Südhof, Edwin Chapman, Reinhard Jahn, and others uncovered the detailed molecular machinery mediating synaptic vesicle fusion and how it is regulated by Synaptotagmins (Chapman 2008; Jahn et al. 2024; Südhof 2013). In summary, synaptic vesicles are small trafficking organelles that are clustered at axons and loaded with neurotransmitters; they undergo calcium-dependent exocytosis in response to action potentials and mediate chemical synaptic transmission. In this chapter, we will cover the molecular aspects of the life of a synaptic vesicle: from biogenesis to fusion and endocytosis at synapses, focusing on the spatial nano-organization of the vesicles and the machinery regulating the synaptic vesicle cycle.

3.2 Biogenesis of Synaptic Vesicles

3.2.1 Sorting from the Golgi Apparatus

Synaptic vesicle constituents are sorted from the trans-Golgi network (TGN) via the secretory pathway, sharing regulatory components with other known transport organelles and plasma membrane proteins. In the model organism *C. elegans*, synaptic vesicle proteins are clustered and sorted from the neuronal Golgi via a mechanism dependent on the scaffolding protein Unc-16/JIP3 (JNK-interacting protein 3) and the protein kinase LRK1/LRRK2/PARK8 (leucine-rich repeat serine/threonine protein kinase 1—involved in polarized sorting from the Golgi) (Choudhary et al. 2017). The adaptors for clathrin-mediated endocytosis AP2 and AP3 are then recruited, leading to the selective inclusion of synaptic vesicle proteins and exclusion from the transport vesicle of proteins resident of the Golgi and other somatic organelles (Choudhary et al. 2017; Edwards et al. 2015). Work in the *Drosophila* neuromuscular junction and in rodent hippocampal neurons showed that these post-TGN transport vesicles not only include synaptic vesicle proteins, such as the Vesicular Glutamate Transporter 1 (VGluT1) and Synaptotagmin-1 (Syt1), but also carry other axonal components, such as the molecules that form the active zone cytomatrix Bruchpilot/Bassoon, and lysosomal proteins, such as the small GTPase Arl8 (Vukoja et al. 2018). Lysosomes, in fact, share a common precursor compartment with synaptic vesicles (Newell-Litwa et al. 2009), suggesting that changes in the sorting of synaptic molecules might alter the lysosome composition or function, and vice versa (this remains to be experimentally demonstrated). These precursor vesicles are transported out of the soma and into the axon by the motor proteins KIF1A/KIF1Bβ, via interaction of the vesicular protein Rab3A and the motor adaptor MADD (Dou et al. 2024; Hummel and Hoogenraad 2021; Niwa et al. 2008). Recent work using iPSC-derived human-induced neurons demonstrated that the local accumulation of precursor vesicles at proto-presynapses is independent of the presence of postsynaptic contacts. These sites are instead defined by increased microtubule dynamics via a mechanism dependent on the severing enzyme Spastin (Aiken and Holzbaur 2023). In conclusion, multiple components necessary to assemble presynapses seem to already be packaged together at the level of the TGN (Dresbach et al. 2006; Ghelani et al. 2021) and then undergo polarized transport toward the axon, where they contribute to the formation, growth, and navigation of the axon (Nichols and Smith 2019; Vukoja et al. 2018).

3.2.2 Generation of Synaptic Vesicles from Precursor Vesicles

Early studies on dissociated hippocampal neurons and the neuroendocrine cell line PC12 proposed that synaptic vesicle proteins, particularly Synaptophysin, cycle multiple times between endosomal intermediates and the plasma membrane of the axon before forming synaptic vesicles (Provoda et al. 2000; Régnier-Vigouroux et al. 1991). Vesicular structures of varying size and shape, larger than synaptic vesicles, were reported in axons (Ahmari et al. 2000). These successive fusion and endocytosis events are mediated by clathrin adaptors including AP-3 (Faundez and Kelly 2000) (but also see Mullins et al. (2000)) and AP-180 (Nonet et al. 1999). The specific interaction of synaptic vesicle proteins with each other leads to clustering and enrichment through these successive constitutive exo–endocytosis cycles, indicating that the synaptic proteins themselves determine the transport pathway. Thus, alterations in synaptic vesicle number and/or shape and recycling in response to neuronal activity have been observed in neurons lacking synaptic vesicle-specific proteins, including Synaptogyrin (a tetraspanin containing the MARVEL domain—a domain involved in membrane transport and apposition), Synaptobrevin-2 (a vesicular SNARE that mediates vesicle fusion), and VGluT1 (Deák et al. 2004; Santos et al. 2009; Stevens et al. 2012). Interactions of synaptic vesicle proteins with motor and cytoskeletal proteins are also responsible for the polarized transport of the precursor carriers (Aiken and Holzbaur 2023; Choudhary et al. 2017; Okada et al. 1995; Song et al. 2009). Moreover, interaction with lipids might also contribute to the clustering of specific proteins and lipids in synaptic vesicles. For example, Synaptophysin (another tetraspanin with the MARVEL domain) binds to cholesterol, and in this way, it might lead to enrichment of synaptic vesicle components and induce membrane curvature necessary to form synaptic vesicles (Thiele et al. 2000). Synaptic vesicles have a larger cholesterol-to-phospholipid ratio than other organelles and smaller than the plasma membrane (ratios are 0.8 for synaptic vesicles, 0.15 for endoplasmic reticulum, 0.5 for endosomes, and 1 for plasma membrane) (Binotti et al. 2021). Synaptic vesicles also have a lower proportion of phosphatidylinositol than the plasma membrane (Binotti et al. 2021), suggesting that not only the proteins but lipids also are enriched or de-enriched through the cycling and trafficking of synaptic vesicle precursors (for a detailed list of synaptic vesicle protein and lipid composition, see Takamori et al. (2006)). Historically, research endeavors have been focused on the transport of synaptic proteins, so how the specific lipid composition of synaptic vesicles is achieved remains largely unknown in neuroscience. Taking advantage of modern methodologies, the polarized delivery of synaptic vesicle proteins to axons has recently been further verified, proposing that the cycling through the plasma membrane may constitute a "depot" for synaptic vesicle biogenesis to be used on demand by the neuron (Watson et al. 2023). Our current knowledge, then, points to a polarized transport toward the axon of TGN-derived presynaptic proteins packaged in precursor vesicles. Besides lysosomal and synaptic vesicle components, the precursor vesicles carry preformed aggregates of active zone proteins (see next section) including Piccolo and Bassoon

(Tao-Cheng 2007). These precursor carriers then fuse with the axonal plasma membrane and cycle through endosomes multiple times before forming mature synaptic vesicles, as defined by their size (40–50 nm diameter) and molecular composition (presence of vesicular neurotransmitter transporters, fusion machinery, and other synaptic vesicle proteins).

3.2.3 *Open Questions*

Numerous laboratories have been investigating the biogenesis of synaptic vesicles for the past 20–30 years and on how the polarized delivery of presynaptic machinery to axons is achieved. However, many questions remain to be answered: Is the transport of synaptic vesicle molecules constitutive or can it be modulated by neuronal activity (by voltage or calcium, for example)? How are non-presynaptic molecules excluded from axons? (see Eichel et al. (2022) and Song et al. (2009)). How are presynaptic components retained at synapses? Regarding this last question, and specifically for synaptic vesicles, clustering at presynaptic boutons is partly mediated by a liquid–liquid phase separation driven by the phosphoprotein Synapsin (Milovanovic et al. 2018). Elimination of Synapsins from neurons leads to redistribution of synaptic vesicles along the axon, reducing the total number of vesicles, but without affecting the number of synaptic vesicles that are docked and ready to fuse at the active zone (Gitler et al. 2004; Siksou et al. 2007). Interaction with the scaffold proteins of the active zone cytomatrix contributes to the retaining of synaptic vesicles, and it also increases their stability by inhibiting the degradation of synaptic proteins and thus synaptic degeneration (Fernández-Busnadiego et al. 2013; Waites et al. 2013). The spatial organization of synaptic vesicles in presynaptic terminals and their functional classification into different pools will be the focus of the next sections.

3.3 Where and When Synaptic Vesicles Fuse

Synaptic transmission is initiated by the arrival of the action potential (AP) which triggers the opening of Voltage-Gated Calcium Channels (VGCC) and the consequent rise in presynaptic calcium levels. This increase in calcium triggers the fusion of neurotransmitter-loaded synaptic vesicles at highly specialized presynaptic release sites in the active zone. Active zones are critical during neurotransmission, since they convert the electrical signals into chemical signals, the neurotransmitters, allowing the passing of information to the efferent neuron. Thus, the size, structure, and composition of the active zone, including calcium channels and synaptic vesicles, determine the efficiency of neurotransmitter release and synaptic transmission in response to APs (Holderith et al. 2012; Südhof 2012).

3.3.1 Composition and Organization of the Active Zone

Electron microscopy studies have defined the active zone as the portion of presynaptic membrane that is next or opposed to the electron-dense synaptic cleft (or synaptic gap) and the postsynaptic density (Emperador-Melero and Kaeser 2020; Siksou et al. 2007). The active zone cytomatrix includes cytoskeletal proteins, scaffolding proteins, and core active zone proteins. Cytoskeletal proteins provide a framework for the transport, recruitment, and stabilization of active zone components, and scaffolding proteins act as intermediaries, linking the core active zone proteins to each other and to the cytoskeleton and thus shaping the spatial organization and function of the active zone (Fig. 3.1). Central scaffolding proteins and the molecular machinery constituting the core of the active zone are evolutionarily conserved (Böhme et al. 2018; Zhai and Bellen 2004), pointing to the importance of active zone organization in synaptic transmission. One key role of active zone scaffolds, including CAST/ELKS in mammals or Bruchpilot (BRP) in *Drosophila*, is to recruit and cluster VGCC and synaptic vesicles at the active zone (Böhme et al. 2018; Held and Kaeser 2018). CAST/ELKS/BRP work via interaction with other central scaffolds, namely RIM (Rab3-interacting molecule) and RIM-binding protein (RIM-BP) (Mittelstaedt et al. 2010; Wang et al. 1997). RIM is essential for the docking of synaptic vesicles and the clustering of VGCC at active zones across multiple species, including the nematode *C. elegans*, *Drosophila*, and rodents (mammals). Docking and priming of synaptic vesicles are achieved via the interaction of RIM with Rab3 on the synaptic vesicles and Unc13/Unc18 (called Munc13 and Munc18 in mammals). Munc13 and Munc18 interact with the synaptic vesicle fusion machinery, the SNARE proteins (for soluble NSF [N-ethylmaleimide-sensitive fusion protein] attachment protein [or SNAP] receptor) orchestrating their assembly into a functional configuration. In this way, Munc13 and Munc18 bridge the active zone plasma membrane and the vesicular membrane and thus prime the synaptic vesicles for fast fusion (Papantoniou et al. 2023; Park et al. 2017; Quade et al. 2019; Sassa et al. 1999). Elimination of Munc13 or Munc18 abrogates all forms of synaptic transmission (without affecting synapse formation) (Richmond et al. 1999; Verhage et al. 2000). Besides docking and priming of synaptic vesicles, CAST/ELKS/BRP scaffolds and RIM/RIM-BP recruit VGCC to active zones (Fig. 3.1). These sites where the active zone scaffolds bring VGCC in close proximity to synaptic vesicles are called release sites and are highly specialized for efficient and fast neurotransmitter release in response to APs. By regulating the number of VGCC and synaptic vesicles recruited to these sites and the distance among them (named the "coupling distance"), active zone scaffolds modulate the presynaptic probability of neurotransmitter release (PR).

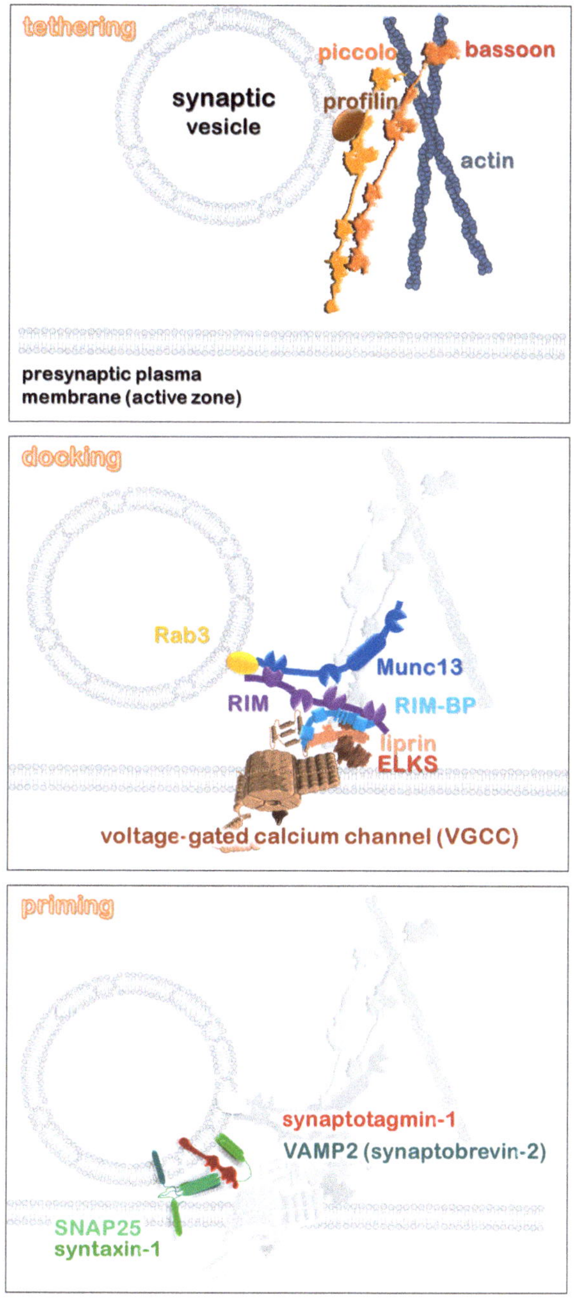

Fig. 3.1 Simplified representation of the molecular machinery involved in synaptic vesicle tethering, docking and priming at release sites in the active zone

3.3.2 Calcium-Driven Synaptic Vesicle Fusion Is Probabilistic

The small group of vesicles that are docked and primed at AZ release sites constitutes the readily releasable pool (RRP). As the name infers, these vesicles are the first ones to fuse and release neurotransmitters in response to APs. The precise timing of release is facilitated by the interaction of the SNARE complex with Synaptotagmins (Syts) and Complexins (Cpx). Syts are highly conserved transmembrane proteins involved in vesicular transport that contain two calcium-binding motifs (C2 domains) in their cytoplasmic region. Syts are a family of 16 genes in mammals (8 in *C. elegans* and 7 in *Drosophila*), with Syt isoforms 1, 2 and 9 mediating fast AP-driven neurotransmitter release in the mammalian nervous system (Chapman 2008; Südhof 2002; Xu et al. 2007). Syt1 C2 domains bind calcium and phospholipids in a coordinated manner, and they also bind to the assembled SNARE complex. Syt1 plays a dual role in neurotransmitter release by clamping the fusion machinery in a partially assembled conformation, thus preventing fusion in resting conditions (i.e., spontaneous release), and allowing the full zippering of the SNAREs upon calcium binding, providing fast (sub-millisecond) time-locked neurotransmitter release in response to APs (Courtney et al. 2021; Südhof 2013; Zhou et al. 2015). To precisely perform both functions, clamping and evoking release, Syt1 works cooperatively with the small cytoplasmic protein Cpx (Ramakrishnan et al. 2020; Zhou et al. 2017). In summary, for synaptic vesicles in the RRP to undergo AP-driven neurotransmitter release, they have to be placed close to VGCC at release sites via the AZ scaffolding machinery and they have to be docked and primed for fusion by Munc13/Munc18, Syt1/Cpx, and the SNARE complex. How efficient and stable is this presynaptic machinery?

The probability of a single synaptic vesicle to fuse in response to an AP, or PR, is highly variable among different types of synapses, and even among individual synapses of the same type and emerging from the same axon (Dürst et al. 2022; Körber and Kuner 2016; Rosenmund et al. 1993) (although evidence from the *Drosophila* neuromuscular junction suggests that PR is stable for individual AZs, see Akbergenova et al. (2018)). So, why is PR variable and what is the evolutionary advantage of having a system that sometimes fails (i.e., PR lower than 1)? One determinant of PR is the number of VGCCs at AZs. At the giant calyx of Held synapse in the auditory system, an increased amount of densely packed VGCCs (i.e., they are clustered, not randomly distributed) leads to a higher rise in calcium concentration around the release site, called calcium nanodomain, thus boosting the probability of calcium-dependent synaptic vesicle fusion (Lübbert et al. 2019; Nakamura et al. 2015). A similar correlation between PR and number of VGCCs has been observed at other synapse types and animal models (Akbergenova et al. 2018; Holderith et al. 2012; Martinez-Dunst et al. 1997). The number of VGCCs at AZs is below saturation, and it can range from tens to a few hundred molecules (Zhang et al. 2022), partially explaining the variability in PR. To note, in small central mammalian synapses, only one VGCC has been proposed to be linked to one synaptic vesicle at release sites, and release sites seem to be more randomly distributed across the area of these small AZs (Scimemi and Diamond 2012). Global PR for a given synapse also scales up with the number of docked and primed vesicles at the AZ, i.e., the RRP size (Holderith et al. 2012; Müller et al. 2022). But for each individual synaptic vesicle, the intrinsic PR is also variable, and one reason for this is the coupling distance between VGCC and synaptic vesicles. At the calyx of Held, for example, the RRP vesicles fuse with two characteristic kinetics in response to APs, fast or slow. Fast-releasing RRP vesicles are located close to VGCC, while slower-releasing RRP vesicles are farther, i.e., have longer coupling distances to VGCC (Chen et al. 2015; Fekete et al. 2019; Keller et al. 2015). The distance between VGCC and synaptic vesicles is regulated by AZ scaffolds, including RIM/ELKS/BRP and other adaptor proteins, and it correlates with PR across different synapse types and species (Böhme et al. 2018, 2019; Butola et al. 2021; Ghelani et al. 2023; Holderith et al. 2012; Müller et al. 2022) (also see Dong et al. (2018); Grabner et al. (2015)) for other interesting VGCC modulatory roles of scaffolds). The copy number of the calcium sensor Syt per vesicle also affects the propensity of individual vesicles to fuse (Bouazza-Arostegui et al. 2022). Moreover, Syt and Cpx help position synaptic vesicles close to VGCC at release sites (Newman et al. 2022; Young and Neher 2009) and thus regulate PR not only by coupling calcium to the SNARE zippering but also by modulating the spatial organization of synaptic vesicles at release sites. As a result, another determinant of PR is the priming state of the synaptic vesicles, i.e., their readiness to fuse (Aldahabi et al. 2022). Collectively, it is the diversity in organization of the AZ molecules and synaptic vesicles, including location, clustering, and amount, that underlies the variability in PR among individual synaptic vesicles and AZs. The possibility of scaling up or down the presynaptic release strength by modulating the structure and composition of the AZ becomes then an interesting substrate for synaptic plasticity. Synaptic plasticity is a general term that refers to how synapses change their efficacy or strength in response to experience. During homeostatic synaptic plasticity, a mechanism that stabilizes neuron activity (Turrigiano 2012), presynaptic actin dynamics in combination with the recruitment of AZ proteins and synaptic vesicles to the RRP mediate the increase in PR and upscaling of synaptic strength (Glebov et al. 2017). Presynaptic homeostatic potentiation of the *Drosophila* neuromuscular junction depends on the recruitment of AZ scaffolds, such as RIM/BRP/Unc13, VGCC, and synaptic vesicles, to the RRP, enhancing neurotransmitter release (Böhme et al. 2019; Ghelani et al. 2023). Similar findings were obtained at the mammalian giant mossy fiber synapse in the hippocampus (Chamberland et al. 2017; Orlando et al. 2021). Interestingly, retrograde signaling from the postsynaptic dendritic spines has been proposed to determine PR in rodent hippocampal synapses (Branco et al. 2008). So, the low and highly variable PR of presynaptic terminals might have been evolutionarily advantageous because it expands the amount of information that synapses can encode and allows for bi-directional plastic changes in synaptic strength (also see Bartol et al. (2015) for an estimation of information storage capacity of the brain). So, we can conclude that one morphological correlate of plasticity is the size and organization of the AZ and the associated RRP of synaptic vesicles.

3.3.3 *Composition and Organization of Synaptic Vesicle Pools*

Functionally, the synaptic vesicles that fuse and release neurotransmitters in response to AP can be divided into two pools: the synchronous release pool and the asynchronous release pool. While synchronous release is time-locked to AP arrival, asynchronous release occurs with a delay of a few tens to hundreds of milliseconds (it can be up to seconds in some giant synapses (Chamberland et al. 2020)). What are the mechanisms underlying these two behaviors of synaptic vesicles? Molecularly, both pools are partially segregated by the composition of the vesicle, mainly the SNARE complex and associated proteins mediating the release (Chamberland and Tóth 2016; Chanaday and Kavalali 2018a). The first synaptic vesicles that fuse synchronously with action potentials have Synaptobrevin-2 (Syb2 or VAMP2) on their membranes which interacts with SNAP25 and Syntaxin-1 at the active zone release site (Brunger et al. 2019; Zhou et al. 2015) (Fig. 3.1). The fusion mediated by this canonical SNARE complex is synchronized by fast Synaptotagmin isoforms, namely Syts 1, 2, and 9, which play the dual role of clamping spontaneous release, thus avoiding stochastic fusion in resting conditions, and coupling neurotransmitter release to the action potential arrival (Südhof 2013; Xu et al. 2007). Asynchronous release, on the other hand, is mediated by synaptic vesicles that contain the vesicular SNARE VAMP4, and it uses a different, slower, and higher affinity calcium sensor, Synaptotagmin-7 (Syt7) (Bacaj et al. 2013; Chen et al. 2017; Raingo et al. 2012). Syt7 is enriched at the plasma membrane compared to the synaptic vesicles (Sugita et al. 2001), so how Syt7 drives the asynchronous release of neurotransmitters from this location remains a mystery. Exocytosis of synaptic vesicles can also occur at rest, independent of action potentials, giving rise to spontaneous neurotransmission (Kavalali 2018). The spontaneously recycling vesicles carry noncanonical SNAREs on their membrane, including Vti-1a, VAMP7, and VAMP4, and can both fuse dependently or independently of calcium (Chamberland and Tóth 2016; Chanaday and Kavalali 2018a). Moreover, several calcium sensing proteins have been found to modulate spontaneous neurotransmission, including the Doc2 family of proteins, suggesting that spontaneously recycling synaptic vesicles are a heterogeneous group of vesicles.

Are the molecular and functional differences related to structural or organizational differences among the pools? As mentioned before, the fast-releasing synaptic vesicles in response to action potentials belong to the RRP, so it is safe to assume that the synchronous pool is part of the RRP (but also see Wang et al. (2016a)). Whether asynchronous vesicles are also part of the RRP or not remains controversial. For mammalian hippocampal glutamatergic synapses, synchronous vesicles seem to fuse at release sites, close to VGCC, while synaptic vesicles fusing asynchronously are more dispersed and show variable distances to VGCC (Malagon et al. 2023; Mendonça et al. 2022) (Fig. 3.2). Thus, while the PR of synchronous release depends on the coupling distance to VGCC, asynchronous release is triggered by the residual, more global increase in presynaptic calcium, and thus it is less dependent on the localization or organization of VGCCs at the active zone

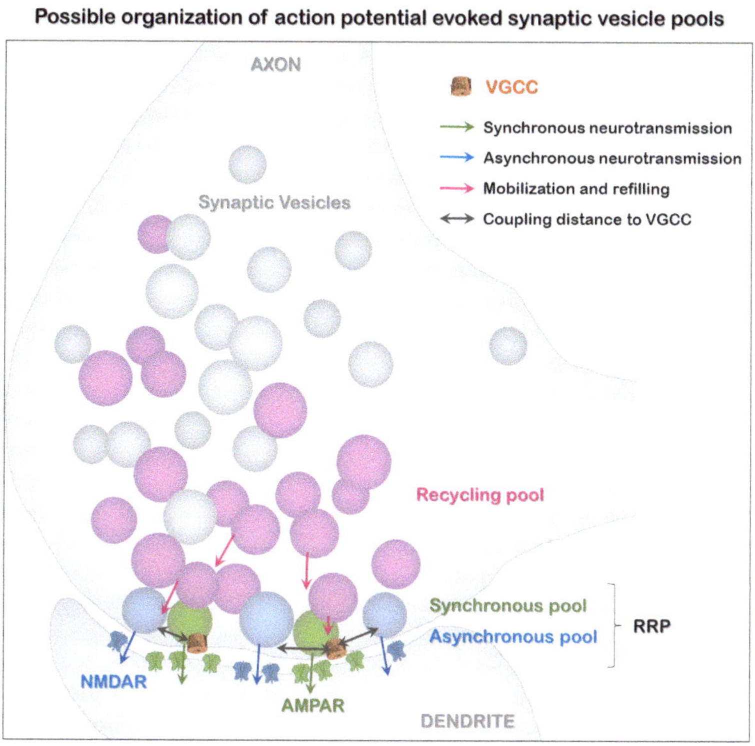

Fig. 3.2 Organization of synaptic vesicle pools for evoked transmission

(Mendonça et al. 2022). This may explain why the probability of detecting synaptic vesicle fusion is higher at the center of the AZ with respect to the periphery at low-frequency firing, while peripheral release sites start to be used at higher frequencies (Maschi and Klyachko 2017, 2020), when presynaptic calcium starts to accumulate. Accordingly, the dynamics of synchronous and asynchronous release plastically change during sustained action potential firing. At the beginning of the stimulus train, the synchronous vesicles in the RRP fuse and release neurotransmitters. Presynaptic calcium accumulation during subsequent stimulations increases the size of presynaptic calcium domains, leading to facilitation of release via the activation of asynchronous vesicles. When the vesicles in the RRP are depleted, new vesicles are recruited to sustain neurotransmission. These vesicles were proposed to belong to the asynchronous pool, thus indicating that synchronous vesicles mediate neurotransmission at the beginning of activity trains, while asynchronous vesicles are the major contributors for the facilitation and maintenance of prolonged release (Chen et al. 2017; Huson and Regehr 2020; Liu et al. 2014; Turecek et al. 2017). At inhibitory synapses, asynchronous release also shows a larger coupling distance to VGCC and dependence on slower and larger calcium transients (Hefft and Jonas 2005). The proportion of asynchronous release varies among different types of synapses, depending on the combination of specific calcium channels, calcium sensors, and calcium buffers expressed (Chamberland et al. 2020; Ermolyuk et al. 2012; Hefft and Jonas 2005). Recent works proposed that synchronous synaptic vesicles are located toward the periphery of active zones, aligned with postsynaptic AMPA-type glutamate receptors, while the asynchronous pool of vesicles may fuse toward the center and are aligned with postsynaptic NMDA-type glutamate receptors (Hruska et al. 2022; Li et al. 2021). Whether VGCC organization correlates with synchronous vesicle location was not explored. This organization of vesicle pools was proposed to maximize the activation of postsynaptic receptors by glutamate and downstream signaling for synaptic plasticity (Hruska et al. 2022; Tang et al. 2016). The size of the synchronous and asynchronous pools of synaptic vesicles is anticorrelated in some synapses (Mendonça et al. 2022; Turecek and Regehr 2019; Wen et al. 2010), and similarly, the number and size of AMPA and NMDA receptor clusters are inversely related (Hruska et al. 2022), suggesting that the trans-synaptic alignment of specific synaptic vesicle pools to postsynaptic receptors might be dynamically scaled during synaptic plasticity, tuning the efficacy and timing of neurotransmission.

Where is the spontaneous pool of synaptic vesicles located at synapses? Besides the RRP, the remaining synaptic vesicles seem to be intermixed at presynaptic boutons, regardless of their molecular composition. However, synaptic vesicles released and recycled during sustained, moderate neuron activity (called the recycling pool) have been reported to be clustered closer to the AZ (Rey et al. 2015; Schikorski and Stevens 2001). Spontaneously recycling synaptic vesicles in particular have been proposed to belong to the resting pool—a pool of vesicles reluctant to fusion in response to action potentials (Chung et al. 2010; Fredj and Burrone 2009); to the recycling pool (Loy et al. 2014; Wilhelm et al. 2010); and to the RRP (Egashira et al. 2022; Grasskamp et al. 2023; Ralowicz et al. 2024). Given the broad variety of fusion proteins and calcium sensors that trigger spontaneous fusion (Chanaday and Kavalali 2018a; Courtney et al. 2018; Kavalali 2015), it is likely that all hypotheses are correct and that the spontaneous pool is actually composed of heterogeneous sub-pools of synaptic vesicles with different molecular and functional properties. Thus, some spontaneous vesicles fuse at the release sites and are activated by the stochastic opening of VGCC (Lee et al. 2022; Williams and Smith 2018), while other vesicles fuse spontaneously at random locations outside release sites (Cork et al. 2016; Wang et al. 2022), independently of the action potential-evoked vesicle pool (Astacio et al. 2022; Horvath et al. 2020; Melom et al. 2013; Sara et al. 2005), and may be activated by alternative calcium sources (Bal et al. 2013; Chanaday et al. 2021; Wang et al. 2023) or even fuse independently of calcium. Why synapses evolved to have redundant mechanisms triggering spontaneous neurotransmission? Stochastic release of neurotransmitters may be important for the growth of axons and dendrites and for the development, formation, and maintenance of synapses (Andreae and Burrone 2015; Banerjee et al. 2021; Choi et al. 2014; Zhang et al. 2017). Spontaneous neurotransmission may also determine basal synaptic properties, including efficacy and excitability, via the activation of homeostatic plasticity mechanisms (Crawford et al. 2017; Horvath et al. 2021; Reese and Kavalali 2015; Sutton et al. 2006). In some cases, the levels of action-potential-evoked neurotransmission can in turn influence the frequency of spontaneous neurotransmission (Lin et al. 2020). Thus, an attractive hypothesis is that the broad variety of molecular mechanisms triggering spontaneous release allows neurons to adapt synaptic properties to different stimuli, activity history, and metabolic states. Testing this hypothesis remains experimentally challenging due to the lack of tools to selectively manipulate spontaneous neurotransmission (but see Crawford et al. (2017)).

3.3.4 *Comment on Synaptic Vesicle Pools in Different Types of Synapses*

The rules discussed for synaptic vesicle pool organization with respect to active zone machinery, including VGCC and release site organizing molecules, are applicable to most types of synaptic contacts. Different types of presynaptic terminals use similar logic to build active zones, and it is the specific molecular composition and subcellular organization they adopt that give rise to the variability in release probabilities, the relative size of each synaptic vesicle pool, and the spatial and temporal dynamics of calcium signals and synaptic vesicle release (for example, see Althof et al. (2015), Banerjee et al. (2022), Boyken et al. (2013), Deguchi-Tawarada et al. (2006), Koppensteiner et al. (2024), Mori et al. (2014), Moulder et al. (2007), Richter et al. (1999), Song and Augustine (2023), and Wu and Cooper (2012)). Our knowledge arises mainly from the research on glutamatergic and GABAergic (i.e., they release γ-aminobutyric acid—GABA) synapses, and we understand less about synaptic vesicle pools and organization of other neurotransmitters (e.g., monoamines) that act on metabotropic receptors and/or perform volume transmission, a slower, large range and tonic form of communication (for a review on volume transmission, see Özçete et al. (2024)). Whether these other neurotransmitters are stored in functionally distinct synaptic vesicle pools needs further investigation.

3.3.5 *Conclusion and Open Questions*

In conclusion, synaptic vesicles are partially segregated into distinct pools with characteristic molecular composition, positioning, and release properties (Chanaday and Kavalali 2018a). Different synaptic vesicle pools fuse and release neurotransmitters at specific locations of the presynaptic membrane that are aligned with segregated clusters of postsynaptic receptors and coupled to different signaling pathways (Guzikowski and Kavalali 2021). Thus, modulating the relative size and organization of individual pools becomes an interesting substrate for synaptic plasticity with the potential to expand the coding capacity of axons. Accordingly, changes in both the RRP and the recycling pool of synaptic vesicles (at the expense of the reserve pool) have been reported during different forms of potentiation or depression of synaptic efficacy (Cazares et al. 2016; Delvendahl et al. 2019; Koppensteiner et al. 2024; Oline and Burger 2014; Rey et al. 2020; Tyler et al. 2006;

Vandael et al. 2020; Wang et al. 2016b). Besides the AZ machinery, synaptic vesicle clustering molecules also regulate the redistribution of synaptic vesicles among the different pools during plasticity, including the phosphoprotein Synapsin (Bruentgens et al. 2024; Moschetta et al. 2022) and its partner Synuclein (Fouke et al. 2021) (see Sansevrino et al. (2023)) for a review on synaptic vesicle clustering mechanisms). The asynchronous pool of synaptic vesicles modulates short forms of presynaptic plasticity via Syt7 (also see Vevea et al. (2021) and Weyrer et al. (2021)), but whether the size of the asynchronous pool of synaptic vesicles changes during synaptic plasticity remains unknown. The spontaneous pool of synaptic vesicles is also capable of plasticity. Muscle denervation at the *Drosophila* neuromuscular junction causes increases in spontaneous release from nearby terminals, for example, contributing to junction outgrowth and recovery (Wang et al. 2021). The spontaneous pool of synaptic vesicles is transiently increased after action potential-evoked activity (Cho et al. 2015; Lin et al. 2020) which contributes to synaptic growth during plasticity (Cho et al. 2015). Transient increases in spontaneous release might also contribute to synapse formation or stabilization during cerebellum development (Le Guellec et al. 2023). Thus, neurons can modify the protein composition and localization of synaptic vesicles depending on their activity levels and needs. So, how do neurons independently regulate and maintain the molecular identity of synaptic vesicle pools during synaptic activity? One likely mechanism is via divergent endocytic and transport pathways for synaptic vesicle proteins. The next section will discuss our current knowledge on the molecular mechanisms, kinetics, and spatial organization of synaptic vesicle recycling.

3.4 Where and When Are Synaptic Vesicles Recycled?

After exocytosis at the AZ membrane, synaptic vesicle components need to be retrieved to reform functional, fusion-competent vesicles. Retrieval can happen at the AZ, via rapid re-closure of the fusion pore, or at the edges of the AZ or peri-AZ, via compensatory endocytosis.

3.4.1 "Kiss-and-Run"

Rapid re-closure of the fusion pore, also termed "kiss-and-run", has been well established in non-neuronal secretory cells that release peptides and other signaling molecules via dense core vesicles (Alabi and Tsien 2013; Shin et al. 2021; Wu et al. 2014). Direct experimental measures of kiss-and-run in neurons have been technically challenging due to the small size of synaptic vesicles, 40–50 nm, and the extremely fast nature of the retrieval process, <300 ms. Rapid vesicle retrieval via pore closure is an attractive mechanism since: (1) it would prevent intermixing of synaptic vesicle and plasma membrane components, thus preserving the molecular identity of the fused vesicle; (2) it would be energetically efficient by avoiding the need to recruit, endocytose, and regenerate synaptic vesicles from the plasma membrane or endosomes; (3) it would be extremely fast, matching the speed of neurotransmission and counteracting synaptic depression; and (4) it would avoid overcrowding of the AZ membrane. Fast retrieval of synaptic vesicles consistent with the closure of a small fusion pore, without intermixing of membrane proteins, has been experimentally evidenced in several neuron types and preparations ranging from cultured hippocampal and dopaminergic neurons to photoreceptors and the calyx of Held synapse in acute slices, or intact neuromuscular junctions (Chanaday and Kavalali 2018b; Henkel et al. 2019; Leitz and Kavalali 2014; Park et al. 2012; Richards 2010; Staal et al. 2004; Sun et al. 2002; Wen et al. 2017) (Fig. 3.3). Recent time-resolved cryo-electron tomography studies in hippocampal neurons provided in situ evidence of "kiss-shrunk-run" happening at synapses (Tao et al. 2025). This mechanism expands on classical "kiss-and-run" and uncovered that most of fused synaptic vesicles maintain a lipidic small fusion pore (~4 nm) and shrink in size during fusion, followed by pore reclosure and migration to the peri-active zone area within 70 ms after fusion, quickly freeing the release sites (Tao et al. 2025). Reconstitution of individual fusion pores using nanodiscs further revealed that SNARE-mediated fusion pores flicker, i.e., they fluctuate between the open and close states before fully collapsing or fully closing (Wu et al. 2016). Adding calcium and Syt1, the main calcium sensor for synchronous release, facilitates the formation and promotes the expansion of the fusion pore (Wu et al. 2021). In this condition, the SNARE complex is not reversible, and pore closure can only happen via disassembly through the action of α-SNAP and NSF (Das et al. 2020). Accordingly, Syt1 and the SNARE protein Syb2 regulate the kinetics of retrieval of individual synaptic vesicles in a calcium-regulated manner (Chanaday and Kavalali 2021; Li et al. 2017a), indicating that the fusion machinery couples release to endocytosis and dictates the kinetics of these processes.

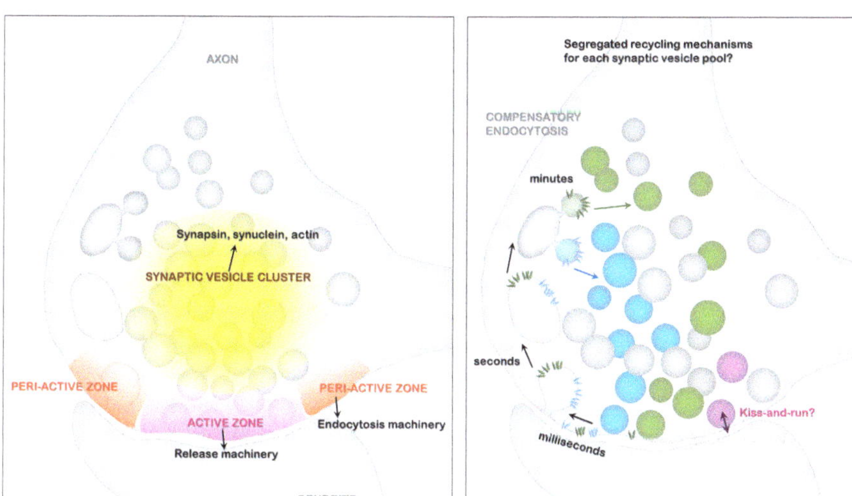

Fig. 3.3 Spatial organization of synaptic vesicle endocytosis machinery at presynaptic boutons

3.4.2 Ultrafast Compensatory Endocytosis

Fusion pore expansion is an energetically favorable process, more so at increasing calcium concentrations; thus, synaptic vesicles can fully collapse with the axonal plasma membrane after fusion. In these cases, the membrane tension generated by the addition of surface area leads to increased membrane curvature in the periphery of the AZ, triggering a compensatory form of endocytosis termed ultrafast endocytosis (Ogunmowo et al. 2023; Watanabe and Boucrot 2017; Watanabe et al. 2013) (Fig. 3.3). The probability of this mode of retrieval increases with the number of vesicles that fuse, adjusting the area of membrane endocytosed accordingly (Ogunmowo et al. 2023). Ultrafast endocytosis is a particularly convenient method to retrieve synaptic vesicle proteins from the plasma membrane since it operates in fast timescales, <300 ms, and it occurs outside the AZ minimizing disturbances of the release machinery (points 3 and 4 in previous paragraph). This endocytic process is regulated by actin, endophilin, and dynamin, independently of clathrin, and relies on the later regeneration of fusion-competent synaptic vesicles from synaptic endosomes (Imoto et al. 2024; Watanabe et al. 2014). Thus, axons use ultrafast endocytosis to swiftly compensate for membrane addition resulting from exocytosis, but synaptic vesicle identity may be lost through intermixing with plasma membrane molecules. Limited diffusion of synaptic vesicle proteins with subsequent re-capture and endocytosis at the peri-AZ has been indeed observed (Gimber et al. 2015). However, there is separate evidence of synaptic vesicle proteins remaining clustered after fusion, in packages that can be later retrieved from the plasma membrane via endocytosis (Opazo et al. 2010) (Fig. 3.3). Synaptic vesicle proteins may even remain clustered in the endosome, facilitating vesicle regeneration (Hoopmann et al. 2010). This observation suggests that molecular identity of the synaptic vesicle, i.e., to what pool that vesicle belongs to, could be maintained after fusion even when there is full collapse fusion and compensatory endocytosis. The fine temporal and spatial regulation of endocytosis at the peri-AZ is achieved via recruitment and clustering of endocytic machinery that is maintained in an inactive and primed state, ready to swiftly start ultrafast endocytosis in response to synaptic activity (Del Signore et al. 2021, 2023; Imoto et al. 2022). At the *Drosophila* larval neuromuscular junction, endocytic machinery is pre-deployed in a radial mesh arrangement around AZs, and the concentration of endocytic proteins correlates with the levels of the AZ protein Bruchpilot (Del Signore et al. 2023). These antecedents suggest that the coupling of exocytosis and endocytosis is both functional and structural. An exciting speculation is that the scaling of synaptic strength during plasticity may not only involve changes in the release machinery but also parallel adjustments of the associated endocytic apparatus. Whether such remodeling of the endocytic mechanisms occurs at mammalian synapses and if it is related to all or some forms of synaptic plasticity remain to be investigated.

3.4.3 *Bulk Endocytosis*

When axons undergo prolonged bursts of high-frequency firing, the large amount of membrane added via synaptic vesicle exocytosis is compensated by activity-dependent bulk endocytosis (ADBE). ADBE is a clathrin-independent, dynamin-dependent form of endocytosis that generates large presynaptic endosomes from which synaptic vesicles are later reformed. Multiple synaptic vesicle proteins have been shown to be retrieved from the plasma membrane via ADBE, including VGluT1, Syb2, and Synaptophysin (Chanaday et al. 2019). ADBE is triggered by calcium influx during synaptic activity, via a mechanism involving the calcium-activated phosphatase Calcineurin and the dephosphins amphiphysin, endophilin, synaptojanin, and dynamin (Cousin and Robinson 2001; Kontaxi et al. 2023; Peng et al. 2021). ADBE shares several effectors and adaptors with ultrafast endocytosis, so one attractive possibility is that both represent extremes of the same pathway, and there is a continuous gradient ranging from ultrafast-retrieved small endosomes to slower (1–2 s)-retrieved large endosomes depending on the size of the exocytic load. However, there are ADBE-specific molecules that might be recruited only during intense activity (Blumrich et al. 2023), suggesting the existence of "activity sensors" that might trigger specific modes of endocytosis on a need basis. In sum, synapses evolved several coexisting mechanisms to recycle their components and regenerate synaptic vesicles. This local recycling is essential for the maintenance of synaptic performance over time and, consequently, disruptions are associated with numerous neurodegenerative diseases (Overhoff et al. 2021).

3.4.4 *Spatiotemporal Coupling of Fusion and Endocytosis*

Are synaptic vesicle pools coupled to specific recycling pathways? The SNARE and associated proteins forming the fusion complex have been shown to likewise modulate the route of endocytosis undertaken by synaptic vesicle proteins. Syb2, for example, regulates calcium-dependent fast endocytosis of synaptic vesicles after synchronous release, but it may not regulate ultrafast endocytosis (Chanaday and Kavalali 2021; Deák et al. 2004). After compensatory endocytosis, Syb2-containing synaptic vesicles are reformed from plasma membrane and endosomes via the adaptor proteins CALM and AP-180 which in turn recruit AP2 and initiate clathrin-mediated endocytosis (Koo et al. 2011, 2015). This recycling pathway depends on the direct interaction of Syb2 with the synaptic vesicle protein Synaptophysin 1 (Kokotos et al. 2019; Rajappa et al. 2016), a tetraspanin that is also involved in synaptic vesicle biogenesis (see the first section of this chapter), suggesting that the regeneration of synaptic vesicles in the recycling pool shares mechanistic features with the biogenesis and transport routes of synaptic vesicles (see Ivanova and Cousin (2022)) for a review on synaptic vesicles and endolysosomal transport interconnections). As discussed above, Syb2 cooperates with the calcium sensor Syt1 for fusion pore formation and expansion. In agreement with its role in fusion pore stabilization, Syt1 slows down the endocytosis of individual synaptic vesicles during mild activity, in a calcium-dependent manner (Leitz and Kavalali 2011; Li et al. 2017a). Syt7, which has higher calcium affinity and membrane penetration than Syt1, promotes more stable and longer lasting fusion pores (Courtney et al. 2023) and accordingly directs synaptic vesicle retrieval to a slower endocytic pathway (Li et al. 2017a), suggesting that synchronous and asynchronous vesicles have segregated fusion endocytosis routes. At high levels of neuron activity, when multiple vesicles are released and compensatory endocytosis is activated, Syt1 retrieval is facilitated by the clathrin adaptor Stonin-2 (Kononenko et al. 2013). Stonin-2 drives endocytosis via AP2 (Kononenko et al. 2013), suggesting a convergent mechanism that brings together Syt1, SV2, Syb2, and Synaptophysin-1 to regenerate synaptic vesicles of the recycling pool and, possibly, maintain the synchronous pool of synaptic vesicles. Release of the asynchronous pool is mediated by the calcium sensor Syt7 and the SNARE protein VAMP4 (Raingo et al. 2012). VAMP4 is essential for ADBE (Nicholson-Fish et al. 2015), and its transport is mediated by the clathrin adaptor AP1 via interaction with VAMP4's dileucine motif (Peden et al. 2001). Thus, synaptic vesicles from the synchronous and asynchronous pools fuse and recycle via two partially segregated pathways. Blockade of one of these pathways seems to redirect all molecules to be trafficked via the other. For example, elimination of Stonin-2 leads to increased ADBE (Kononenko et al. 2013), while mutations in the VAMP4 dileucine motif cause VAMP4 to recycle similar to Syb2 (Raingo et al. 2012). As mentioned above, spontaneous neurotransmission is mediated by several SNARE proteins, including Vti-1a, VAMP7, and VAMP4. These SNAREs are also involved in post-Golgi transport of endosomes and lysosomes, further supporting the existence of overlaps between the synaptic vesicle cycle and the endolysosomal trafficking pathways (Ivanova and Cousin 2022). VAMP7 belongs to the longin family of SNAREs because it has a long N-terminal region (absent in Syb2, for example). This N-terminal region or longin domain has an auto-inhibitory role on fusion complex formation (Daste et al. 2015; Vivona et al. 2010), maybe explaining why VAMP7-containing vesicles are highly resilient to fuse (constituting the resting pool). The longing domain also contains a binding motif for the clathrin adaptor AP3 (Martinez-Arca et al. 2003), suggesting that these synaptic vesicles may be recycled independently of the action potential-driven recycling pool. AP3-mediated recycling might be a more global regulator of the resting pool of synaptic vesicles, since a recent report uncovered a novel AP3-dependent synaptic vesicle mobilization mechanism that is only activated during high-frequency firing (Xu et al. 2023). VAMP4 and Vti-1a colocalize to the same synaptic vesicles (Kreykenbohm et al. 2002). Besides its role in driving spontaneous neurotransmission, Vti-1a regulates post-Golgi sorting and transport of synaptic vesicles (Emperador-Melero et al. 2018) and the growth and development of neuronal processes (Bollmann et al. 2022). The fact that VAMP4-mediated fusion contributes to both asynchronous and spontaneous release of neurotransmitters suggests certain level of overlap in the recycling pathways for different synaptic vesicle pools. Synaptic vesicles generated by VAMP4-mediated ADBE have indeed a higher propensity to fuse spontaneously (Lin et al. 2020). This overlap could provide a molecular link whereby the intensity of synaptic activity modulates the levels of spontaneous neurotransmission, influencing neuron excitability and synapse stability. Thus, while spatial, functional, and molecular segregation of recycling of the different synaptic vesicle pools may expand the coding capacity and fidelity of information transfer in neurons, a partial overlap among these pathways may be a substrate for synaptic plasticity by which axons can tune neurotransmitter release in response to their activity history. Experimental evidence following the same synapses for prolonged periods of time is necessary to demonstrate these points.

3.4.5 *Synaptic Vesicle Recycling for Different Neurotransmitters*

Most of the research on synaptic vesicle pools was performed using excitatory, glutamatergic synapses as a model. VGluT, the transporter responsible for loading glutamate into synaptic vesicles, has several endocytic signals including multiple dileucine-like and polyproline motifs that interact with AP1, AP2, AP3, and endophilins (De Gois et al. 2006; Foss et al. 2013; Li et al. 2017b; Weston et al. 2011). This redundancy might ensure that all synaptic vesicle pools, regardless of endocytic route, receive VGluT and are loaded with neurotransmitters. Not all VGluT isoforms are similar, however. Cortical neurons have mainly VGluT1 in their synaptic vesicles and show a very low probability of neurotransmitter release. Neurons in subcortical areas, such as the thalamus, midbrain, and brainstem, express mostly VGluT2 and/or VGluT3 and have high probability of glutamate release. This variability in the propensity of synaptic vesicles to fuse is attributed to differences in the affinities among each VGluT isoform and the various endocytic adaptors, and it influences presynaptic performance including short-term plasticity (facilitation versus depression) (Foss et al. 2013; Weston et al. 2011). Vesicular transporters for other neurotransmitters also contain dileucine or dileucine-like motifs, including the GABA, monoamine, and acetylcholine transporters (VGAT, VMAT, and VAChT, respectively) (Fei et al. 2008; Santos et al. 2013; Tan et al. 1998). Accordingly, their post-Golgi sorting and local trafficking at synapses seem to be regulated by clathrin adaptors and dynamin (Fei et al. 2008). Many presynaptic varicosities can co-release two neurotransmitters (Hnasko and Edwards 2012; Svensson et al. 2018; Vaaga et al. 2014; Wallace and Sabatini 2023), for example, a combination of glutamate with either GABA, a monoamine (e.g., dopamine or serotonin), or acetylcholine. Historically, it has been intensely debated whether both neurotransmitters are stored in the same synaptic vesicle, in different vesicles residing in the same synapse, or in different synapses arising from the same axon. Behavioral and electrophysiological studies on brain circuits that use co-transmission have shown that, in some cases, the probabilities of release of the two co-released neurotransmitters are correlated and both encode the same behavioral output (Fattorini et al. 2015; Gras et al. 2008; Jonas et al. 1998; Kim et al. 2022; Melani and Tritsch 2022). These results suggested a model of co-packaging of the different neurotransmitters into the same synaptic vesicles. Possible benefits of co-packaging and co-signaling include facilitation of synaptic vesicle filling by influencing the electrochemical gradient across its membrane and expanding the signaling and plasticity repertoire of the synapse. Molecularly, co-packaging may result from the fact that many of these transporters contain dileucine motifs and might use overlapping clathrin adaptors, thus being transported and endocytosed together. VGluT1, VMAT2, and VGAT, for example, all interact with AP3 (Fei et al. 2008; Santos et al. 2013; Tan et al. 1998). In other cases, the two neurotransmitters at co-transmitting circuits encode dissimilar, nonredundant information, are released with independent probabilities, and signal with different kinetics (Kljakic et al. 2022; Silm et al. 2019; Zych and Ford 2022), indicating these co-releasing synapses have synaptic vesicles that only carry one type of neurotransmitter. As proposed for the different VGluT isoforms, these differences arise, at least in part, from the use of segregated endocytic adaptors and trafficking routes by the vesicular transporters (Bae et al. 2020; Silm et al. 2019), which in turn influences the efficacy of neurotransmitter release, short-term plasticity, and the amount of information one synapse can encode. Whether endocytosis and recycling of synaptic vesicles carrying different neurotransmitters occur at the same or different locations of the presynaptic bouton remains unknown.

3.4.6 Conclusion

In summary, the molecular identity of synaptic vesicle pools, either defined by release kinetics—synchronous, asynchronous, or spontaneous—or by neurotransmitter, is maintained by partially segregated recycling pathways at presynaptic terminals. This segregation is molecular and spatial, and it may increase the coding capacity and flexibility of synapses. There is also partial overlap among the pools, and this in turn may offer the advantage of cross-communication and maintenance of global balance (homeostasis) among the different forms of neurotransmission. Regardless of the synaptic vesicle pool, all forms of neurotransmitter release are either triggered or modulated by calcium. In the next section, we will summarize our current knowledge on presynaptic calcium sources, their spatial and temporal organization, and putative links to the different synaptic vesicle pools.

3.5 Spatial Organization of Calcium Sources and the Synaptic Vesicle Pools

Fast, synchronous release of neurotransmitters is initiated by the action-potential-driven opening of VGCC at the AZ. The efficiency of the stimulation–release coupling is modulated by the AZ scaffolds that determine the relative number and organization of VGCC and synaptic vesicles. As discussed above, the vesicles that are docked and primed at release sites are believed to belong to the RRP, and at least they include the synchronous pool of vesicles. There are three major families of VGCC: Cav1 (also called L-type or Dmca1D in *Droshophila*); Cav2 (P/Q-, N- and R-type, or Dmca1A); and Cav3 (T-type) (Dolphin and Lee 2020; Mochida 2018; Zhang et al. 2022). Cav2 channels have fast activation and inactivation kinetics and are essential for most of synchronous neurotransmission in multiple synapse types and organisms. Cav1 channels have larger conductance and slower inactivation kinetics ("L" means long opening). In muscle, Cav1 mediates excitation–contraction coupling together with the endoplasmic reticulum (ER) calcium channel Ryanodine Receptor (RyR). In neurons, it has been proposed to mainly be involved in different forms of short- and long-term synaptic plasticity. Cav1 channels also trigger neurotransmitter release at a special type of presynaptic specification called ribbon synapse, where secretion is graded (i.e., scales linearly with the membrane potential—different from an action potential). Cav3 channels are activated at low voltage and are primarily involved in generating rhythmic firing patterns (e.g., in neurons in the thalamus and in the heart) (Dolphin and Lee 2020; Mochida 2018; Zhang et al. 2022).

3.5.1 Are Specific Calcium Channels Associated with Different Synaptic Vesicle Pools?

The giant mossy fiber synapse in the hippocampus contains three types of VGCC: P/Q-, N-, and R-type. These channels have different gating properties, activation and deactivation kinetics, and subcellular distribution along the mossy fiber axon and boutons (Li et al. 2007; Pelkey et al. 2006). Thus, it was proposed that regulation of the relative amount and organization of each VGCC subtype allows mossy fibers to independently tune neurotransmitter release properties at presynaptic contacts with different targets (Pelkey et al. 2006). At individual presynaptic boutons, N-type VGCCs seem to generate highly localized and fast calcium signals (nanodomains) that trigger synchronous fusion of RRP vesicles at a few release sites. During repetitive firing, P/Q-type VGCCs are activated; these channels generate larger calcium increases (microdomains), thus reaching more release sites and increasing synchronous and asynchronous synaptic vesicle release, causing short-term facilitation (Chamberland et al. 2017, 2020) (Fig. 3.4). Thus, the synchronous and asynchronous pools of synaptic vesicles are associated with separate calcium channels.

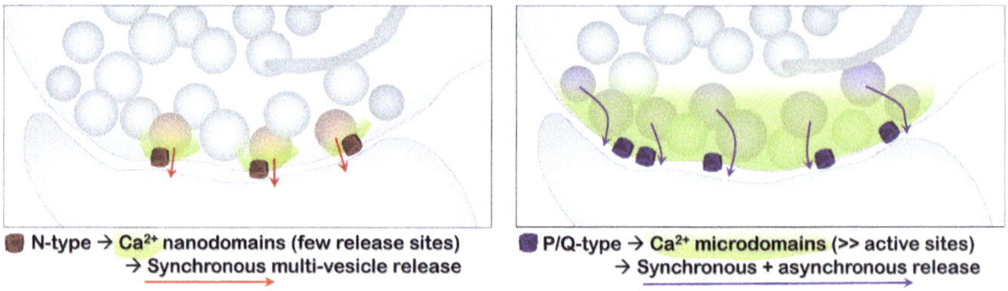

Fig. 3.4 Simplified representation of calcium channel organization at the mossy fiber bouton and their associated synaptic vesicle pools

While N- and P/Q-type VGCC channels mediate most of synaptic transmission, the slower R-type channels are essential for long-term potentiation of mossy fiber presynaptic boutons (a long-lasting form of plasticity that increases presynaptic efficacy) (Castillo et al. 1994). Then, at least at this specialized synapse of the hippocampus, different VGCCs have distinct roles in presynaptic function. At a different synapse, the mouse neuromuscular junction, neurotransmitter release at low levels of activity is almost exclusively dependent on Cav2 channels (N-type), suggesting that the RRP and subsequent synchronous neurotransmission are triggered by this type of calcium channel (Fig. 3.5). At high-frequency stimulation, however, calcium entry via Cav1 channels (L-type) promotes synaptic vesicle endocytosis and directs them to a fast recycling pathway, possibly the recycling pool, that also shows high PR (Perissinotti et al. 2008). In this way, Cav1/L-type channels may maintain neurotransmitter release during intense, sustained activity. Similarly, Cav1 channels are necessary for short-term facilitation of neurotransmitter release at central mammalian synapses and participate in the establishment of presynaptic long-term potentiation (Fourcaudot et al. 2009; Jensen and Mody 2001). These Cav1 functions are strikingly similar to the proposed roles for Syt7, the calcium sensor for asynchronous neurotransmission. Indeed, it was proposed that Cav1 channels are linked to the asynchronous pool of synaptic vesicles (Eggers et al. 2013; Few et al. 2012) and, in hair cells, Cav1 do not interact with Syt1, the calcium sensor for synchronous release, and instead binds to Otoferlin (the main protein coupling calcium to graded release for hearing (Hams et al. 2017)). Separate locations and functions of Cav1 and Cav2 have also been found in nonmammalian synapses. At the neuromuscular junction of the nematode *C. elegans*, Cav2 is localized to tight clusters associated with AZ scaffold proteins, including ELKS, and it mediates around 50% of all acetylcholine release (Mueller et al. 2023). The remainder neurotransmission is driven by the joined activation of Cav1 and RyR, resembling their role in muscle cells (Fig. 3.5). Cav1 is distributed more diffusely in the active zone, does not co-localize with ELKS, and is more prone to trigger multivesicular release (Mueller et al. 2023). Cav1 and Cav2 seem to be associated with different components of the synaptic fusion machinery too (Mueller et al. 2023), resembling the differences observed in the mammalian auditory system (Wichmann 2015). Similar organization was found at the *Drosophila* neuromuscular junction, where Cav2 channels

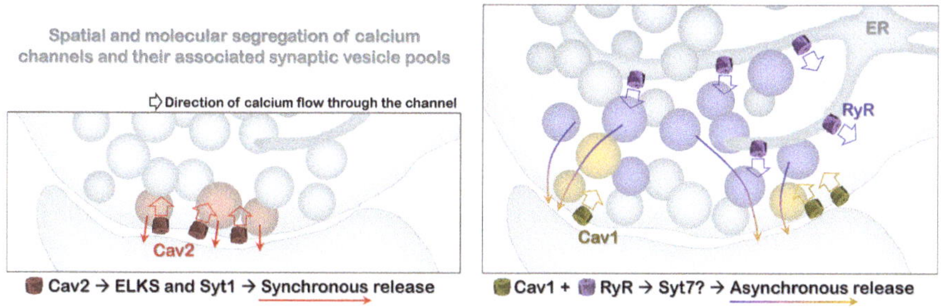

Fig. 3.5 Simplified representation of calcium channel organization at neuromuscular active zones and their associated synaptic vesicle pools

localize at the center of the AZ and mediate the majority of action potential-driven synchronous release (Krick et al. 2021). Cav1 channels are excluded from the center, do not regulate the RRP size, and localize around the AZ (Krick et al. 2021). This more peripheral distribution coincides with the sites of endocytosis (Del Signore et al. 2023), suggesting that Cav1 may regulate synaptic vesicle recycling as it was proposed for the mammalian neuromuscular junction (Perissinotti et al. 2008). In agreement, Cav1 is necessary for short-term facilitation at these synapses via acceleration of synaptic vesicle endocytosis (Krick et al. 2021). Given that Cav1 channels have slower kinetics and larger conductance than Cav2, how does calcium influx from Cav1 remain constrained to the periphery of the AZ? The plasma membrane calcium ATPase (PMCA) co-localizes with Cav1 in the periphery of the AZ, and it counterbalances Cav1 activity, negatively regulating synaptic vesicle fusion. The authors proposed that PMCA protects the RRP from being depleted by the activity of Cav1 (Krick et al. 2021). Similarly, in some neurons, PMCA levels are enriched at axons, where they act as negative regulators of cytoplasmic calcium levels, limiting the amount of neurotransmitter release (Erhardt et al. 2021; Fujii et al. 1996). Interestingly, PMCA is also located in synaptic vesicles where it regulates calcium and proton gradient (Fujii et al. 1996; Ono et al. 2019), suggesting it might have a role in neurotransmitter loading. In conclusion, the organization of plasma membrane calcium channels at the active zone seems to mirror the spatial distribution of the fusion and endocytosis machinery, ensuring the temporal fidelity of neurotransmitter release, the prompt and efficient retrieval and regeneration of vesicles, and the independent modulation of the different synaptic vesicle pools. Many open questions remain, including whether the organization of calcium channel types at small mammalian synapses also shows spatial segregation, the detailed docking and fusion machinery associated with each channel type, and the mechanisms determining the recruitment of the channels to their specific locations. Moreover, VGCC alone is not sufficient to maintain synaptic transmission, especially in response to sustained activity or during presynaptic forms of plasticity, and the contribution from other calcium channels and transporters at the plasma membrane, the ER, and mitochondria is necessary.

3.5.2 Alternative Calcium Sources and the Synaptic Vesicle Cycle

The ER is the main intracellular store of calcium, and it serves two functions: it clears calcium from the cytoplasm by pumping it into the lumen of the ER, and it amplifies cytoplasmic calcium signals via release through ER-resident calcium channels. The buffering capacity of the ER depends on the sarco-/endoplasmic reticulum calcium ATPase (SERCA). Computational models predict that SERCA activity at axons influences calcium-dependent short-term facilitation of neurotransmitter release (Singh et al. 2021). According to the model, at low firing rates, the SERCA would clear calcium from the presynaptic cytoplasm, but at increasing frequencies, SERCA saturates, and thus calcium would accumulate facilitating synaptic vesicle fusion (Singh et al. 2021). Indeed, in the hair cells of the auditory system, the efficiency of high-frequency firing depends on fast calcium clearance via the combined action of PMCA, SERCA, and the mitochondrial calcium uniporter (MCU) (Cuadra et al. 2021). Mitochondrial calcium uptake is a limiting factor in calcium accumulation at presynaptic boutons that contain mitochondria, modulating PR of synaptic vesicle release, RRP depletion and refilling rates, asynchronous neurotransmission, and synaptic vesicle endocytosis (Devine et al. 2022; Lopez-Manzaneda et al. 2022; Marland et al. 2016; Singh et al. 2018; Smith et al. 2016; Thomas et al. 2019; Vos et al. 2010). Thus, fast calcium buffering by PMCA at the plasma membrane, SERCA at the ER, and MCU in mitochondria are all necessary to ensure the temporal precision of neurotransmitter release and information transfer at synapses (Fig. 3.6). Since these calcium transporters are located in different organelles, at varying relative distances to the synaptic vesicle pools and release sites, it remains unclear whether each extrusion mechanism is activated by similar or different types of stimuli or associated with specific synaptic vesicle pools. The ER is also a source of calcium for neurotransmitter release, amplifying VGCC-driven calcium increase (Fig. 3.6). Calcium influx through VGCC triggers a second wave of calcium release from the ER via two calcium channels: RyR and inositol 1,4,5-trisphosphate receptors (IP3R), through a pathway called calcium-induced calcium release (CICR). RyR may be primarily activated by Cav1 channels, as discussed above. IP3R are coincidence detectors, since they are activated by the concerted binding of calcium and IP3 and thus usually require signaling from metabotropic neurotransmitter or neuropeptide receptors, besides VGCC. CICR has been widely shown to potentiate neurotransmitter release at multiple synapse types (e.g., glutamate, GABA, dopamine, and acetylcholine) at various nervous system regions and animal models (Chanaday et al. 2021; Ghatpande et al. 2006; Mathew and Hablitz 2008; Meadows et al. 2021; Patel et al. 2009; Ramesh et al. 2021; Sharma and Hasan 2020; Szikra et al. 2008; Warrier et al. 2005; Wen et al. 2013). At rod photoreceptors, CICR triggers fusion of synaptic vesicles outside of ribbons which leads to slow tonic neurotransmitter release (Chen et al. 2014), suggesting it may be activating a synaptic vesicle pool different to the RRP. Similarly, CICR can selectively augment spontaneous glutamate release from the VAMP7-containing

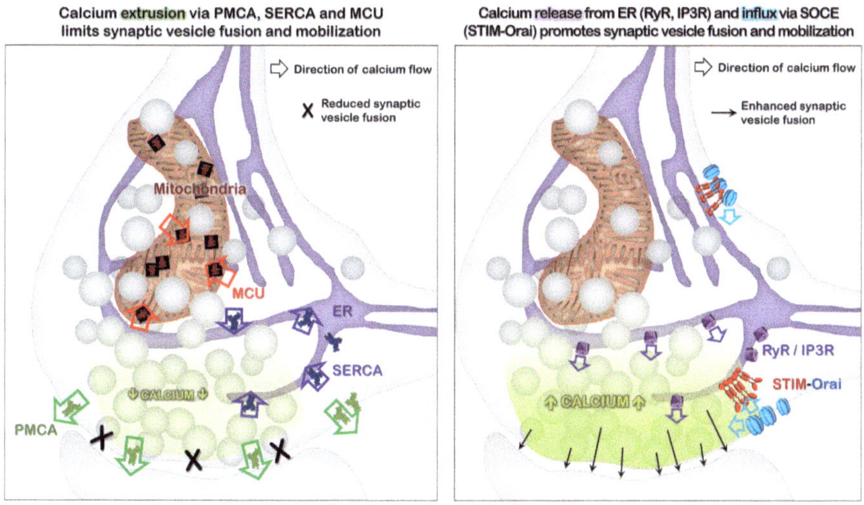

Fig. 3.6 *Left: Mechanisms of calcium clearance at axons*. Reduction in cytoplasmic calcium is achieved by transport against its electrochemical gradient by: PMCA (calcium flows towards the extracellular space); SERCA (pumping into the ER); MCU (calcium load into mitochondria). *Right: Mechanisms of calcium amplification at axons*. Cytoplasmic calcium is increased by: release from the ER via IP3Rs and RyRs, and via influx from the extracellular space through Orai channels. The interplay between extrusion and amplification will determine cytoplasmic calcium concentration, influencing the probability of synaptic vesicle fusion

resting pool at hippocampal neurons (Bal et al. 2013), and it drives the tonic activity of auditory hair cells (Castellano-Munoz et al. 2016). At the zebrafish neuromuscular junction, synchronous neurotransmission driven by the RRP strictly depends on VGCC as in the mammalian, *Drosophila,* and nematode motorneuron presynaptic boutons. But asynchronous neurotransmission is maintained by a "propagating intracellular calcium source" (Wen et al. 2013), which resembles IP3R-dependent CICR ("calcium waves") (Ross 2012). Nicotinic acetylcholine receptors trigger action potential-independent glutamate release from hippocampal neurons also via the activation of RyR-driven CICR; this response is significant enough to trigger firing of the postsynaptic neuron (Sharma and Vijayaraghavan 2003). CICR may also be essential for somatic release of serotonin, driving volume transmission (De-Miguel 2022). Thus, calcium release from intracellular stores is a central player in neurotransmission, modulating the intensity and kinetics of synaptic vesicle mobilization, fusion, and recycling. Depending on the synapse type, stimulus intensity, and external signals, specific forms of CICR are activated, and they seem to be coupled to segregated synaptic vesicle pools and modes of neurotransmitter release.

Another source of calcium influx into axons is through store-operated calcium entry (SOCE). SOCE is triggered by calcium depletion in the ER. Calcium in the lumen of the ER is sensed by the ER-resident Stromal Interaction Molecules (STIM1 and STIM2) (Liou et al. 2005; Roos et al. 2005) which, upon calcium unbinding, translocate and aggregate at ER–plasma membrane junctions where they interact with and gate the opening of plasma membrane calcium channels, mainly Orai channels (also known as CRAC—calcium release-activated calcium channels) (Feske et al. 2006; Penner et al. 1988). As a consequence, there is calcium influx into the cytosol (Fig. 3.6), followed by calcium pumping into the ER by the SERCA, restoring calcium levels in the ER (Soboloff et al. 2012). While the existence and abundance of ER–plasma membrane junctions at presynaptic terminals remains controversial (Wu et al. 2017), there is recent evidence supporting a role for STIM and Orai proteins in axon maturation (Kasturacharya et al. 2022; Mitchell et al. 2012; Pathak et al. 2015; Pavez et al. 2019; Shim et al. 2013) and function. For example, SOCE modulates neuron excitability by enhancing axonal membrane depolarization (Dou et al. 2018; Wei et al. 2017). STIM1 may influence synaptic vesicle PR via modulation of calcium levels in the ER (de Juan-Sanz et al. 2017). A neuron-specific STIM1 splice variant, STIM1B, facilitates calcium accumulation in the axon during sustained action potential firing, supporting synaptic vesicle fusion and avoiding short-term depression (Ramesh et al. 2021). STIM2, on the other hand, selectively drives spontaneous neurotransmitter release at glutamatergic but not GABAergic synapses (Chanaday et al. 2021), suggesting that different forms of SOCE may be activated depending on the context and trigger the release of specific synaptic vesicle pools. The subcellular organization of STIM and Orai proteins in the axon, and specifically their localization relative to the synaptic vesicle pools, remains to be uncovered.

What are the calcium sources for spontaneous neurotransmission? As mentioned, the spontaneous pool of synaptic vesicles appears to be heterogeneous, and thus its release can be triggered by calcium from VGCC, from the ER, or even be calcium-independent (Bal et al. 2013; Kavalali 2015; Williams and Smith 2018). The calcium sensors for spontaneous neurotransmission may also differ when compared to the action potential-evoked pool or even among synaptic vesicles that carry different neurotransmitters (e.g., GABA versus glutamate) (Chanaday et al. 2021; Courtney et al. 2018; Ermolyuk et al. 2013; Groffen et al. 2010; Ramirez et al. 2017; Vyleta and Smith 2011). Considering the various functions spontaneous neurotransmission plays in axon and synapse development, synapse maintenance, and setting circuit properties and excitability (e.g., homeostatic synaptic efficacy), these multiple calcium sources might ensure that regardless of the specific activity history or metabolic state of the synapse, they all converge on scaling spontaneous neurotransmitter release to ultimately fine-tune the synaptic strength and the signal-to-noise ratio of information transfer in the circuit.

3.6 Final Thoughts

We have come a long way since the discovery of synaptic vesicles and the molecules regulating the synaptic vesicle cycle. Several synaptic vesicle pools have been found that differ in their molecular composition, spatial organization, and dependence on calcium sources, and which mediate distinct forms of neurotransmission and synaptic plasticity processes. As it is common in scientific endeavors, these discoveries opened new questions: How are calcium channels organized at different types of synapses? Are all release sites similar or does the active zone machinery change depending on neurotransmitter, neuron type, and brain region? How are active zone and plasma membrane components at release sites recycled and protected from disassembly by the continuous insertion of synaptic vesicle membrane? How are the different synaptic vesicle pools generated during biogenesis? What are the molecular signals that determine the specific pools a given synapse will have? Is the molecular and spatial segregation of synaptic vesicle pools complete or is it a gradient? How stable are synaptic vesicle pools over time? Do they change with synaptic plasticity in terms of location, composition, and clustering? What about the morphology of the endoplasmic reticulum at presynaptic terminals and the nano-organization of calcium channels and calcium transporters in internal organelles? How do all these calcium sources and synaptic vesicle pools achieve temporal and spatial compartmentalization? How many different messages (information) can one synapse really encode? These are challenging questions, since synaptic vesicles are very small and tightly clustered in groups of hundreds of vesicles at each presynaptic bouton. Moreover, synaptic vesicles fuse and are endocytosed in just milliseconds, challenging both the spatial and temporal limits of resolution of current microscopy techniques. The collaborative work of multidisciplinary teams of scientists, applying complementary techniques from different fields (e.g., electrophysiology, molecular biology, microscopy, animal behavior, and computational modeling) will be necessary to answer these questions.

Acknowledgments This work was supported by the University of Pennsylvania and the Margaret Landenberger Foundation. The author thanks Jiangnan Xiao, Robert T. Moul, and Dr. A. Alejandro Vilcaes for their constructive criticism and feedback on the manuscript and figures.

References

Ahmari SE, Buchanan J, Smith SJ. Assembly of presynaptic active zones from cytoplasmic transport packets. Nat Neurosci. 2000;3(5):445–51.

Aiken J, Holzbaur ELF. Spastin locally amplifies microtubule dynamics to pattern the axon for presynaptic cargo delivery. bioRxiv. 2023. https://doi.org/10.1101/2023.08.08.552320.

Akbergenova Y, et al. Characterization of developmental and molecular factors underlying release heterogeneity at Drosophila synapses. elife. 2018;7:e38268.

Alabi AA, Tsien RW. Perspectives on kiss-and-run: role in exocytosis, endocytosis, and neurotransmission. Annu Rev Physiol. 2013;75:393–422.

Aldahabi M, et al. Different priming states of synaptic vesicles underlie distinct release probabilities at hippocampal excitatory synapses. Neuron. 2022;110(24):4144–4161.e7.

Althof D, et al. Inhibitory and excitatory axon terminals share a common nano-architecture of their Cav2.1 (P/Q-type) Ca(2+) channels. Front Cell Neurosci. 2015;9:315.

Andreae LC, Burrone J. Spontaneous neurotransmitter release shapes dendritic arbors via long-range activation of NMDA receptors. Cell Rep. 2015;10(6):873–82.

Astacio H, Vasin A, Bykhovskaia M. Stochastic properties of spontaneous synaptic transmission at individual active zones. J Neurosci. 2022;42(6):1001–19.

Augustine GJ, Eckert R. Divalent cations differentially support transmitter release at the squid giant synapse. J Physiol. 1984;346:257–71.

Augustine GJ, Charlton MP, Smith SJ. Calcium entry and transmitter release at voltage-clamped nerve terminals of squid. J Physiol. 1985;367:163–81.

Bacaj T, et al. Synaptotagmin-1 and synaptotagmin-7 trigger synchronous and asynchronous phases of neurotransmitter release. Neuron. 2013;80(4):947–59.

Bae JR, et al. Distinct synaptic vesicle recycling in inhibitory nerve terminals is coordinated by SV2A. Prog Neurobiol. 2020;194:101879.

Bal M, et al. Reelin mobilizes a VAMP7-dependent synaptic vesicle pool and selectively augments spontaneous neurotransmission. Neuron. 2013;80(4):934–46.

Banerjee S, et al. Miniature neurotransmission is required to maintain Drosophila synaptic structures during ageing. Nat Commun. 2021;12(1):4399.

Banerjee A, et al. Molecular and functional architecture of striatal dopamine release sites. Neuron. 2022;110(2):248–265.e9.

Bartol TM, et al. Nanoconnectomic upper bound on the variability of synaptic plasticity. elife. 2015;4:e10778.

Binotti B, Jahn R, Pérez-Lara Á. An overview of the synaptic vesicle lipid composition. Arch Biochem Biophys. 2021;709:108966.

Blumrich EM, et al. Phosphatidylinositol 4-kinase IIα is a glycogen synthase kinase 3-regulated interaction hub for activity-dependent bulk endocytosis. Cell Rep. 2023;42(6):112633.

Böhme MA, Grasskamp AT, Walter AM. Regulation of synaptic release-site Ca^{2+} channel coupling as a mechanism to control release probability and short-term plasticity. FEBS Lett. 2018;592(21):3516–31.

Böhme MA, et al. Rapid active zone remodeling consolidates presynaptic potentiation. Nat Commun. 2019;10(1):1085.

Bollmann C, et al. Primary neurons lacking the SNAREs vti1a and vti1b show altered neuronal development. Neural Dev. 2022;17(1):12.

Bouazza-Arostegui B, et al. Deconstructing synaptotagmin-1's distinct roles in synaptic vesicle priming and neurotransmitter release. J Neurosci. 2022;42(14):2856–71.

Boyken J, et al. Molecular profiling of synaptic vesicle docking sites reveals novel proteins but few differences between glutamatergic and GABAergic synapses. Neuron. 2013;78(2):285–97.

Branco T, et al. Local dendritic activity sets release probability at hippocampal synapses. Neuron. 2008;59(3):475–85.

Brose N, et al. Synaptotagmin: a calcium sensor on the synaptic vesicle surface. Science. 1992;256(5059):1021–5.

Bruentgens F, et al. The lack of synapsin alters presynaptic plasticity at hippocampal mossy fibers in male mice. eNeuro. 2024;11:ENEURO.0330.

Brunger AT, et al. The pre-synaptic fusion machinery. Curr Opin Struct Biol. 2019;54:179–88.

Butola T, et al. RIM-binding protein 2 organizes Ca2+ channel topography and regulates release probability and vesicle replenishment at a fast central synapse. J Neurosci. 2021;41(37):7742–67.

Castellano-Munoz M, Schnee ME, Ricci AJ. Calcium-induced calcium release supports recruitment of synaptic vesicles in auditory hair cells. J Neurophysiol. 2016;115(1):226–39.

Castillo PE, Weisskopf MG, Nicoll RA. The role of Ca2+ channels in hippocampal mossy fiber synaptic transmission and long-term potentiation. Neuron. 1994;12(2):261–9.

Cazares VA, et al. Dynamic partitioning of synaptic vesicle pools by the SNARE-binding protein Tomosyn. J Neurosci. 2016;36(44):11208–22.

Chamberland S, Tóth K. Functionally heterogeneous synaptic vesicle pools support diverse synaptic signalling. J Physiol. 2016;594(4):825–35.

Chamberland S, Evstratova A, Tóth K. Short-term facilitation at a detonator synapse requires the distinct contribution of multiple types of voltage-gated calcium channels. J Neurosci. 2017;37(19):4913–27.

Chamberland S, et al. Slow-decaying presynaptic calcium dynamics gate long-lasting asynchronous release at the hippocampal mossy fiber to CA3 pyramidal cell synapse. Synapse. 2020;74(12):e22178.
Chanaday NL, Kavalali ET. Presynaptic origins of distinct modes of neurotransmitter release. Curr Opin Neurobiol. 2018a;51:119–26.
Chanaday NL, Kavalali ET. Optical detection of three modes of endocytosis at hippocampal synapses. elife. 2018b;7:e36097.
Chanaday NL, Kavalali ET. Synaptobrevin-2 dependent regulation of single synaptic vesicle endocytosis. Mol Biol Cell. 2021;32(19):1818–23.
Chanaday NL, et al. The synaptic vesicle cycle revisited: new insights into the modes and mechanisms. J Neurosci. 2019;39(42):8209–16.
Chanaday NL, et al. Presynaptic store-operated Ca(2+) entry drives excitatory spontaneous neurotransmission and augments endoplasmic reticulum stress. Neuron. 2021;109(8):1314–1332.e5.
Chapman ER. How does synaptotagmin trigger neurotransmitter release? Annu Rev Biochem. 2008;77:615–41.
Chen M, Križaj D, Thoreson WB. Intracellular calcium stores drive slow non-ribbon vesicle release from rod photoreceptors. Front Cell Neurosci. 2014;8:20.
Chen Z, et al. Ca2+ channel to synaptic vesicle distance accounts for the readily releasable pool kinetics at a functionally mature auditory synapse. J Neurosci. 2015;35(5):2083–100.
Chen C, et al. Triple function of synaptotagmin 7 ensures efficiency of high-frequency transmission at central GABAergic synapses. Cell Rep. 2017;21(8):2082–9.
Cho RW, et al. Phosphorylation of complexin by PKA regulates activity-dependent spontaneous neurotransmitter release and structural synaptic plasticity. Neuron. 2015;88(4):749–61.
Choi BJ, et al. Miniature neurotransmission regulates Drosophila synaptic structural maturation. Neuron. 2014;82(3):618–34.
Choudhary B, et al. UNC-16/JIP3 regulates early events in synaptic vesicle protein trafficking via LRK-1/LRRK2 and AP complexes. PLoS Genet. 2017;13(11):e1007100.
Chung C, et al. Acute dynamin inhibition dissects synaptic vesicle recycling pathways that drive spontaneous and evoked neurotransmission. J Neurosci. 2010;30(4):1363–76.
Cork KM, Van Hook MJ, Thoreson WB. Mechanisms, pools, and sites of spontaneous vesicle release at synapses of rod and cone photoreceptors. Eur J Neurosci. 2016;44(3):2015–27.
Courtney NA, et al. Excitatory and inhibitory neurons utilize different Ca(2+) sensors and sources to regulate spontaneous release. Neuron. 2018;98(5):977–991.e5.
Courtney KC, et al. Synaptotagmin 1 oligomerization via the juxtamembrane linker regulates spontaneous and evoked neurotransmitter release. Proc Natl Acad Sci USA. 2021;118(48):e2113859118.
Courtney KC, et al. Synaptotagmin-7 outperforms synaptotagmin-1 to promote the formation of large, stable fusion pores via robust membrane penetration. Nat Commun. 2023;14(1):7761.
Cousin MA, Robinson PJ. The dephosphins: dephosphorylation by calcineurin triggers synaptic vesicle endocytosis. Trends Neurosci. 2001;24(11):659–65.
Crawford DC, et al. Selective molecular impairment of spontaneous neurotransmission modulates synaptic efficacy. Nat Commun. 2017;8:14436.
Cuadra AE, et al. Phase-locking requires efficient Ca^{2+} extrusion at the auditory hair cell ribbon synapse. J Neurosci. 2021;41(8):1625–35.
Das D, et al. Resolving kinetic intermediates during the regulated assembly and disassembly of fusion pores. Nat Commun. 2020;11(1):231.
Daste F, Galli T, Tareste D. Structure and function of longin SNAREs. J Cell Sci. 2015;128(23):4263–72.
De Gois S, et al. Identification of endophilins 1 and 3 as selective binding partners for VGLUT1 and their co-localization in neocortical glutamatergic synapses: implications for vesicular glutamate transporter trafficking and excitatory vesicle formation. Cell Mol Neurobiol. 2006;26(4–6):679–93.

de Juan-Sanz J, et al. Axonal endoplasmic reticulum Ca(2+) content controls release probability in CNS nerve terminals. Neuron. 2017;93(4):867–881.e6.
De Robertis ED, Bennett HS. Some features of the submicroscopic morphology of synapses in frog and earthworm. J Biophys Biochem Cytol. 1955;1(1):47–58.
De Robertis E, et al. Isolation of synaptic vesicles and structural organization of the acetycholine system within brain nerve endings. J Neurochem. 1963;10:225–35.
Deák F, et al. Synaptobrevin is essential for fast synaptic-vesicle endocytosis. Nat Cell Biol. 2004;6(11):1102–8.
Deguchi-Tawarada M, et al. Active zone protein CAST is a component of conventional and ribbon synapses in mouse retina. J Comp Neurol. 2006;495(4):480–96.
Del Castillo J, Katz B. Quantal components of the end-plate potential. J Physiol. 1954;124(3):560–73.
Del Signore SJ, et al. An autoinhibitory clamp of actin assembly constrains and directs synaptic endocytosis. elife. 2021;10:e69597.
Del Signore SJ, et al. An approach for quantitative mapping of synaptic periactive zone architecture and organization. Mol Biol Cell. 2023;34(6):ar51.
Delvendahl I, Kita K, Müller M. Rapid and sustained homeostatic control of presynaptic exocytosis at a central synapse. Proc Natl Acad Sci USA. 2019;116(47):23783–9.
De-Miguel FF. The thermodynamically expensive contribution of three calcium sources to somatic release of serotonin. Int J Mol Sci. 2022;23(3):1495.
Devine MJ, et al. Mitochondrial Ca2+ uniporter haploinsufficiency enhances long-term potentiation at hippocampal mossy fibre synapses. J Cell Sci. 2022;135(22):jcs259823.
Dolphin AC, Lee A. Presynaptic calcium channels: specialized control of synaptic neurotransmitter release. Nat Rev Neurosci. 2020;21(4):213–29.
Dong W, et al. CAST/ELKS proteins control voltage-gated Ca2+ channel density and synaptic release probability at a mammalian central synapse. Cell Rep. 2018;24(2):284–293.e6.
Dou Y, et al. Orai1 plays a crucial role in central sensitization by modulating neuronal excitability. J Neurosci. 2018;38(4):887–900.
Dou D, Aiken J, Holzbaur ELF. RAB3 phosphorylation by pathogenic LRRK2 impairs trafficking of synaptic vesicle precursors. J Cell Biol. 2024;223(6):e202307092.
Dresbach T, et al. Assembly of active zone precursor vesicles: obligatory trafficking of presynaptic cytomatrix proteins Bassoon and Piccolo via a trans-Golgi compartment. J Biol Chem. 2006;281(9):6038–47.
Dürst CD, et al. Vesicular release probability sets the strength of individual Schaffer collateral synapses. Nat Commun. 2022;13(1):6126.
Edwards SL, et al. UNC-16 (JIP3) acts through synapse-assembly proteins to inhibit the active transport of cell soma organelles to Caenorhabditis elegans motor neuron axons. Genetics. 2015;201(1):117–41.
Egashira Y, et al. Spontaneously recycling synaptic vesicles constitute readily releasable vesicles in intact neuromuscular synapses. J Neurosci. 2022;42(17):3523–36.
Eggers ED, Klein JS, Moore-Dotson JM. Slow changes in Ca2(+) cause prolonged release from GABAergic retinal amacrine cells. J Neurophysiol. 2013;110(3):709–19.
Eichel K, et al. Endocytosis in the axon initial segment maintains neuronal polarity. Nature. 2022;609(7925):128–35.
Emperador-Melero J, Kaeser PS. Assembly of the presynaptic active zone. Curr Opin Neurobiol. 2020;63:95–103.
Emperador-Melero J, et al. Vti1a/b regulate synaptic vesicle and dense core vesicle secretion via protein sorting at the Golgi. Nat Commun. 2018;9(1):3421.
Erhardt B, et al. Plasma membrane calcium ATPase downregulation in dopaminergic neurons alters cellular physiology and motor behaviour in Drosophila melanogaster. Eur J Neurosci. 2021;54(6):5915–31.
Ermolyuk YS, et al. Independent regulation of basal neurotransmitter release efficacy by variable Ca^{2+} influx and bouton size at small central synapses. PLoS Biol. 2012;10(9):e1001396.

Ermolyuk YS, et al. Differential triggering of spontaneous glutamate release by P/Q-, N- and R-type Ca2+ channels. Nat Neurosci. 2013;16(12):1754–63.

Fatt P, Katz B. Some observations on biological noise. Nature. 1950;166(4223):597–8.

Fatt P, Katz B. Spontaneous subthreshold activity at motor nerve endings. J Physiol. 1952;117(1):109–28.

Fattorini G, et al. Co-expression of VGLUT1 and VGAT sustains glutamate and GABA co-release and is regulated by activity in cortical neurons. J Cell Sci. 2015;128(9):1669–73.

Faundez VV, Kelly RB. The AP-3 complex required for endosomal synaptic vesicle biogenesis is associated with a casein kinase Ialpha-like isoform. Mol Biol Cell. 2000;11(8):2591–604.

Fei H, et al. Trafficking of vesicular neurotransmitter transporters. Traffic. 2008;9(9):1425–36.

Fekete A, et al. Underpinning heterogeneity in synaptic transmission by presynaptic ensembles of distinct morphological modules. Nat Commun. 2019;10(1):826.

Fernández-Busnadiego R, et al. Cryo-electron tomography reveals a critical role of RIM1α in synaptic vesicle tethering. J Cell Biol. 2013;201(5):725–40.

Feske S, et al. A mutation in Orai1 causes immune deficiency by abrogating CRAC channel function. Nature. 2006;441(7090):179–85.

Few AP, et al. Asynchronous Ca2+ current conducted by voltage-gated Ca2+ (CaV)-2.1 and CaV2.2 channels and its implications for asynchronous neurotransmitter release. Proc Natl Acad Sci USA. 2012;109(7):E452–60.

Foss SM, et al. Multiple dileucine-like motifs direct VGLUT1 trafficking. J Neurosci. 2013;33(26):10647–60.

Fouke KE, et al. Synuclein regulates synaptic vesicle clustering and docking at a vertebrate synapse. Front Cell Dev Biol. 2021;9:774650.

Fourcaudot E, et al. L-type voltage-dependent Ca(2+) channels mediate expression of presynaptic LTP in amygdala. Nat Neurosci. 2009;12(9):1093–5.

Fredj NB, Burrone J. A resting pool of vesicles is responsible for spontaneous vesicle fusion at the synapse. Nat Neurosci. 2009;12(6):751–8.

Fujii JT, et al. Plasma membrane calcium ATPase in synaptic terminals of chick Edinger-Westphal neurons. Brain Res. 1996;734(1–2):193–202.

Ghatpande AS, Sivaraaman K, Vijayaraghavan S. Store calcium mediates cholinergic effects on mIPSCs in the rat main olfactory bulb. J Neurophysiol. 2006;95(3):1345–55.

Ghelani T, et al. Nanoscopical analysis reveals an orderly arrangement of the presynaptic scaffold protein bassoon at the Golgi-apparatus. Front Mol Neurosci. 2021;14:744034.

Ghelani T, et al. Interactive nanocluster compaction of the ELKS scaffold and Cacophony Ca2+ channels drives sustained active zone potentiation. Sci Adv. 2023;9(7):eade7804.

Gimber N, et al. Diffusional spread and confinement of newly exocytosed synaptic vesicle proteins. Nat Commun. 2015;6:8392.

Gitler D, et al. Different presynaptic roles of synapsins at excitatory and inhibitory synapses. J Neurosci. 2004;24(50):11368–80.

Glebov OO, et al. Nanoscale structural plasticity of the active zone matrix modulates presynaptic function. Cell Rep. 2017;18(11):2715–28.

Grabner CP, et al. RIM1/2-mediated facilitation of Cav1.4 channel opening is required for Ca2+-stimulated release in mouse rod photoreceptors. J Neurosci. 2015;35(38):13133–47.

Gras C, et al. The vesicular glutamate transporter VGLUT3 synergizes striatal acetylcholine tone. Nat Neurosci. 2008;11(3):292–300.

Grasskamp AT, et al. Spontaneous neurotransmission at evocable synapses predicts their responsiveness to action potentials. Front Cell Neurosci. 2023;17:1129417.

Groffen AJ, et al. Doc2b is a high-affinity Ca2+ sensor for spontaneous neurotransmitter release. Science. 2010;327(5973):1614–8.

Guzikowski NJ, Kavalali ET. Nano-organization at the synapse: segregation of distinct forms of neurotransmission. Front Synaptic Neurosci. 2021;13:796498.

Hams N, et al. Otoferlin is a multivalent calcium-sensitive scaffold linking SNAREs and calcium channels. Proc Natl Acad Sci USA. 2017;114(30):8023–8.

Hefft S, Jonas P. Asynchronous GABA release generates long-lasting inhibition at a hippocampal interneuron-principal neuron synapse. Nat Neurosci. 2005;8(10):1319–28.
Held RG, Kaeser PS. ELKS active zone proteins as multitasking scaffolds for secretion. Open Biol. 2018;8(2):170258.
Henkel AW, Mouihate A, Welzel O. Differential release of exocytosis marker dyes indicates stimulation-dependent regulation of synaptic activity. Front Neurosci. 2019;13:1047.
Heuser JE, Reese TS. Evidence for recycling of synaptic vesicle membrane during transmitter release at the frog neuromuscular junction. J Cell Biol. 1973;57(2):315–44.
Heuser JE, Reese TS, Landis DM. Functional changes in frog neuromuscular junctions studied with freeze-fracture. J Neurocytol. 1974;3(1):109–31.
Heuser JE, et al. Synaptic vesicle exocytosis captured by quick freezing and correlated with quantal transmitter release. J Cell Biol. 1979;81(2):275–300.
Hnasko TS, Edwards RH. Neurotransmitter corelease: mechanism and physiological role. Annu Rev Physiol. 2012;74:225–43.
Hodgkin AL, Huxley AF. Currents carried by sodium and potassium ions through the membrane of the giant axon of Loligo. J Physiol. 1952a;116(4):449–72.
Hodgkin AL, Huxley AF. The dual effect of membrane potential on sodium conductance in the giant axon of Loligo. J Physiol. 1952b;116(4):497–506.
Hodgkin AL, Huxley AF, Katz B. Measurement of current-voltage relations in the membrane of the giant axon of Loligo. J Physiol. 1952;116(4):424–48.
Holderith N, et al. Release probability of hippocampal glutamatergic terminals scales with the size of the active zone. Nat Neurosci. 2012;15(7):988–97.
Hoopmann P, et al. Endosomal sorting of readily releasable synaptic vesicles. Proc Natl Acad Sci USA. 2010;107(44):19055–60.
Horvath PM, et al. Spontaneous and evoked neurotransmission are partially segregated at inhibitory synapses. elife. 2020;9:e52852.
Horvath PM, et al. A subthreshold synaptic mechanism regulating BDNF expression and resting synaptic strength. Cell Rep. 2021;36(5):109467.
Hruska M, Cain RE, Dalva MB. Nanoscale rules governing the organization of glutamate receptors in spine synapses are subunit specific. Nat Commun. 2022;13(1):920.
Hummel JJA, Hoogenraad CC. Specific KIF1A-adaptor interactions control selective cargo recognition. J Cell Biol. 2021;220(10):e202105011.
Huson V, Regehr WG. Diverse roles of synaptotagmin-7 in regulating vesicle fusion. Curr Opin Neurobiol. 2020;63:42–52.
Imoto Y, et al. Dynamin is primed at endocytic sites for ultrafast endocytosis. Neuron. 2022;110(17):2815–2835.e13.
Imoto Y, et al. Dynamin 1xA interacts with Endophilin A1 via its spliced long C-terminus for ultrafast endocytosis. EMBO J. 2024;43:3327.
Ivanova D, Cousin MA. Synaptic vesicle recycling and the endolysosomal system: a reappraisal of form and function. Front Synaptic Neurosci. 2022;14:826098.
Jahn R, Cafiso DC, Tamm LK. Mechanisms of SNARE proteins in membrane fusion. Nat Rev Mol Cell Biol. 2024;25(2):101–18.
Jensen K, Mody I. L-type Ca2+ channel-mediated short-term plasticity of GABAergic synapses. Nat Neurosci. 2001;4(10):975–6.
Jonas P, Bischofberger J, Sandkühler J. Corelease of two fast neurotransmitters at a central synapse. Science. 1998;281(5375):419–24.
Kasturacharya N, Dhall JK, Hasan G. A STIM dependent dopamine-insulin axis maintains the larval drive to feed and grow in Drosophila. bioRxiv. 2022. https://doi.org/10.1101/2022.09.20.508641.
Katz B, Miledi R. The effect of calcium on acetylcholine release from motor nerve terminals. Proc R Soc Lond B Biol Sci. 1965;161:496–503.
Katz B, Miledi R. The timing of calcium action during neuromuscular transmission. J Physiol. 1967;189(3):535–44.
Katz B, Miledi R. Further study of the role of calcium in synaptic transmission. J Physiol. 1970;207(3):789–801.

Kavalali ET. The mechanisms and functions of spontaneous neurotransmitter release. Nat Rev Neurosci. 2015;16(1):5–16.

Kavalali ET. Spontaneous neurotransmission: a form of neural communication comes of age. J Neurosci Res. 2018;96(3):331–4.

Keller D, et al. An exclusion zone for Ca2+ channels around docked vesicles explains release control by multiple channels at a CNS synapse. PLoS Comput Biol. 2015;11(5):e1004253.

Kim S, et al. Co-packaging of opposing neurotransmitters in individual synaptic vesicles in the central nervous system. Neuron. 2022;110(8):1371–1384.e7.

Kljakic O, et al. Functional dissociation of behavioral effects from acetylcholine and glutamate released from cholinergic striatal interneurons. FASEB J. 2022;36(2):e22135.

Kokotos AC, et al. Synaptophysin sustains presynaptic performance by preserving vesicular synaptobrevin-II levels. J Neurochem. 2019;151(1):28–37.

Kononenko NL, et al. Compromised fidelity of endocytic synaptic vesicle protein sorting in the absence of stonin 2. Proc Natl Acad Sci USA. 2013;110(6):E526–35.

Kontaxi C, Kim N, Cousin MA. The phospho-regulated amphiphysin/endophilin interaction is required for synaptic vesicle endocytosis. J Neurochem. 2023;166(2):248–64.

Koo SJ, et al. SNARE motif-mediated sorting of synaptobrevin by the endocytic adaptors clathrin assembly lymphoid myeloid leukemia (CALM) and AP180 at synapses. Proc Natl Acad Sci USA. 2011;108(33):13540–5.

Koo SJ, et al. Vesicular synaptobrevin/VAMP2 levels guarded by AP180 control efficient neurotransmission. Neuron. 2015;88(2):330–44.

Koppensteiner P, et al. $GABA_B$ receptors induce phasic release from medial habenula terminals through activity-dependent recruitment of release-ready vesicles. Proc Natl Acad Sci USA. 2024;121(8):e2301449121.

Körber C, Kuner T. Molecular machines regulating the release probability of synaptic vesicles at the active zone. Front Synaptic Neurosci. 2016;8:5.

Kreykenbohm V, et al. The SNAREs vti1a and vti1b have distinct localization and SNARE complex partners. Eur J Cell Biol. 2002;81(5):273–80.

Krick N, et al. Separation of presynaptic Ca_v2 and Ca_v1 channel function in synaptic vesicle exo$^-$ and endocytosis by the membrane anchored Ca^{2+} pump PMCA. Proc Natl Acad Sci USA. 2021;118(28):e2106621118.

Le Guellec B, et al. Depolarization-induced bursts of miniature synaptic currents in individual synapses of developing cerebellum. J Gen Physiol. 2023;155(5):e202213212.

Lee BJ, et al. Voltage-gated calcium channels contribute to spontaneous glutamate release directly via nanodomain coupling or indirectly via calmodulin. Prog Neurobiol. 2022;208:102182.

Leitz J, Kavalali ET. Ca^{2+} influx slows single synaptic vesicle endocytosis. J Neurosci. 2011;31(45):16318–26.

Leitz J, Kavalali ET. Fast retrieval and autonomous regulation of single spontaneously recycling synaptic vesicles. elife. 2014;3:e03658.

Li L, Bischofberger J, Jonas P. Differential gating and recruitment of P/Q-, N-, and R-type Ca2+ channels in hippocampal mossy fiber boutons. J Neurosci. 2007;27(49):13420–9.

Li YC, et al. Synaptotagmin-1- and synaptotagmin-7-dependent fusion mechanisms target synaptic vesicles to kinetically distinct endocytic pathways. Neuron. 2017a;93(3):616–631.e3.

Li H, et al. VGLUT2 trafficking is differentially regulated by adaptor proteins AP-1 and AP-3. Front Cell Neurosci. 2017b;11:324.

Li S, et al. Asynchronous release sites align with NMDA receptors in mouse hippocampal synapses. Nat Commun. 2021;12(1):677.

Lin PY, et al. VAMP4 maintains a Ca^{2+}-sensitive pool of spontaneously recycling synaptic vesicles. J Neurosci. 2020;40(28):5389–401.

Liou J, et al. STIM is a Ca2+ sensor essential for Ca2+-store-depletion-triggered Ca2+ influx. Curr Biol. 2005;15(13):1235–41.

Liu H, et al. Synaptotagmin 7 functions as a Ca2+-sensor for synaptic vesicle replenishment. elife. 2014;3:e01524.

Lopez-Manzaneda M, Fuentes-Moliz A, Tabares L. Presynaptic mitochondria communicate with release sites for spatio-temporal regulation of exocytosis at the motor nerve terminal. Front Synaptic Neurosci. 2022;14:858340.

Loy K, et al. Common strength and localization of spontaneous and evoked synaptic vesicle release sites. Mol Brain. 2014;7:23.

Lübbert M, et al. CaV2.1 α1 subunit expression regulates presynaptic CaV2.1 abundance and synaptic strength at a central synapse. Neuron. 2019;101(2):260–273.e6.

Malagon G, Myeong J, Klyachko VA. Two forms of asynchronous release with distinctive spatiotemporal dynamics in central synapses. elife. 2023;12:e84041.

Marland JR, et al. Mitochondrial calcium uptake modulates synaptic vesicle endocytosis in central nerve terminals. J Biol Chem. 2016;291(5):2080–6.

Martinez-Arca S, et al. A dual mechanism controlling the localization and function of exocytic v-SNAREs. Proc Natl Acad Sci USA. 2003;100(15):9011–6.

Martinez-Dunst C, Michaels RL, Fuchs PA. Release sites and calcium channels in hair cells of the chick's cochlea. J Neurosci. 1997;17(23):9133–44.

Maschi D, Klyachko VA. Spatiotemporal regulation of synaptic vesicle fusion sites in central synapses. Neuron. 2017;94(1):65–73.e3.

Maschi D, Klyachko VA. Spatiotemporal dynamics of multi-vesicular release is determined by heterogeneity of release sites within central synapses. elife. 2020;9:e55210.

Mathew SS, Hablitz JJ. Calcium release via activation of presynaptic IP3 receptors contributes to kainate-induced IPSC facilitation in rat neocortex. Neuropharmacology. 2008;55(1):106–16.

Meadows MA, et al. Glycine release is potentiated by cAMP via EPAC2 and Ca(2+) stores in a retinal interneuron. J Neurosci. 2021;41(46):9503–20.

Melani R, Tritsch NX. Inhibitory co-transmission from midbrain dopamine neurons relies on presynaptic GABA uptake. Cell Rep. 2022;39(3):110716.

Melom JE, et al. Spontaneous and evoked release are independently regulated at individual active zones. J Neurosci. 2013;33(44):17253–63.

Mendonça PRF, et al. Asynchronous glutamate release is enhanced in low release efficacy synapses and dispersed across the active zone. Nat Commun. 2022;13(1):3497.

Milovanovic D, et al. A liquid phase of synapsin and lipid vesicles. Science. 2018;361(6402):604–7.

Mitchell CB, et al. STIM1 is necessary for store-operated calcium entry in turning growth cones. J Neurochem. 2012;122(6):1155–66.

Mittelstaedt T, Alvaréz-Baron E, Schoch S. RIM proteins and their role in synapse function. Biol Chem. 2010;391(6):599–606.

Mochida S. Presynaptic calcium channels. Neurosci Res. 2018;127:33–44.

Mori M, Tanifuji S, Mochida S. Kinetic organization of Ca2+ signals that regulate synaptic release efficacy in sympathetic neurons. Mol Pharmacol. 2014;86(3):297–305.

Moschetta M, et al. Ca2+ binding to synapsin I regulates resting Ca2+ and recovery from synaptic depression in nerve terminals. Cell Mol Life Sci. 2022;79(12):600.

Moulder KL, et al. Vesicle pool heterogeneity at hippocampal glutamate and GABA synapses. J Neurosci. 2007;27(37):9846–54.

Mueller BD, et al. CaV1 and CaV2 calcium channels mediate the release of distinct pools of synaptic vesicles. elife. 2023;12:e81407.

Müller JA, et al. A presynaptic phosphosignaling hub for lasting homeostatic plasticity. Cell Rep. 2022;39(3):110696.

Mullins C, Hartnell LM, Bonifacino JS. Distinct requirements for the AP-3 adaptor complex in pigment granule and synaptic vesicle biogenesis in Drosophila melanogaster. Mol Gen Genet. 2000;263(6):1003–14.

Nakamura Y, et al. Nanoscale distribution of presynaptic Ca(2+) channels and its impact on vesicular release during development. Neuron. 2015;85(1):145–58.

Newell-Litwa K, et al. Roles of BLOC-1 and adaptor protein-3 complexes in cargo sorting to synaptic vesicles. Mol Biol Cell. 2009;20(5):1441–53.

Newman ZL, et al. Determinants of synapse diversity revealed by super-resolution quantal transmission and active zone imaging. Nat Commun. 2022;13(1):229.

Nichols EL, Smith CJ. Synaptic-like vesicles facilitate pioneer axon invasion. Curr Biol. 2019;29(16):2652–2664.e4.

Nicholson-Fish JC, et al. VAMP4 is an essential cargo molecule for activity-dependent bulk endocytosis. Neuron. 2015;88(5):973–84.

Niwa S, Tanaka Y, Hirokawa N. KIF1Bbeta- and KIF1A-mediated axonal transport of presynaptic regulator Rab3 occurs in a GTP-dependent manner through DENN/MADD. Nat Cell Biol. 2008;10(11):1269–79.

Nonet ML, et al. UNC-11, a Caenorhabditis elegans AP180 homologue, regulates the size and protein composition of synaptic vesicles. Mol Biol Cell. 1999;10(7):2343–60.

Ogunmowo TH, et al. Membrane compression by synaptic vesicle exocytosis triggers ultrafast endocytosis. Nat Commun. 2023;14(1):2888.

Okada Y, et al. The neuron-specific kinesin superfamily protein KIF1A is a unique monomeric motor for anterograde axonal transport of synaptic vesicle precursors. Cell. 1995;81(5):769–80.

Oline SN, Burger RM. Short-term synaptic depression is topographically distributed in the cochlear nucleus of the chicken. J Neurosci. 2014;34(4):1314–24.

Ono Y, et al. Expression of plasma membrane calcium ATPases confers Ca^{2+}/H^{+} exchange in rodent synaptic vesicles. Sci Rep. 2019;9(1):4289.

Opazo F, et al. Limited intermixing of synaptic vesicle components upon vesicle recycling. Traffic. 2010;11(6):800–12.

Orlando M, et al. Recruitment of release sites underlies chemical presynaptic potentiation at hippocampal mossy fiber boutons. PLoS Biol. 2021;19(6):e3001149.

Overhoff M, De Bruyckere E, Kononenko NL. Mechanisms of neuronal survival safeguarded by endocytosis and autophagy. J Neurochem. 2021;157(2):263–96.

Özçete Ö, Banerjee A, Kaeser PS. Mechanisms of neuromodulatory volume transmission. Mol Psychiatry. 2024;29:3680.

Papantoniou C, et al. Munc13- and SNAP25-dependent molecular bridges play a key role in synaptic vesicle priming. Sci Adv. 2023;9(25):eadf6222.

Park H, Li Y, Tsien RW. Influence of synaptic vesicle position on release probability and exocytotic fusion mode. Science. 2012;335(6074):1362–6.

Park S, et al. UNC-18 and Tomosyn antagonistically control synaptic vesicle priming downstream of UNC-13 in Caenorhabditis elegans. J Neurosci. 2017;37(36):8797–815.

Patel JC, et al. Mobilization of calcium from intracellular stores facilitates somatodendritic dopamine release. J Neurosci. 2009;29(20):6568–79.

Pathak T, et al. Store-operated calcium entry through Orai is required for transcriptional maturation of the flight circuit in drosophila. J Neurosci. 2015;35(40):13784–99.

Pavez M, et al. STIM1 is required for remodeling of the endoplasmic reticulum and microtubule cytoskeleton in steering growth cones. J Neurosci. 2019;39(26):5095–114.

Peden AA, Park GY, Scheller RH. The Di leucine motif of vesicle-associated membrane protein 4 is required for its localization and AP-1 binding. J Biol Chem. 2001;276(52):49183–7.

Pelkey KA, et al. Compartmentalized Ca(2+) channel regulation at divergent mossy-fiber release sites underlies target cell-dependent plasticity. Neuron. 2006;52(3):497–510.

Peng YJ, et al. Minibrain kinase and calcineurin coordinate activity-dependent bulk endocytosis through synaptojanin. J Cell Biol. 2021;220(12):e202011028.

Penner R, Matthews G, Neher E. Regulation of calcium influx by second messengers in rat mast cells. Nature. 1988;334(6182):499–504.

Perin MS, et al. Phospholipid binding by a synaptic vesicle protein homologous to the regulatory region of protein kinase C. Nature. 1990;345(6272):260–3.

Perissinotti PP, Giugovaz Tropper B, Uchitel OD. L-type calcium channels are involved in fast endocytosis at the mouse neuromuscular junction. Eur J Neurosci. 2008;27(6):1333–44.

Provoda CJ, Waring MT, Buckley KM. Evidence for a primary endocytic vesicle involved in synaptic vesicle biogenesis. J Biol Chem. 2000;275(10):7004–12.

Quade B, et al. Membrane bridging by Munc13-1 is crucial for neurotransmitter release. elife. 2019;8:e42806.

Raingo J, et al. VAMP4 directs synaptic vesicles to a pool that selectively maintains asynchronous neurotransmission. Nat Neurosci. 2012;15(5):738–45.
Rajappa R, et al. Synaptophysin 1 clears synaptobrevin 2 from the presynaptic active zone to prevent short-term depression. Cell Rep. 2016;14(6):1369–81.
Ralowicz AJ, Hokeness S, Hoppa MB. Frequency of spontaneous neurotransmission at individual boutons corresponds to the size of the readily releasable pool of vesicles. J Neurosci. 2024;44(18):e1253232024.
Ramakrishnan S, et al. Synergistic roles of synaptotagmin-1 and complexin in calcium-regulated neuronal exocytosis. elife. 2020;9:e54506.
Ramesh G, et al. A short isoform of STIM1 confers frequency-dependent synaptic enhancement. Cell Rep. 2021;34(11):108844.
Ramirez DMO, et al. Loss of Doc2-dependent spontaneous neurotransmission augments glutamatergic synaptic strength. J Neurosci. 2017;37(26):6224–30.
Reese AL, Kavalali ET. Spontaneous neurotransmission signals through store-driven Ca(2+) transients to maintain synaptic homeostasis. elife. 2015;4:e09262.
Régnier-Vigouroux A, Tooze SA, Huttner WB. Newly synthesized synaptophysin is transported to synaptic-like microvesicles via constitutive secretory vesicles and the plasma membrane. EMBO J. 1991;10(12):3589–601.
Rey SA, et al. Ultrastructural and functional fate of recycled vesicles in hippocampal synapses. Nat Commun. 2015;6:8043.
Rey S, et al. Nanoscale remodeling of functional synaptic vesicle pools in Hebbian plasticity. Cell Rep. 2020;30(6):2006–2017.e3.
Richards DA. Regulation of exocytic mode in hippocampal neurons by intra-bouton calcium concentration. J Physiol. 2010;588(Pt 24):4927–36.
Richmond JE, Davis WS, Jorgensen EM. UNC-13 is required for synaptic vesicle fusion in C. elegans. Nat Neurosci. 1999;2(11):959–64.
Richter K, et al. Presynaptic cytomatrix protein bassoon is localized at both excitatory and inhibitory synapses of rat brain. J Comp Neurol. 1999;408(3):437–48.
Roos J, et al. STIM1, an essential and conserved component of store-operated Ca2+ channel function. J Cell Biol. 2005;169(3):435–45.
Rosenmund C, Clements JD, Westbrook GL. Nonuniform probability of glutamate release at a hippocampal synapse. Science. 1993;262(5134):754–7.
Ross WN. Understanding calcium waves and sparks in central neurons. Nat Rev Neurosci. 2012;13(3):157–68.
Sabatini BL, Regehr WG. Timing of neurotransmission at fast synapses in the mammalian brain. Nature. 1996;384(6605):170–2.
Sansevrino R, Hoffmann C, Milovanovic D. Condensate biology of synaptic vesicle clusters. Trends Neurosci. 2023;46(4):293–306.
Santos MS, Li H, Voglmaier SM. Synaptic vesicle protein trafficking at the glutamate synapse. Neuroscience. 2009;158(1):189–203.
Santos MS, et al. Sorting of the vesicular GABA transporter to functional vesicle pools by an atypical dileucine-like motif. J Neurosci. 2013;33(26):10634–46.
Sara Y, et al. An isolated pool of vesicles recycles at rest and drives spontaneous neurotransmission. Neuron. 2005;45(4):563–73.
Sassa T, et al. Regulation of the UNC-18-Caenorhabditis elegans syntaxin complex by UNC-13. J Neurosci. 1999;19(12):4772–7.
Schikorski T, Stevens CF. Morphological correlates of functionally defined synaptic vesicle populations. Nat Neurosci. 2001;4(4):391–5.
Scimemi A, Diamond JS. The number and organization of Ca2+ channels in the active zone shapes neurotransmitter release from Schaffer collateral synapses. J Neurosci. 2012;32(50):18157–76.
Sharma A, Hasan G. Modulation of flight and feeding behaviours requires presynaptic IP3Rs in dopaminergic neurons. elife. 2020;9:e62297.
Sharma G, Vijayaraghavan S. Modulation of presynaptic store calcium induces release of glutamate and postsynaptic firing. Neuron. 2003;38(6):929–39.

Shim S, Zheng JQ, Ming GL. A critical role for STIM1 in filopodial calcium entry and axon guidance. Mol Brain. 2013;6:51.

Shin W, et al. Preformed Ω-profile closure and kiss-and-run mediate endocytosis and diverse endocytic modes in neuroendocrine chromaffin cells. Neuron. 2021;109(19):3119–3134.e5.

Siksou L, et al. Three-dimensional architecture of presynaptic terminal cytomatrix. J Neurosci. 2007;27(26):6868–77.

Silm K, et al. Synaptic vesicle recycling pathway determines neurotransmitter content and release properties. Neuron. 2019;102(4):786–800.e5.

Simon SM, Llinás RR. Compartmentalization of the submembrane calcium activity during calcium influx and its significance in transmitter release. Biophys J. 1985;48(3):485–98.

Singh M, et al. Presynaptic loss of dynamin-related protein 1 impairs synaptic vesicle release and recycling at the mouse calyx of Held. J Physiol. 2018;596(24):6263–87.

Singh N, et al. Presynaptic endoplasmic reticulum regulates short-term plasticity in hippocampal synapses. Commun Biol. 2021;4(1):241.

Smith SJ, et al. The spatial distribution of calcium signals in squid presynaptic terminals. J Physiol. 1993;472:573–93.

Smith HL, et al. Mitochondrial support of persistent presynaptic vesicle mobilization with age-dependent synaptic growth after LTP. elife. 2016;5:e15275.

Soboloff J, et al. STIM proteins: dynamic calcium signal transducers. Nat Rev Mol Cell Biol. 2012;13(9):549–65.

Song SH, Augustine GJ. Different mechanisms of synapsin-induced vesicle clustering at inhibitory and excitatory synapses. Cell Rep. 2023;42(8):113004.

Song AH, et al. A selective filter for cytoplasmic transport at the axon initial segment. Cell. 2009;136(6):1148–60.

Staal RG, Mosharov EV, Sulzer D. Dopamine neurons release transmitter via a flickering fusion pore. Nat Neurosci. 2004;7(4):341–6.

Stevens RJ, et al. Abnormal synaptic vesicle biogenesis in Drosophila synaptogyrin mutants. J Neurosci. 2012;32(50):18054–67, 18067a.

Südhof TC. Synaptotagmins: why so many? J Biol Chem. 2002;277(10):7629–32.

Südhof TC. The presynaptic active zone. Neuron. 2012;75(1):11–25.

Südhof TC. Neurotransmitter release: the last millisecond in the life of a synaptic vesicle. Neuron. 2013;80(3):675–90.

Sugita S, et al. Synaptotagmin VII as a plasma membrane Ca(2+) sensor in exocytosis. Neuron. 2001;30(2):459–73.

Sun JY, Wu XS, Wu LG. Single and multiple vesicle fusion induce different rates of endocytosis at a central synapse. Nature. 2002;417(6888):555–9.

Sutton MA, et al. Miniature neurotransmission stabilizes synaptic function via tonic suppression of local dendritic protein synthesis. Cell. 2006;125(4):785–99

Svensson E, et al. General principles of neuronal Co-transmission: insights from multiple model systems. Front Neural Circuits. 2018;12:117.

Szikra T, et al. Depletion of calcium stores regulates calcium influx and signal transmission in rod photoreceptors. J Physiol. 2008;586(20):4859–75.

Takamori S, et al. Molecular anatomy of a trafficking organelle. Cell. 2006;127(4):831–46.

Tan PK, et al. A leucine-based motif mediates the endocytosis of vesicular monoamine and acetylcholine transporters. J Biol Chem. 1998;273(28):17351–60.

Tang AH, et al. A trans-synaptic nanocolumn aligns neurotransmitter release to receptors. Nature. 2016;536(7615):210–4.

Tao CL, Tian CL, Liu YT, Lu ZH, Qi L, Li XW, Li C, Shen X, Gu ML, Huang WL, Liu S, Yang LQ, Liao Z, Ma X, Wu J, Sun J, Wang P, Lau PM, Zhou ZH, Bi GQ. "Kiss-shrink-run" unifies mechanisms for synaptic vesicle exocytosis and hyperfast recycling. Science. 2025 Oct 16;390(6770):eads7954. https://doi.org/10.1126/science.ads7954. Epub 2025 Oct 16. PMID: 41100620.

Tao-Cheng JH. Ultrastructural localization of active zone and synaptic vesicle proteins in a preassembled multi-vesicle transport aggregate. Neuroscience. 2007;150(3):575–84.
Thiele C, et al. Cholesterol binds to synaptophysin and is required for biogenesis of synaptic vesicles. Nat Cell Biol. 2000;2(1):42–9.
Thomas CI, et al. Presynaptic mitochondria volume and abundance increase during development of a high-fidelity synapse. J Neurosci. 2019;39(41):7994–8012.
Turecek J, Regehr WG. Neuronal regulation of fast synaptotagmin isoforms controls the relative contributions of synchronous and asynchronous release. Neuron. 2019;101(5):938–949.e4.
Turecek J, Jackman SL, Regehr WG. Synaptotagmin 7 confers frequency invariance onto specialized depressing synapses. Nature. 2017;551(7681):503–6.
Turrigiano G. Homeostatic synaptic plasticity: local and global mechanisms for stabilizing neuronal function. Cold Spring Harb Perspect Biol. 2012;4(1):a005736.
Tyler WJ, et al. BDNF increases release probability and the size of a rapidly recycling vesicle pool within rat hippocampal excitatory synapses. J Physiol. 2006;574(Pt 3):787–803.
Vaaga CE, Borisovska M, Westbrook GL. Dual-transmitter neurons: functional implications of co-release and co-transmission. Curr Opin Neurobiol. 2014;29:25–32.
Vandael D, et al. Short-term plasticity at hippocampal mossy fiber synapses is induced by natural activity patterns and associated with vesicle pool engram formation. Neuron. 2020;107(3):509–521.e7.
Verhage M, et al. Synaptic assembly of the brain in the absence of neurotransmitter secretion. Science. 2000;287(5454):864–9.
Vevea JD, et al. Synaptotagmin 7 is targeted to the axonal plasma membrane through γ-secretase processing to promote synaptic vesicle docking in mouse hippocampal neurons. elife. 2021;10:e67261.
Vivona S, et al. The longin SNARE VAMP7/TI-VAMP adopts a closed conformation. J Biol Chem. 2010;285(23):17965–73.
Vos M, Lauwers E, Verstreken P. Synaptic mitochondria in synaptic transmission and organization of vesicle pools in health and disease. Front Synaptic Neurosci. 2010;2:139.
Vukoja A, et al. Presynaptic biogenesis requires axonal transport of lysosome-related vesicles. Neuron. 2018;99(6):1216–1232.e7.
Vyleta NP, Smith SM. Spontaneous glutamate release is independent of calcium influx and tonically activated by the calcium-sensing receptor. J Neurosci. 2011;31(12):4593–606.
Waites CL, et al. Bassoon and Piccolo maintain synapse integrity by regulating protein ubiquitination and degradation. EMBO J. 2013;32(7):954–69.
Wallace ML, Sabatini BL. Synaptic and circuit functions of multitransmitter neurons in the mammalian brain. Neuron. 2023;111(19):2969–83.
Wang Y, et al. Rim is a putative Rab3 effector in regulating synaptic-vesicle fusion. Nature. 1997;388(6642):593–8.
Wang SSH, et al. Fusion competent synaptic vesicles persist upon active zone disruption and loss of vesicle docking. Neuron. 2016a;91(4):777–91.
Wang X, Pinter MJ, Rich MM. Reversible recruitment of a homeostatic reserve pool of synaptic vesicles underlies rapid homeostatic plasticity of quantal content. J Neurosci. 2016b;36(3):828–36.
Wang Y, et al. Structural and functional synaptic plasticity induced by convergent synapse loss in the Drosophila neuromuscular circuit. J Neurosci. 2021;41(7):1401–17.
Wang CS, et al. Probing the segregation of evoked and spontaneous neurotransmission via photobleaching and recovery of a fluorescent glutamate sensor. elife. 2022;11:e76008.
Wang CS, Monteggia LM, Kavalali ET. Spatially non-overlapping Ca^{2+} signals drive distinct forms of neurotransmission. Cell Rep. 2023;42(10):113201.
Warrier A, et al. Calcium from internal stores triggers GABA release from retinal amacrine cells. J Neurophysiol. 2005;94(6):4196–208.
Watanabe S, Boucrot E. Fast and ultrafast endocytosis. Curr Opin Cell Biol. 2017;47:64–71.
Watanabe S, et al. Ultrafast endocytosis at mouse hippocampal synapses. Nature. 2013;504(7479):242–7.
Watanabe S, et al. Clathrin regenerates synaptic vesicles from endosomes. Nature. 2014;515(7526):228–33.

Watson ET, et al. Synaptic vesicle proteins are selectively delivered to axons in mammalian neurons. elife. 2023;12:e82568.
Wei D, et al. Orai1 and Orai3 mediate store-operated calcium entry contributing to neuronal excitability in dorsal root ganglion neurons. Front Cell Neurosci. 2017;11:400.
Wen H, et al. Distinct roles for two synaptotagmin isoforms in synchronous and asynchronous transmitter release at zebrafish neuromuscular junction. Proc Natl Acad Sci USA. 2010;107(31):13906–11.
Wen H, et al. Synchronous and asynchronous modes of synaptic transmission utilize different calcium sources. elife. 2013;2:e01206.
Wen X, Saltzgaber GW, Thoreson WB. Kiss-and-run is a significant contributor to synaptic exocytosis and endocytosis in photoreceptors. Front Cell Neurosci. 2017;11:286.
Weston MC, et al. Interplay between VGLUT isoforms and endophilin A1 regulates neurotransmitter release and short-term plasticity. Neuron. 2011;69(6):1147–59.
Weyrer C, et al. Introduction of synaptotagmin 7 promotes facilitation at the climbing fiber to Purkinje cell synapse. Cell Rep. 2021;36(12):109719.
Whittaker VP, Sheridan MN. The morphology and acetylcholine content of isolated cerebral cortical synaptic vesicles. J Neurochem. 1965;12:363–72.
Whittaker VP, Michaelson IA, Kirkland RJ. The separation of synaptic vesicles from nerve-ending particles ("synaptosomes"). Biochem J. 1964;90(2):293–303.
Wichmann C. Molecularly and structurally distinct synapses mediate reliable encoding and processing of auditory information. Hear Res. 2015;330(Pt B):178–90.
Wilhelm BG, Groemer TW, Rizzoli SO. The same synaptic vesicles drive active and spontaneous release. Nat Neurosci. 2010;13(12):1454–6.
Williams CL, Smith SM. Calcium dependence of spontaneous neurotransmitter release. J Neurosci Res. 2018;96(3):335–47.
Wu WH, Cooper RL. The regulation and packaging of synaptic vesicles as related to recruitment within glutamatergic synapses. Neuroscience. 2012;225:185–98.
Wu LG, et al. Exocytosis and endocytosis: modes, functions, and coupling mechanisms. Annu Rev Physiol. 2014;76:301–31.
Wu Z, et al. Nanodisc-cell fusion: control of fusion pore nucleation and lifetimes by SNARE protein transmembrane domains. Sci Rep. 2016;6:27287.
Wu Y, et al. Contacts between the endoplasmic reticulum and other membranes in neurons. Proc Natl Acad Sci USA. 2017;114(24):E4859–67.
Wu Z, et al. The neuronal calcium sensor synaptotagmin-1 and SNARE proteins cooperate to dilate fusion pores. elife. 2021;10:e68215.
Xu J, Mashimo T, Südhof TC. Synaptotagmin-1, -2, and -9: Ca(2+) sensors for fast release that specify distinct presynaptic properties in subsets of neurons. Neuron. 2007;54(4):567–81.
Xu H, et al. Adaptor protein AP-3 produces synaptic vesicles that release at high frequency by recruiting phospholipid flippase ATP8A1. Nat Neurosci. 2023;26(10):1685–700.
Young SM, Neher E. Synaptotagmin has an essential function in synaptic vesicle positioning for synchronous release in addition to its role as a calcium sensor. Neuron. 2009;63(4):482–96.
Zhai RG, Bellen HJ. The architecture of the active zone in the presynaptic nerve terminal. Physiology (Bethesda). 2004;19:262–70.
Zhang YV, et al. The KIF1A homolog Unc-104 is important for spontaneous release, postsynaptic density maturation and perisynaptic scaffold organization. Sci Rep. 2017;7:38172.
Zhang W, Jiang HH, Luo F. Diverse organization of voltage-gated calcium channels at presynaptic active zones. Front Synaptic Neurosci. 2022;14:1023256.
Zhou Q, et al. Architecture of the synaptotagmin-SNARE machinery for neuronal exocytosis. Nature. 2015;525(7567):62–7.
Zhou Q, et al. The primed SNARE-complexin-synaptotagmin complex for neuronal exocytosis. Nature. 2017;548(7668):420–5.
Zych SM, Ford CP. Divergent properties and independent regulation of striatal dopamine and GABA co-transmission. Cell Rep. 2022;39(7):110823.

Chapter 4
Active Zone: Linking Resolution Levels of Microscopic Modalities

Maksim Galkov, Paulina Nemcova, Dirk Dietrich, and Susanne Schoch

Abstract The presynaptic active zone (AZ) is a precisely organized nanoscale domain where synaptic vesicle exocytosis and neurotransmitter release are governed by tightly regulated protein networks. This review synthesizes recent insights from electron microscopy (EM) and super-resolution fluorescence microscopy that have deepened our understanding of active zone architecture in mammalian central nervous system synapses and at the Drosophila neuromuscular junction (NMJ). These imaging techniques have elucidated the spatial organization of key active zone proteins relative to one another and to the plasma membrane, which is notably well-ordered at the Drosophila NMJ. Here, we present a detailed overview of the nanometer-scale positioning of AZ proteins across the two types of synapses. In parallel, the idea that active zone nanostructures may form through liquid–liquid phase separation has emerged as a potential organizing principle. The transient and dynamic interactions characteristic of phase-separated protein condensates contrast with models that attribute nanodomain organization to specific, stable protein–protein interactions, raising the question of how the active zone's stable core architecture is reconciled with its capacity for dynamic plasticity.

Keywords Electron microscopy · Superresolution microscopy · Phase condensates · Presynaptic ultrastructure · Dense projections

4.1 Introduction

In the nervous system, information is transmitted between neurons and their target cells at specialized contact sites known as synapses, a concept introduced by Sir Charles Sherrington in 1897 (Foster and Sherrington 1897). The morphological visualization of synapses was first achieved in the 1940s by Couteaux (1946, 1947),

M. Galkov · P. Nemcova · D. Dietrich · S. Schoch (✉)
Institute of Cellular Neurosciences II, Medical Faculty, University of Bonn, Bonn, Germany
e-mail: susanne.schoch@uni-bonn.de

E. T. Kavalali (ed.), *Nano-organization of the Synapse*, Advances in Neurobiology 48,
https://doi.org/10.1007/978-3-032-12594-1_4

and their ultrastructure was described using electron microscopy (EM) images in the 1950s by Palade and Palay (1955), De Robertis and Bennett (1955), and Robertson (1956). Based on the observations of synaptic ultrastructure, Palay formulated the concept of "synaptic complex" in 1956, defining it as a discrete synaptic subregion characterized by a cluster of synaptic vesicles in close proximity to the dense part of the presynaptic membrane, opposed to the thickening of the postsynaptic membrane (Palay 1956, 1958). The identification of these "synaptic complexes" as the site of synaptic vesicle exocytosis was first demonstrated by Couteaux & Pécot-Dechavassine through electron microscopy examination of the frog neuromuscular junction (NMJ) (Couteaux and Pécot-Dechavassine 1970), with Couteaux introducing the term "zone active" to describe the area of the presynaptic plasma membrane where fusion of neurotransmitter-filled synaptic vesicles occurs and neurotransmitter is released. Further evidence supporting the existence of an "active zone" was provided by freeze-fracture studies of the frog NMJ by Dreyer et al. (1973) and Heuser et al. (1974). Subsequent ultrastructural analyses have revealed the presence of active zones (AZs) in various neuron types in a wide range of organisms, from *C. elegans* and *Drosophila* to humans (reviewed in (Zhai and Bellen 2004; Südhof 2012; Zhai 2023)). Although the exact ultrastructural organization of AZs may vary depending on the neuron type and postsynaptic partner, common morphological features have been identified: (1) a small number of synaptic vesicles directly contact the plasma membrane of the AZ and fuse there, (2) an electron dense material, consisting of proteins, is proximally lined up in a segregated fashion near the AZ plasma membrane, (3) the synaptic cleft maintains a uniform size (20–30 nm), and (4) the AZ is aligned with the electron dense material of the postsynaptic partner, known as the postsynaptic density (PSD).

The proteins found at the AZ fall into distinct categories such as the core complex regulating synaptic vesicle fusion (e.g., SNAREs, Munc18, Munc13, Complexin, Synaptotagmin) and endocytosis (e.g., Dynamin, Amphiphysin, Intersectin), regulators of AZ assembly (e.g., Syd-2/Liprin-α and Syd-1), adaptor and scaffolding proteins, cell adhesion molecules, cytoskeletal elements, and voltage-gated calcium channels (VGCCs). Over the past two decades, significant progress has been made in identifying the AZ proteome (reviewed in (Laßek et al. 2015; Jin and Zhai 2023)) and understanding how AZ components participate in the regulation of synaptic vesicle fusion (reviewed by Jin and Zhai (2023)). While most of the proteins that constitute the scaffolding network, called the cytomatrix at the active zone (CAZ), are not exclusive to presynaptic AZs but also occur along the axon or at high concentrations throughout the presynaptic bouton, six evolutionary conserved CAZ protein families that are AZ-specific/-enriched have been identified: RIM, RIM-BP, Munc13/Unc13, ELKS (and ELKS-like protein Bruchpilot in *Drosophila*), Liprin-α, Bassoon and Piccolo (reviewed in (Schoch and Gundelfinger 2006; Jin and Zhai 2023)). These AZ proteins are large (from ~100 to over 500 kDa) and contain multiple structural domains involved in protein–protein, protein–lipid and protein–ion binding as well as long intrinsically disordered regions (IDRs). They engage in extensive homo- and heteromeric interactions with each other, as well as with the components of the synaptic vesicle (SV) release and recycling machinery, the cytoskeleton, cell adhesion molecules, and the plasma membrane (PM), resulting in AZ proteins being largely insoluble in biochemical purifications (Phillips et al. 2001; Boyken et al. 2013).

Despite advances in imaging techniques, such as tissue preparation without fixatives, electron tomography, and super-resolution microscopy, which have provided insights into the three-dimensional (3D) organization of the overall CAZ material, our knowledge about the molecular arrangement and the spatial organization of AZ proteins is still limited. To gain a comprehensive understanding of how AZ proteins regulate SV fusion and endocytosis as well as presynaptic strength and plasticity at the molecular level, it will be essential to obtain a detailed knowledge about the submicroscopic organization of individual proteins and their spatial relationship with each other, which ultimately defines their possible interactions. Here, we will focus on and summarize the currently available data about the spatial organization of AZ molecules.

Two contrasting concepts of spatial organization of AZ-specific proteins have emerged: (1) AZ proteins form a protein scaffold, the CAZ, with a defined stoichiometry based on highly specific and stable protein–protein interactions (PPIs), which is connected to SVs and VGCCs (reviewed in (Schoch and Gundelfinger 2006; Jin and Zhai 2023), and (2) the formation of membrane-less fluidic subcompartments of varying composition, called liquid–liquid phases (LLPS) or phase condensates that are characterized by less specific and more transient interactions and high local protein concentrations in the phase condensate compared to the surrounding solution (McDonald and Shen 2021; Lautenschläger 2022). With the currently available technological approaches, it remains challenging, if not impossible, to resolve the position and orientation of individual molecules (scaffold model), and definitive evidence for or against LLPS in presynaptic boutons or at the AZ is lacking.

4.2 Model Synapses with Different Dimensions: Neuromuscular Junction and Mammalian Central Small Synapses

The ultrastructure of the AZ has been examined using EM in various organisms, ranging from *C. elegans* over frogs to humans, and regions, for example, at the neuromuscular junction, the inner ear, the retina, and the hippocampus, revealing synapse types with strongly diverging architectures (reviewed in (Zhai and Bellen 2004)). Here, we will review the current knowledge of the AZ nanoscale organization of small mammalian central nervous system (CNS) and *Drosophila* NMJ synapses. These synapses differ in their overall architecture, for example, their size, the number of docked and total SVs, and the number of AZs per synaptic contact.

In rodents, small CNS synapses (~1 μm bouton diameter) in the hippocampus, cerebellum, and cortex have AZs ranging from 0.2 to 0.4 μm in diameter, 0.03 to 0.14 μm^2 in area, and containing about 4–10 docked SVs/AZ or ~100 docked SVs/μm^2 (Carlin et al. 1980; Schikorski and Stevens 1997, 1999; Xu-Friedman et al. 2001; Paulussen et al. 2023).

The *Drosophila* larval NMJ is another but a much larger example of a synaptic contact, which is formed by a branched chain of synaptic boutons of the motor neuron. On average, one NMJ includes 20–50 boutons, and each bouton having a diameter of ~2–4 μm maintains ~10–30 AZs (~0.8 AZs per μm^2 of the bouton surface area), which are defined as protein-enriched areas where SVs cluster and fuse (Sigrist et al. 2002; Prokop and Meinertzhagen 2006; Collins and DiAntonio 2007; Paul et al. 2015; Nijhof et al. 2016; Hur et al. 2018). The area of the dense material forming an individual AZ in NMJ has a diameter of 0.6–0.8 μm, covers 0.3–0.5 μm^2 (Sigrist et al. 2002; Song et al. 2002; Dawson-Scully et al. 2007; Barber et al. 2018; Hur et al. 2018), and contains approximately 30–50 docked SVs/AZ at a similar density compared to small rodent synapses of ~100 docked SVs/μm^2 (Song et al. 2002; Dawson-Scully et al. 2007).

4.3 Light Microscopic Views of AZ Material

4.3.1 Small CNS Synapses

Due to the diffraction limit, confocal microscopy resolves only the outline of small mammalian CNS synapses, and proteins appear uniformly distributed inside the synaptic volume (Fig. 4.1a). Accordingly, presynapses appear as puncta that partially overlap with the postsynaptic signal.

Stimulated emission depletion (STED) microscopy and stochastic optical reconstruction microscopy (STORM), with their enhanced resolution capabilities, offer deeper insights into the distribution of AZ proteins in presynaptic nerve terminals of small mammalian CNS synapses. STED and STORM imaging revealed that the AZ protein material segregates into two zones: approximately 0–50 nm and 50–100 nm from the PM. Within the zone proximal to the PM, fluorescence peaks are observed for RIM1/2, Munc13, RIM-BP2, ELKS2, Liprin-α3, and the C-terminal region of Bassoon and Piccolo. The N-terminal regions of Bassoon and Piccolo are found in the distal zone (50–100 nm) (Dani et al. 2010; Tang et al. 2016; Wong et al. 2018; Held et al. 2020; Wiesner et al. 2020; Emperador-Melero et al. 2021). Solely, the fluorescent signal of Liprin-α2 was broadly distributed in the presynaptic bouton without forming a sharp intensity peak. Quantification of the lateral distribution of AZ proteins within the presynaptic bouton showed a high overlap (73–90%) of the area stained for RIM1/2, ELKS2, RIM-BP2, and Munc13, with the AZ area defined by labeling against Bassoon, whereas a large fraction of Liprin-α2 and -α3 was present outside this area (Wong et al. 2018).

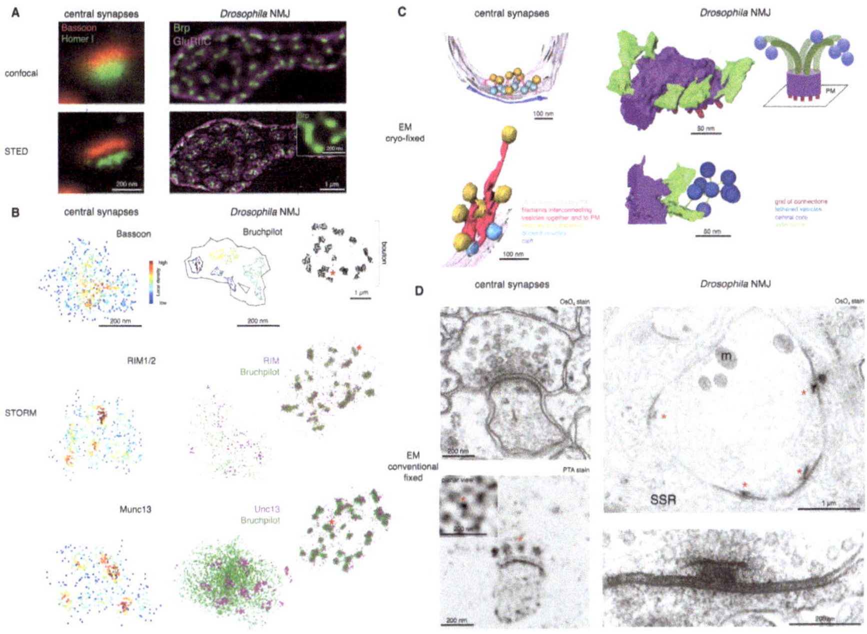

Fig. 4.1 Central synapses and *Drosophila* NMJ visualized with different imaging modalities. Active zone appearance as revealed by light microscopy: Confocal and STED microscopy (**a**, enlarged Bruchpilot (Brp) array at the *Drosophila* NMJ is shown in the inset), and STORM (**b**, in *Drosophila* NMJ boutons, asterisks mark individual active zones, enlarged in the insets; for Bruchpilot, colors indicate distinct protein clusters). Images were adapted for central synapses from (Wiesner et al. 2020) (confocal and STED) and (Tang et al. 2016) (STORM); for *Drosophila* NMJ from (Muttathukunnel et al. 2022) (confocal and STED), (Mrestani et al. 2021; Dannhäuser et al. 2022; Mrestani et al. 2023) (STORM). Active zone appearance, as revealed by electron microscopy: Reconstructions of filamentous structures at central synapses and the T-shaped structure at the *Drosophila* NMJ in cryo-fixed specimens (**c**, the cartoon's color coding of the T-shape structure corresponds to the cryo-tomogram); conventionally fixed synapses contrasted with phosphotungstic acid (PTA) or osmium tetroxide (OsO_4) (**d**, for PTA-stained central synapse, asterisk indicates a single dense projection shown in cross-section and planar views; for OsO_4-stained *Drosophila* NMJ, a bouton cross section with several active zones denoted by asterisks (M, mitochondria; SSR, subsynaptic reticulum) and an individual active zone at higher magnification appearing as an electron-dense area with the T-shaped structure are shown). Images were adapted for central synapses from (Siksou et al. 2007) (cryo-fixed preparations), own data (conventionally fixed); for *Drosophila* NMJ from (Jiao et al. 2010) (cryo-electron tomography) and (McCabe et al. 2003) (conventionally fixed); scale bar size for cryo-reconstructions (**c**) and conventionally fixed *Drosophila* NMJ (**d**) was estimated based on original electron micrographs or the vesicle diameter, respectively

Moreover, STORM data provided evidence for the idea that some AZ proteins are arranged in a specific orientation relative to the PM (Dani et al. 2010). Antibodies targeting the N- and C-terminal regions of Piccolo and Bassoon detected their N-termini approximately 70–80 nm from the PM and their C-termini around 40 nm from the PM. STED microscopy also revealed the polarization of Bassoon molecules, showing that the distance between the N- and C-terminal regions of the protein in cultured hippocampal neurons ranges from 30 to 68 nm (Wong et al. 2018; Held et al. 2020; Ghelani et al. 2021).

To better understand the microanatomy of the CAZ, the two-dimensional (2D) distribution of CAZ components within the AZ plane is of particular interest. Super-resolution microscopy techniques like STORM and DNA-PAINT have shown that AZ proteins such as Bassoon, RIM1/2, Munc13, and VGCCs (Tang et al. 2016; Glebov et al. 2017; Sakamoto et al. 2018; Müller et al. 2022; Dharmasri et al. 2024; Emperador-Melero et al. 2024) are not uniformly distributed across the AZ but form local protein "densities" called nanoclusters (Fig. 4.1b). Precise measurements of nanocluster size and abundance were possible due to the high resolution of these techniques. RIM1/2, Munc13, and VGCCs form nanoclusters that are less than 80 nm in diameter (Tang et al. 2016; Sakamoto et al. 2018; Müller et al. 2022; Emperador-Melero et al. 2024; Anderson et al. 2025), while Bassoon forms larger nanoclusters exceeding 100 nm (Tang et al. 2016). The number of nanoclusters per synapse is also different for each AZ protein: RIM1/2 forms 2–3 nanoclusters per synapse (~30–40 clusters per μm^2 of AZ area) (Tang et al. 2016; Held et al. 2020; Müller et al. 2022; Sun et al. 2022; Lloyd et al. 2023; Guzikowski and Kavalali 2024), Munc13 forms 2–7 (~30–100 per μm^2) (Sakamoto et al. 2018; Karlocai et al. 2021; Dharmasri et al. 2024; Emperador-Melero et al. 2024; Guzikowski and Kavalali 2024, Anderson et al. 2025), VGCCs form 1-2 clusters (~15–30 per μm^2) partially segregated from Munc13 (Emperador-Melero et al. 2024), while Bassoon shows a more homogeneous distribution with 1–2 larger clusters (15–30 clusters per μm^2) (Tang et al. 2016). Furthermore, STORM analysis in cultured hippocampal neurons revealed that RIM1/2 nanoclusters in the AZ are aligned with postsynaptic PSD-95 nanoclusters, leading to the concept of trans-synaptic nanocolumns (Tang et al. 2016).

While data regarding the radial distribution of postsynaptic receptors indicate that NMDA receptors are concentrated in the center and AMPA receptors are enriched at the periphery of the PSD (reviewed in (Scheefhals and MacGillavry 2018; Maynard et al. 2023)), little is known about the detailed radial distribution of presynaptic proteins within the AZ. This limitation is partly due to the difficulty in definitely establishing the borders of the AZ in most super-resolution microscopy studies. Tang et al. reported that RIM1/2, Bassoon, and Munc13 clusters are more frequently found in the inner part of the AZ area (a central area of 100 nm diameter) (Tang et al. 2016). Cell adhesion molecules, particularly Neurexin-1, show a more peripheral distribution, located ~280 nm from the center, while Neurexin-3 is found closer to the synaptic center at ~180 nm from it (Lloyd et al. 2023). Collectively, super-resolution microscopy data suggests that the distribution of AZ nanoclusters at small CNS synapses is not homogeneous.

4.3.2 Drosophila *NMJ*

Confocal microscopy combined with immunolabeling of glutamatergic NMJ boutons in *Drosophila* revealed that a major AZ component, Bruchpilot (Brp), forms approximately two puncta per 1 μm^2 of the bouton's planar area. Depending on the bouton size, this results in 5–20 Brp puncta per bouton, each with a diameter of ~500 nm. These Brp-positive puncta are also referred to as AZs or synapses and are closely aligned with postsynaptic glutamate receptor clusters (Paul et al. 2015; Hoover et al. 2019; Guan et al. 2020; Mrestani et al. 2021). Other AZ proteins, such as RIM-BP, RIM, Unc13A and B, Syd-1, and Liprin-α, similarly appear as distinct puncta of similar size (Owald et al. 2010; Liu et al. 2011; Graf et al. 2012; Böhme et al. 2016; Reddy-Alla et al. 2017; Petzoldt et al. 2020; Mrestani et al. 2023). In contrast, the adhesion molecule Neurexin shows a less organized distribution, forming discontinuous patches that extend into the Brp-negative regions (Li et al. 2007; Owald et al. 2012). Notably, confocal microscopy, followed by image deconvolution revealed that Brp, stained against its C-terminal region, forms ring-shaped structures with a pronounced decrease in fluorescence intensity at their centers, rather than blurry round spots (Hoover et al. 2019). Taken together, the data obtained by confocal microscopy suggests a radial organizational pattern, with Brp being a central structure of the *Drosophila* AZ.

STED imaging provided deeper insight into the spatial structure of the *Drosophila* NMJ (Fig 4.1A). Data from side-viewed synapses revealed that Brp molecules extend up to 150 nm from the PM into the presynaptic bouton (Fouquet et al. 2009; Liu et al. 2011), defining a volume that also contains Unc13A and RIM-BP (Liu et al. 2011; Reddy-Alla et al. 2017). Additionally, STED imaging also showed that Brp molecules are polarized: their N-terminus is proximal to the PM, while the C-terminus projects into the presynaptic bouton, with an average distance of 70 nm between them (Fouquet et al. 2009). Brp is not oriented perpendicular to the PM but is slightly tilted, with the N-terminus positioned centrally in the AZ and the C-terminus 90–120 nm from the AZ center (Fouquet et al. 2009; Böhme et al. 2016; Petzoldt et al. 2020). For RIM-BP, the N-terminal epitope is located closer to the AZ center compared to the C-terminal epitope (Petzoldt et al. 2020), and it also shows polarization relative to the PM, with the C-terminus positioned closer to the PM than the N-terminus (Liu et al. 2011). Similarly, Unc13A is oriented with the C-terminus slightly closer to the PM and the N-terminus extending into the nerve terminal (Reddy-Alla et al. 2017). However, STED data on the orientation of other AZ proteins such as RIM, Syd-1, and Liprin-α are not yet available.

STED microscopy also corroborates confocal data, providing clear evidence of a radial pattern in the 2D distribution of the CAZ material. Unlike mammalian CNS synapses, the AZ of the *Drosophila* NMJ exhibits a distinct organization: protein clusters form circular, ring-like arrays (Fouquet et al. 2009; Liu et al. 2011; Matkovic et al. 2013; Reddy-Alla et al. 2017; Petzoldt et al. 2020; Muttathukunnel et al. 2022). The central region of the AZ is occupied by VGCCs, closely associated with the N-termini of Brp molecules (Fouquet et al. 2009; Liu et al. 2011; Böhme et al. 2016). Surrounding the Brp core, the AZ proteins RIM-BP and Unc13A are located at distances of 30–40 and 70 nm from the AZ center, respectively (Böhme et al. 2016; Petzoldt et al. 2020). Unc13B, another Unc13 isoform, is positioned further from the AZ center, at about 120 nm, occasionally occupying Brp-negative areas (Böhme et al. 2016). Syd-1 and Liprin-α co-localize strongly, forming distinct clusters at the AZ periphery around the Brp core, about 150–170 nm from the AZ center (Fouquet et al. 2009; Owald et al. 2010). STED microscopy also shows a wide distribution of Neurexin across the NMJ bouton, with limited co-localization with Brp (Chen et al. 2010; Owald et al. 2012), supporting previous confocal findings. Notably, the circular protein arrays observed by STED are further segmented into 3–6 "densities" per AZ for proteins like Brp (Matkovic et al. 2013; Böhme et al. 2019; Muttathukunnel et al. 2022), Unc13A (Böhme et al. 2019; Petzoldt et al. 2020), RIM-BP (Böhme et al. 2019), Syd-1, and Liprin-α (Owald et al. 2010). Thus, the CAZ material is not only organized radially but also exhibits additional spatial segregation.

The nanoarchitecture of *Drosophila* NMJ, as revealed by STORM, shows that the protein arrays of Brp, RIM, RIM-BP, and Unc13 observed with STED microscopy consist of distinct nanoclusters, each less than 50 nm in diameter (Fig. 4.1b) (Ehmann et al. 2014; Mrestani et al. 2021; Dannhäuser et al. 2022; Mrestani et al. 2023). For Brp, 15 clusters per AZ were identified (Ehmann et al. 2014; Mrestani et al. 2021), 11 for Unc13 (Dannhäuser et al. 2022), and 10 for RIM (Mrestani et al. 2023), amounting to approximately 30 clusters per μm^2 of AZ for each protein. The average distance of Brp (C-terminus), RIM, and Unc13 clusters to the estimated center of the AZ was 120–140 nm (Ehmann et al. 2014; Mrestani et al. 2021; Dannhäuser et al. 2022; Mrestani et al. 2023), consistent with the STED microscopy findings. Furthermore, STORM revealed that individual protein clusters can form "superclusters", with around 50% of Unc13 and RIM clusters into 2–4 "superclusters" per AZ on average (Mrestani et al. 2021; Dannhäuser et al. 2022). These "superclusters" likely correspond to the protein "densities" seen in STED imaging, given their similar number per AZ. Overall, the ultrastructural organization of the *Drosophila* NMJ AZ is markedly different from that of small CNS synapses, with a greater abundance of nanoclusters for individual AZ molecules per AZ detected by STORM.

In summary, key aspects of the spatial molecular organization of the AZ at the *Drosophila* NMJ have been uncovered through various light microscopy techniques. A defined pattern has emerged, with AZ proteins primarily arranged in co-aligned, circular arrays around a Brp core scaffold that positions most AZ components. While mammalian synapses share some features with NMJs, such as cluster and vesicle numbers per AZ area, light microscopy does not reveal a similar molecular organization of the CAZ in mammalian central synapses. Super-resolution microscopy identifies nanoclusters of specific AZ proteins, but no central organizing protein or symmetry has been observed. Molecular arrangements in mammalian synapses may be too localized to be resolved by light microscopy, and the three-dimensional structure of small synapse—often bent or invaginated—further complicates the detection of regular patterns (Petralia et al. 2018).

4.4 Electron Microscopic View of the AZ

4.4.1 Small Central Synapses

Conventional osmium/uranyl acetate-stained transmission electron micrographs clearly show small central synapses, marked by an accumulation of synaptic vesicles near a narrow synaptic cleft. However, since osmium primarily stains lipids and visualizes cellular membranes and organelles, these micrographs do not clearly reveal the AZ material (Fig. 4.1d). In glutamatergic synapses, postsynaptic proteinaceous material appears as an electron-dense correlate of the PSD, but presynaptic proteins remain inconspicuous. When samples are treated differently and fixed by rapid freezing, followed by freeze-substitution ("cryo-fixed" as opposed to conventional chemical fixation), and viewed as reconstructions after electron tomography (which enhances contrast and resolution), presynaptic proteins appear as supramolecular densities near the presynaptic membrane and as filaments (Fig. 4.1c). In hippocampal synapses, AZ material has been shown to contain filaments of varying length (5-120 nm) that interconnect SVs within the presynaptic bouton or tether them to the PM (Siksou et al. 2007; Cole et al. 2016). Near the presynaptic membrane and adjacent to docked SVs, electron-dense structures have been observed, from which filaments extend further into the bouton. While their molecular nature has not been fully characterized, it seems likely that these near-membrane filaments and densities correspond to macromolecular super-assemblies and reflect the spatial organization of AZ proteins. This idea is supported by studies combining cryo-fixation with immunogold labeling, which demonstrated that Bassoon and ELKS2 co-localize to filaments. ELKS2 was found to reside very close to the membrane (~30 nm), whereas the N-terminus of Bassoon was mapped at a distance of ~70 nm (Siksou et al. 2007). Proteins like RIMs, Bassoon, and Piccolo may contribute to these filaments as well (reviewed in (Chakrabarti and Wichmann 2019)). Cryo-electron tomography has also reported these filaments and supramolecular densities near the membrane (Fernández-Busnadiego et al. 2010; Held et al. 2024). In this approach, samples are viewed in a frozen and vitrified state without additional staining, chemical fixation, dehydration, or embedding (Fernández-Busnadiego et al. 2010; Held et al. 2024), and AZ densities were observed at ~50/μm^2 (equivalent to ~5 per AZ, Held et al. 2024), similar to the number of docked SVs.

For decades, prominent electron-dense material in the AZ of small central synapses has been described in transmission electron micrographs stained with phosphotungstic acid (PTA). PTA binds to amino acids with basic side chains, making proteins visible in the electron microscope, while lipids and membranes do not stain. PTA-stained electron dense material is typically found in large protein assemblies, such as in the mitochondrial matrix, endoplasmic reticulum, and filaments. In synapses, PTA reveals a characteristic structure: regular arrays of the so-called presynaptic dense projections (DPs) along the AZ membrane and a pronounced, rather homogeneous staining of the postsynaptic complex (Fig. 4.1d). Each synapse contains around 12–18 DPs (120-180/μm^2), each with a diameter of ~50 nm (Jones et al. 1976, Hillman and Chen 1985, Vrensen et al. 1980). When a synapse is viewed in tangential synaptic slices, DPs appear arranged in a hexagonal pattern across the AZ, spaced ~45 nm apart (Gray 1963, 1964; Bloom and Aghajanian 1968; Akert et al. 1969; Pfenninger et al. 1969; Jones et al. 1976; Vrensen et al. 1980; Igarashi et al. 1988; Phillips et al. 2001). A few studies using alternative protein-staining methods visualized DPs and synaptic vesicles in the same sample and showed that DPs are located between vesicles (Burette et al. 2012; Akert et al. 1969; Jones et al. 1976; Rowland et al. 2000), suggesting that DPs are formed by AZ proteins involved in organizing vesicles and the release machinery.

Indeed, combining PTA staining with immunogold labeling and transmission EM not only provided direct evidence that DPs contain AZ proteins but also strongly indicated a specific arrangement of these proteins (Limbach et al. 2011). In this study, antibodies against specific domains of Piccolo and Bassoon, as well as Munc13 and RIM1/2, were used in combination with PTA staining to visualize DPs. Based on quantitative analysis, a model was proposed in which Piccolo is polarized relative to the PM: its N-terminal region was found farthest from the PM (~80 nm from the AZ), its central region was located near the DP tip (~60 nm from the AZ), and its C-terminal region, which co-localizes with the C-terminus of Bassoon, run parallel to the PM at ~40 nm (Limbach et al. 2011). In addition, immunogold labeling identified the N-termini of RIM1/2 and Munc13 near the PM (~20 nm from the AZ) and at the periphery of the DPs (~20 nm from the DP center), co-localized with the C-termini of Piccolo and Bassoon (Limbach et al. 2011). Genetic deletion of RIM1/2 did not affect DPs (Kaeser et al. 2011), but DPs are sensitive to a combined deletion of RIM1/2 and RIM-BP1/2 (Acuna et al. 2016).

Cryo-electron tomography of synapses together with 3D reconstruction of supramolecular assemblies in the vicinity of synaptic vesicles revealed that SVs at different stages of docking are tethered to the PM by protein densities of variable length (~10–20 nm) without apparent organization. However, a circular array of protein clusters was observed beneath primed SVs (≤6 nm to the PM) (Radhakrishnan et al. 2021). Aligned and averaged 3D reconstructions revealed six rod-like protein densities forming a circle with an outer diameter of ~40 nm in the PM plane. These findings suggest that these protein clusters may represent six SNAREpins organized by Munc13 molecules (Radhakrishnan et al. 2021). This is further supported by the intermembrane crystal structures of the Munc13 core domain, which show that six Munc13 molecules can interact to form a hexamer and assemble SNAREpins(Grushin et al. 2022; Bera et al. 2025). To date, such detailed data are not available for any of the other AZ-enriched proteins, such as RIM, RIM-BP, ELKS, and Liprin-α,

4.4.2 Drosophila *NMJ*

At the *Drosophila* NMJ, lipid (osmium) staining of side-view NMJs reveals a distinct electron-dense structure at the center of most AZs, known as T-bar, which has a characteristic shape resembling the letter "T" or a bouquet of flowers (Fig. 4.1d) (Reist et al. 1998; Koenig and Ikeda 1999; Sigrist et al. 2002, Wichmann and Sigrist 2010). The T-shaped structure has an average height of 60 nm, with a base width of about 100 nm. The roof, or "platform," is approximately twice as wide, extending to a total width of 200 nm (Matkovic et al. 2013; Bruckner et al. 2017; Wagner 2017; Petzoldt et al. 2020). EM data using rapid cryo-fixation followed by freeze-substitution show that most docked SVs at the *Drosophila* larval NMJ are located within 150 nm of the central region of the T-bar, with the highest density observed between 30 and 60 nm (Böhme et al. 2016; Bruckner et al. 2017; Reddy-Alla et al. 2017; Kobbersmed et al. 2020). If the periphery of the AZ is defined by Liprin-α and Syd-1 clusters, the estimated planar radius of the CAZ protein scaffold is approximately 200 nm, as shown by STED microscopy. These findings indicate that the area occupied by docked SVs is comparable to the dimensions of the AZ protein complex. Interestingly, the peak density of docked SVs corresponds to the location of the Unc13A fluorescence signal, highlighting its critical role in SV docking (Böhme et al. 2016; Kobbersmed et al. 2020). Furthermore, the peak density of total SVs relative to the vertical PM plane is found between 100 and 200 nm (Zhan et al. 2016). This supports the idea that Brp filaments, extending into the bouton, play a key role in forming a "platform" that captures and tethers SVs, which was also supported by cryo-electron tomography data (Fig. 4.1c) (Jiao et al. 2010). Notably, the *Drosophila* larval NMJs also exhibit an exclusion zone between VGCCs and SV docking sites, similar to what is seen in mammalian central synapses (Keller et al. 2015; Rebola et al. 2019; Eguchi et al. 2022). The peak density of docked SVs is observed at a distance from the center of the T-bar, where VGCCs are located, with simulation data estimating the exclusion zone width at around 30 nm (Kobbersmed et al. 2020).

At *Drosophila* NMJs, EM data confirmed the polarized structure of the large Brp protein, which is considered a "topological" analog of Bassoon and Piccolo. The data revealed a 25 nm distance between the N- and C-terminal regions of Brp, with the N-terminal epitope positioned close to the PM (Fouquet et al. 2009).

4.5 Integrated View of the AZ Nano-organization

In summary, the combination of super-resolution, electron microscopy, fluorescently tagged lines, and domain-specific antibodies has greatly advanced our understanding of the submicroscopic spatial organization of AZ proteins at the *Drosophila* NMJ. At this synapse, the spatial orientation of all major AZ-enriched proteins has been resolved in relation to the PM, the Brp scaffold, and SVs, revealing a highly precise arrangement of these proteins (Fig. 4.2). From a planar perspective, this organization follows a "ring-in-ring" pattern, where the filamentous nature of Brp bundles serves as a defining structural frame, delineating the spatial distribution of the other AZ proteins and of SVs at the AZ (Fig. 4.2a, c, and references in the figure). The Brp N-terminus and VGCCs occupy the central region of the AZ, surrounded by docked SVs and by circular arrays of the AZ proteins in the following order: RIM-BP N-terminus, RIM-BP C-terminus, Unc13A, RIM, and Unc13B. Liprin-α, and Syd-1 are positioned at the periphery of the AZ. From a lateral viewpoint, Brp, along with possibly RIM-BP and Unc13A, shows a molecular polarization relative to the PM plane. The Brp N-terminal region is close to the PM, while its C-terminal region extends in a curved manner into the presynaptic

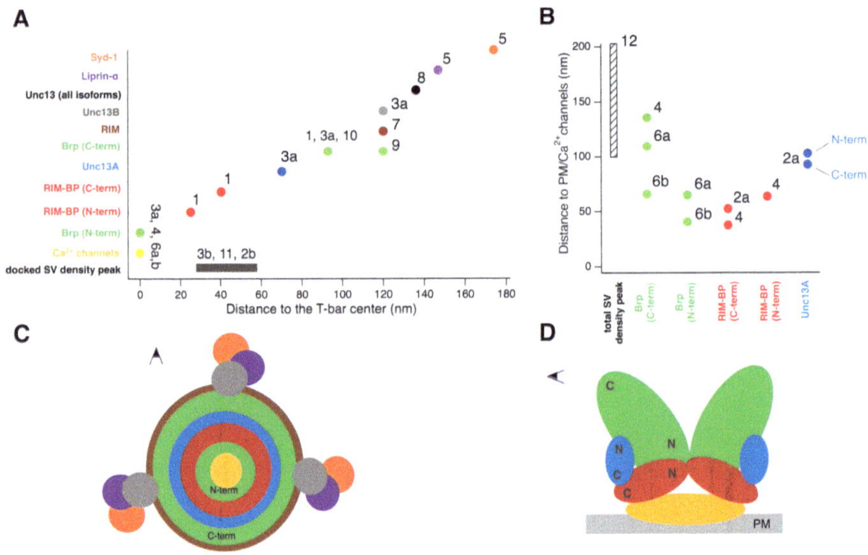

Number	Reference	Method
1	Petzoldt et al., 2020	STED
2	Reddy-Alla et al., 2017	a. STED b. EM
3	Böhme et al., 2016	a. STED b. EM
4	Liu et al., 2011	STED
5	Owald et al., 2010	STED
6	Fouquet et al., 2009	a. STED b. EM
7	Mrestani et al., 2023	STORM
8	Dannhäuser et al., 2022	STORM
9	Mrestani et al., 2021	STORM
10	Ehmann et al., 2014	STROM
11	Bruckner et al., 2017	EM
12	Zhan et al., 2016	EM

Fig. 4.2 Positioning of AZ members in the *Drosophila* larval NMJ: Distances to the AZ center (defined by Ca^{2+}-channels and the N-terminal part of Bruchpilot) shown from a top view (**a**) and from a side view (**b**), relative to the plasma membrane (based on EM data) or to the Ca^{2+}-channel signal peak (based on STED data). (**c**, **d**) General scheme of the AZ topology, with colors corresponding to the AZ members from panels a and b (all references for the plotted values are listed in the table; approximate numerical data are shown without variability ranges. Abbreviations: Brp, Bruchpilot; N-/C-term, N-/C-terminal protein regions; PM, plasma membrane; SV, synaptic vesicle)

bouton, forming the T-bar structure. The C-termini of RIM-BP and Unc13A are positioned slightly closer to the PM plane compared to their N-termini, suggesting that these molecules also have an elongated structure (Fig. 4.2b, d, and references in the figure).

In contrast, no unified model has been established for the organization of AZ proteins in small mammalian CNS synapses. The diameter of protein nanoclusters identified by super-resolution light microscopy, ranging between 30 and 100 nm, is comparable to the diameter of DPs detected using transmission EM (~50 nm), though this has to be considered within the context of methodological differences. The number of DPs per AZ, typically 12–18 observed by EM of PTA-stained tissue, is much higher than the number of nanoclusters identified by super-resolution techniques. This discrepancy currently is unresolved.

Notably, STORM data reveal a higher number of clusters formed by individual AZ components at the *Drosophila* NMJ compared to mammalian central synapses. The differences can be attributed to the fact that the area of AZ at NMJs is 3–5 times larger than that of CNS synapses (Fig. 4.1d). Thus, the organizational principle of the CAZ material appears to be "isometric", with the number of protein clusters per unit of AZ area remaining consistent across different synapses.

Both light microscopy and EM data suggest that most AZ proteins, such as RIM1/2, ELKS2, RIM-BP, Munc13, Liprin-α3, the cytoplasmic loop of P/Q-type VGCCs, and the C-terminal regions of Piccolo and Bassoon, are located within a ~50 nm wide layer directly adjacent to the PM (Fig. 4.3a, and references in the figure). The width of this layer corresponds to the height of the DPs and the SV diameter, supporting the notion that AZ proteins may form nanoclusters visible as DPs in EM images that facilitate the coupling of VGCCs and SVs due to their close proximity. Findings from the STED and STORM imaging of side-view synapses show that AZ proteins are highly enriched in the proximity of the AZ PM, supporting the idea that DPs are composed of CAZ material. This was further confirmed by a combination of pan-protein staining and immunolabeling. The defined and differential localization of Piccolo, Bassoon, Munc13, and RIM1/2 relative to DPs was demonstrated by immunogold labeling (Limbach et al. 2011), and overlap between immunolabeled Bassoon and DPs was further shown using expansion microscopy (M'Saad et al. 2025). Super-resolution techniques like STED and STORM, as well as EM data, indicate distinct localizations for the AZ proteins Piccolo and Bassoon, which likely organize and frame the CAZ scaffold. The C-termini of these proteins are situated close to the PM, while their N-termini extend into the presynaptic terminal. Immunogold EM data also suggest that Piccolo is bent, with its N-terminus projecting into the presynaptic bouton volume, its central regions located at the tips of DPs and its C-terminus positioned near SVs at the DP periphery. RIM1/2 and Munc13 are localized at the edges of DPs, where they may play a role in SV docking (Fig. 4.3b, c, and references in the figure). These findings align with the *in silico* models of Piccolo and Bassoon, which predict that these proteins are approximately 80 nm in length (Gundelfinger et al. 2016). According to the model, the proteins may be tilted or undergo spatial folding at the AZ, reducing the distance between their N- and C-termini to the experimentally observed 30–40 nm. As previously

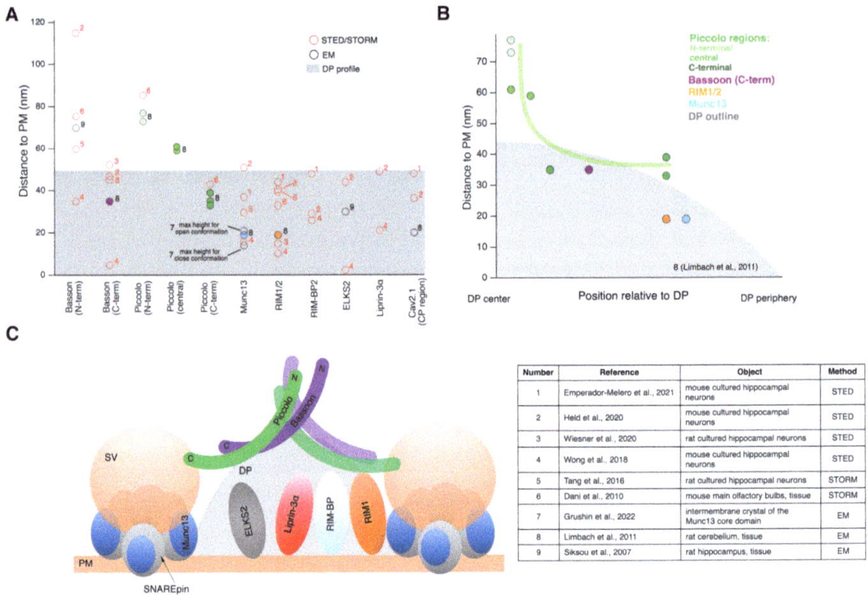

Number	Reference	Object	Method
1	Emperador-Melero et al., 2021	mouse cultured hippocampal neurons	STED
2	Held et al., 2020	mouse cultured hippocampal neurons	STED
3	Wiesner et al., 2020	rat cultured hippocampal neurons	STED
4	Wong et al., 2018	mouse cultured hippocampal neurons	STED
5	Tang et al., 2016	rat cultured hippocampal neurons	STORM
6	Dani et al., 2010	mouse main olfactory bulbs, tissue	STORM
7	Grushin et al., 2022	intermembrane crystal of the Munc13 core domain	EM
8	Limbach et al., 2011	rat cerebellum, tissue	EM
9	Siksou et al., 2007	rat hippocampus, tissue	EM

Fig. 4.3 Positioning of AZ members in the mammalian central synapse: Lateral distances from the plasma membrane plane (PM) determined by STED or STORM and EM techniques, with a 50 nm high dense projection (DP) profile indicated (Jones et al. 1976; Limbach et al. 2011) (**a**). Protein positions relative to the center-to-periphery axis of the DP (**b**), and a hypothetical scheme of CAZ organization (**c**) (all references for the plotted values are listed in the table; approximate numerical data are shown without variability ranges; for STED and STORM data, all distances were initially obtained using PSD-95 intensity peak as a common reference and later recalculated relative to the PM plane, assuming that PSD-95 distributed at 45 nm from the presynaptic membrane: ~20 nm between PSD-95 and the postsynaptic membrane (Chen et al. 2008; Chen et al. 2011; Jeyifous et al. 2016) with added synaptic cleft width equal to ~25 nm (Lučić et al. 2005; Zuber et al. 2005; Chen et al. 2008). Abbreviations: CP loop, cytoplasmic loop; DP, dense projection; N-/C-term and central, N-/C-terminal and central protein regions; PM, plasma membrane; SV, synaptic vesicle)

mentioned, Munc13 molecules are thought to form hexagonal rings around docked SVs and appear to be spatially segregated from VGCCs.

4.6 Liquid–Liquid Phase Separation (LLPS) as a Potential Organizational Mechanism of the CAZ

Recently, the formation of membrane-less subcompartments of varying composition, the so-called phase condensates, has been suggested as one mechanism to organize AZ proteins in the presynaptic nerve terminal (reviewed in (McDonald and Shen 2021; Lautenschläger 2022; Wu et al. 2023; Wu et al. 2025). These phase condensates contain particular proteins at high concentrations compared to the surrounding solution (Hyman et al. 2014; Banani et al. 2017; Alberti et al. 2019; Gorsheneva et al. 2024). The two phases stably coexist, even though their components are dynamically exchanged. Phase separation is strongly affected by even minute changes in environmental parameters like pH, temperature, osmotic pressure, and ATP levels in the cell (Li et al. 2022). As a result, phase separation enables cells to quickly adapt to changes in their environment.

Using purified protein fragments and/or overexpression in HEK293T cells, it was shown that RIM, RIM-BP, ELKS, Liprin-α, Munc13, and Piccolo can undergo phase separation (Emperador-Melero et al. 2021; Liang et al. 2021; Wu et al. 2019, 2021; Qiu et al. 2024). When purified protein fragments were combined with small unilamellar vesicles (SUVs) and giant unilamellar vesicles (GUVs), which mimic presynaptic membranes, it was found that SUVs, together with Synapsin/Intersectin, coated the surface of condensed droplets formed by AZ proteins on the GUV membranes through phase separation. This process, involving RIM and RIM-BP, enabled these phase-separated condensates to recruit the cytoplasmic tail of VGCCs that were tethered to the GUV membrane (Wu et al. 2021). Considering that RIM1 and RIM-BP co-condense with Liprin-α3 and Munc13 (Emperador-Melero et al. 2021), as well as with ELKS1 (Wu et al. 2021; Qiu et al. 2024) and the C-terminal region of Piccolo (Qiu et al. 2024), it appears possible to reconstitute the CAZ network as a liquid phase on the surface of the GUV membrane.

So far, only one study aimed at examining the occurrence of phase separation at synapses *in vivo* (McDonald et al. 2020). McDonald et al. identified sequences in Liprin-α and ELKS1 that were essential for phase separation *in vitro*. When a mutant Liprin-α variant lacking the ability to undergo phase condensation was expressed in *C. elegans,* they were enriched at synapses but showed defects in the AZ assembly and synapse function. These deficits could be rescued by expressing mutants, in which the sequences driving phase separation were replaced by a well-characterized phase condensation-promoting motif. However, these results do not directly proof the formation of AZ phase condensates. The Liprin-α phase separation-promoting region that was deleted and replaced as well as the two well-studied replacement motifs are intrinsically disordered regions (IDRs). IDRs mainly engage in transient and often promiscuous interactions. Therefore, it cannot be excluded that the observed defects were caused by the failure of the proteins to engage in regular protein protein interactions or due to other still unknown functions of the deleted sequences. Furthermore, also the observation that proteins rapidly recover after photobleaching is not sufficient to conclude that phase condensation has occurred, in particular, as the resolution of conventional fluorescence microscopy cannot resolve subsynaptic structures (McSwiggen et al. 2019; Alberti et al. 2019; Deng and Wan 2024). A problem with the genetic approach of expressing mutant proteins to deduce that an observed phenotype is the result of altered phase separation can also be the time scale. The time between expression and observation leaves room for many indirect processes to occur that could cause the phenotype (Alberti et al. 2019), in this case AZ assembly and synapse function. In addition, pharmacological treatment of cultured hippocampal neurons with an aliphatic alcohol capable of dissolving liquid condensates did not result in a complete disassembly of AZ nanoclusters. Instead, the sizes of RIM1/2 and Munc13 clusters decreased by ~20%, without any change in their number per synapse (Guzikowski and Kavalali 2024). Taken together, there is strong evidence that the majority of AZ proteins are capable of undergoing phase condensation. However, unequivocal proof for the existence of AZ phase condensate is still lacking.

This raises the question if the requirements for AZ proteins to undergo phase separation in the presynaptic nerve terminal are met. Biochemical and biophysical studies have shown that phase condensation is driven by a combination of key factors (reviewed in (Hyman et al. 2014; Banani et al. 2017; Alberti et al. 2019; André and Spruijt 2020; Li et al. 2022)):

1. *Concentration of molecules*: For phase separation to occur, the concentration of a protein or a set of interacting molecules needs to exceed a critical threshold. Below this threshold, the molecules remain evenly dispersed; above it, they spontaneously form condensates. Local increases in concentration, such as those created at the sites of active signaling, can drive the formation of phase-separated compartments.
2. *Molecular crowding*: A high concentration of macromolecules, the so-called crowding agents, can enhance LLPS by several mechanisms. Crowding agents occupying a certain environmental volume raise the effective concentration of interacting partners and promote intermolecular attractions. Moreover, crowding agents can interact with phase-separating biomolecules or change their solubility, also affecting phase separation.
3. *Multivalent interactions:* Proteins containing multiple domains or motifs, known as "stickers", can interact with other proteins through multivalent interactions, leading to the formation of a network of weak, reversible bonds, thereby promoting phase condensation.
4. *Low complexity and intrinsically disordered regions (IDRs)*: IDRs lack a stable secondary or tertiary structure. They are highly flexible and can engage in a variety of weak interactions, e.g., charge–charge, dipole–dipole interaction, hydrogen bonds, π–π staking, with other IDRs or structured domains, thereby contributing to the required multivalency.
5. *Post-translational modifications (PTMs):* PTMs, such as phosphorylation, methylation, and acetylation, can modulate the ability of proteins to engage in phase separation by changing their charge and by regulating protein interactions. Thereby, PTMs allow cells to dynamically regulate the formation and dissolution of condensates in response to signaling events or environmental changes.

AZ proteins fulfill the biophysical criteria for phase condensation listed above. They are multivalent as they contain multiple domains and motifs, which are able to engage in a large number of protein interactions, in particular also among each other (Schoch and Gundelfinger 2006; Jin and Zhai 2023). In addition, a significant portion of their amino acid sequence is predicted to be intrinsically disordered and of low complexity (Lautenschläger 2022). Phosphoproteomic screens have also shown that AZ proteins are strongly regulated by phosphorylation in response to neuronal activity (Kohansal-Nodehi et al. 2016; Engholm-Keller et al. 2019). Accordingly, they have been identified in a bioinformatics analysis as droplet-driver proteins (Hardenberg et al. 2020) and as mentioned above form phase condensates *in vitro* and in heterologous cell systems.

Micron-scale phase-separated condensates in the cell nucleus contain between 15,000 and 85,000 molecules per 1 μm^3 of condensate (Ali et al. 2025). However, at the synapse, due to the size limitations of the compartment, condensates will be a lot smaller and therefore contain significantly fewer molecules. To date, only limited data is available with regard to the concentration of AZ proteins in the presynapse. Wilhelm et al. reported the number of molecules per synapse for Bassoon (~446), Munc13 (~1551), Piccolo (~100), and RIM1 (~39) (Wilhelm et al. 2014). As the critical concentration for a protein or a protein mixture is specific to these molecules, it is difficult to judge if the concentration of AZ proteins in the presynaptic nerve terminal would allow for phase condensation. However, molecular crowding promotes phase separation, and EM images show the presynaptic bouton to be highly crowded, as it is filled with SVs, mitochondria, and filaments (Shepherd and Harris 1998; Harris and Weinberg 2012). Also, a 3D model of the presynaptic compartment, including proteins, organelles, and filaments, based on data obtained by quantitative mass spectrometry, STED imaging, and EM data depicts a highly crowded environment (Wilhelm et al. 2014). Therefore, environmental crowding may promote the formation of AZ phase condensates. This raises the question of how the AZ-phase condensate concept fits together with the data obtained by EM and super-resolution microscopy.

It is still unresolved how many AZ-phase condensates could be present in one presynaptic bouton. Furthermore, the arrangement of proteins in AZ-phase condensates is also still unresolved. In general, phase condensates resemble liquids and are dynamic with biomolecules constantly moving within them and exchanging with the cytoplasm. This dynamic nature is based on the underlying protein interactions in the phase condensate which are mainly of lower affinity, transient, and promiscuous, resulting in a random arrangement of molecules in the phase condensate (Boeynaems et al. 2018; Abyzov et al. 2022; Brangwynne 2013).

In contrast, multiple lines of evidence suggest that the release machinery, those proteins that bring SVs into proximity of the VGCCs and that are needed to fuse the SVs, like SNARE proteins, Munc13, and Synaptogmin-1, are arranged in a highly ordered fashion, forming ring-like structures at the base of SVs (Wang et al. 2014; Radhakrishnan et al. 2021; Grushin et al. 2022). This ordered arrangement would argue that release sites with readily releasable SVs are not part of phase condensates and are not visible as distinct units in super-resolution images, as the protein content is too low. Radhakrishnan and colleagues propose that each of the six protein densities observed under docked SVs consists of one "exocytosis module" likely including only a small number of protein molecules: SNAP25, Syntaxin-1, VAMP2, Complexin (one protein copy for each of them), and Synaptotagmin-1 C2B domains (Radhakrishnan et al. 2021), a model that is consistent with later experimental findings (Papantoniou et al. 2023). On the other hand, nanoclusters and DPs could represent phase condensates that act as molecular reservoirs of AZ-enriched proteins that keep them close to the AZ membrane. These AZ phase condensates could be tethered to the PM by specific interactions of AZ proteins with VGCCs and cell adhesion molecules, such as Neurexin and Receptor Tyrosine Phosphatases (RTPs), and connect to SVs on the side facing the presynaptic bouton. There are multiple nanoclusters of individual AZ proteins and DPs per AZ; however, so far, no data is available showing an enrichment of different AZ proteins in one nanocluster. This could be interpreted in a way that DPs represent all phase condensates present at an AZ and that the abundance of individual AZ proteins in the phase condensates differs strongly and that STORM only detects the phase condensates/nanoclusters with a high abundance of specific proteins. Such AZ-phase condensates of diverging composition serving as molecular reservoirs could also be a mean to regulate the availability, localization, and overall abundance of specific AZ proteins. If nanoclusters are in fact DPs with diverging composition, this could explain why EM of PTA-stained tissue and after immunogold labeling suggests the AZ material to be distributed throughout the AZ in a fairly regular pattern (Bloom and Aghajanian 1968; Jones et al. 1976; Vrensen et al. 1980; Igarashi et al. 1988; Phillips et al. 2001; Limbach et al. 2011), while nanoclusters appear more randomly distributed (Tang et al. 2016; Glebov et al. 2017; Sakamoto et al. 2018; Müller et al. 2022; Dharmasri et al. 2024; Emperador-Melero et al. 2024), as this could be due to the fact that not all AZ-phase condensates/nanoclusters are visible.

However, one has to keep in mind that the EM, STORM, and STED images were obtained using fixed tissue and that fixation may have introduced artifacts. Studies examining the impact of fixation on protein localization in cells have shown that the artifacts introduced by fixation are complex and differ between proteins and fixatives (Yoshida et al. 2023). Fixation can increase/decrease the size and number of signals detected and cause the appearance/disappearance of new signals. Fixation-induced artifacts depend in part on the time constant of the fixation and of the molecule binding–unbinding rate but can also be caused by other mechanisms like disruption of sub-cellular structures and organelle morphology and incomplete fixation. The above listed fixation artifacts have also been observed for phase condensates in cells, ranging from the appearance of phase condensates in cells that in the live condition do not have droplets to the opposite, the complete disappearance of droplets that had been detected in live cells (Irgen-Gioro et al. 2022).

Images obtained from fixed tissue and cells always only represent a snapshot of the live situation and do not capture the dynamics of physiological systems. Single-particle tracking of AMPA, NMDA, GABA, and Glycine receptors in the postsynaptic density (reviewed in (Maynard et al. 2023)) and photoactivated localization microscopy (PALM) of P/Q-type VGCCs in the AZ (Heck et al. 2019) in synapses of live primary hippocampal neurons have revealed that the majority of molecules are mobile. Similarly, EM immunogold labeling has shown that RIM1 molecules redistribute into the inner synaptic volume following high-K^+ stimulation (Tao-Cheng 2006). However, even in live neurons, P/Q-type VGCCs assemble into clusters of approximately 80 nm diameter, where individual P/Q-type VGCCs transiently reside for around 100 ms. Notably, these clusters demonstrate dynamic behaviors, including formation, deformation, and drifting within the PM (Heck et al. 2019). It is conceivable that these clusters represent phase condensates.

Irrespective of higher-order protein complexes formed by specific PPIs or LLPS, the question arises what might be the mechanism causing this clustered versus a general homogeneous distribution. One possible explanation could be that certain rigid protein complexes, like SNAREpins, or cell adhesion molecules might form a pillar on which other proteins build upon. A clustered distribution could be advantageous, for example, to segregate different modes of neurotransmitter release like evoked, spontaneous, and asynchronous. It also provides more possibilities for fine-tuning the efficiency of neurotransmission, e.g., by forming pre- and postsynaptic nanocolumns.

To unequivocally resolve the question if and how phase condensation is involved in organizing the AZ ultrastructure and function, more experimental data will be required. To this end, it will also be necessary to develop more specific ways to interfere with phase condensates, e.g., drugs/inhibitors, as the existing approaches are too unspecific (McSwiggen et al. 2019; Alberti et al. 2019; Lyon et al. 2021). To specifically interfere with the phase condensation of individual proteins, it would be necessary to better understand the role of their IDRs in this process in order to design mutant proteins, with only few amino acid exchanges, that would still be fully functional but unable to undergo phase separation. It will be important to resolve the link between phase condensates and DPs to determine if DPs indeed represent phase condensates in EM images. Finally, *in vivo* experiments for the general characterization of LLPS of AZ proteins were so far mainly performed in HEK293T cells as they allow the visualization of droplets. However, it remains unclear whether the results observed in HEK293T cells after overexpression are comparable with the situation at the presynapse with the endogenous levels of AZ proteins.

4.7 Outlook

Much progress has been made since the invention of super-resolution microscopy and the revival of EM in recent years in resolving the submicroscopic organization of the presynaptic AZ and the arrangement of individual AZ proteins. However, many open questions and challenges remain that need to be addressed and might need the development of novel tools:

1. It will be essential to obtain experimental evidence that allows to differentiate between the formation of AZ-phase condensates or transient hubs with a local high concentration of biomolecules. Can LLPS occur at the AZ taking into account the abundance of individual molecules and the compartment size?
2. What is the molecular stoichiometry of DPs, nanoclusters, and phase condensates at the AZ? Does the stoichiometry change during synaptic plasticity?
3. How do the concepts developed based on fixed tissue hold up to data obtained from live neurons? How dynamic is the AZ ultrastructure?

4. Considering the structured spatial organization of *Drosophila* NMJs compared to the less structured small CNS synapses, does the high degree of global organizational pattern dominated by the T-bar at the NMJ reflect specialization for high-throughput vesicle release, whereas at central synapses, smaller release units, possibly around DPs, allow more room for functional plasticity and fine-tuning?

Acknowledgments Our work was supported by the Deutsche Forschungsgemeinschaft (SS: SFB1089, SCHO 820/4-1, 5-2, 6-2, 7-1, 8-1, and 9-1, SPP1757; DD: SFB1089, SPP1757, INST1172 15, DI853/3-5&7, INST 217/785-1), the Federal State of Nordrhein-Westfalen (iBehave, SS), and the BONFOR program of the University of Bonn Medical Center (SS, DD). We thank Dr. Hanes Beckert and the Microscopy Core Facility of the Medical Faculty of the University of Bonn for the osmium-stained central synapse electron micrograph presented in Figure 4.1D.

References

Abyzov A, Blackledge M, Zweckstetter M. Conformational dynamics of intrinsically disordered proteins regulate biomolecular condensate chemistry. Chem Rev. 2022;122(6):6719–48. https://doi.org/10.1021/acs.chemrev.1c00774.

Acuna C, Liu X, Südhof TC. How to make an active zone: unexpected universal functional redundancy between RIMs and RIM-BPs. Neuron. 2016;91(4):792–807. https://doi.org/10.1016/j.neuron.2016.07.042.

Akert K, Moor H, Pfenninger K, Sandri C. Contributions of new impregnation methods and freeze etching to the problems of synaptic fine structure. Prog Brain Res. 1969;31:223–40. https://doi.org/10.1016/s0079-6123(08)63241-0.

Alberti S, Gladfelter A, Mittag T. Considerations and challenges in studying liquid-liquid phase separation and biomolecular condensates. Cell. 2019;176(3):419–34. https://doi.org/10.1016/j.cell.2018.12.035.

Ali C, Muzzopappa F, Erdel F. A molecular census to elucidate the demixing mechanism of membraneless organelles Abstract Genome Biology. 2025;26(1). https://doi.org/10.1186/s13059-025-03806-0.

Anderson MC, Dharmasri PA, Damenti M, Metzbower SR, Laghaei R, Blanpied TA, Levy AD. Trans-synaptic molecular context of NMDA receptor nanodomains Nature Communications. 2025;16(1). https://doi.org/10.1038/s41467-025-62766-y.

André AAM, Spruijt E. Liquid–liquid phase separation in crowded environments. Int J Mol Sci. 2020;21(16):5908. https://doi.org/10.3390/ijms21165908.

Banani SF, Lee HO, Hyman AA, Rosen MK. Biomolecular condensates: organizers of cellular biochemistry. Nat Rev Mol Cell Biol. 2017;18(5):285–98. https://doi.org/10.1038/nrm.2017.7.

Barber KR, Hruska M, Bush KM, Martinez JA, Fei H, Levitan IB, Dalva MB, Wairkar YP. Levels of Par-1 kinase determine the localization of Bruchpilot at the Drosophila neuromuscular junction synapses. Sci Rep. 2018;8(1):16099. https://doi.org/10.1038/s41598-018-34250-9.

Bera M, Grushin K, Kalyana Sundaram RV, Hinzen JS, Chen J, Chatterjee A, Radhakrishnan A, Lee S, Padmanarayana M, Coleman J, Pincet F, Rothman JE, Dittman JS. Two successive oligomeric Munc13 assemblies scaffold vesicle docking and SNARE assembly to support neurotransmitter release Nature Communications. 2025;16(1). https://doi.org/10.1038/s41467-025-62420-7.

Bloom FE, Aghajanian GK. Fine structural and cytochemical analysis of the staining of synaptic junctions with phosphotungstic acid. J Ultrastruct Res. 1968;22(5–6):361–75. https://doi.org/10.1016/s0022-5320(68)90027-0.

Boeynaems S, Alberti S, Fawzi NL, Mittag T, Polymenidou M, Rousseau F, Schymkowitz J, Shorter J, Wolozin B, Bosch LVD, Tompa P, Fuxreiter M. Protein phase separation: a new phase in cell biology. Trends Cell Biol. 2018;28(6):420–35. https://doi.org/10.1016/j.tcb.2018.02.004.

Böhme MA, Beis C, Reddy-Alla S, Reynolds E, Mampell MM, Grasskamp AT, Lützkendorf J, Bergeron DD, Driller JH, Babikir H, Göttfert F, Robinson IM, O'Kane CJ, Hell SW, Wahl MC, Stelzl U, Loll B, Walter AM, Sigrist SJ. Active zone scaffolds differentially accumulate Unc13 isoforms to tune Ca^{2+} channel–vesicle coupling. Nat Neurosci. 2016;19(10):1311–20. https://doi.org/10.1038/nn.4364.

Böhme MA, McCarthy AW, Grasskamp AT, Beuschel CB, Goel P, Jusyte M, Laber D, Huang S, Rey U, Petzoldt AG, Lehmann M, Göttfert F, Haghighi P, Hell SW, Owald D, Dickman D, Sigrist SJ, Walter AM. Rapid active zone remodeling consolidates presynaptic potentiation. Nat Commun. 2019;10(1):1085. https://doi.org/10.1038/s41467-019-08977-6.

Boyken J, Grønborg M, Riedel D, Urlaub H, Jahn R, Chua JJE. Molecular profiling of synaptic vesicle docking sites reveals novel proteins but few differences between glutamatergic and GABAergic synapses. Neuron. 2013;78(2):285–97. https://doi.org/10.1016/j.neuron.2013.02.027.

Brangwynne CP. Phase transitions and size scaling of membrane-less organelles. J Cell Biol. 2013;203(6):875–81. https://doi.org/10.1083/jcb.201308087.

Bruckner JJ, Zhan H, Gratz SJ, Rao M, Ukken F, Zilberg G, O'Connor-Giles KM. Fife organizes synaptic vesicles and calcium channels for high-probability neurotransmitter release. J Cell Biol. 2017;216(1):231–46. https://doi.org/10.1083/jcb.201601098.

Burette AC, Lesperance T, Crum J, Martone M, Volkmann N, Ellisman MH, Weinberg RJ Electron tomographic analysis of synaptic ultrastructure Abstract Journal of Comparative Neurology. 2012;520(12):2697–2711. https://doi.org/10.1002/cne.v520.1210.1002/cne.23067.

Carlin RK, Grab DJ, Cohen RS, Siekevitz P. Isolation and characterization of postsynaptic densities from various brain regions: enrichment of different types of postsynaptic densities. J Cell Biol. 1980;86(3):831–45. https://doi.org/10.1083/jcb.86.3.831.

Chakrabarti R, Wichmann C. Nanomachinery organizing release at neuronal and ribbon synapses. Int J Mol Sci. 2019;20(9):2147. https://doi.org/10.3390/ijms20092147.

Chen X, Winters C, Azzam R, Li X, Galbraith JA, Leapman RD, Reese TS. Organization of the core structure of the postsynaptic density. Proc Natl Acad Sci U S A. 2008;105(11):4453–8. https://doi.org/10.1073/pnas.0800897105.

Chen K, Gracheva EO, Yu S-C, Sheng Q, Richmond J, Featherstone DE. Neurexin in embryonic Drosophila neuromuscular junctions. PLoS One. 2010;5(6):e11115. https://doi.org/10.1371/journal.pone.0011115.

Chen X, Nelson CD, Li X, Winters CA, Azzam R, Sousa AA, Leapman RD, Gainer H, Sheng M, Reese TS. PSD-95 is required to sustain the molecular organization of the postsynaptic density. J Neurosci. 2011;31(17):6329–38. https://doi.org/10.1523/jneurosci.5968-10.2011

Cole AA, Chen X, Reese TS. A network of three types of filaments organizes synaptic vesicles for storage, mobilization, and docking. J Neurosci. 2016;36(11):3222–30. https://doi.org/10.1523/jneurosci.2939-15.2016.

Collins CA, DiAntonio A. Synaptic development: insights from Drosophila. Curr Opin Neurobiol. 2007;17(1):35–42. https://doi.org/10.1016/j.conb.2007.01.001.

Couteaux R. Sur Les Gouttieres Synaptiques Du Muscle Strie. Comptes Rendus Des Seances De La Societe De Biologie Et De Ses Filiales. 1946;140(4):270–1.

Couteaux R. Contribution à l'étude de la synapse myoneurale. Rev Canad Biol. 1947;6:563–711.

Couteaux R, Pécot-Dechavassine M. Synaptic vesicles and pouches at the level of "active zones" of the neuromuscular junction. C R Hebd des seances l'Acad des Sci Ser D: Sci Nat. 1970;271(25):2346–9.

Dani A, Huang B, Bergan J, Dulac C, Zhuang X. Superresolution imaging of chemical synapses in the brain. Neuron. 2010;68(5):843–56. https://doi.org/10.1016/j.neuron.2010.11.021.

Dannhäuser S, Mrestani A, Gundelach F, Pauli M, Komma F, Kollmannsberger P, Sauer M, Heckmann M, Paul MM. Endogenous tagging of Unc-13 reveals nanoscale reorganization at active zones during presynaptic homeostatic potentiation. Front Cell Neurosci. 2022;16:1074304. https://doi.org/10.3389/fncel.2022.1074304.

Dawson-Scully K, Lin Y, Imad M, Zhang J, Marin L, Horne JA, Meinertzhagen IA, Karunanithi S, Zinsmaier KE, Atwood HL. Morphological and functional effects of altered cysteine string protein at the Drosophila larval neuromuscular junction. Synapse. 2007;61(1):1–16. https://doi.org/10.1002/syn.20335.

Deng B, Wan G. Technologies for studying phase-separated biomolecular condensates. Adv Biotechnol. 2024;2(1):10. https://doi.org/10.1007/s44307-024-00020-0.

Dharmasri PA, Levy AD, Blanpied TA. Differential nanoscale organization of excitatory synapses onto excitatory vs. inhibitory neurons. Proc Natl Acad Sci U S A. 2024;121(17):e2315379121. https://doi.org/10.1073/pnas.2315379121.

Dreyer F, Peper K, Akert K, Sandri C, Moor H. Ultrastructure of the 'active zone' in the frog neuromuscular junction. Brain Res. 1973;62(2):373–80. https://doi.org/10.1016/0006-8993(73)90699-9.

Eguchi K, Montanaro J, Monnier EL, Shigemoto R. The number and distinct clustering patterns of voltage-gated calcium channels in nerve terminals. Front Neuroanat. 2022;16:846615. https://doi.org/10.3389/fnana.2022.846615.

Ehmann N, van de Linde S, Alon A, Ljaschenko D, Keung XZ, Holm T, Rings A, DiAntonio A, Hallermann S, Ashery U, Heckmann M, Sauer M, Kittel RJ. Quantitative super-resolution imaging of Bruchpilot distinguishes active zone states. Nat Commun. 2014;5(1):4650. https://doi.org/10.1038/ncomms5650.

Emperador-Melero J, Wong MY, Wang SSH, de Nola G, Nyitrai H, Kirchhausen T, Kaeser PS. PKC-phosphorylation of Liprin-α3 triggers phase separation and controls presynaptic active zone structure. Nat Commun. 2021;12(1):3057. https://doi.org/10.1038/s41467-021-23116-w.

Emperador-Melero J, Andersen JW, Metzbower SR, Levy AD, Dharmasri PA, de Nola G, Blanpied TA, Kaeser PS. Distinct active zone protein machineries mediate Ca^{2+} channel clustering and vesicle priming at hippocampal synapses. Nat Neurosci. 2024;27(9):1–15. https://doi.org/10.1038/s41593-024-01720-5.

Engholm-Keller K, Waardenberg AJ, Müller JA, Wark JR, Fernando RN, Arthur JW, Robinson PJ, Dietrich D, Schoch S, Graham ME. The temporal profile of activity-dependent presynaptic phospho-signalling reveals long-lasting patterns of poststimulus regulation. PLoS Biol. 2019;17(3):e3000170. https://doi.org/10.1371/journal.pbio.3000170.

Fernández-Busnadiego R, Zuber B, Maurer UE, Cyrklaff M, Baumeister W, Lučić V. Quantitative analysis of the native presynaptic cytomatrix by cryoelectron tomography. J Cell Biol. 2010;188(1):145–56. https://doi.org/10.1083/jcb.200908082.

Foster M, Sherrington CS. A textbook of physiology. 7th ed. New York: MacMillan; 1897.

Fouquet W, Owald D, Wichmann C, Mertel S, Depner H, Dyba M, Hallermann S, Kittel RJ, Eimer S, Sigrist SJ. Maturation of active zone assembly by Drosophila Bruchpilot. J Cell Biol. 2009;186(1):129–45. https://doi.org/10.1083/jcb.200812150.

Ghelani T, Montenegro-Venegas C, Fejtova A, Dresbach T. Nanoscopical analysis reveals an orderly arrangement of the presynaptic scaffold protein Bassoon at the Golgi-apparatus. Front Mol Neurosci. 2021;14:744034. https://doi.org/10.3389/fnmol.2021.744034.

Glebov OO, Jackson RE, Winterflood CM, Owen DM, Barker EA, Doherty P, Ewers H, Burrone J. Nanoscale structural plasticity of the active zone matrix modulates presynaptic function. Cell Rep. 2017;18(11):2715–28. https://doi.org/10.1016/j.celrep.2017.02.064.

Gorsheneva NA, Sopova JV, Azarov VV, Grizel AV, Rubel AA. Biomolecular condensates: structure, functions, methods of research. Biochemistry (Moscow). 2024;89(Suppl 1):S205–23. https://doi.org/10.1134/s0006297924140116.

Graf ER, Valakh V, Wright CM, Wu C, Liu Z, Zhang YQ, DiAntonio A. RIM promotes calcium channel accumulation at active zones of the Drosophila neuromuscular junction. J Neurosci. 2012;32(47):16586–96. https://doi.org/10.1523/jneurosci.0965-12.2012.

Gray EG. Electron microscopy of presynaptic organelles of the spinal cord. J Anat. 1963;97:101–6.

Gray E. Tissue of the central nervous system. In: Kurtz S, editor. Electron microscopic anatomy. New York: Academic Press; 1964. p. 369–417.

Grushin K, Sundaram RVK, Sindelarb CV, Rothman JE. Munc13 structural transitions and oligomers that may choreograph successive stages in vesicle priming for neurotransmitter release. Proc Natl Acad Sci U S A. 2022;119(7) https://doi.org/10.1073/pnas.2121259119.

Guan Z, Quiñones-Frías MC, Akbergenova Y, Littleton JT. Drosophila Synaptotagmin 7 negatively regulates synaptic vesicle release and replenishment in a dosage-dependent manner. eLife. 2020;9:e55443. https://doi.org/10.7554/elife.55443.

Gundelfinger ED, Reissner C, Garner CC. Role of Bassoon and Piccolo in assembly and molecular organization of the active zone. Front Synaptic Neurosci. 2016;7:19. https://doi.org/10.3389/fnsyn.2015.00019.

Guzikowski NJ, Kavalali ET. Functional specificity of liquid-liquid phase separation at the synapse Nature Communications. 2024;15(1). https://doi.org/10.1038/s41467-024-54423-7.

Hardenberg M, Horvath A, Ambrus V, Fuxreiter M, Vendruscolo M. Widespread occurrence of the droplet state of proteins in the human proteome. Proc Natl Acad Sci U S A. 2020;117(52):33254–62. https://doi.org/10.1073/pnas.2007670117.

Harris KM, Weinberg RJ. Ultrastructure of synapses in the mammalian brain. Cold Spring Harb Perspect Biol. 2012;4(5):a005587. https://doi.org/10.1101/cshperspect.a005587.

Heck J, Parutto P, Ciuraszkiewicz A, Bikbaev A, Freund R, Mitlöhner J, Andres-Alonso M, Fejtova A, Holcman D, Heine M. Transient confinement of $Ca_V2.1$ Ca^{2+}-channel splice variants shapes synaptic short-term plasticity. Neuron. 2019;103(1):66–79.e12. https://doi.org/10.1016/j.neuron.2019.04.030.

Held RG, Liu C, Ma K, Ramsey AM, Tarr TB, Nola GD, Wang SSH, Wang J, Maagdenberg AMJM van den, Schneider T, Sun J, Blanpied TA, Kaeser PS. Synapse and active zone assembly in the absence of presynaptic Ca^{2+} channels and Ca^{2+} entry. Neuron. 2020;107(4):667–683.e9. https://doi.org/10.1016/j.neuron.2020.05.032.

Held RG, Liang J, Brunger AT. Nanoscale architecture of synaptic vesicles and scaffolding complexes revealed by cryo-electron tomography. Proc Natl Acad Sci U S A. 2024;121(27):e2403136121. https://doi.org/10.1073/pnas.2403136121.

Heuser JE, Reese TS, Landis DMD. Functional changes in frog neuromuscular junctions studied with freeze-fracture. J Neurocytol. 1974;3(1):109–31. https://doi.org/10.1007/bf01111936.

Hillman DE, Chen S. Compensation in the number of presynaptic dense projections and synaptic vesicles in remaining parallel fibres following cerebellar lesions Journal of Neurocytology. 1985;14(4):673–687. https://doi.org/10.1007/BF01200804.

Hoover KM, Gratz SJ, Qi N, Herrmann KA, Liu Y, Perry-Richardson JJ, Vanderzalm PJ, O'Connor-Giles KM, Broihier HT. The calcium channel subunit $\alpha_2\delta$-3 organizes synapses via an activity-dependent and autocrine BMP signaling pathway. Nat Commun. 2019;10(1):5575. https://doi.org/10.1038/s41467-019-13165-7.

Hur JH, Lee S-H, Kim A-Y, Koh YH. Regulation of synaptic architecture and synaptic vesicle pools by Nervous wreck at Drosophila Type 1b glutamatergic synapses. Exp Mol Med. 2018;50(3):e462–e462. https://doi.org/10.1038/emm.2017.303.

Hyman AA, Weber CA, Jülicher F. Liquid-liquid phase separation in biology. Annu Rev Cell Dev Biol. 2014;30(1):39–58. https://doi.org/10.1146/annurev-cellbio-100913-013325.

Igarashi H, Nasu F, Inomata K, Arii T. High-voltage electron microscopy of E-PTA-stained synaptic junctions in the rat frontal cortex. Brain Res. 1988;473(2):365–8. https://doi.org/10.1016/0006-8993(88)90867-0.

Irgen-Gioro S, Yoshida S, Walling V, Chong S. Fixation can change the appearance of phase separation in living cells. eLife. 2022;11:e79903. https://doi.org/10.7554/elife.79903.

Jeyifous O, Lin EI, Chen X, Antinone SE, Mastro R, Drisdel R, Reese TS, Green WN. Palmitoylation regulates glutamate receptor distributions in postsynaptic densities through control of PSD95 conformation and orientation. Proc Natl Acad Sci U S A. 2016;113(52):E8482–91. https://doi.org/10.1073/pnas.1612963113.

Jiao W, Masich S, Franzén O, Shupliakov O. Two pools of vesicles associated with the presynaptic cytosolic projection in Drosophila neuromuscular junctions. J Struct Biol. 2010;172(3):389–94. https://doi.org/10.1016/j.jsb.2010.07.007.

Jin Y, Zhai RG. Presynaptic cytomatrix proteins. Adv Neurobiol. 2023;33:23–42. https://doi.org/10.1007/978-3-031-34229-5_2.

Jones DG, Ellison LT, Reading LC, Dittmer MM. A critical evaluation of the relationship between the presynaptic network, synaptic vesicles and dense projections in central synapses. Cell Tissue Res. 1976;169(1):49–66. https://doi.org/10.1007/bf00219307.

Kaeser PS, Deng L, Wang Y, Dulubova I, Liu X, Rizo J, Südhof TC. RIM proteins tether Ca^{2+} channels to presynaptic active zones via a direct PDZ-domain interaction. Cell. 2011;144(2):282–95. https://doi.org/10.1016/j.cell.2010.12.029.

Karlocai MR, Heredi J, Benedek T, Holderith N, Lorincz A, Nusser Z. Variability in the Munc13-1 content of excitatory release sites. eLife. 2021;10:e67468. https://doi.org/10.7554/elife.67468.

Keller D, Babai N, Kochubey O, Han Y, Markram H, Schürmann F, Schneggenburger R. An exclusion zone for Ca^{2+} channels around docked vesicles explains release control by multiple channels at a CNS synapse. PLoS Comput Biol. 2015;11(5):e1004253. https://doi.org/10.1371/journal.pcbi.1004253.

Kobbersmed JR, Grasskamp AT, Jusyte M, Böhme MA, Ditlevsen S, Sørensen JB, Walter AM. Rapid regulation of vesicle priming explains synaptic facilitation despite heterogeneous vesicle: Ca^{2+} channel distances. eLife. 2020;9:e51032. https://doi.org/10.7554/elife.51032.

Koenig JH, Ikeda K. Contribution of active zone subpopulation of vesicles to evoked and spontaneous release. J Neurophysiol. 1999;81(4):1495–505. https://doi.org/10.1152/jn.1999.81.4.1495.

Kohansal-Nodehi M, Chua JJ, Urlaub H, Jahn R, Czernik D. Analysis of protein phosphorylation in nerve terminal reveals extensive changes in active zone proteins upon exocytosis. eLife. 2016;5:e14530. https://doi.org/10.7554/elife.14530.

Laßek M, Weingarten J, Volknandt W. The synaptic proteome. Cell Tissue Res. 2015;359(1):255–65. https://doi.org/10.1007/s00441-014-1943-4.

Lautenschläger J. Protein phase separation hotspots at the presynapse. Open Biol. 2022;12(2):210334. https://doi.org/10.1098/rsob.210334.

Li J, Ashley J, Budnik V, Bhat MA. Crucial role of Drosophila neurexin in proper active zone apposition to postsynaptic densities, synaptic growth, and synaptic transmission. Neuron. 2007;55(5):741–55. https://doi.org/10.1016/j.neuron.2007.08.002.

Li J, Zhang M, Ma W, Yang B, Lu H, Zhou F, Zhang L. Post-translational modifications in liquid-liquid phase separation: a comprehensive review Abstract Molecular Biomedicine. 2022;3(1). https://doi.org/10.1186/s43556-022-00075-2.

Liang M, Jin G, Xie X, Zhang W, Li K, Niu F, Yu C, Wei Z. Oligomerized liprin-α promotes phase separation of ELKS for compartmentalization of presynaptic active zone proteins. Cell Rep. 2021;34(12):108901. https://doi.org/10.1016/j.celrep.2021.108901.

Limbach C, Laue MM, Wang X, Hu B, Thiede N, Hultqvist G, Kilimann MW. Molecular in situ topology of Aczonin/Piccolo and associated proteins at the mammalian neurotransmitter release site. Proc Natl Acad Sci U S A. 2011;108(31):E392–401. https://doi.org/10.1073/pnas.1101707108.

Liu KSY, Siebert M, Mertel S, Knoche E, Wegener S, Wichmann C, Matkovic T, Muhammad K, Depner H, Mettke C, Bückers J, Hell SW, Müller M, Davis GW, Schmitz D, Sigrist SJ. RIM-binding protein, a central part of the active zone, is essential for neurotransmitter release. Science. 2011;334:1565–9. https://doi.org/10.1126/science.1212991.

Lloyd BA, Han Y, Roth R, Zhang B, Aoto J. Neurexin-3 subsynaptic densities are spatially distinct from Neurexin-1 and essential for excitatory synapse nanoscale organization in the hippocampus. Nat Commun. 2023;14(1):4706. https://doi.org/10.1038/s41467-023-40419-2.

Lučić V, Yang T, Schweikert G, Förster F, Baumeister W. Morphological characterization of molecular complexes present in the synaptic cleft. Structure. 2005;13(3):423–34. https://doi.org/10.1016/j.str.2005.02.005.

Lyon AS, Peeples WB, Rosen MK. A framework for understanding the functions of biomolecular condensates across scales. Nat Rev Mol Cell Biol. 2021;22(3):215–35. https://doi.org/10.1038/s41580-020-00303-z.

Matkovic T, Siebert M, Knoche E, Depner H, Mertel S, Owald D, Schmidt M, Thomas U, Sickmann A, Kamin D, Hell SW, Bürger J, Hollmann C, Mielke T, Wichmann C, Sigrist SJ. The Bruchpilot cytomatrix determines the size of the readily releasable pool of synaptic vesicles. J Cell Biol. 2013;202(4):667–83. https://doi.org/10.1083/jcb.201301072.

Maynard SA, Ranft J, Triller A. Quantifying postsynaptic receptor dynamics: insights into synaptic function. Nat Rev Neurosci. 2023;24(1):4–22. https://doi.org/10.1038/s41583-022-00647-9.

McCabe BD, Marqués G, Haghighi AP, Fetter RD, Crotty ML, Haerry TE, Goodman CS, O'Connor MB. The BMP homolog Gbb provides a retrograde signal that regulates synaptic growth at the Drosophila neuromuscular junction Neuron 2003:39(2) 241–254. https://doi.org/10.1016/S0896-6273(03)00426-4.

McDonald NA, Shen K. Finding functions of phase separation in the presynapse. Curr Opin Neurobiol. 2021;69:178–84. https://doi.org/10.1016/j.conb.2021.04.001.

McDonald NA, Fetter RD, Shen K. Assembly of synaptic active zones requires phase separation of scaffold molecules. Nature. 2020;588(7838):454–8. https://doi.org/10.1038/s41586-020-2942-0.

McSwiggen DT, Mir M, Darzacq X, Tjian R. Evaluating phase separation in live cells: diagnosis, caveats, and functional consequences. Genes Dev. 2019;33(23–24):1619–34. https://doi.org/10.1101/gad.331520.119.

Mrestani A, Pauli M, Kollmannsberger P, Repp F, Kittel RJ, Eilers J, Doose S, Sauer M, Sirén A-L, Heckmann M, Paul MM. Active zone compaction correlates with presynaptic homeostatic potentiation. Cell Rep. 2021;37(1):109770. https://doi.org/10.1016/j.celrep.2021.109770.

Mrestani A, Dannhäuser S, Pauli M, Kollmannsberger P, Hübsch M, Morris L, Langenhan T, Heckmann M, Paul MM. Nanoscaled RIM clustering at presynaptic active zones revealed by endogenous tagging. Life Sci Alliance. 2023;6(12):e202302021. https://doi.org/10.26508/lsa.202302021.

M'Saad O, Cairns A, Gulcicek J, Kasula RK, Liao J, Kondratiuk I, Bewersdorf EH, Kidd P, Falahati H, Gentile JE, Niescier RF. Bewersdorf All-optical visualization of specific molecules in the ultrastructural context of brain tissue Nature Biotechnology. 2025: https://doi.org/10.1038/s41587-025-02905-4.

Müller JA, Betzin J, Santos-Tejedor J, Mayer A, Oprişoreanu A-M, Engholm-Keller K, Paulußen I, Gulakova P, McGovern TD, Gschossman LJ, Schönhense E, Wark JR, Lamprecht A, Becker AJ, Waardenberg AJ, Graham ME, Dietrich D, Schoch S. A presynaptic phosphosignaling hub for lasting homeostatic plasticity. Cell Rep. 2022;39(3):110696. https://doi.org/10.1016/j.celrep.2022.110696.

Muttathukunnel P, Frei P, Perry S, Dickman D, Müller M. Rapid homeostatic modulation of transsynaptic nanocolumn rings. Proc Natl Acad Sci. 2022;119(45) https://doi.org/10.1073/pnas.2119044119.

Nijhof B, Castells-Nobau A, Wolf L, Gooyert JMS, Monedero I, Torroja L, Coromina L, van der Laak JAWM, Schenck A. A new Fiji-based algorithm that systematically quantifies nine synaptic parameters provides insights into Drosophila NMJ morphometry. PLoS Comput Biol. 2016;12(3):e1004823. https://doi.org/10.1371/journal.pcbi.1004823.

Owald D, Fouquet W, Schmidt M, Wichmann C, Mertel S, Depner H, Christiansen F, Zube C, Quentin C, Körner J, Urlaub H, Mechtler K, Sigrist SJ. A Syd-1 homologue regulates pre- and postsynaptic maturation in Drosophila. J Cell Biol. 2010;188(4):565–79. https://doi.org/10.1083/jcb.200908055.

Owald D, Khorramshahi O, Gupta VK, Banovic D, Depner H, Fouquet W, Wichmann C, Mertel S, Eimer S, Reynolds E, Holt M, Aberle H, Sigrist SJ. Cooperation of Syd-1 with Neurexin synchronizes pre- with postsynaptic assembly. Nat Neurosci. 2012;15(9):1219–26. https://doi.org/10.1038/nn.3183.

Palay SL. Structure and function in the neuron. Prog Neurobiol. 1956;1:64–82.

Palay SL. The morphology of synapses in the central nervous system. Exp cell Res. 1958;14(Suppl 5):275–93.

Palay SL, Palade GE. The fine structure of neurons. J Cell Biol. 1955;1(1):69–88. https://doi.org/10.1083/jcb.1.1.69.

Papantoniou C, Laugks U, Betzin J, Capitanio C, Ferrero JJ, Sánchez-Prieto J, Schoch S, Brose N, Baumeister W, Cooper BH, Imig C, Lučić V. Munc13- and SNAP25-dependent molecular bridges play a key role in synaptic vesicle priming Science Advances 2023:9(25). https://doi.org/10.1126/sciadv.adf6222.

Paul MM, Pauli M, Ehmann N, Hallermann S, Sauer M, Kittel RJ, Heckmann M. Bruchpilot and Synaptotagmin collaborate to drive rapid glutamate release and active zone differentiation. Front Cell Neurosci. 2015;9:29. https://doi.org/10.3389/fncel.2015.00029.

Paulussen I, Beckert H, Musial TF, Gschossmann LJ, Wolf J, Schmitt M, Clasadonte J, Mairet-Coello G, Wolff C, Schoch S, Dietrich D. SV2B defines a subpopulation of synaptic vesicles. J Mol Cell Biol. 2023:mjad054. https://doi.org/10.1093/jmcb/mjad054.

Petralia RS, Wang Y-X, Mattson MP, Yao PJ. Invaginating structures in mammalian synapses. Front Synaptic Neurosci. 2018;10:4. https://doi.org/10.3389/fnsyn.2018.00004.

Petzoldt AG, Götz TWB, Driller JH, Lützkendorf J, Reddy-Alla S, Matkovic-Rachid T, Liu S, Knoche E, Mertel S, Ugorets V, Lehmann M, Ramesh N, Beuschel CB, Kuropka B, Freund C, Stelzl U, Loll B, Liu F, Wahl MC, Sigrist SJ. RIM-binding protein couples synaptic vesicle recruitment to release sites. J Cell Biol. 2020;219(7):e201902059. https://doi.org/10.1083/jcb.201902059.

Pfenninger K, Sandri C, Akert K, Eugster CH. Contribution to the problem of structural organization of the presynaptic area. Brain Res. 1969;12(1):10–8. https://doi.org/10.1016/0006-8993(69)90051-1.

Phillips GR, Huang JK, Wang Y, Tanaka H, Shapiro L, Zhang W, Shan W-S, Arndt K, Frank M, Gordon RE, Gawinowicz MA, Zhao Y, Colman DR. The presynaptic particle web ultrastructure, composition, dissolution, and reconstitution. Neuron. 2001;32(1):63–77. https://doi.org/10.1016/s0896-6273(01)00450-0.

Prokop A, Meinertzhagen IA. Development and structure of synaptic contacts in Drosophila. Semin Cell Dev Biol. 2006;17(1):20–30. https://doi.org/10.1016/j.semcdb.2005.11.010

Qiu H, Wu X, Ma X, Li S, Cai Q, Ganzella M, Ge L, Zhang H, Zhang M. Short-distance vesicle transport via phase separation. Cell. 2024;187(9):2175–2193.e21. https://doi.org/10.1016/j.cell.2024.03.003.

Radhakrishnan A, Li X, Grushin K, Krishnakumar SS, Liu J, Rothman JE. Symmetrical arrangement of proteins under release-ready vesicles in presynaptic terminals. Proc Natl Acad Sci U S A. 2021;118(5):1–7. https://doi.org/10.1073/pnas.2024029118.

Rebola N, Reva M, Kirizs T, Szoboszlay M, Lőrincz A, Moneron G, Nusser Z, DiGregorio DA. Distinct nanoscale calcium channel and synaptic vesicle topographies contribute to the diversity of synaptic function. Neuron. 2019;104(4):693–710.e9. https://doi.org/10.1016/j.neuron.2019.08.014.

Reddy-Alla S, Böhme MA, Reynolds E, Beis C, Grasskamp AT, Mampell MM, Maglione M, Jusyte M, Rey U, Babikir H, McCarthy AW, Quentin C, Matkovic T, Bergeron DD, Mushtaq Z, Göttfert F, Owald D, Mielke T, Hell SW, Sigrist SJ, Walter AM. Stable positioning of Unc13 restricts synaptic vesicle fusion to defined release sites to promote synchronous neurotransmission. Neuron. 2017;95(6):1350–1364.e12. https://doi.org/10.1016/j.neuron.2017.08.016.

Reist NE, Buchanan J, Li J, DiAntonio A, Buxton EM, Schwarz TL. Morphologically docked synaptic vesicles are reduced in synaptotagmin mutants of Drosophila. J Neurosci. 1998;18(19):7662–73. https://doi.org/10.1523/jneurosci.18-19-07662.1998.

Robertis EDPD, Bennett HS. Some features of the submicroscopic morphology of synapses in frog and earthworm. J Cell Biol. 1955;1(1):47–58. https://doi.org/10.1083/jcb.1.1.47.

Robertson JD. The ultrastructure of a reptilian myoneural junction. J Cell Biol. 1956;2(4):381–94. https://doi.org/10.1083/jcb.2.4.381.

Rowland KC, Irby NK, Spirou GA. Specialized synapse-associated structures within the calyx of Held. J Neurosci. 2000;20(24):9135–44. https://doi.org/10.1523/jneurosci.20-24-09135.2000.
Sakamoto H, Ariyoshi T, Kimpara N, Sugao K, Taiko I, Takikawa K, Asanuma D, Namiki S, Hirose K. Synaptic weight set by Munc13-1 supramolecular assemblies. Nat Neurosci. 2018;21(1):41–9. https://doi.org/10.1038/s41593-017-0041-9.
Scheefhals N, MacGillavry HD. Functional organization of postsynaptic glutamate receptors. Mol Cell Neurosci. 2018;91:82–94. https://doi.org/10.1016/j.mcn.2018.05.002.
Schikorski T, Stevens CF. Quantitative ultrastructural analysis of hippocampal excitatory synapses. J Neurosci. 1997;17(15):5858–67. https://doi.org/10.1523/jneurosci.17-15-05858.1997.
Schikorski T, Stevens CF. Quantitative fine-structural analysis of olfactory cortical synapses. Proc Natl Acad Sci U S A. 1999;96(7):4107–12. https://doi.org/10.1073/pnas.96.7.4107.
Schoch S, Gundelfinger ED. Molecular organization of the presynaptic active zone. Cell Tissue Res. 2006;326(2):379–91. https://doi.org/10.1007/s00441-006-0244-y.
Shepherd GMG, Harris KM. Three-dimensional structure and composition of CA3→CA1 axons in rat hippocampal slices: implications for presynaptic connectivity and compartmentalization. J Neurosci. 1998;18(20):8300–10. https://doi.org/10.1523/jneurosci.18-20-08300.1998.
Sigrist SJ, Thiel PR, Reiff DF, Schuster CM. The postsynaptic glutamate receptor subunit DGluR-IIA mediates long-term plasticity in Drosophila. J Neurosci. 2002;22(17):7362–72. https://doi.org/10.1523/jneurosci.22-17-07362.2002.
Siksou L, Rostaing P, Lechaire J-P, Boudier T, Ohtsuka T, Fejtova A, Kao H-T, Greengard P, Gundelfinger ED, Triller A, Marty S. Three-dimensional architecture of presynaptic terminal cytomatrix. J Neurosci. 2007;27(26):6868–77. https://doi.org/10.1523/jneurosci.1773-07.2007.
Song W, Ranjan R, Dawson-Scully K, Bronk P, Marin L, Seroude L, Lin Y-J, Nie Z, Atwood HL, Benzer S, Zinsmaier KE. Presynaptic regulation of neurotransmission in Drosophila by the G protein-coupled receptor methuselah. Neuron. 2002;36(1):105–19. https://doi.org/10.1016/s0896-6273(02)00932-7.
Südhof TC. The presynaptic active zone. Neuron. 2012;75(1):11–25. https://doi.org/10.1016/j.neuron.2012.06.012.
Sun S-Y, Li X-W, Cao R, Zhao Y, Sheng N, Tang A-H. Correlative assembly of subsynaptic nanoscale organizations during development. Front Synaptic Neurosci. 2022;14:748184. https://doi.org/10.3389/fnsyn.2022.748184.
Tang A-H, Chen H, Li TP, Metzbower SR, MacGillavry HD, Blanpied TA. A trans-synaptic nanocolumn aligns neurotransmitter release to receptors. Nature. 2016;536(7615):210–4. https://doi.org/10.1038/nature19058.
Tao-Cheng J-H. Activity-related redistribution of presynaptic proteins at the active zone. Neuroscience. 2006;141(3):1217–24. https://doi.org/10.1016/j.neuroscience.2006.04.061.
Vrensen G, Cardozo JN, Müller L, Want JVD. The presynaptic grid: a new approach. Brain Res. 1980;184(1):23–40. https://doi.org/10.1016/0006-8993(80)90585-5.
Wagner N. Ultrastructural comparison of the Drosophila larval and adult ventral abdominal neuromuscular junction. J Morphol. 2017;278(7):987–96. https://doi.org/10.1002/jmor.20692.
Wang J, Bello O, Auclair SM, Wang J, Coleman J, Pincet F, Krishnakumar SS, Sindelar CV, Rothman JE. Calcium sensitive ring-like oligomers formed by synaptotagmin. Proc Natl Acad Sci U S A. 2014;111(38):13966–71. https://doi.org/10.1073/pnas.1415849111.
Wichmann C, Sigrist SJ The active zone T-bar—A plasticity module? Journal of Neurogenetics . 2010;24(3):133–145. https://doi.org/10.3109/01677063.2010.489626.
Wiesner T, Bilodeau A, Bernatchez R, Deschênes A, Raulier B, Koninck PD, Lavoie-Cardinal F. Activity-dependent remodeling of synaptic protein organization revealed by high throughput analysis of STED nanoscopy images. Front Neural Circuits. 2020;14:57. https://doi.org/10.3389/fncir.2020.00057.
Wilhelm BG, Mandad S, Truckenbrodt S, Kröhnert K, Schäfer C, Rammner B, Koo SJ, Claßen GA, Krauss M, Haucke V, Urlaub H, Rizzoli SO. Composition of isolated synaptic boutons reveals the amounts of vesicle trafficking proteins. Science. 2014;344(6187):1023–8. https://doi.org/10.1126/science.1252884.

Wong MY, Liu C, Wang SSH, Roquas ACF, Fowler SC, Kaeser PS. Liprin-α3 controls vesicle docking and exocytosis at the active zone of hippocampal synapses. Proc Natl Acad Sci U S A. 2018;115(9):2234–9. https://doi.org/10.1073/pnas.1719012115.

Wu X, Cai Q, Shen Z, Chen X, Zeng M, Du S, Zhang M. RIM and RIM-BP form presynaptic active-zone-like condensates via phase separation. Mol Cell. 2019;73(5):971–984.e5. https://doi.org/10.1016/j.molcel.2018.12.007.

Wu X, Ganzella M, Zhou J, Zhu S, Jahn R, Zhang M. Vesicle tethering on the surface of phase-separated active zone condensates. Mol Cell. 2021;81(1):13–24.e7. https://doi.org/10.1016/j.molcel.2020.10.029.

Wu X, Qiu H, Zhang M. Interactions between membraneless condensates and membranous organelles at the presynapse: a phase separation view of synaptic vesicle cycle. J Mol Biol. 2023;435(1):167629. https://doi.org/10.1016/j.jmb.2022.167629.

Wu X, Shen Z, Zhang M. Phase separation–mediated compartmentalization underlies synapse formation and plasticity Annual Review of Neuroscience. 2025;48(1):149–168. https://doi.org/10.1146/annurev-neuro-112723-040159.

Xu-Friedman MA, Harris KM, Regehr WG. Three-dimensional comparison of ultrastructural characteristics at depressing and facilitating synapses onto cerebellar Purkinje cells. J Neurosci. 2001;21(17):6666–72. https://doi.org/10.1523/jneurosci.21-17-06666.2001.

Yoshida SR, Maity BK, Chong S. Visualizing protein localizations in fixed cells: caveats and the underlying mechanisms. J Phys Chem B. 2023;127(19):4165–73. https://doi.org/10.1021/acs.jpcb.3c01658.

Zhai RG. The architecture of the presynaptic release site. Adv Neurobiol. 2023;33:1–21. https://doi.org/10.1007/978-3-031-34229-5_1.

Zhai RG, Bellen HJ. The architecture of the active zone in the presynaptic nerve terminal. Physiology. 2004;19(5):262–70. https://doi.org/10.1152/physiol.00014.2004.

Zhan H, Bruckner J, Zhang Z, O'Connor-Giles K. Three-dimensional imaging of Drosophila motor synapses reveals ultrastructural organizational patterns. J Neurogenet. 2016;30(3–4):237–46. https://doi.org/10.1080/01677063.2016.1253693.

Zuber B, Nikonenko I, Klauser P, Muller D, Dubochet J. The mammalian central nervous synaptic cleft contains a high density of periodically organized complexes. Proc Natl Acad Sci U S A. 2005;102(52):19192–7. https://doi.org/10.1073/pnas.0509527102.

Chapter 5
Synaptic Cell Adhesion: A Functional Perspective

Kelly Honkanen and Richard C. Sando

Abstract Synaptic cell adhesion molecules are critical components of the molecular programs underlying synapse formation and neural circuit assembly. Here, we discuss our current understanding of the functional roles of several of these synaptic adhesion molecules in the mammalian central nervous system. Emerging evidence, driven by advances in super-resolution approaches, supports that pre- and postsynaptic machinery are highly organized at the nanoscale level and that these precise sub-synaptic positions are important for synaptic transmission. We subsequently describe the nano-organization of several synaptic cell adhesion molecules and how these trans-synaptic complexes align release machinery to shape the synaptic function. Collectively, an understanding of the mechanistic roles of synaptic cell adhesion complexes will provide insights into how neural circuits assemble by vast numbers of diverse synaptic connections in the brain.

Keywords Synapse formation · Synapse specificity · Cell adhesion molecule · Trans-synaptic signaling · Adhesion GPCR

5.1 Introduction

The mammalian brain is composed of a remarkable number of diverse synaptic connections that form highly stereotyped circuits during development (Colón-Ramos 2009; Südhof 2018). The organization of these diverse synapses helps shape the brain's ability to process information and generate behaviors (Luo 2021). After their initial establishment, synaptic circuits are continually refined by our experiences and are maintained throughout life (Attardo et al. 2015). What are the mechanisms

K. Honkanen · R. C. Sando (✉)
Department of Pharmacology, Vanderbilt Brain Institute, Vanderbilt University, Nashville, TN, USA
e-mail: richard.sando@Vanderbilt.Edu

E. T. Kavalali (ed.), *Nano-organization of the Synapse*, Advances in Neurobiology 48,
https://doi.org/10.1007/978-3-032-12594-1_5

that guide the assembly of the synaptic circuitry within the mammalian brain? The human brain is composed of an estimated >1 × 10^{12} synaptic connections, and individual neurons can receive thousands of synaptic inputs. During early postnatal development, it is estimated that ~700 remarkable new synaptic connections are formed every second in the human brain (Mikhailova and McAllister 2020). The complexity of the brain wiring problem has been acknowledged for some time, and the mechanisms still remain unclear. Over five decades ago, Roger Sperry proposed the chemoaffinity hypothesis (Meyer 1998; Vanderhaeghen and Polleux 2004; Berlucchi and Marzi 2024; Sperry 1951; Sperry 1963), which postulated the existence of chemical codes that specify precise synaptic connectivity. Since then, evidence has supported that networks of synaptic cell adhesion molecules contribute to the fundamental aspects of synapse formation and specification in neural circuits (Zipursky and Sanes 2010; Sanes and Zipursky 2020; Siddiqui and Craig 2011; Südhof 2018). Such combinatorial molecular codes are further diversified via mechanisms including alternative splicing, expression gradients, and post-translational modifications (e.g., glycosylation profiles of extracellular proteins). These processes expand the capacity of relevant genetic composites to help generate a molecular "blueprint" underlying synapse formation and brain wiring (Nusser 2018; Cortés et al. 2023; Hassan and Hiesinger 2015; Jain and Zipursky 2023; Xie et al. 2022; Luo 2006; Sanes and Yamagata 2009; Shen and Cowan 2010; Südhof 2021). The development of other multicellular structures into precise patterns also requires highly specific cell–cell recognition (Honig and Shapiro 2020). Thus, brain wiring can be viewed as an incredibly complex example of multicellular assembly, where insights from the development of other tissues can be used as examples. Here, we summarize several aspects of our current understanding of how synaptic cell adhesion contributes to synapse formation, function, and circuit assembly in the mammalian central nervous system.

5.2 Signaling Mechanisms of Synapse Formation

Intricate molecular networks of secreted cues and adhesion complexes contribute to circuit assembly and the functional specification of distinct synapse subtypes (de Wit and Ghosh 2016; Zipursky et al. 2006; Wu et al. 2011; Yogev and Shen 2014). However, extracellular interactions alone cannot fully account for the complexities of synapse formation and specificity (Wolterhoff & Hiesinger 2024; Margeta and Shen 2010). For instance, bi-directional signaling is also required to guide the necessary biological processes to establish synapses (Biederer & Stagi 2008, Biederer 2005, Dalva et al. 2007). These signals modulate the cytoskeleton and recruit the proper functional components to nascent synaptic compartments (Jin and Garner 2008). Do these signals emanate from the complexes of cell adhesion molecules during synapse formation or from other sources such as neurotransmitter release from nascent synapses?

If intrinsic molecular codes help shape circuit assembly, what is the role of signaling via trans-synaptic adhesion complexes? Alternatively, can trans-synaptic signaling via neurotransmission guide synapse formation at nascent synapses independently of or preceding adhesion? The contribution of neuronal activity, such as synaptic transmission, compared to intrinsic molecular mechanisms has been studied in many contexts. For example, experience-driven activity is important for circuit assembly during critical periods of development (Akin and Zipursky 2020). Moreover, activity has been shown to modulate a vast array of functional and structural plasticity processes in the developing and adult brain. However, does synaptic transmission drive synapse formation in the mammalian CNS, or is synapse formation orchestrated by intrinsic molecular programs that precede transmission?

Studies investigating the presynaptic machinery essential for synaptic release provided insights into this question. Constitutive deletion of Munc18-1 abolishes all forms of neurotransmitter release from presynaptic terminals throughout development. Remarkably, despite the absence of synaptic transmission during early development, Munc18-1 mutants display overtly normal brain development, including assembly of synapses containing characteristic ultrastructural features (Verhage et al. 2000). Deletion of the presynaptic active zone proteins Munc13-1/Munc13-2, which are essential for synaptic vesicle priming, ablates both spontaneous and evoked release. However, neurons from these models form synapses at a normal density with typical synaptic ultrastructural features, such as vesicle pools and postsynaptic densities (Varoqueaux et al. 2002). Likewise, loss of the major SNARE protein on synaptic vesicles essential for fast calcium-driven synaptic vesicle exocytosis, Synaptobrevin-2 (Syb2/VAMP2), results in morphologically normal synapses, including preserved vesicle pools (Schoch et al. 2001; Deák et al. 2004). Importantly, these constitutive KOs are lethal immediately after birth, which precluded the analysis of synapses at later phases of synaptogenesis. Constitutive treatment of primary neurons with the tetanus toxin, which cleaves Syb2, during their maturation also resulted in no alterations in synapse density or organization (Harms and Craig 2005). These results highlight that synapses in the CNS can be assembled without synaptic transmission. Activity-dependent mechanisms are essential for the refinement of circuits after this initial assembly (Zhang and Poo 2001).

More recent studies used similar approaches to ablate presynaptic release yet devised strategies to enable the analysis of synapses at later phases of synaptogenesis. These studies cultured hippocampal slices from Munc13-1/2 double KO mice for up to 3 weeks postnatally (Sigler et al. 2017). This time frame encompasses a major period of postnatal synaptogenesis. Thus, neurons from these hippocampal slices undergo early synaptogenesis virtually devoid of synaptic transmission. Nevertheless, CA1 pyramidal neurons from Munc13Munc13 dKO slices developed normal dendritic arbors and functional spines at a density comparable to controls. Complementary studies examined the overall contribution of presynaptic glutamate release to synapse assembly by expressing tetanus toxin, which cleaves the presynaptic SNARE Syb2, in glutamatergic neurons of the forebrain in vivo (Sando et al. 2017). A subset of these mice could survive into postnatal development despite the lack of glutamate release from cortical and hippocampal pyramidal neurons.

Examination of these brains at the level of overall circuit architecture, single synapse morphology, and synaptic ultrastructure supported that synaptic glutamate release is largely dispensable for many aspects of synapse formation and circuit assembly during early postnatal development. Namely, overall stereotyped connectivity was preserved throughout the hippocampus. While neurons displayed a loss of a subset of dendritic spines, loss of presynaptic glutamate release did not dramatically alter the development of presynaptic terminals nor spine morphology. Moreover, synapses formed in the absence of presynaptic glutamate release assembled stereotypical ultrastructural features, including presynaptic vesicular pools, the presynaptic active zone, and postsynaptic density. Altogether, these studies further emphasized the role of intrinsic programs that precede or are independent from neurotransmission in driving the central aspects of synapse and circuit assembly.

Does postsynaptic signaling via neurotransmitter receptors instruct synapse formation? Cell autonomous deletion of AMPA and NMDA receptors helped illuminate this question (Lu et al. 2013). Hippocampal CA1 neurons which lack postsynaptic glutamatergic transmission displayed normal dendritic morphology and developed postsynaptic spines at a density indistinguishable from control cells. These spines contained normal postsynaptic densities and received presynaptic inputs harboring normal vesicular pools. Importantly, these manipulations were conducted in a sparse, cell-autonomous manner to allow the analysis of glutamate receptor-deficient and control CA1 neurons in the same sample. These results support the functional segregation of postsynaptic excitatory neurotransmission and postsynaptic neurodevelopment, including the formation of spines (Lu et al. 2013).

This scenario may vary across different synapse subtypes. For example, recent studies have demonstrated that ectopic expression of GABA synthesis enzymes and GABA presynaptic vesicular transporters in purely glutamatergic neurons can support presynaptic GABA release which can drive the formation of functional GABAergic postsynaptic apparatus (Burlingham et al. 2022). This suggests that presynaptic GABA release alone can organize postsynaptic GABAergic responses. However, given the ectopic expression in these experimental paradigms, it remains unclear if presynaptic neurotransmitter release alone can drive postsynaptic assembly *in vivo*.

However, glutamate uncaging experiments support that localized release of a neurotransmitter on a dendritic shaft can induce postsynaptic spine formation, albeit with varying efficacies (Kwon and Sabatini 2011). Are subsets of new synapses susceptible to induction via synaptic transmission? Mounting evidence also supports that learning and memory can support new synapse formation, supporting that experience-driven activity can promote synapse formation. These observations, along with studies examining how ablation of synaptic transmission affects synapse formation and brain development, suggest there is likely an intricate interplay between neuronal activity and molecular codes including adhesion molecules during synapse formation.

Which signals initially guide synapse formation in the CNS, and how are these signals transduced by cell surface receptors? Synapse formation likely involves several canonical signaling pathways that are spatially compartmentalized at nascent synapses (Ellisdon & Halls 2016). For example, recent studies illustrate that JNK and PKA are important for the development of both pre- and postsynaptic specializations, whereas AKT is exclusively important for postsynaptic development (Jiang et al. 2021). Moreover, compartmentalized postsynaptic cAMP signaling has been shown to be important for glutamatergic synapse formation (Kwon and Sabatini 2011; Sando et al. 2022). Aside from these pathways in mammalian neurons, protein ubiquitination via ubiquitin ligase HighWire/RPM-1 is critical in the signaling pathways mediating synapse assembly in invertebrates (Chang & Balice-Gordon 2000, Jin and Garner 2008; Tian et al. 2011). Therefore, these intrinsic adhesion systems, coupled to several downstream signaling pathways, participate in synapse formation, regulate synapse maturation, and recruit fundamental functional components to nascent synapses (Chia et al. 2013). Our current understanding of functional roles of several of these adhesion modules will be further discussed in the subsequent section.

5.3 Functional Roles of Cell Adhesion Molecules (CAMs) in Synapse Formation

5.3.1 Adhesion GPCRs

Bi-directional trans-synaptic signaling by cell adhesion molecules directs the critical aspects of circuit assembly and function (Dalva et al. 2007; Gomez et al. 2021; Südhof 2021). Increasing evidence supports that adhesion G protein-coupled receptors (aGPCRs) control the central facets of synapse formation and circuit development via trans-cellular adhesion and intracellular signaling. aGPCRs are a large GPCR class, with 33 members in humans divided into 9 families, yet their biological functions and signal transduction mechanisms remain poorly understood (Paavola & Hall 2012). While certain aGPCR families are virtually unstudied in the brain and may have currently unreported roles at synapses, three families have well-documented roles in synapse formation and circuit assembly, namely, the Latrophilins (Adgrl), BAIs (brain angiogenesis inhibitor; Adgrb), and CELSRs (Flamingo in invertebrates; Adgrc). We describe several of these roles in the following subsections.

5.3.2 aGPCR Latrophilins

α-Latrotoxin, the potent neurotoxin present in black widow spider venom that targets vertebrates, is a relatively large (130 kD) toxin that induces a massive increase in spontaneous synaptic vesicle release. A screen using immobilized α-Latrotoxin to identify its endogenous receptor in mammalian neurons identified Latrophilin-1

(Lphn1) together with Neurexin-1 (Nrxn1) as candidate α-Latrotoxin receptors (Südhof 2001, Krasnoperov et al. 1997; Lelianova et al. 1997). Nrxn binds to latrotoxin in a Ca^{2+}-dependent manner, whereas Lphn1 binding to latrotoxin is Ca^{2+}-independent. Lphns are conserved aGPCRs that exhibit three paralogs in mammals (Lphn1-3) which are derived from (*Drosophila melanogaster*) or LAT-1/LAT-2 (*C. elegans*). Constitutive Lphn1 KO mice are viable yet display breeding deficits. Lphn1 KO neurons have a partial loss of α-latrotoxin binding and diminished α-latrotoxin-induced glutamate release from synaptosomes, which initially alluded to a putative presynaptic localization (Tobaben et al. 2000).

Lphns are multidomain adhesion receptors composed of a large extracellular N-terminus that harbors several domains mediating cell adhesion, an aGPCR signature GAIN domain which is followed by a 7-transmembrane domain, and an extensive intracellular C-terminal tail. The Lphn N-terminus exhibits a signal peptide, followed by a Lectin domain, olfactomedin domain, and hormone-binding domain (Fig. 5.1). Within this extracellular region, Lphns interact with at least three previously identified adhesion molecules: neurexins (Boucard et al. 2012), teneurins (Araç & Li 2019, Silva et al. 2011; Boucard et al. 2014), and FLRTs (fibronectin leucine-rich repeat transmembrane proteins) (O'Sullivan et al. 2012). However, Lphns may engage additional unidentified interactors given their extensive N-terminal region.

Neurexins interact with the Lphn1 olfactomedin domain to generate a high-affinity intercellular adhesion complex (Boucard et al. 2012). Interestingly, this interaction affinity is regulated by alternative splicing. Namely, Neurexin lacking splice site 4 (SS4) binds to Lphn1 with nanomolar affinity, while the inclusion of SS4 diminishes Lphn1 binding. This adds to the growing body of evidence supporting that alternative splicing in the extracellular region of adhesion molecules expands the diversity of possible adhesion complex compositions by changing their affinities for different interactors. Neurexin–Lphn1 binding occurs in a *trans*-configuration, which indicates a postsynaptic Lphn localization. While the functional significance of Lphn–Teneurin and Lphn–FLRT interaction has been studied, the role of Lphn–Nrxn interaction at synapses is currently unclear.

Teneurins interact in a *trans*-configuration with Lphns via their Lectin domain. Structural studies determined that the Teneurin barrel region interacts with the Lphn lectin domain (Li et al. 2018). Interestingly, an alternative splice site proximal to the Lphn Lectin domain has been shown to switch Lphn affinity to Teneurins, with the inclusion of this splice insert reducing the affinity for Teneurin binding by ~five-fold (Boucard et al. 2014). This further supports that alternative mRNA splicing modulates the composition of synaptic cell adhesion molecule complexes by altering the affinity. FLRTs (FLRT1-3) have been shown to interact with the Lphn olfactomedin domain, similar to Nrxns. Thus, Lphns participate in large multiprotein combinatorial adhesion complexes that are likely cell type- and synapse-specific

The GPCR activation mechanisms of Lphns and other aGPCRs remains incompletely understood (Hamann et al. 2015). The most intensively studied Lphn aGPCR

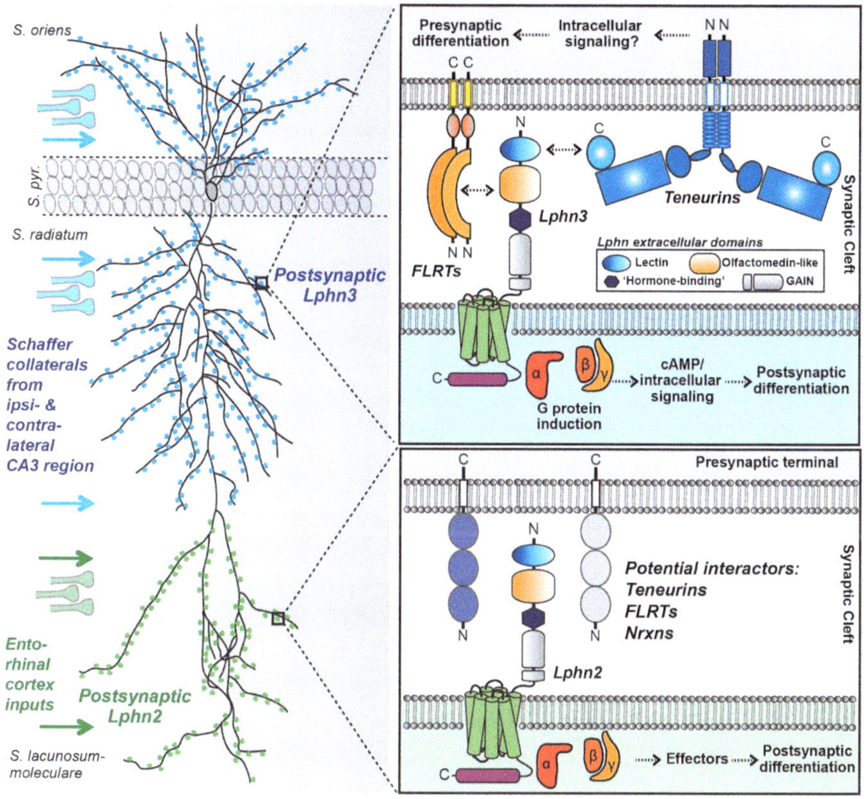

Fig. 5.1 *Model of aGPCR Lphn function at excitatory synapses in hippocampal CA1 pyramidal neurons***.** Two different Lphn paralogs, Lphn2 and Lphn3, are postsynaptic and targeted to distinct dendritic sub-regions of hippocampal CA1 neurons. Lphn3 is restricted to the *stratum oriens* and *radiatum* regions, which receive CA3 inputs via the Schaffer collateral pathway, while Lphn2 is anchored to *stratum lacunosum-moleculare* region dendrites which receive entorhinal cortical inputs. Each Lphn paralog is important for the formation of excitatory inputs into their respective dendritic subregions. Lphn3 function at Schaffer collateral synapses requires simultaneous binding to two unrelated presynaptic interactors, FLRTs and Teneurins. This function also requires GPCR-mediated signal transduction, which likely induces intracellular signaling pathways to organize postsynaptic differentiation. While Teneurin-FLRT presynaptic interactors likely also induce concomitant presynaptic signaling to orchestrate presynaptic differentiation, these signaling mechanisms remain to be explored. For Lphn2 at *stratum lacunosum-moleculare region* synapses, while evidence implicates a potential role of FLRTs in Lphn2 function in certain contexts, the precise interactors involved in its postsynaptic role remain unresolved. Moreover, the function of a third putative presynaptic Lphn ligand Nrxns in this complex in vivo remains unexplored. Similar to Lphn3, postsynaptic Lphn2 likely regulates compartmentalized GPCR signaling in nascent perforant pathway spines, although the signaling pathways and whether they differ from Lphn3 are unclear. (Figure modified from Sando et al. 2019)

induction paradigms involve the GAIN (GPCR Autoproteolysis-Inducing Domain) domain and tethered agonism (Lala and Hall 2022; Vizurraga et al. 2020). The GAIN domain of Lphns and several other aGPCRs exhibit an autoproteolytic cleavage site which is constitutively cleaved during protein trafficking prior to Lphn reaching the plasma membrane surface (Araç et al. 2012). Once at the surface, the N-terminal fragment remains associated with the Lphn C-terminal fragment. Autoproteolytic cleavage in the GAIN generates a membrane-proximal tethered agonist (TA), which can activate the 7-TM GPCR to transduce G protein-mediated signaling. Currently, there are two primary models for how this could occur. The first involves the removal of the N-terminal fragment, which thereby exposes the TA to induce signaling. The second is a more tunable model in which the NTF remains associated and induces TA-mediated signaling via conformational changes (Purcell and Hall 2018). Studies have shown that deletion of the Lphn N-terminal fragment generates constitutive GPCR activity. Moreover, inducible approaches have been utilized based on the replacement of the endogenous NTF with an experimentally cleavable sequence, such as a PAR1 recognition site or enterokinase cleavage site (Mathiasen et al. 2020; Lizano et al. 2021; Bui et al. 2023). These strategies allow for acute treatment with cleavage enzymes, which exposes the TA and activates Lphn-dependent G protein coupling. While these strategies support the model that the removal and exposure of the TA can increase GPCR activation, several lines of evidence support that aGPCR signaling may also involve TA exposure-independent and TA-independent mechanisms. For example, a Lphn3 point mutant that does not undergo autoproteolysis is still fully capable of rescuing deficits in excitatory synapse formation in Lphn3-deficient neurons, suggesting that the full-length uncleaved receptor can fulfill Lphn3 function at synapses (Sando et al. 2019). Importantly, recent structural studies of the Lphn3 holoreceptor consisting of the GAIN-7TM supports interplay between the extracellular region and 7TM that modulates signaling (Kordon et al. 2024). Thus, multiple induction mechanisms exist for aGPCRs, and direct mechanical binding, such as during synapse formation, may also influence GPCR signaling at nascent synapses (Lala and Hall 2022; Vizurraga et al. 2020).

Lphns have been shown to modulate a number of different signal transduction pathways and G proteins. *C. elegans* LAT-1 has displayed transduction of the $G\alpha_s$ pathway, while *Drosophila* ortholog CIRL (calcium-independent receptor for latrotoxin) interacts with a $G\alpha_i$ pathway (Müller et al. 2015; Scholz et al. 2017). Mammalian Lphn1 has displayed interaction with $G\alpha_s$, $G\alpha_i$, $G\alpha_o$, and $G\alpha_q$ pathways (Ovando-Zambrano et al. 2019; Nazarko et al. 2018; Lelianova et al. 1997). Lphn2 is currently less examined but has exhibited coupling to $G\alpha_s$ and $G\alpha_{12/13}$ (Camillo et al. 2021; Sando and Südhof 2021; Pederick et al. 2023). Lphn3 has shown several G protein interactions, including $G\alpha_s$, $G\alpha_i$, $G\alpha_q$, and $G\alpha_{12/13}$ (Qian et al. 2022; Barros-Alvarez et al. 2022; Sando and Südhof 2021; Nazarko et al. 2018; Moreno-Salinas et al. 2022; Mathiasen et al. 2020). As with other GPCRs, which G proteins are transduced may be context-dependent or biased based on the specific ligand or mode of activation.

The C-terminal tail of Lphns is an extensive intracellular sequence that harbors a very C-terminal PDZ binding motif (TSL-COOH). This PDZ-binding motif binds to the postsynaptic PDZ domain protein SHANK (Tobaben et al. 2000). SHANKs are postsynaptic scaffolding proteins, which further support a postsynaptic Lphn localization. This interaction is likely important for anchoring Lphns at postsynaptic sites and/or recruiting additional postsynaptic machinery to nascent synapses. Lphn3 rescue experiments found that the presence of the C-terminal tail was essential for its role in excitatory synapse formation (Sando and Südhof 2021). Given its extensive size (~350 amino acids for LPHN3), this C-terminal tail may serve as a scaffold for various other protein–protein interactions and may help anchor different Lphns at different synapses. However, the precise composition of these interactions at synapses, together with putative Lphn–Arrestin interactions in the C-terminus, remains unresolved.

Functionally, increasing evidence supports that Lphns and their interaction partners mediate synapse formation and circuit wiring. Lphn2 loss-of-function studies revealed that Lphn2 acts as a postsynaptic target recognition receptor for incoming entorhinal cortical axons onto the *stratum lacunosum-moleculare* region of CA1 pyramidal neurons (Anderson et al. 2017). Cultured hippocampal neurons lacking Lphn2 displayed a ~ 50% decrease in excitatory synapse density with no alterations in inhibitory synapses. Considering the lack of reliable antibodies to detect endogenous Lphns, tagged knock-in mice have been instrumental toward detecting endogenous Lphns in the brain. These genetic tools revealed that Lphn2 is postsynaptic in the hippocampal CA1 region and targeted specifically to the *stratum lacunosum-moleculare* region of the CA1 pyramidal neuron dendritic arbor (Anderson et al. 2017). This subcellular dendritic domain receives presynaptic inputs from the entorhinal cortical pyramidal neurons. In vivo loss-of-function studies found that postsynaptic Lphn2 is essential for the formation of these inputs into the CA1 region. Thus, Lphn2 functions as a postsynaptic adhesion receptor for entorhinal cortical–hippocampal synapse assembly (Anderson et al. 2017). Ten3 was found to be expressed in the complementary regions of hippocampal networks relative to Lphn2, suggesting topographical expression gradients of Teneurin/Lphn may help guide their functions (Pederick et al. 2021). Further studies focused on the parahippocampal circuit determined that Lphn2 is located in the interconnected regions of the medial entorhinal cortex (MEC), presubiculum (PrS), and parasubiculum (PaS) and selectively regulates MEC inputs arising from the ipsilateral PrS (Donohue et al. 2021). These results further support the fundamental role of Lphn2 as an adhesion receptor in synaptic circuit assembly.

As compared to Lphn2, studies of Lphn3 in the hippocampus revealed that this paralog localizes to non-overlapping, complementary dendritic subregions of CA1 pyramidal neurons, where it selectively directs the formation of inputs into these subregions (Sando et al. 2019). Again, employing tag knock-in Lphn3 mouse models to detect endogenous Lphn3, these experiments found that Lphn3 is postsynaptic and selectively targeted to the *stratum oriens* and *stratum radiatum* of CA1 pyramidal neurons. These regions receive inputs from the CA3 Schaffer collateral pathway, distinct from the perforant pathway inputs into the *stratum lacunosum-moleculare*. Similar to Lphn2, Lphn3 deletion in hippocampal neurons resulted in ~50% decrease in excitatory synapse density with no observable changes in inhibitory synapses. In vivo, postsynaptic Lphn3 mediates the formation of excitatory Schaffer collateral inputs into these specific dendritic compartments of CA1 pyramidal neurons. Thus, the adhesion receptors Lphn2 and Lphn3 selectively direct the formation of perforant pathway and Schaffer collateral pathway synapses, respectively, onto hippocampal CA1 neurons. These findings revealed that different paralogs of a postsynaptic adhesion receptor can be targeted to distinct dendritic subdomains where they function in different synapses within the same neuron.

Mechanistically, the role of Lphn3 in hippocampal synapse formation requires binding to both Teneurins in the Lphn lectin domain and FLRTs in the Lphn olfactomedin domain (Sando et al. 2019). These studies support a coincidence detection mechanism for synapse specificity controlled by combinations of adhesion molecules. Coincidence detection by synaptic cell adhesion molecule complexes can help explain the specificity and diversity of synaptic connectivity in circuits. Input-specific synapse formation via extracellular combinatorial adhesion interactions also supports the notion of molecular codes directing synapse formation. Therefore, combinatorial codes are a critical component of the mechanisms that can help explain how trillions of connections are made with specificity in the mammalian brain.

Further rescue experiments revealed that Lphn3 function in hippocampal circuit assembly requires GPCR function as well as the C-terminal intracellular tail region (Sando and Südhof 2021). This supports that Lphns function as both adhesion molecules and signaling receptors at synapses. While several adhesion molecules important for synapse formation and function have been identified, the intracellular signaling pathways emanating from these receptors remain relatively poorly understood. Further research is required to elucidate the complex intracellular signal transduction mechanisms that drive the synapse assembly. One model is that nascent CA3-CA1 circuits contain Lphn3 targeted intracellularly to the *stratum radiatum* region, while Lphn2 is targeted to the *stratum lacunosum-moleculare* (Fig. 5.1). While intracellular tethering to differential sets of postsynaptic scaffolds is a potential scenario, the precise sets of interaction partners that differentially target Lphn3 vs. Lphn2 remain unclear. Incoming axons from either the CA3 Schaffer collateral or Entorhinal cortex perforant pathway then target their respective subregions. For Schaffer collateral inputs, presynaptic Teneurin and FLRT specifically recognize postsynaptic Lphn3, which transduces signals that guide the postsynaptic assembly. Trans-synaptic binding to both Teneurin and FLRT is essential, supporting a coincidence detection mechanism. The Lphn3–FLRT–Teneurin trans-synaptic complex nucleates a trans-cellular signaling apparatus that involves Lphn3 GPCR signaling in the nascent postsynaptic compartment. The corresponding presynaptic signaling component remains unclear but may involve Teneurin-mediated signaling (Bagutti et al. 2003).

Lphns exhibit context-dependent functions in different circuits. For example, in the cerebellar climbing fiber-to-Purkinje cell and parallel fiber-to-Purkinje cell system, Lphn2 and Lphn3 are redundantly important for parallel fiber-specific synapse function (Zhang et al. 2020). This redundant function in the cerebellar system is distinct from the roles of Lphn2 and Lphn3 in the hippocampal CA1 where they are individually important for different synaptic inputs onto the same postsynaptic cell. Contrary to Lphn2 and Lphn3, Lphn1 is dispensable for excitatory synapse formation but is important for perisomatic inhibitory synapse assembly in hippocampal neurons (Matúš et al. 2024). Lphns have also displayed distinct roles in the mammalian retina. Lphn3 was identified in horizontal cell dendrites where it engages a trans-synaptic mechanism that selectively controls cone synapse function (Wang et al. 2021a, 2021b). Interestingly, this function was dependent upon alternative splicing in a region that negatively modulates trans-synaptic interactions. These studies support context-specific functions in specific circuits.

A multitude of unanswered questions remain regarding Lphns and the Lphn–FLRT–Teneurin complex. First, the precise roles of this ternary complex in synapse formation vs. elimination remains uncertain. Addressing this question will likely require new live imaging approaches to help disentangle these processes. Lphn signaling dynamics during synapse assembly and aGPCR signaling dynamics, in general, are poorly characterized. The presynaptic signaling components of this complex are unclear. Future studies are required to determine these aspects of Lphn–Teneurin–FLRT signaling. Extensive human genetics studies have linked Lphn3 to ADHD, and several constitutive Lphn3 KO models ranging from *Drosophila*, zebrafish, to rodents display hyperactivity, which supports the robustness of this effect (Arcos-Burgos et al. 2010; Bruxel et al. 2021; Domené et al. 2011; Huang et al. 2019; Hwang et al. 2015; Regan et al. 2021). While this is likely due to deficits in striatal circuit connectivity and function, the mechanistic basis of the role of Lphn3 in this disease remains unknown.

5.3.3 Latrophilin Adhesion Complex—Teneurins

Teneurins encode large single-pass type II transmembrane proteins with an unusual extracellular region displaying a toxin-like domain conserved with bacterial Tc-toxins, a large cylindrical barrel domain, beta-propeller domain, Ig-like domain, and EGF repeats that mediate *cis*-dimerization. As type II transmembrane proteins, Teneurins exhibit an ~350-residue intracellular N-terminal region and ~ 2400-residue extracellular C-terminal region (Woelfle et al. 2016). Evidence suggests that the Teneurin intracellular region may undergo proteolytic processing and translocate to the nucleus to regulate the transcriptional activity (Bagutti et al. 2003; Kenzelmann et al. 2008). This raises the possibility that the Lphn–FLRT–Teneurin complex signals in a bidirectional manner, with Lphns modulating postsynaptic GPCR signaling and Teneurins regulating presynaptic signaling.

Teneurins were originally identified as pair rule genes in *Drosophila* and termed Oz (odz). Pair rule genes are involved in the segmentation of insect embryos. As most pair rule genes encode transcription factors, Oz was unusual in its role as a transmembrane adhesion molecule with a pair gene function (Levine et al. 1994). Teneurins are highly conserved and present throughout multicellular organisms. Since their discovery, increasing evidence supports that Teneurins have roles in nervous system wiring in a range of systems, from *Drosophila*, to *C. elegans*, to mice. In *Drosophila*, Ten-m and Ten-a have functions in synaptic partner matching between the axons of defined classes of olfactory sensory neurons connecting to postsynaptic projection neurons (Hong et al. 2012). *Drosophila* Ten-m and Ten-a have shown a similar target selection function in the neuromuscular junction and important roles in visual system connectivity (Mosca et al. 2012; Hunyara et al. 2023).

Mammals contain four Teneurin paralogs (Ten1-4). Ten-3 displays expression in topographically interconnected areas of the parahippocampal circuit. Loss of Ten-3 from these interconnected areas results in aberrant axonal targeting to the respective target areas, supporting a role in circuit assembly (Berns et al. 2018). Super-resolution microscopy revealed that Ten-3 localizes to presynaptic nanoclusters present within most excitatory synapses of the hippocampus (Zhang et al. 2022). Presynaptic Ten-3 and Ten-4 deletion from the medial entorhinal cortex impaired the synapse assembly in the hippocampal CA1, where postsynaptic Ten-3/Ten-4 deletion had no effect. Collectively, these studies support that presynaptic Teneurins underlie circuit wiring specificity and synapse formation via trans-interaction within a heterophilic adhesion complex, likely involving postsynaptic Lphns (Zhang et al. 2022).

In addition to the roles of full-length Teneurins at synapses, studies have shown that Teneurins contain a C-terminal sequence termed TCAP (Teneurin C-terminal-associated peptide) that may work independently from Teneurins via proteolytic processing, perhaps at a furin-like site, or by transcription at an independent transcriptional start site in the Teneurin last exon (Dodsworth & Lovejoy 2022, Woelfle et al. 2016). TCAP is a 40- to 41-residue peptide that interacts with Lphns (Husić et al. 2019), suggesting it may have a role in parallel or in competition with Teneurins. Future work is required to dissect the function of TCAP at synapses and its interplay with full-length Teneurin and Lphns.

5.3.4 Latrophilin Adhesion Complex—FLRTs (Fibronectin Leucine Rich Repeat Transmembrane Proteins)

Mammals encode three fibronectin leucine-rich repeat transmembrane (FLRT) proteins, FLRT1-3, that contain a single-pass transmembrane, ten extracellular leucine-rich repeat (LRR) domains, and a membrane proximal fibronectin type 3 (FN3) domain. FLRTs were identified as interaction partners for Unc5 proteins (Karaulanov et al. 2009; Söllner and Wright 2009), which regulate axon repulsion via Netrin-1 (Yamagishi et al. 2011). Correspondingly, FLRT2 and FLRT3 have shown to act in the repulsion of axons and soma of Unc5-positive neurons in the developing mammalian cortex.

FLRT3 was initially found as an interactor for Lphn3 at synapses using proteomics approaches (O'Sullivan et al. 2012). Depletion of FLRT3 in hippocampal neurons reduced the excitatory synapse density, supporting a role in excitatory synapse formation (O'Sullivan et al. 2012). FLRT3 binds to the olfactomedin domain of Lphn3 in a *trans*-configuration, where it forms a ternary complex involving FLRT/Teneurin/Lphn (Lu et al. 2015). Interestingly, FLRTs can also interact in *trans* with Unc5 proteins, suggesting a putative role of these proteins in the FLRT/Teneurin/Lphn complex (Karaulanov et al. 2009; Yamagishi et al. 2011). Future studies are required to explore the interplay between these molecules and pathways during the circuit assembly.

5.3.5 aGPCR CELSRs (Cadherin EGF LAG Seven Transmembrane Receptors) and Cadherins

The cadherin family of adhesion proteins is evolutionarily ancient and likely played critical roles in the generation of multicellular organisms (Friedman et al. 2015; Honig and Shapiro 2020). The superfamily became extensive and diverse in vertebrates, composed of type I–IV cadherins, clustered/nonclustered protocadherins, the aGPCR Flamingo/CELSRs, desmocollins, desmogelins, Ret, Dachsous/Fat, calsyntenins, and primitive cadherins. The proliferation of cadherins in vertebrates brought highly specialized functions, including roles at synapses which are briefly outlined below (Brusés 2006, Bamji 2005, Basu et al. 2015, Ranscht 2000, Shapiro & Colman 1999).

Like Latrophilins, CELSR aGPCRs are highly conserved from invertebrate to vertebrates, suggesting they underlie evolutionarily ancient biological functions (Boutin et al. 2012; Goffinet and Tissir 2017). Mammalian CELSRs (CELSR1-3) exhibit an extensive N-terminal extracellular region which harbors nine cadherin repeats, followed by EGF-like repeats, laminin G-like and laminin EGF-like domains, a hormone-binding domain, followed by a GAIN domain. Similar to LPHNs, CELSRs also exhibit an intracellular C-terminal tail. Notably, CELSRs display an atypical GAIN region, and studies found that mouse CELSR1-3 display differential autoproteolysis profiles and signaling mechanisms (Bui et al. 2023). The CELSR2 GAIN domain contains several canonical residues known to exhibit efficient autoproteolysis and correspondingly is cleaved at this site. However, CELSR1 and CELSR3 have atypical GAINs lacking these residues and thereby lack GAIN cleavage, exemplifying a subset of aGPCRs that do not undergo autoproteolysis and remain full length receptors. CELSR1-3 all engage Gα_s, yet through different mechanisms. While CELSR2 requires several canonical TA residues for Gα_s induction, it did not exhibit TA-exposed induction, suggesting that it requires N-terminal adhesion sequences upstream of the GAIN. In contrast, CELSR1 and CELSR3 lacked TA-dependent transduction altogether and maintained Gα_s transduction capabilities despite mutations in essential TA residues (Bui et al. 2023). These results align with recent screens suggesting that approximately half of the aGPCRs lack TA-dependent transduction, suggesting several TA-independent mechanisms of aGPCR activation may be at play (Dates et al. 2024). For example, the push and pull of adhesion may result in conformational changes that directly modulate aGPCR activation for certain aGPCRs lacking TA-exposure-dependent or TA-dependent G protein induction, such as CELSRs. This model is supported by recent evidence revealing interplay between the NTF and 7TM in the structure of an aGPCR holoreceptor (Kordon et al. 2024). These studies help inform how aGPCRs critical for synapse formation and circuit assembly function coordinate adhesion with signaling.

Initial studies into CELSR function found that the *Drosophila* CELSR ortholog Flamingo (also known as *Starry night—stan*) was a central component of the planar cell polarity (PCP) pathway (Chae et al. 1999; Lu et al. 1999; Usui et al. 1999). Since these initial results, extensive literature has illustrated CELSRs' role in the core PCP pathway. Studies of *stan* in *Drosophila* have shown a critical role in the stereotyped development of sensory bristles and wing hairs (Formstone 2010; Devenport and Fuchs 2008). As part of the PCP pathway, CELSRs interact genetically with core PCP components including Frizzled, Van Gogh, Dishevelled, Prickle, and Diego (Boutin et al. 2012; Tissir and Goffinet 2013; Yates et al. 2010). Developmentally regulated spatial organization of these components helps guide the establishment of polarity in nascent tissues (Shi 2023). While CELSRs interact genetically with core PCP proteins, their direct adhesion partners remain less characterized. Evidence suggests that the 7-TM region of CELSRs may interact with Frizzled, which is another GPCR that binds Wnt to regulate Wnt signaling, thus potentially forming GPCR heterodimers (Wang et al. 2014). As with other cadherin-containing proteins, the N-terminal cadherin repeats of CELSRs are likely involved with homophilic CELSR interactions (Basta et al. 2022), and recent structural studies have helped to begin resolving this question for CELSR1 (Nishiguchi et al. 2023; Tamilselvan and Sotomayor 2024; Bandekar et al. 2024). However, given the CELSR's extensive size (~300 kD) and multidomain architecture, it likely participates in several currently undiscovered interactions.

Functionally in neural circuits, seminal studies found that Flamingo mutants exhibited disruptions in *Drosophila* visual system topographic maps, suggesting a role in neuronal target selection during circuit assembly (Lee et al. 2003; Senti et al. 2003). The single *C. elegans* CELSR ortholog, Flamingo (cFMI-1), underlies a crucial role in axonal pathfinding and neuronal development (Loveless and Hardin 2012; Steimel et al. 2010). cFMI-1 mutants exhibited substantial deficits in GABAergic motor neuron axonal pathfinding, decreased synapse density, and accumulation of synaptic vesicles at nonsynaptic regions (Najarro et al. 2012). cFMI-1 has been shown to genetically interact with the Wnt signaling pathway for proper neurite targeting functions (Ackley 2014; Huarcaya Najarro & Ackley 2013). Further invertebrate studies showed that *Drosophila* Flamingo functions as a short-range signal to influence the growth cone choice of correct postsynaptic partners, further supporting a role in circuit wiring during development (Chen and Clandinin 2008).

Previous characterization of several CELSR loss-of-function mouse models also supports important roles in neurodevelopment (Berger-Müller and Suzuki 2011; Qu et al. 2010). ENU mutagenesis screens found two CELSR1 mutants, *Crash* and *Spin Cycle*, which have major developmental anomalies (Curtin et al. 2003). CELSR1 *Crash* mice contain a single point mutation in the eighth cadherin repeat, while CELSR1 *Spin Cycle* mice harbor a single point mutation between cadherin repeats 8 and 9. Mice heterozygous for these mutations display head-shaking behavior and altered hair cilia bundles in inner ear cells due to deficits in the PCP pathway. Homozygous constitutive CELSR1 KO mice are mostly viable but sterile and display deficits in neural tube closure that lead to embryonic lethality in a significant fraction of mice. Interestingly, constitutive CELSR3 KO mice die shortly after birth and show major deficits in axonal patterning (Tissir et al. 2005; Zhou et al. 2008). Despite these intriguing studies in mammals, most previous works have been done on constitutive CELSR KO models or with the heterozygous point mutations described above. Therefore, cell type-specific function using conditional deletions, as well as examination of potential redundancy between mammalian CELSR paralogs in a cell type-specific manner, remains less explored.

While the role of CELSR3 has been examined in the context of early axonal targeting, CELSRs' role during postnatal synaptogenesis remains less explored. Interestingly, CELSR3 and other PCP components have also been shown to regulate glutamatergic synapse density in mice (Ban et al. 2021; Thakar et al. 2017). Several CELSR3 missense mutations are associated with seizures and epilepsy (Li et al. 2022). One CELSR3 missense mutation in epilepsy is within intracellular loop 3 (ICL3) of the 7-transmembrane GPCR, a region critical for G protein coupling, and two different mutations are present within the extracellular region. These studies support that alterations in CELSR3 adhesion or signaling functions contribute to disease.

Aside from aGPCR CELSRs, other cadherins (calcium-dependent adherent proteins) are important during neural development and participate in collaborative extracellular interactions to modulate synaptic architecture, signaling, and plasticity (Nern et al. 2008; Pecot et al. 2013; Kurusu et al. 2012). More generally, cadherins play critical roles in cellular adhesion and tissue morphogenesis across various systems and have been described in several detailed and comprehensive reviews (e.g. Hulpiau and van Roy 2009; Yagi and Takeichi 2000). Interestingly, studies have shown that N-cadherin is initially localized throughout the synaptic cleft but becomes sequestered into discrete clusters at the periphery of the synapse as synapses mature (Elste and Benson 2006; Uchida et al. 1996; Yamagata et al. 1995). This raises the possibility that these classical Cadherins may participate in the nano-organization of synapses during development.

5.3.6 aGPCR BAIs (Brain Angiogenesis Inhibitors)

Mammalian BAI1-3 (ADGRB1-3) are aGPCRs comprised of an N-terminal extracellular region containing several thrombospondin type 1 repeats (TSRs) involved in cellular adhesion, a hormone-binding domain, and a GAIN domain, followed by the 7-transmembrane GPCR and intracellular C-terminal tail possessing a PDZ-binding motif. BAI3, a predominant paralog in the brain, also displays a very N-terminal CUB domain. BAI1 was originally identified in a screen for p53-inducible genes and inhibits neovascularization (Nishimori et al. 1997; Duda et al. 2002; Kaur et al. 2005; Kang et al. 2006; Kudo et al. 2007). BAI2 and BAI3 do not display p53-dependent induction but have also been shown to inhibit angiogenesis (Shiratsuchi et al. 1997; Kee et al. 2002; Kee et al. 2004; Jeong et al. 2006). Correspondingly, BAIs have been implicated in tumorigenesis and cancer (Nair et al. 2021; Mathema and Na-Bangchang 2017; Cork and Van Meir 2011).

In addition to BAIs' roles in angiogenesis, neovascularization, and tumorigenesis, increasing evidence also supports a crucial role in neural circuit development. BAI1 and BAI3 have shown enrichment at the postsynaptic density, suggesting a role as postsynaptic receptors. Constitutive BAI1 KO mice display alterations in the postsynaptic ultrastructure of CA1 pyramidal neurons and deficits in spatial learning (Zhu et al. 2015). BAI3 is present at excitatory synapses in the brain where it localizes to dendritic spines. BAI2 appears broadly expressed throughout the brain, but its function and localization at synapses remain understudied compared to other paralogs (Duman et al. 2016).

C1ql proteins have been identified as extracellular interaction partners for BAI3, and this interaction is important for its role in circuit assembly (Martinelli et al. 2016). C1ql1-4 are part of the secreted C1q family that includes C1ql and Cbln (precerebellins) (Iijima et al. 2010). In the cerebellum, climbing fiber connections onto Purkinje cells display a "single-winner" mechanism during development, where a subset of connections "win" and are maintained, while the rest are pruned. C1ql1 was found as an important anterograde signal to maintain single-winner climbing fiber connectivity via interaction with postsynaptic BAI3 on Purkinje cells (Sigoillot et al. 2015; Kakegawa et al. 2015; Aimi et al. 2023). C1ql1 was reported to interact with BAI3 via the BAI3 N-terminal CUB domain, while C1ql3 was shown to interact with the TSRs of BAI3 (Bolliger et al. 2011; Park et al. 2018). This study also found that all four C1ql proteins are capable of interacting with the TSRs of BAI3 via the globular C1q domains of C1ql.

5.3.7 Neurexins and Neuroligins

Work over the past several decades reveal that neurexins (Nrxns) are presynaptic adhesion molecules that nucleate a trans-synaptic adhesion network that controls the functional specification of synapses. Three Nrxn genes (Nrxn1-3) encode both long α-neurexins and short β-neurexins (and Nrxn1 specific γ-neurexin) and are subject to extensive alternative splicing at six canonical splice sites. Nrxns bind an array of distinct postsynaptic partners including Nlgn, LRRTM, Lphn, and Cblns to regulate synaptic organization and function (Varoqueax et al. 2006). Given recent reviews have comprehensively summarized neurexins and their postsynaptic ligands, we will only briefly describe certain aspects of the neurexin function to avoid redundancy (Südhof 2017).

Neurexins (Nrxn1-3) are extensively alternatively spliced presynaptic CAMs. Alternative splice isoforms help generate molecular diversity that encodes specific functional properties. While several postsynaptic Nrxn ligands are now known, the first discovered Nrxn interactor was postsynaptic neuroligins (Nlgn1-4) (Ichtchenko et al. 1998). One of the features initially documented for Nrxns and Nlgns was their ability to induce synapse formation onto non-neuronal cells when expressed on the cell surface. For example, overexpression of neuroligin in non-neuronal cells, such as HEK293T or COS-7 cells, was found to be sufficient to induce presynaptic organization onto the heterologous cell (Gokce & Südhof 2013, Scheiffele et al. 2000). Likewise, overexpression of neurexin in heterologous cells induces postsynaptic induction onto that cell (Graf et al. 2004). While surprising, previous pioneering studies found that poly-lysine-coated beads alone are sufficient to induce synapse formation (Burry 1982; Burry and Hayes 1986). Over time, many cell adhesion molecules were shown to exhibit the ability to induce synapse formation when overexpressed in heterologous cells, a property also known as artificial synapse formation. In fact, over 20 different adhesion molecules from an array of gene families have exhibited this property. While seemingly nonspecific, many adhesion molecules display directionality in artificial synapse induction (i.e., a presynaptic molecule will induce postsynaptic specializations and vice versa). Moreover, these observations highlight that at least in certain scenarios, pre- and postsynaptic specializations can assemble independently of one another, which has also been observed in invertebrates (Kurshan and Shen 2019).

The observation that many adhesion molecules that lack any sequence homology can all induce pre- or postsynaptic differentiation suggests an important role of adhesion in synapse formation. Indeed, studies which engineered novel synaptic adhesion molecules support this conclusion (Hale et al. 2023). These studies utilized the extremely high affinity between bacterial Barnstar and Barnase, which exhibits one of the highest affinity interactions known. When used to replace the extracellular region of neurexin or neuroligin, these fusions, referred to as Barnoligin (Barnase/Neuroligin fusion) and Starexin (Barnstar/Neurexin fusion) generate a high-affinity trans-cellular adhesion complex. Importantly, because Barnstar/Barnase have no orthologous sequence in mammals, this system removes all endogenous neurexin–neuroligin extracellular interactions and tests for nonselective extracellular adhesion. However, Barnstar/Barnase were still capable of inducing synaptic differentiation in heterologous cells (Hale et al. 2023). These studies highlight the importance of cellular adhesion in synapse formation.

Furthermore, these studies highlight the potential utility of engineering molecular tools based on synaptic adhesion molecules to probe or alter circuit function. Recent studies developed a novel synthetic adhesion protein that was capable of altering synaptic function and restoring a number of disease relevant deficits. This synthetic adhesion molecule, termed CPTX, is a fusion between components of two structurally unrelated endogenous adhesion molecules, Neuronal Pentraxin 1 (NP1) and Cerebellin-1 (Cbln1), and capitalizes upon two distinct features of these molecules (Suzuki et al. 2020). CPTX contains the N-terminal CRR domains of Cbln1 to enable binding to presynaptic Nrxn and induction of presynaptic differentiation but lacks the C-terminal gCLq region of native Cbln1 which binds to the non-ionotropic glutamate receptor GluD2. Instead, CPTX contains the C-terminal PTX domain of NP1, which interacts with AMPARs. This fusion enables both presynaptic induction via Cbln1 features as well as recruitment of functional postsynaptic apparatus via NP1 features. By functioning as a synthetic bridge across the excitatory synapse, CPTX was shown to rescue numerous disease-relevant deficits in excitatory function in mouse models. These approaches highlight the power of utilizing novel adhesion systems to manipulate circuit development and function. Importantly, as evidenced by the structure-guided design of CPTX, understanding the fundamental biological functions of these synaptic adhesion systems is essential prior to using this knowledge to design novel synaptic regulators.

5.3.8 LRR (Leucine-Rich Repeat) Proteins

Aside from FLRTs, numerous LRR-containing proteins have emerged as critical organizers of synapse development and function (de Wit et al. 2011). Other examples include LRRTMs, Slitrks, ELFN1, and NGLs. One extensively studied example includes the Leucine-rich repeat transmembrane neuronal proteins (LRRTM1-4), which are type I transmembrane proteins with extracellular LRR domains and intracellular PDZ-binding motifs (Ko 2012). Postsynaptic LRRTMs have been shown to interact with Nrxns in a Nrxn SS4-dependent manner (Aoto et al. 2013) and recruit AMPARs to the postsynaptic membrane, thereby supporting LTP (Roppongi et al. 2017, Bhouri et al. 2018).

ELFN1 (Extracellular Leucine-Rich Repeat and Fibronectin Type III Domain Containing 1), which contains extracellular LRR repeats, a single-pass transmembrane domain, and a cytoplasmic tail, is expressed in oriens-lacunosum moleculare interneurons (OLM), a class of somatostatin-positive hippocampal interneurons that reside in the *stratum oriens* of the hippocampal CA1 region yet project specifically to the dendritic subdomain of the *stratum lacunosum-moleculare* (Sylwestrak and Ghosh 2012). OLM interneurons receive excitatory input from CA1 pyramidal neurons, and these synapses contain ELFN1. CA1 neurons also project to PV-positive neurons which lack ELFN1. These two inputs onto interneurons formed by CA1 pyramidal neurons have distinct presynaptic release probabilities, with CA1–OLM synapses being facilitating while CA1–PV synapses depressing. Reduction of ELFN1 in OLM interneurons increases the release probability of the CA1–OLM synapse, suggesting that ELFN1 is involved in specifying this synapse-specific functional parameter (Sylwestrak and Ghosh 2012).

Different gene families of LRR containing proteins may play synapse-specific modular roles in a given neural circuit. For example, three classes of LRR-containing adhesion molecules, FLRTs, LRRTMs and Slitrks, were found to contribute to input-specific synapse properties in the hippocampal CA1 region (Schroeder et al.

2018). FLRTs regulated synapse density, consistent with their roles in excitatory synapse formation, while LRRTM1 and Slitrk organized presynaptic vesicle pools. These results support that partially overlapping modules of cell adhesion molecules are differentially targeted to synapse-specific inputs via their trans-synaptic binding partners and/or intracellular scaffolding interactions, where they contribute to context-specific functions in synapse formation or the functional parameters of synapses.

5.3.9 SynCAM

SynCAM1-4 exhibit three extracellular immunoglobulin (Ig) domains, a single-pass transmembrane sequence, followed by an intracellular tail with a C-terminal PDZ binding motif (Biederer et al. 2002). SynCAM localizes to synapses, where it was initially reported to be both pre- and postsynaptic via homophilic adhesion. SynCAM has important roles in synapse assembly and induces presynaptic specializations in the artificial synapse formation assay in a manner dependent upon its Ig domains (Biederer et al. 2002). Presynaptic specializations induced by SynCAM were functional and indistinguishable to terminals induced via Nlgn (Sara et al. 2005). Further studies found that while SynCAM1-4 can interact in a homophilic manner with the same paralog, they preferentially interact in a heterophilic manner with different paralogs (Fogel et al. 2007). Moreover, the same paralogs can form *cis* oligomers that increase adhesive capacity (Fogel et al. 2011). Interestingly, more recent studies found cooperativity between SynCAM and LRRTM1, with double deletion resulting in a more profound loss of dendritic spines compared to single deletions. This effect was observed in the prefrontal cortex but not the hippocampus, implicating context-specific functions of this cooperative interaction (Perez de Arce et al. 2023).

5.4 Nano-organization of Adhesion Complexes at Synapses

5.4.1 Overview of Synaptic Organization

Neurotransmission is intricately organized and occurs at specific regions of the synapse rather than being uniformly distributed across the synaptic terminal. Central to this process are three compartments: the presynaptic active zone, the postsynaptic density (PSD), and the synaptic cleft. Over the past two decades, advancements in cryo-EM and super-resolution microscopy (SRM) have revealed that many critical proteins within each synaptic compartment have distinct subsynaptic distributions and often organize into discrete regions of higher density at the nanometer scale. These nanostructures are commonly referred to as nanoclusters or nanodomains (reviewed in Fukata et al. 2024 and Biederer et al. 2017). For instance, seminal SRM analyses of the PSD in cultured hippocampal neurons demonstrate that AMPA receptors (AMPARs) form 1–3 nanodomains, each containing up to 25 receptors and spanning approximately 80–100 nm (Fukata et al. 2013; MacGillavry et al. 2013; Nair et al. 2013). These AMPAR nanodomains distribute across the synapse but tend to localize toward the periphery of the PSD (Racca et al. 2000; Dani et al. 2010; Chen et al. 2008; Tang et al. 2016). Similarly, NMDA receptors (NMDARs) form 1–3 nanodomains per synapse with a median diameter of 60 nm. NMDARs co-organize with AMPARs within a 300 nm area of the PSD of most excitatory synapses, yet the nanodomains remain spatially segregated, with NMDARs typically occupying more central locations (Racca et al. 2000; Chen et al. 2008; Goncalves et al. 2020; Li et al. 2021). However, the specific organizational properties of NMDARs are subunit-specific and can vary across synapses (Kellermayer et al. 2018; Dani et al. 2010). Postsynaptic scaffolding protein PSD-95 assembles into 1–3 nanodomains, averaging ~80 nm in diameter, that frequently co-organize with AMPAR nanodomains in both hippocampal neuron cultures (MacGillavry et al. 2013; Nair et al. 2013; Tang et al. 2016) and tissue samples (Fukata et al. 2013; Broadhead et al. 2016; Tang et al. 2016). The presynaptic active zone also displays complex nano-organization. For example, presynaptic vesicle docking and priming protein Munc13-1 forms nanoassemblies of about 45 nm in diameter, and active zone organizing protein RIM forms ~2 high-density nanoassemblies spanning ~80 nm in hippocampal neuron cultures. Interestingly, aspects of both Munc13-1 and RIM nano-organization are correlated with vesicular release, but there is significant variation in their co-organization with respect to one another and other key components of synaptic release machinery (Sakamoto et al. 2018; Tang et al. 2016). Within the synaptic cleft, synaptic cell adhesion molecules also form discrete nanodomains which are thought to modulate the nano-organization and function of synapses—detailed below.

Mathematical modeling suggests that the alignment of presynaptic vesicular release with postsynaptic receptors strongly impacts transmission efficiency and thus synaptic strength (outlined in Biederer et al. 2017). These models are supported by experimental results indicating that not all receptors contribute to a postsynaptic response (Nimchinsky et al. 2004; Liu et al. 1999; McAllister and Stevens 2000), and additional models demonstrate that the efficiency of receptor activation diminishes with increasing distance from the vesicular release site (Raghavachar & Lisman 2004; MacGillavry et al. 2011). Indeed, a transformative study using SRM provided direct evidence that presynaptic release machinery can align with postsynaptic receptors on a nanometer scale. Specifically, presynaptic RIM1 nanodomains align with postsynaptic PSD-95 and AMPAR nanodomains in hippocampal neuron cultures and tissue sections, forming a trans-synaptic "nanocolumn" (Tang et al. 2016). Since this time, nanocolumns have been identified in several species, ranging from mice to *Drosophila* (Cole & Reese 2023, Muttathukunnel et al. 2022), indicating evolutionary conservation. Additionally, nanocolumns have proven to be dynamic structures that change in density and heterogeneity during synapse maturation (Sun et al. 2022) and in response to synaptic activity (Muttathukunnel et al. 2022).

Although important for transmission efficiency, it is not completely understood what regulates nanodomain and nanocolumn formation and retention at the synapse. Several studies suggest clustering of postsynaptic scaffolding proteins, particularly PSD-95, initiates the formation of postsynaptic receptor nanodomains (Chen et al. 2011a, 2011b; Ehrlich et al. 2007; El-Husseini et al. 2000; Elias et al. 2006; Nair et al. 2013), likely through receptor-trapping mechanisms (Opazo and Choquet 2011). If so, what initiates prior clustering of PSD-95 into nanodomains? Evidence indicates that it likely involves reversible palmitoylation of PSD-95 (Fukata et al. 2013). It is also plausible that receptor nanodomains are organized by interactions with proteins other than scaffolding molecules. Glutamate receptors are known to bind in *cis* with ion channels, other receptors, and several cell adhesion molecules. Additionally, the NTD of AMPARs extend into the synaptic cleft (García-Nafría et al. 2016), providing the opportunity for extracellular interactions. Alternatively, postsynaptic receptors themselves could drive the nanodomain organization. The mechanisms of presynaptic nano-organization are even more unclear. Presynaptic active zone proteins RIM, Munc13, and Bassoon organize into nanoclusters near the membrane, yet they lack a transmembrane domain. Thus, presynaptic organization must be mediated by interactions with other transmembrane proteins. It is hypothesized that active zone protein organization may be mediated by interactions with Ca^{2+} channels, which directly bind RIM, (Chen et al. 2011a, 2011b; Hibino et al. 2002; Kaeser et al. 2011), Liprin-α which binds RIM and transmembrane protein LAR (Schoch et al. 2002 ; Sara et al. 2005), or cell adhesion molecules (Luo et al. 2020, Südhof 2018). Finally, what regulates the alignment of pre- and postsynaptic nanodomains across the synapse to form nanocolumns? Increasing evidence suggests alignment is mediated by cell adhesion molecules (Fukata et al. 2024).

Cell adhesion molecules are strong candidates as regulators of nano-organization and nanocolumn dynamics given that they are transmembrane proteins which link the pre- and postsynaptic compartments, have precise nonhomogeneous organization, and/or directly interact with nanocolumn-associated proteins in *cis* and *trans.* It has been hypothesized that cell adhesion molecules may facilitate nanocolumn formation through recruitment of associated proteins, or they may occupy portions of the synapse surrounding nanostructures, resulting in molecular crowding and protein trapping within the nanostructures (Biederer et al. 2017). Here, we will describe nano-organization of cell adhesion molecules at the synapse and discuss evidence suggesting that cell adhesion molecules are critical mediators of pre- and postsynaptic nano-organization.

5.4.2 *Nano-organization of CAMs*

Determining nano-organization of CAMs is a challenging problem due to a lack of reliable antibodies. Additionally, the small size of most CNS synapses renders the resolution of standard light microscopy insufficient to determine if an adhesion molecule is on the pre- or postsynaptic side, especially given their extracellular regions are present within the cleft. It is also difficult to dissect the functional role of CAM nanostructures due to significant redundancy across the numerous isoforms and protein families. The introduction of super-resolution microscopy and advanced genetic and protein engineering approaches has provided significant new opportunities to characterize the nanoscale organization of adhesion complexes at synapses and determine how these complexes organize synaptic junctions to help ensure the fidelity of synaptic communication.

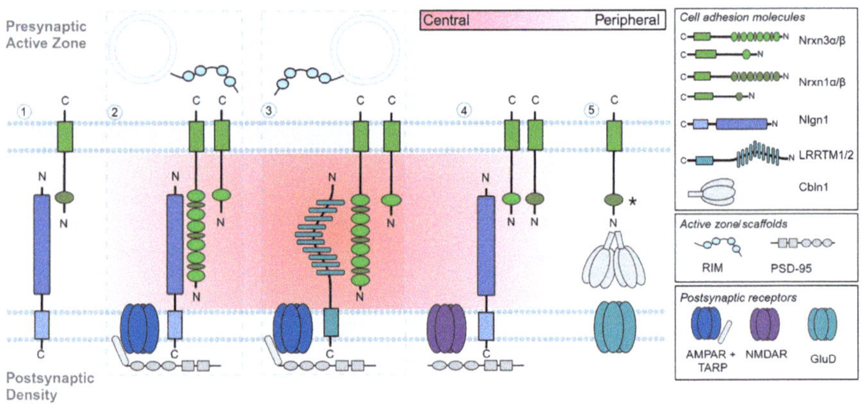

Fig. 5.2 *Current model of the nano-organization of Nrxn, Nlgn, and LRRTM adhesion complexes at excitatory hippocampal synapses***.** (**1**) Nrxn1 localizes toward the periphery of PSD in mature excitatory synapses, and forms ~1 nanodomain per excitatory synapse (Trotter et al. 2019). Nlgn1 distributes throughout the synapse, but forms transcellular complexes with Nrxn1β (Chamma et al. 2016) which are peripherally located. (**2**) Alignment of RIM1/2 and AMPAR nanodomains within nanocolumns is dependent on Nlgn1–PSD-95 interactions (Haas et al. 2018). This study did not specify the specific form of Nrxn involved. However, Nrxn3 nanodomains were shown to associate with nanocolumns, whereas Nrxn1 nanodomains do not (Lloyd et al. 2023). Thus, it is likely that Nlgn binds Nrxn3 in these nanocolumns. (**3**) LRRTM2 forms ~2 centrally located nanodomains and co-organize with RIM and AMPA receptor nanodomains into nanocolumns (Chamma et al. 2016; Ramsey et al. 2021). Centrally localized Nrxn3 is enriched within nanocolumns opposing LRRTM2 (Lloyd et al. 2023). Nrxn-LRRTM1 complexes were found to preferentially regulate nano-organization of AMPARs (Nozawa et al. 2022). (**4**) Nrxn-Nlgn1 complexes preferentially regulate nano-organization of NMDA receptors, and these complexes have a more peripheral localization in some brain regions (Nozawa et al. 2022). This particular study showed that Nrxn3β–Nlgn1 interactions recruit NMDARs, but it is likely that Nrxn1 also binds Nlgn1 to regulate NMDARs given that Nrxn1 is a well-established Nlgn1 ligand, localizes peripherally, and preferentially regulates NMDAR function (Chamma et al. 2016; Trotter et al. 2019; Lloyd et al. 2023; Dai et al. 2019). (**5**) Nrxn1 is peripherally enriched around GluD1 nanodomains (Lloyd et al. 2023) and Nrxn-Cbln1 complexes preferentially regulate nano-organization of GluD receptors (Nozawa et al. 2022). *Nrxnα/β isoforms may be participating in this interaction. Gray shading is indicative of nanodomains. Gray dotted outline is indicative of a nanocolumn. Both Nrxn α and β are shown together if the isoform was not specified. Central (red) and peripheral (white) localizations are estimates based on the literature. (Created in BioRender.) https://BioRender.com/f74n346)

5.4.3 *Neurexins*

As briefly described above, neurexins (Nrxn1-3 α/β) are presynaptic CAMs that undergo extensive alternative splicing and bind an array of postsynaptic ligands to modulate synaptogenesis and synaptic transmission (reviewed in Südhof 2017). Specifically, Nrxn1 and Nrxn3 regulate NMDAR- and AMPAR-mediated responses, respectively, and are involved in trans-synaptic recruitment of these glutamate receptors. This implicates them as synaptic organizers and provides a mechanism by which they modulate synaptic function (Aoto et al. 2013; Dai et al. 2019; Südhof 2017; Südhof 2018). How are Nrxns organized at the synapse? Using STORM, Nrxn1 molecules were found to frequently assemble into a single, discrete nanocluster at the periphery of excitatory pre-synapses in cultured hippocampal neurons (Trotter et al. 2019) (Fig. 5.2(1)). Synapses containing Nrxn1 nanocluster(s) display higher GluA1 AMPAR surface expression and increased vesicular release when compared to synapses lacking Nrxn1 nanoclusters, suggesting an association between increased synaptic activity and the presence of Nrxn1 nanoclusters. However, conditional knockout of Nrxn1 did not have an effect on synapse density or activity, likely due to redundancy with Nrxn2/3 which were also identified in Nrxn1 nanoclusters (Trotter et al. 2019). A similar study using custom Nrxn mouse models and dSTORM found that Nrxn1 and Nrxn3 assemble into distinct nanoclusters in hippocampal excitatory synapses. Endogenous Nrxn3 forms on average two centrally located nanoclusters per excitatory synapse (Fig. 5.2(2) and (3)), often enriched within trans-synaptic nanocolumns opposing LRRTM2. Deletion of Nrxn3 decreases the average number of GluA1 nanodomains as well as receptor density within the nanodomains. Furthermore, Nrxn3 deletion impairs GluA1 alignment with RIM1 nanodomains, which decreases spontaneous and evoked excitatory responses, but does not affect RIM1-PSD-95 alignment. This suggests a trans-synaptic mechanism in which Nrxn3 selectively governs the organization of GluA1 nanodomains and remodeling of RIM1–GluA1 nanocolumns (Lloyd et al. 2023). Conversely, Nrxn1 is enriched at the periphery of the synapse opposing GluD1 nanoclusters (Fig. 5.2(5)) and is not associated with nanocolumns (Lloyd et al. 2023). Together, these analyses illustrate the differential expression and function of Nrxn homologs and exemplify how super-resolution approaches combined with functional analyses such as electrophysiology can be used to determine how Nrxn nano-organization at the synapse affects synaptic organization and function.

5.4.4 *Neurexin-Binding Partners: LRRTMs (Leucine Rich Repeat Transmembrane Proteins), Neuroligins, and Cblns (Cerebellins)*

Neurexin interactors LRRTM and Nlgn bind PSD-95 and have roles in regulating and recruiting glutamate receptors at the PSD (Soler-Llavina et al. 2011; Südhof 2017; Sudhof 2018), making them candidate modulators of formation and/or alignment of nanodomains. Novel SRM approaches in fixed and live hippocampal samples revealed that Nlgn1 and LRRTM2 exhibit distinct spatial distribution at excitatory synapses of hippocampal neuron cultures. LRRTMs form compact, stable nanoclusters that match the size and number of AMPAR nanodomains (Chamma et al. 2016). A separate study using dSTORM confirmed that LRRTM2 nanoclusters co-organize with AMPAR and PSD-95 nanodomains as well as presynaptic RIM1/2 (Ramsey et al. 2021) and Nrxn3 (Lloyd et al. 2023), forming nanocolumns (Fig. 5.2(3)). Acute proteolysis at an engineered cleavage site within the extracellular domain of LRRTM2 resulted in rapid declustering of AMPAR nanodomains within the nanocolumn and diminished evoked, but not spontaneous, excitatory responses (Ramsey et al. 2021). In contrast, Nlgn1 is relatively dispersed at the synapse but does form trans-synaptic complexes with opposing Nrxn1β nanodomains (Fig. 5.2(1)). Chemically induced synaptic depression largely disrupted Nlgn1/Nrxn1β association, indicating that Nlgn1 nanostructures are activity-dependent and dynamic (Chamma et al. 2016). Furthermore, Nlgn1 co-organizes with GluA2-containing AMPAR nanodomains, and transient expression of mutant Nlgn1 lacking a large portion of the C-terminus disrupted this co-organization, likely by preventing PSD-95-mediated scaffolding. Notably, the expression of this mutant also disrupts AMPAR alignment with RIM (Haas et al. 2018), indicating that Nlgn1 is also associated with nanocolumns (Fig. 5.2(2)). Accordingly, disruption of Nlgn1–PSD-95 interactions via acute treatment with Nlgn C-terminal peptide decreased spontaneous and evoked AMPAR-mediated neurotransmission in cultured hippocampal neurons by 30%. Modeling suggests that this disruption was caused by a ~ 100 nm shift between presynaptic vesicular release and postsynaptic AMPAR nanodomains (Haas et al. 2018; Fukata et al. 2024). Altogether, this indicates that LRRTM2 and Nlgn1 are important for maintaining trans-synaptic nano-organization of AMPARs, and this precise organization is important for synaptic function. It must be noted that within each family, there are four different LRRTM and Nlgn genes that are differentially expressed and have different binding preferences. Future studies are necessary to parse out the specific role of each LRRTM and Nlgn gene in organizing different synaptic types.

The plethora of Nrxn postsynaptic ligands together with the array of Nrxn splice variants increases the potential ways Nrxns can contribute to synaptic nano-organization. For example, different Nrxn splice variants may be present at the same or distinct synapses, where they control distinct nanostructures via differential binding of postsynaptic ligands. Studies employing expansion microscopy in mouse brain tissue support this notion (Sachs et al. 2024, Nozawa et al. 2022). These studies examined the nano-organization of three Nrxn ligands, Nlgn1, LRRTM1, and the extracellular scaffolding protein Cerebellin 1 (Cbln1) (Yuzaki 2018) in different brain regions including the cerebellum, hippocampus, and cortex. These brain regions express different compositions of Nrxn +/-SS4, a splice site which regulates binding to LRRTMs and Cblns. Interestingly, Cbln1, Nlgn1, and LRRTM1 assemble into overlapping or distinct nanodomains depending upon the composition of presynaptic Nrxn SS4. Moreover, the Cbln1, Nlgn1, and LRRTM1 nanodomains preferentially regulate the size and distribution of GluD1, NMDARs, and AMPARs nanodomains, respectively (Nozawa et al. 2022) (Fig. 5.2(5), (4), and (3), respectively). These findings are consistent with functional studies showing that LRRTM and Nlgn primarily regulate AMPAR-mediated and NMDAR-mediated synaptic activity, respectively (Südhof 2017), further connecting nano-organization with synaptic activity. The study also highlights a mechanism through which Nrxns, and potentially other CAMs, precisely regulate synaptic nano-organization in-trans: via preferential binding of certain splice variants to coordinating or competing ligands.

5.4.5 SynCAM1, Ephrins, and Cadherins

SRM revealed that other important cell adhesion molecules and receptors, Synaptic Cell Adhesion Molecule 1 (SynCAM1), and postsynaptic tyrosine kinase EphB2 receptors, also have distinct nano-organization. By combining STORM with cryo-EM, it was found that SynCAM1 is typically located near the periphery of excitatory synaptic clefts, whereas EphB2 receptors are centrally localized, supporting distinct nano-organization of different adhesion molecules (Perez de Arce et al. 2015). Interestingly, EphB receptors bind GluN1 subunits of NMDARs in an interaction stimulated by presynaptic EphrinB (Dalva et al. 2000; Grunwald et al. 2001). Co-organization of EphrinB/EphB2 receptor nanodomains with NMDAR nanodomains requires further investigation, as their organization and activity may be associated.

As previously mentioned, N-cadherin CAMs are initially localized throughout the synaptic cleft but become sequestered into discrete clusters at the periphery of the synapse as synapses mature (Elste and Benson 2006; Uchida et al. 1996). The dynamic and nonhomogeneous organization of cadherins raises the possibility that they may regulate nano-organization of the synapse during development. Intriguingly, N-Cadherin has been shown to bind GluA subunits of AMPARs both in *cis* and *trans* (Saglietti et al. 2007), suggesting they may have a role in recruiting AMPARs to the periphery of the synapse. Cadherin localization in mature synapses corresponds with SynCAM1, which often marks the edge of the synaptic cleft (Perez de Arce et al. 2015). Thus, it has been proposed that high-density nanodomains of SynCAM1 and Cadherin at the edges of the synapse may reserve these subregions for purposes other than protein clustering, driving nanodomain formation in more central locations (Biederer et al. 2017). This "center-surround inhibition" mode of organization may be occurring throughout the synapse, encouraging nanostructures to form in specific zones.

5.4.6 Latrophilins

It is likely that synaptic adhesion GPCRs, including Lphns and BAIs, are also spatially organized into nanodomains and that this organization spatially restricts their signaling events. Recent studies utilizing genetic models combined with STED microscopy found that aGPCR Lphn1 localizes to nanoclusters in both excitatory and inhibitory synapses of cultured hippocampal neurons. Consistent with other synaptic CAMs, these synapses contained approximately 1–3 Lphn1 nanoclusters which on average were ~ 90–100 nm in diameter. Interestingly, deletion of Lphn1 results in impairment of select inhibitory synapses at the soma (Matúš et al. 2024). Additional studies are needed to address the importance of Lphn1 nano-organization and signaling activity at these inhibitory synapses. Moreover, studies describing the nano-organization of other synaptic aGPCRs, including Lphn2 and Lphn3 which are selectively localized to distinct excitatory synapses, will further our understanding of how aGPCR-mediated adhesion and signaling events are regulated at the synapse.

Evidence also suggests that presynaptic Lphn interactors assemble into organized nanoclusters. Extensive STORM SRM determined that in most hippocampal excitatory synapses, endogenous Ten-3 forms one presynaptic nanocluster of approximately 80 nm. Presynaptic deletion of Ten-3 and Ten-4 revealed deficits in entorhinal cortical–hippocampal excitatory synapse assembly, while postsynaptic deletions had no effect (Zhang et al. 2022). Therefore, presynaptic Teneurins likely function in circuit assembly via direct binding to postsynaptic Lphns with high nanomolar affinity. However, it is currently unknown whether these Ten-3 nanodomains align with postsynaptic Lphn nanodomains. Moreover, whether other Teneurin paralogs, Ten-1 and Ten-2, or other Lphn presynaptic ligands, such as FLRTs, assemble into synaptic nanostructures remains to be determined.

5.4.7 Nano-organization and CAMs at Inhibitory Synapses

The majority of studies addressing the importance of nano-organization is limited to glutamatergic synapses. However, SRM approaches have shown that GABAergic synapses also organize into trans-synaptic nanocolumns, aligning presynaptic release sites with postsynaptic GABA receptors and Gephyrin scaffolding proteins. Disruption of $GABA_A$ receptor nano-alignment alters the synaptic physiological function, underscoring the significance of precise spatial organization in inhibitory synapse function (Olah et al. 2023). Recent studies begin to emphasize the importance of CAM nanoscale organization at GABAergic synapses. Using a combination of super-resolution imaging, electrophysiology, and protein engineering, studies have shown that rapid cleavage and removal of Neuroligin 2 from the cell surface impair spontaneous miniature inhibitory postsynaptic currents due to disrupted nano-alignment of RIM1/2 and $GABA_A$ receptors. Thus, postsynaptic

Neuroligin-2 has a crucial role in organizing GABAergic inhibitory synapses in cultured hippocampal neurons (Xu et al. 2024). Additionally, as described above, Lphn1 nanodomains are important for the formation of perisomatic inhibitory synapses. However, additional studies are needed to analyze Lphn1 co-organization and alignment with other synaptic nanostructures which modulate how Lphn1 nanodomains affect inhibitory synaptic function.

Considering the relatively recent advent of super-resolution techniques, our understanding of CAM nano-organization at synapses and how these CAMs control synaptic properties remains in its infancy, and several important questions remain. For example, how are CAMs within the same family differentially organized in different brain regions and across diverse synaptic types? What are the shared vs. distinct roles of CAM complexes in organizing synaptic subtypes? How do distinct CAM complexes preferentially organize specific components of functional machinery within the same synapse? Relatedly, do different CAM complexes differentially organize spontaneous vs. evoked release sites? What is the temporal interplay between pre- and postsynaptic CAMs during synaptic nanoassembly? How do synaptic CAMs organize compartmentalized intracellular signaling machinery at the nanoscale level? These are just a few of the numerous unanswered questions in this area of neuronal cell biology.

5.5 Conclusions and Perspectives

The mammalian brain is intricately wired into stereotyped neural circuits that govern all cognitive processes and behavior. The synapse is the fundamental unit of neural communication within these circuits, yet many aspects of synapse specificity, formation, and organization are not fully understood. It is generally recognized that many aspects of synaptogenesis are directed by intrinsic molecular programs, but the exact nature and mechanisms of these programs remain elusive. However, evidence highlights the crucial role of cell adhesion molecules in these processes. Artificial synapse formation assays have shown that numerous cell adhesion molecules can induce pre- or postsynaptic specializations in opposing neurons, with adhesion being necessary for this phenomenon. Adhesive interactions at initial synaptic contacts are thought to trigger bi-directional signaling cascades that recruit the necessary synaptic machinery, stabilize the synapse, and shape synaptic properties. However, loss-of-function studies have failed to identify a single "master regulator" molecule or pathway for synapse formation, suggesting multiple synaptogenic components work in parallel. These components likely function as combinatorial modules, and the precise composition dictates synapse specificity and synapse-specific properties. Moreover, this introduces redundancy into the mechanisms of synapse formation, ensuring that impairment of one component does not ablate synaptogenesis. This redundancy explains why deletion of most synaptic adhesion molecules does not affect synapse density. Additionally, there is a remarkable diversity of synaptic subtypes in the brain, each likely with different compositions of cell adhesion molecules. It is likely that some cell adhesion modules mediate fundamental aspects of synapse assembly and are shared across most synaptic types, while others form non-overlapping molecular codes that specify synapse specificity and diversity.

Advancements in super-resolution microscopy are unveiling the precise nanoscale composition and organization of synapses. Presynaptic vesicle release machinery and postsynaptic receptors form distinct nanodomains that occasionally align into nanocolumns, enhancing synaptic transmission. However, the mechanisms behind nanodomain formation, localization, and retention at the synapse remain unclear. Nevertheless, cell adhesion molecules are emerging as strong candidate regulators. Recent studies reveal that cell adhesion molecules also assemble into nanodomains that often co-organize with other synaptic nanodomains and/or trans-synaptic nanocolumns. Disrupting CAM expression and/or function alters the composition and alignment of these nanodomains, thereby disrupting nanocolumns and impairing synaptic function. This suggests that cell adhesion complexes may recruit the appropriate nanodomain components and link pre- and postsynaptic nanodomains to form nanocolumns. Conversely, some CAM complexes do not co-organize with other nanodomains nor associate with nanocolumns. The primary purpose of these adhesion complexes may simply be to adhere to the pre- and post-synaptic junctions. Alternatively, they may regulate synaptic nano-organization by occupying certain regions of synaptic space, thus inducing precise nano-organization in other, unoccupied regions through molecular crowding.

The role of cell adhesion molecules (CAMs) in synapse formation, organization, and function is a rapidly growing field, yet many critical areas remain unanswered. Several fundamental questions remain, including interplay between pre- and post-synaptic specializations during synapse formation. Does the presynaptic specialization instruct the postsynaptic structure, or vice versa? The organization and function of CAMs have been extensively studied within excitatory glutamatergic synapses in the mammalian forebrain, particularly the hippocampus. However, numerous diverse synaptic types throughout the brain have yet to be analyzed. Evidence suggests that CAMs are differentially expressed across various synaptic subtypes, and if CAMs function combinatorially, this differential expression could impact adhesive interactions, signaling activity, and the nano-organization of CAM complexes. Therefore, it is essential to explore the expression and function of CAMs in different synaptic types. Additionally, while the adhesive roles of certain CAMs during synapse formation are becoming clearer, significant gaps remain in our understanding of the bi-directional signaling pathways they control, the mechanisms by which they induce signaling, and how this signaling regulates synapse formation, maturation, and function. Adhesion GPCRs (aGPCRs) present a promising avenue for exploration, as they are examples of CAMs shown to be necessary for proper synaptogenesis in specific synaptic types and are known to signal via G-protein coupling. However, aGPCRs possess an unusual mode of activation, and the majority remain orphan receptors uncharacterized in the brain. Thus, further investigation into their signaling roles at the synapse, as well as other CAMs, is required to elucidate the pathways underlying circuit wiring. Regarding the organization of synapses, it remains unclear what initiates and regulates the nano-organization and alignment of synaptic proteins. Although CAMs are likely candidates, our understanding of their nano-organization is still incompletely understood, with only several molecules analyzed in a few synaptic models. In these models, certain CAMs appear crucial for the formation and alignment of nanodomains, affecting transmission efficiency. However, how the nano-organization of CAMs differs across synaptic types and how the assembly of CAMs into distinct nanodomains influences their function have not been thoroughly investigated. Collectively, an understanding of CAM modules across diverse synaptic types, detailed analyses of CAM signaling pathways, and thorough investigations into the nano-organization of CAMs are essential to fully understand their roles in neural circuit wiring and synaptic function.

References

Ackley BD. Wnt-signaling and planar cell polarity genes regulate axon guidance along the anteroposterior axis in C. elegans. Developmental neurobiology. 2014;74(8):781–796. https://doi.org/10.1002/dneu.v74.810.1002/dneu.22146.

Aimi T, Matsuda K, Yuzaki M. C1ql1-Bai3 signaling is necessary for climbing fiber synapse formation in mature Purkinje cells in coordination with neuronal activity. Mol Brain. 2023;16:61. https://doi.org/10.1186/s13041-023-01048-4.

Akin O, Zipursky SL. Activity regulates brain development in the fly. Curr Opin Genet Dev. 2020;65:8–13. https://doi.org/10.1016/j.gde.2020.04.005.

Anderson GR, Maxeiner S, Sando R, Tsetsenis T, Malenka RC, Südhof TC. Postsynaptic adhesion GPCR latrophilin-2 mediates target recognition in entorhinal-hippocampal synapse assembly. J Cell Biol. 2017;216:3831–46. https://doi.org/10.1083/jcb.201703042.

Aoto J, Martinelli DC, Malenka RC, Tabuchi K, Südhof TC. Presynaptic neurexin-3 alternative splicing trans-synaptically controls postsynaptic AMPA receptor trafficking. Cell. 2013;154:75–88. https://doi.org/10.1016/j.cell.2013.05.060.

Araç D, Li J. Teneurins and latrophilins: two giants meet at the synapse. Curr Opin Struct Biol. 2019;54:141–51. https://doi.org/10.1016/j.sbi.2019.01.028.

Araç D, Boucard AA, Bolliger MF, Nguyen J, Soltis SM, Südhof TC, Brunger AT. A novel evolutionarily conserved domain of cell-adhesion GPCRs mediates autoproteolysis. EMBO J. 2012;31:1364–78. https://doi.org/10.1038/emboj.2012.26.

Arcos-Burgos M, Jain M, Acosta MT, Shively S, Stanescu H, Wallis D, Domené S, Vélez JI, Karkera JD, Balog J, Berg K, Kleta R, Gahl WA, Roessler E, Long R, Lie J, Pineda D, Londoño AC, Palacio JD, Arbelaez A, Lopera F, Elia J, Hakonarson H, Johansson S, Knappskog PM, Haavik J, Ribases M, Cormand B, Bayes M, Casas M, Ramos-Quiroga JA, Hervas A, Maher BS, Faraone SV, Seitz C, Freitag CM, Palmason H, Meyer J, Romanos M, Walitza S, Hemminger U, Warnke A, Romanos J, Renner T, Jacob C, Lesch K-P, Swanson J, Vortmeyer A, Bailey-Wilson JE, Castellanos FX, Muenke M. A common variant of the latrophilin 3 gene, LPHN3, confers susceptibility to ADHD and predicts effectiveness of stimulant medication. Mol Psychiatry. 2010;15:1053–66. https://doi.org/10.1038/mp.2010.6.

Attardo A, Fitzgerald JE, Schnitzer MJ. Impermanence of dendritic spines in live adult CA1 hippocampus. Nature. 2015;523:592–6. https://doi.org/10.1038/nature14467.

Bagutti C, Forro G, Ferralli J, Rubin B, Chiquet-Ehrismann R. The intracellular domain of teneurin-2 has a nuclear function and represses zic-1-mediated transcription. J Cell Sci. 2003;116:2957–66. https://doi.org/10.1242/jcs.00603.

Bamji SX. Cadherins: actin with the cytoskeleton to form synapses. Neuron. 2005;47:175–8. https://doi.org/10.1016/j.neuron.2005.06.024.

Ban Y, Yu T, Feng B, Lorenz C, Wang X, Baker C, Zou Y. Prickle promotes the formation and maintenance of glutamatergic synapses by stabilizing the intercellular planar cell polarity complex. Sci Adv. 2021;7:eabh2974. https://doi.org/10.1126/sciadv.abh2974.

Bandekar SJ, Garbett K, Kordon SP, Dintzner E, Shearer T, Sando RC, Araç D. Structure of the extracellular region of the adhesion GPCR CELSR1 reveals a compact module which regulates G protein-coupling. bioRxiv. 2024:2024.01.26.577439. https://doi.org/10.1101/2024.01.26.577439.

Barros-Álvarez X, Nwokonko RM, Vizurraga A, Matzov D, He F, Papasergi-Scott MM, Robertson MJ, Panova O, Yardeni EH, Seven AB, Kwarcinski FE. The tethered peptide activation mechanism of adhesion GPCRs. Nature. 2022;28;604(7907):757–762. https://doi.org/10.1038/s41586-022-04575-7.

Basta LP, Sil P, Jones RA, Little KA, Hayward-Lara G, Devenport D. Celsr1 and Celsr2 exhibit distinct adhesive interactions and contributions to planar cell polarity. Front Cell Dev Biol. 2022;10:1064907. https://doi.org/10.3389/fcell.2022.1064907.

Basu R, Taylor MR, Williams ME. The classic cadherins in synaptic specificity. Cell Adhes Migr. 2015;9:193–201. https://doi.org/10.1080/19336918.2014.1000072.

Berger-Müller S, Suzuki T. Seven-pass transmembrane cadherins: roles and emerging mechanisms in axonal and dendritic patterning. Mol Neurobiol. 2011;44:313–20. https://doi.org/10.1007/s12035-011-8201-5.

Berlucchi G, Marzi CA. Roger Sperry, the maverick brain scientist who was haunted by psyche. Front Hum Neurosci. 2024;18 https://doi.org/10.3389/fnhum.2024.1392660.

Berns DS, DeNardo LA, Pederick DT, Luo L. Teneurin-3 controls topographic circuit assembly in the hippocampus. Nature. 2018;554:328–33. https://doi.org/10.1038/nature25463.

Bhouri M, Morishita W, Temkin P, Goswami D, Kawabe H, Brose N, Südhof TC, Craig AM, Siddiqui TJ, Malenka R. Deletion of LRRTM1 and LRRTM2 in adult mice impairs basal AMPA receptor transmission and LTP in hippocampal CA1 pyramidal neurons. Proc Natl Acad Sci USA. 2018;115:E5382–9. https://doi.org/10.1073/pnas.1803280115.

Biederer T. Progress from the postsynaptic side: signaling in synaptic differentiation. Sci STKE. 2005;2005:pe9. https://doi.org/10.1126/stke.2742005pe9.

Biederer T, Stagi M. Signaling by synaptogenic molecules. Curr Opin Neurobiol. 2008;18:261–9. https://doi.org/10.1016/j.conb.2008.07.014.

Biederer T, Sara Y, Mozhayeva M, Atasoy D, Liu X, Kavalali ET, Südhof TC. SynCAM, a synaptic adhesion molecule that drives synapse assembly. Science. 2002;297:1525–31. https://doi.org/10.1126/science.1072356.

Biederer T, Kaeser PS, Blanpied TA. Transcellular Nanoalignment of synaptic function. Neuron. 2017;96:680–96. https://doi.org/10.1016/j.neuron.2017.10.006.

Bolliger MF, Martinelli DC, Südhof TC. The cell-adhesion G protein-coupled receptor BAI3 is a high-affinity receptor for C1q-like proteins. Proc Natl Acad Sci USA. 2011;108:2534–9. https://doi.org/10.1073/pnas.1019577108.

Boucard AA, Ko J, Südhof TC. High affinity neurexin binding to cell adhesion G-protein-coupled receptor CIRL1/latrophilin-1 produces an intercellular adhesion complex. J Biol Chem. 2012;287:9399–413. https://doi.org/10.1074/jbc.M111.318659.

Boucard AA, Maxeiner S, Südhof TC. Latrophilins function as Heterophilic cell-adhesion molecules by binding to Teneurins: regulation by alternative splicing. J Biol Chem. 2014;289:387–402. https://doi.org/10.1074/jbc.M113.504779.

Boutin C, Goffinet AM, Tissir F. Celsr1-3 cadherins in PCP and brain development. Curr Top Dev Biol. 2012;101:161–83. https://doi.org/10.1016/B978-0-12-394592-1.00010-7.

Broadhead MJ, Horrocks MH, Zhu F, Muresan L, Benavides-Piccione R, DeFelipe J, Fricker D, Kopanitsa MV, Duncan RR, Klenerman D, Komiyama NH, Lee SF, Grant SGN. PSD95 nanoclusters are postsynaptic building blocks in hippocampus circuits. Sci Rep. 2016;6:24626. https://doi.org/10.1038/srep24626.

Brusés JL. N-cadherin signaling in synapse formation and neuronal physiology. Mol Neurobiol. 2006;33:237–52. https://doi.org/10.1385/MN:33:3:237.

Bruxel EM, Moreira-Maia CR, Akutagava-Martins GC, Quinn TP, Klein M, Franke B, Ribasés M, Rovira P, Sánchez-Mora C, Kappel DB, Mota NR, Grevet EH, Bau CHD, Arcos-Burgos M, Rohde LA, Hutz MH. Meta-analysis and systematic review of ADGRL3 (LPHN3) polymorphisms in ADHD susceptibility. Mol Psychiatry. 2021;26:2277–85. https://doi.org/10.1038/s41380-020-0673-0.

Bui DLH, Roach A, Li J, Bandekar SJ, Orput E, Raghavan R, Araç D, Sando RC. The adhesion GPCRs CELSR1-3 and LPHN3 engage G proteins via distinct activation mechanisms. Cell Rep. 2023;42:112552. https://doi.org/10.1016/j.celrep.2023.112552.

Burlingham SR, Wong NF, Peterkin L, Lubow L, Dos Santos Passos C, Benner O, Ghebrial M, Cast TP, Xu-Friedman MA, Südhof TC, Chanda S. Induction of synapse formation by de novo neurotransmitter synthesis. Nat Commun. 2022;13:3060. https://doi.org/10.1038/s41467-022-30756-z.

Burry RW. Development of apparent presynaptic elements formed in response to polylysine coated surfaces. Brain research. 1982:9;247(1):1–16. https://doi.org/10.1016/0006-8993(82)91022-8.

Burry RW, Hayes DM. Development and elimination of presynaptic elements on polylysine-coated beads implanted in neonatal rat cerebellum. Journal of neuroscience research. 1986;15(1): 67–78. https://doi.org/10.1002/jnr.v15:110.1002/jnr.490150107.

Camillo C, Facchinello N, Villari G, Mana G, Gioelli N, Sandri C, Astone M, Tortarolo D, Clapero F, Gays D, Oberkersch RE, Arese M, Tamagnone L, Valdembri D, Santoro MM, Serini G. LPHN2 inhibits vascular permeability by differential control of endothelial cell adhesion. J Cell Biol. 2021;220:e202006033. https://doi.org/10.1083/jcb.202006033.

Chae J, Kim MJ, Goo JH, Collier S, Gubb D, Charlton J, Adler PN, Park WJ. The drosophila tissue polarity gene starry night encodes a member of the protocadherin family. Development. 1999;126:5421–9. https://doi.org/10.1242/dev.126.23.5421.

Chamma I, Letellier M, Butler C, Tessier B, Lim K-H, Gauthereau I, Choquet D, Sibarita J-B, Park S, Sainlos M, Thoumine O. Mapping the dynamics and nanoscale organization of synaptic adhesion proteins using monomeric streptavidin. Nat Commun. 2016;7:10773. https://doi.org/10.1038/ncomms10773.

Chang Q, Balice-Gordon RJ. Highwire, rpm-1, and futsch: balancing synaptic growth and stability. Neuron. 2000;26:287–90. https://doi.org/10.1016/s0896-6273(00)81161-7.

Chen P-L, Clandinin TR. The cadherin flamingo mediates level-dependent interactions that guide photoreceptor target choice in drosophila. Neuron. 2008;58:26–33. https://doi.org/10.1016/j.neuron.2008.01.007.

Chen X, Winters C, Azzam R, Li X, Galbraith JA, Leapman RD, Reese TS. Organization of the core structure of the postsynaptic density. Proc Natl Acad Sci USA. 2008;105:4453–8. https://doi.org/10.1073/pnas.0800897105.

Chen J, Billings SE, Nishimune H. Calcium channels link the muscle-derived synapse organizer laminin β2 to bassoon and CAST/Erc2 to organize presynaptic active zones. J Neurosci. 2011a;31:512–25. https://doi.org/10.1523/JNEUROSCI.3771-10.2011.

Chen X, Nelson CD, Li X, Winters CA, Azzam R, Sousa AA, Leapman RD, Gainer H, Sheng M, Reese TS. PSD-95 is required to sustain the molecular organization of the postsynaptic density. J Neurosci. 2011b;31:6329–38. https://doi.org/10.1523/JNEUROSCI.5968-10.2011.

Chia PH, Li P, Shen K. Cell biology in neuroscience: cellular and molecular mechanisms underlying presynapse formation. J Cell Biol. 2013;203:11–22. https://doi.org/10.1083/jcb.201307020.

Cole AA, Reese TS. Transsynaptic assemblies link domains of presynaptic and postsynaptic intracellular structures across the synaptic cleft. J Neurosci. 2023;43:5883–92. https://doi.org/10.1523/JNEUROSCI.2195-22.2023.

Colón-Ramos DA. Synapse formation in developing neural circuits. Curr Top Dev Biol. 2009;87:53–79. https://doi.org/10.1016/S0070-2153(09)01202-2.

Cork SM, Van Meir EG. Emerging roles for the BAI1 protein family in the regulation of phagocytosis, synaptogenesis, neurovasculature, and tumor development. J Mol Med (Berl). 2011;89:743–52. https://doi.org/10.1007/s00109-011-0759-x.

Cortés E, Pak JS, Özkan E. Structure and evolution of neuronal wiring receptors and ligands. Dev Dyn. 2023;252:27–60. https://doi.org/10.1002/dvdy.512.

Curtin JA, Quint E, Tsipouri V, Arkell RM, Cattanach B, Copp AJ, Henderson DJ, Spurr N, Stanier P, Fisher EM, Nolan PM, Steel KP, Brown SDM, Gray IC, Murdoch JN. Mutation of Celsr1 disrupts planar polarity of inner ear hair cells and causes severe neural tube defects in the mouse. Curr Biol. 2003;13:1129–33. https://doi.org/10.1016/s0960-9822(03)00374-9.

Dai J, Aoto J, Südhof TC. Alternative splicing of presynaptic neurexins differentially controls postsynaptic NMDA and AMPA receptor responses. Neuron. 2019;102:993–1008.e5. https://doi.org/10.1016/j.neuron.2019.03.032.

Dalva MB, Takasu MA, Lin MZ, Shamah SM, Hu L, Gale NW, Greenberg ME. EphB receptors interact with NMDA receptors and regulate excitatory synapse formation. Cell. 2000;103:945–56. https://doi.org/10.1016/s0092-8674(00)00197-5.

Dalva MB, McClelland AC, Kayser MS. Cell adhesion molecules: signalling functions at the synapse. Nat Rev Neurosci. 2007;8:206–20. https://doi.org/10.1038/nrn2075.

Dani A, Huang B, Bergan J, Dulac C, Zhuang X. Superresolution imaging of chemical synapses in the brain. Neuron. 2010;68:843–56. https://doi.org/10.1016/j.neuron.2010.11.021.

Dates AN, Jones DTD, Smith JS, Skiba MA, Rich MF, Burruss MM, Kruse AC, Blacklow SC. Heterogeneity of tethered agonist signaling in adhesion G protein-coupled receptors. Cell Chem Biol. 2024;31(8):1542–1553.e4. https://doi.org/10.1016/j.chembiol.2024.03.004.

de Arce KP, Ribic A, Chowdhury D, Watters K, Thompson GJ, Sanganahalli BG, Lippard ETC, Rohlmann A, Strittmatter SM, Missler M, Hyder F, Biederer T. Concerted roles of LRRTM1 and SynCAM 1 in organizing prefrontal cortex synapses and cognitive functions. Nat Commun. 2023;14:459. https://doi.org/10.1038/s41467-023-36042-w.

de Wit J, Ghosh A. Specification of synaptic connectivity by cell surface interactions. Nat Rev Neurosci. 2016;17:4–4. https://doi.org/10.1038/nrn.2015.3.

de Wit J, Hong W, Luo L, Ghosh A. Role of leucine-rich repeat proteins in the development and function of neural circuits. Annu Rev Cell Dev Biol. 2011;27:697–729. https://doi.org/10.1146/annurev-cellbio-092910-154111.

Deák F, Schoch S, Liu X, Südhof TC, Kavalali ET. Synaptobrevin is essential for fast synaptic-vesicle endocytosis. Nat Cell Biol. 2004;6:1102–8. https://doi.org/10.1038/ncb1185.

Devenport D, Fuchs E. Planar polarization in embryonic epidermis orchestrates global asymmetric morphogenesis of hair follicles. Nat Cell Biol. 2008;10:1257–68. https://doi.org/10.1038/ncb1784.

Dodsworth TL, Lovejoy DA. Role of Teneurin C-terminal associated peptides (TCAP) on intercellular adhesion and communication. Front Neurosci. 2022;16:868541. https://doi.org/10.3389/fnins.2022.868541.

Domené S, Stanescu H, Wallis D, Tinloy B, Pineda DE, Kleta R, Arcos-Burgos M, Roessler E, Muenke M. Screening of human LPHN3 for variants with a potential impact on ADHD susceptibility. Am J Med Genet B Neuropsychiatr Genet. 2011;156B:11–8. https://doi.org/10.1002/ajmg.b.31141.

Donohue JD, Amidon RF, Murphy TR, Wong AJ, Liu ED, Saab L, King AJ, Pae H, Ajayi MT, Anderson GR. Parahippocampal latrophilin-2 (ADGRL2) expression controls topographical presubiculum to entorhinal cortex circuit connectivity. Cell Rep. 2021;37:110031. https://doi.org/10.1016/j.celrep.2021.110031.

Duda DG, Sunamura M, Lozonschi L, Yokoyama T, Yatsuoka T, Motoi F, Horii A, Tani K, Asano S, Nakamura Y, Matsuno S. Overexpression of the p53-inducible brain-specific angiogenesis inhibitor 1 suppresses efficiently tumour angiogenesis. Br J Cancer. 2002;86:490–6. https://doi.org/10.1038/sj.bjc.6600067.

Duman JG, Tu YK, Tolias KF. Emerging roles of BAI adhesion-GPCRs in synapse development and plasticity. Neural plasticity. 2016;(1)9. https://doi.org/10.1155/2016/8301737.

Ehrlich I, Klein M, Rumpel S, Malinow R. PSD-95 is required for activity-driven synapse stabilization. Proc Natl Acad Sci USA. 2007;104:4176–81. https://doi.org/10.1073/pnas.0609307104.

El-Husseini AE, Schnell E, Chetkovich DM, Nicoll RA, Bredt DS. PSD-95 involvement in maturation of excitatory synapses. Science. 2000;290:1364–8.

Elias GM, Funke L, Stein V, Grant SG, Bredt DS, Nicoll RA. Synapse-specific and developmentally regulated targeting of AMPA receptors by a family of MAGUK scaffolding proteins. Neuron. 2006;52:307–20. https://doi.org/10.1016/j.neuron.2006.09.012.

Ellisdon AM, Halls ML. Compartmentalization of GPCR signalling controls unique cellular responses. Biochem Soc Trans. 2016;44:562–7. https://doi.org/10.1042/BST20150236.

Elste AM, Benson DL. Structural basis for developmentally regulated changes in cadherin function at synapses. J Comp Neurol. 2006;495:324–35. https://doi.org/10.1002/cne.20876.

Fogel AI, Akins MR, Krupp AJ, Stagi M, Stein V, Biederer T. SynCAMs organize synapses through heterophilic adhesion. J Neurosci. 2007;27:12516–30. https://doi.org/10.1523/JNEUROSCI.2739-07.2007.

Fogel AI, Stagi M, Perez de Arce K, Biederer T. Lateral assembly of the immunoglobulin protein SynCAM 1 controls its adhesive function and instructs synapse formation. EMBO J. 2011;30:4728–38. https://doi.org/10.1038/emboj.2011.336.

Formstone CJ. 7TM-Cadherins: developmental roles and future challenges. Adv Exp Med Biol. 2010;706:14–36. https://doi.org/10.1007/978-1-4419-7913-1_2.

Friedman LG, Benson DL, Huntley GW. Cadherin-based transsynaptic networks in establishing and modifying neural connectivity. Curr Top Dev Biol. 2015;112:415–65. https://doi.org/10.1016/bs.ctdb.2014.11.025.

Fukata Y, Dimitrov A, Boncompain G, Vielemeyer O, Perez F, Fukata M. Local palmitoylation cycles define activity-regulated postsynaptic subdomains. J Cell Biol. 2013;202:145–61. https://doi.org/10.1083/jcb.201302071.

Fukata Y, Fukata M, MacGillavry HD, Nair D, Hosy E. Celebrating the birthday of AMPA receptor Nanodomains: illuminating the nanoscale organization of Excitatory Synapses with 10 nanocandles. J Neurosci. 2024;44:e2104232024. https://doi.org/10.1523/JNEUROSCI.2104-23.2024.

García-Nafría J, Herguedas B, Watson JF, Greger IH. The dynamic AMPA receptor extracellular region: a platform for synaptic protein interactions. J Physiol. 2016;594:5449–58. https://doi.org/10.1113/JP271844.

Goffinet AM, Tissir F. Seven pass Cadherins CELSR1-3. Semin Cell Dev Biol. 2017;69:102–10. https://doi.org/10.1016/j.semcdb.2017.07.014.

Gokce O, Südhof TC. Membrane-tethered monomeric neurexin LNS-domain triggers synapse formation. J Neurosci. 2013;33:14617–28. https://doi.org/10.1523/JNEUROSCI.1232-13.2013.

Gomez AM, Traunmüller L, Scheiffele P. Neurexins: molecular codes for shaping neuronal synapses. Nat Rev Neurosci. 2021;22:137–51. https://doi.org/10.1038/s41583-020-00415-7.

Goncalves J, Bartol TM, Camus C, Levet F, Menegolla AP, Sejnowski TJ, Sibarita J-B, Vivaudou M, Choquet D, Hosy E. Nanoscale co-organization and coactivation of AMPAR, NMDAR, and mGluR at excitatory synapses. Proc Natl Acad Sci USA. 2020;117:14503–11. https://doi.org/10.1073/pnas.1922563117.

Graf ER, Zhang X, Jin S-X, Linhoff MW, Craig AM. Neurexins induce differentiation of GABA and glutamate postsynaptic specializations via neuroligins. Cell. 2004;119:1013–26. https://doi.org/10.1016/j.cell.2004.11.035.

Grunwald IC, Korte M, Wolfer D, Wilkinson GA, Unsicker K, Lipp HP, Bonhoeffer T, Klein R. Kinase-independent requirement of EphB2 receptors in hippocampal synaptic plasticity. Neuron. 2001;32:1027–40. https://doi.org/10.1016/s0896-6273(01)00550-5.

Haas KT, Compans B, Letellier M, Bartol TM, Grillo-Bosch D, Sejnowski TJ, Sainlos M, Choquet D, Thoumine O, Hosy E. Pre-post synaptic alignment through neuroligin-1 tunes synaptic transmission efficiency. eLife. 2018;7:e31755. https://doi.org/10.7554/eLife.31755.

Hale WD, Südhof TC, Huganir RL. Engineered adhesion molecules drive synapse organization. Proc Natl Acad Sci USA. 2023;120:e2215905120. https://doi.org/10.1073/pnas.2215905120.

Hamann J, Aust G, Araç D, Engel FB, Formstone C, Fredriksson R, Hall RA, Harty BL, Kirchhoff C, Knapp B, Krishnan A, Liebscher I, Lin H-H, Martinelli DC, Monk KR, Peeters MC, Piao

X, Prömel S, Schöneberg T, Schwartz TW, Singer K, Stacey M, Ushkaryov YA, Vallon M, Wolfrum U, Wright MW, Xu L, Langenhan T, Schiöth HB. International union of basic and clinical pharmacology. XCIV. Adhesion G protein-coupled receptors. Pharmacol Rev. 2015;67:338–67. https://doi.org/10.1124/pr.114.009647.

Harms KJ, Craig AM. Synapse composition and organization following chronic activity blockade in cultured hippocampal neurons. J Comp Neurol. 2005;490:72–84. https://doi.org/10.1002/cne.20635.

Hassan BA, Hiesinger PR. Beyond molecular codes: simple rules to wire complex brains. Cell. 2015;163:285–91. https://doi.org/10.1016/j.cell.2015.09.031.

Hibino H, Pironkova R, Onwumere O, Vologodskaia M, Hudspeth AJ, Lesage F. RIM binding proteins (RBPs) couple Rab3-interacting molecules (RIMs) to voltage-gated ca(2+) channels. Neuron. 2002;34:411–23. https://doi.org/10.1016/s0896-6273(02)00667-0.

Hong W, Mosca TJ, Luo L. Teneurins instruct synaptic partner matching in an olfactory map. Nature. 2012;484:201–7. https://doi.org/10.1038/nature10926.

Honig B, Shapiro L. Adhesion protein structure, molecular affinities, and principles of cell-cell recognition. Cell. 2020;181:520–35. https://doi.org/10.1016/j.cell.2020.04.010.

Huang X, Zhang Q, Gu X, Hou Y, Wang M, Chen X, Wu J. LPHN3 gene variations and susceptibility to ADHD in Chinese Han population: a two-stage case-control association study and gene-environment interactions. Eur Child Adolesc Psychiatry. 2019;28:861–73. https://doi.org/10.1007/s00787-018-1251-8.

Huarcaya Najarro E, Ackley BD. C. elegans fmi-1/flamingo and Wnt pathway components interact genetically to control the anteroposterior neurite growth of the VD GABAergic neurons. Dev Biol. 2013;377:224–35. https://doi.org/10.1016/j.ydbio.2013.01.014.

Hulpiau P, Van Roy F. Molecular evolution of the cadherin superfamily. The international journal of biochemistry & cell biology. 2009; 1;41(2): 349–369.

Hunyara JL, Daly KM, Torres K, Yurgel ME, Komal R, Hattar S, Kolodkin AL. Teneurin-3 regulates the generation of non-image-forming visual circuitry and responsiveness to light in the suprachiasmatic nucleus. PLoS Biol. 2023;21:e3002412. https://doi.org/10.1371/journal.pbio.3002412.

Husić M, Barsyte-Lovejoy D, Lovejoy DA. Teneurin C-Terminal Associated Peptide (TCAP)-1 and Latrophilin interaction in HEK293 cells: evidence for modulation of intercellular adhesion. Front Endocrinol (Lausanne). 2019;10:22. https://doi.org/10.3389/fendo.2019.00022.

Hwang IW, Lim MH, Kwon HJ, Jin HJ. Association of LPHN3 rs6551665 A/G polymorphism with attention deficit and hyperactivity disorder in Korean children. Gene. 2015;566:68–73. https://doi.org/10.1016/j.gene.2015.04.033.

Ichtchenko K, Khvotchev M, Kiyatkin N, Simpson L, Sugita S, Südhof TC. alpha-latrotoxin action probed with recombinant toxin: receptors recruit alpha-latrotoxin but do not transduce an exocytotic signal. EMBO J. 1998;17:6188–99. https://doi.org/10.1093/emboj/17.21.6188.

Iijima T, Miura E, Watanabe M, Yuzaki M. Distinct expression of C1q-like family mRNAs in mouse brain and biochemical characterization of their encoded proteins. Eur J Neurosci. 2010;31:1606–15. https://doi.org/10.1111/j.1460-9568.2010.07202.x.

Jain S, Zipursky SL. Temporal control of neuronal wiring. Semin Cell Dev Biol., Special Issue: temporal patterning in the CNS. 2023;142:81–90. https://doi.org/10.1016/j.semcdb.2022.05.012.

Jeong BC, Kim M-Y, Lee JH, Kee HJ, Kho DH, Han KE, Qian YR, Kim J-K, Kim KK. Brain-specific angiogenesis inhibitor 2 regulates VEGF through GABP that acts as a transcriptional repressor. FEBS Lett. 2006;580:669–76. https://doi.org/10.1016/j.febslet.2005.12.086.

Jiang X, Sando R, Südhof TC. Multiple signaling pathways are essential for synapse formation induced by synaptic adhesion molecules. Proc Natl Acad Sci U S A. 2021;118:e2000173118. https://doi.org/10.1073/pnas.2000173118.

Jin Y, Garner CC. Molecular mechanisms of presynaptic differentiation. Annu Rev Cell Dev Biol. 2008;24:237–62. https://doi.org/10.1146/annurev.cellbio.23.090506.123417.

Kaeser PS, Deng L, Wang Y, Dulubova I, Liu X, Rizo J, Südhof TC. RIM proteins tether Ca2+ channels to presynaptic active zones via a direct PDZ-domain interaction. Cell. 2011;144:282–95. https://doi.org/10.1016/j.cell.2010.12.029.

Kakegawa W, Mitakidis N, Miura E, Abe M, Matsuda K, Takeo YH, Kohda K, Motohashi J, Takahashi A, Nagao S, Muramatsu S, Watanabe M, Sakimura K, Aricescu AR, Yuzaki M. Anterograde C1ql1 signaling is required in order to determine and maintain a single-winner climbing fiber in the mouse cerebellum. Neuron. 2015;85:316–29. https://doi.org/10.1016/j.neuron.2014.12.020.

Kang X, Xiao X, Harata M, Bai Y, Nakazaki Y, Soda Y, Kurita R, Tanaka T, Komine F, Izawa K, Kunisaki R, Setoyama M, Nishimori H, Natsume A, Sunamura M, Lozonshi L, Saitoh I, Tokino T, Asano S, Nakamura Y, Tani K. Antiangiogenic activity of BAI1 in vivo: implications for gene therapy of human glioblastomas. Cancer Gene Ther. 2006;13:385–92. https://doi.org/10.1038/sj.cgt.7700898.

Karaulanov E, Böttcher RT, Stannek P, Wu W, Rau M, Ogata S, Cho KWY, Niehrs C. Unc5B interacts with FLRT3 and Rnd1 to modulate cell adhesion in Xenopus embryos. PLoS One. 2009;4:e5742. https://doi.org/10.1371/journal.pone.0005742.

Kaur B, Brat DJ, Devi NS, Van Meir EG. Vasculostatin, a proteolytic fragment of brain angiogenesis inhibitor 1, is an antiangiogenic and antitumorigenic factor. Oncogene. 2005;24:3632–42. https://doi.org/10.1038/sj.onc.1208317.

Kee HJ, Koh JT, Kim M-Y, Ahn KY, Kim JK, Bae CS, Park SS, Kim KK. Expression of brain-specific angiogenesis inhibitor 2 (BAI2) in normal and ischemic brain: involvement of BAI2 in the ischemia-induced brain angiogenesis. J Cereb Blood Flow Metab. 2002;22:1054–67. https://doi.org/10.1097/00004647-200209000-00003.

Kee HJ, Ahn KY, Choi KC, Won Song J, Heo T, Jung S, Kim J-K, Bae CS, Kim KK. Expression of brain-specific angiogenesis inhibitor 3 (BAI3) in normal brain and implications for BAI3 in ischemia-induced brain angiogenesis and malignant glioma. FEBS Lett. 2004;569:307–16. https://doi.org/10.1016/j.febslet.2004.06.011.

Kellermayer B, Ferreira JS, Dupuis J, Levet F, Grillo-Bosch D, Bard L, Linarès-Loyez J, Bouchet D, Choquet D, Rusakov DA, Bon P, Sibarita J-B, Cognet L, Sainlos M, Carvalho AL, Groc L. Differential nanoscale topography and functional role of GluN2-NMDA receptor subtypes at glutamatergic synapses. Neuron. 2018;100:106–119.e7. https://doi.org/10.1016/j.neuron.2018.09.012.

Kenzelmann D, Chiquet-Ehrismann R, Leachman NT, Tucker RP. Teneurin-1 is expressed in interconnected regions of the developing brain and is processed in vivo. BMC Dev Biol. 2008;8:30. https://doi.org/10.1186/1471-213X-8-30.

Ko J. The leucine-rich repeat superfamily of synaptic adhesion molecules: LRRTMs and Slitrks. Mol Cells. 2012;34:335–40. https://doi.org/10.1007/s10059-012-0113-3.

Kordon SP, Cechova K, Bandekar SJ, Leon K, Dutka P, Siffer G, Kossiakoff AA, Vafabakhsh R, Araç D. Conformational coupling between extracellular and transmembrane domains modulates holo-adhesion GPCR function. Nature communications. 2024:15(1). https://doi.org/10.1038/s41467-024-54836-4.

Krasnoperov VG, Bittner MA, Beavis R, Kuang Y, Salnikow KV, Chepurny OG, Little AR, Plotnikov AN, Wu D, Holz RW, Petrenko AG. alpha-Latrotoxin stimulates exocytosis by the interaction with a neuronal G-protein-coupled receptor. Neuron. 1997;18:925–37. https://doi.org/10.1016/s0896-6273(00)80332-3.

Kudo S, Konda R, Obara W, Kudo D, Tani K, Nakamura Y, Fujioka T. Inhibition of tumor growth through suppression of angiogenesis by brain-specific angiogenesis inhibitor 1 gene transfer in murine renal cell carcinoma. Oncol Rep. 2007;18:785–91.

Kurshan PT, Shen K. Synaptogenic pathways. Curr Opin Neurobiol. 2019;57:156–62. https://doi.org/10.1016/j.conb.2019.03.005.

Kurusu M, Katsuki T, Zinn K, Suzuki E. Developmental changes in expression, subcellular distribution, and function of Drosophila N-cadherin, guided by a cell-intrinsic program during neuronal differentiation. Dev Biol. 2012;366:204–17. https://doi.org/10.1016/j.ydbio.2012.04.006.

Kwon H-B, Sabatini BL. Glutamate induces de novo growth of functional spines in developing cortex. Nature. 2011;474:100–4. https://doi.org/10.1038/nature09986.

Lala T, Hall RA. Adhesion G protein-coupled receptors: structure, signaling, physiology, and pathophysiology. Physiol Rev. 2022;102:1587–624. https://doi.org/10.1152/physrev.00027.2021.

Lee RC, Clandinin TR, Lee C-H, Chen P-L, Meinertzhagen IA, Zipursky SL. The protocadherin flamingo is required for axon target selection in the drosophila visual system. Nat Neurosci. 2003;6:557–63. https://doi.org/10.1038/nn1063.

Lelianova VG, Davletov BA, Sterling A, Rahman MA, Grishin EV, Totty NF, Ushkaryov YA. Alpha-latrotoxin receptor, latrophilin, is a novel member of the secretin family of G protein-coupled receptors. J Biol Chem. 1997;272:21504–8. https://doi.org/10.1074/jbc.272.34.21504.

Levine A, Bashan-Ahrend A, Budai-Hadrian O, Gartenberg D, Menasherow S, Wides R. Odd Oz: a novel Drosophila pair rule gene. Cell. 1994;77:587–98. https://doi.org/10.1016/0092-8674(94)90220-8.

Li J, Shalev-Benami M, Sando R, Jiang X, Kibrom A, Wang J, Leon K, Katanski C, Nazarko O, Lu YC, Südhof TC, Skiniotis G, Araç D. Structural basis for Teneurin function in circuit-wiring: a toxin motif at the synapse. Cell. 2018;173:735–748.e15. https://doi.org/10.1016/j.cell.2018.03.036.

Li S, Raychaudhuri S, Lee SA, Brockmann MM, Wang J, Kusick G, Prater C, Syed S, Falahati H, Ramos R, Bartol TM, Hosy E, Watanabe S. Asynchronous release sites align with NMDA receptors in mouse hippocampal synapses. Nat Commun. 2021;12:677. https://doi.org/10.1038/s41467-021-21004-x.

Li J, Lin S-M, Qiao J-D, Liu X-R, Wang J, Jiang M, Zhang J, Zhong M, Chen X-Q, Zhu J, He N, Su T, Shi Y-W, Yi Y-H, Liao W-P, China Epilepsy Gene 1.0 Project. CELSR3 variants are associated with febrile seizures and epilepsy with antecedent febrile seizures. CNS Neurosci Ther. 2022;28:382–9. https://doi.org/10.1111/cns.13781.

Liu G, Choi S, Tsien RW. Variability of neurotransmitter concentration and nonsaturation of postsynaptic AMPA receptors at synapses in hippocampal cultures and slices. Neuron. 1999;22:395–409. https://doi.org/10.1016/s0896-6273(00)81099-5.

Lizano E, Hayes JL, Willard FS. A synthetic method to assay adhesion-family G-protein coupled receptors. Determination of the G-protein coupling profile of ADGRG6(GPR126). Biochem Biophys Res Commun. 2021;534:317–22. https://doi.org/10.1016/j.bbrc.2020.11.086.

Lloyd BA, Han Y, Roth R, Zhang B, Aoto J. Neurexin-3 subsynaptic densities are spatially distinct from Neurexin-1 and essential for excitatory synapse nanoscale organization in the hippocampus. Nat Commun. 2023;14:4706. https://doi.org/10.1038/s41467-023-40419-2.

Loveless T, Hardin J. Cadherin complexity: recent insights into cadherin superfamily function in C. Elegans. Curr Opin Cell Biol. 2012;24:695–701. https://doi.org/10.1016/j.ceb.2012.06.008.

Lu B, Usui T, Uemura T, Jan L, Jan YN. Flamingo controls the planar polarity of sensory bristles and asymmetric division of sensory organ precursors in Drosophila. Curr Biol. 1999;9:1247–50. https://doi.org/10.1016/s0960-9822(99)80505-3.

Lu W, Bushong EA, Shih TP, Ellisman MH, Nicoll RA. The cell-autonomous role of excitatory synaptic transmission in the regulation of neuronal structure and function. Neuron. 2013;78:433–9. https://doi.org/10.1016/j.neuron.2013.02.030.

Lu YC, Nazarko OV, Sando R, Salzman GS, Li N-S, Südhof TC, Araç D. Structural basis of Latrophilin-FLRT-UNC5 interaction in cell adhesion. Structure. 2015;23:1678–91. https://doi.org/10.1016/j.str.2015.06.024.

Luo L. Developmental neuroscience: two gradients are better than one. Nature. 2006;439:23–4. https://doi.org/10.1038/439023a.

Luo L. Architectures of neuronal circuits. Science. 2021;373:eabg7285. https://doi.org/10.1126/science.abg7285.

Luo F, Sclip A, Jiang M, Südhof TC. Neurexins cluster Ca2+ channels within the presynaptic active zone. EMBO J. 2020;39:e103208. https://doi.org/10.15252/embj.2019103208.

MacGillavry HD, Kerr JM, Blanpied TA. Lateral organization of the postsynaptic density. Mol Cell Neurosci. 2011;48:321–31. https://doi.org/10.1016/j.mcn.2011.09.001.

MacGillavry HD, Song Y, Raghavachari S, Blanpied TA. Nanoscale scaffolding domains within the postsynaptic density concentrate synaptic AMPA receptors. Neuron. 2013;78:615–22. https://doi.org/10.1016/j.neuron.2013.03.009.

Margeta MA, Shen K. Molecular mechanisms of synaptic specificity. Mol Cell Neurosci. 2010;43:261–7. https://doi.org/10.1016/j.mcn.2009.11.009.

Martinelli DC, Chew KS, Rohlmann A, Lum MY, Ressl S, Hattar S, Brunger AT, Missler M, Südhof TC. Expression of C1ql3 in discrete neuronal populations controls efferent synapse numbers and diverse behaviors. Neuron. 2016;91:1034–51. https://doi.org/10.1016/j.neuron.2016.07.002.

Mathema VB, Na-Bangchang K. Regulatory roles of brain-specific angiogenesis inhibitor 1(BAI1) protein in inflammation, tumorigenesis and phagocytosis: a brief review. Crit Rev Oncol Hematol. 2017;111:81–6. https://doi.org/10.1016/j.critrevonc.2017.01.006.

Mathiasen S, Palmisano T, Perry NA, Stoveken HM, Vizurraga A, McEwen DP, Okashah N, Langenhan T, Inoue A, Lambert NA, Tall GG, Javitch JA. G12/13 is activated by acute tethered agonist exposure in the adhesion GPCR ADGRL3. Nat Chem Biol. 2020;16:1343–50. https://doi.org/10.1038/s41589-020-0617-7.

Matúš D, Lopez JM, Sando RC, Südhof TC. Essential role of Latrophilin-1 adhesion GPCR nanoclusters in inhibitory synapses. J Neurosci. 2024;44(23):e1978232024. https://doi.org/10.1523/JNEUROSCI.1978-23.2024.

McAllister AK, Stevens CF. Nonsaturation of AMPA and NMDA receptors at hippocampal synapses. Proc Natl Acad Sci USA. 2000;97:6173–8. https://doi.org/10.1073/pnas.100126497.

Meyer RL. Roger Sperry and his chemoaffinity hypothesis. Neuropsychologia. 1998;36:957–80. https://doi.org/10.1016/s0028-3932(98)00052-9.

Mikhailova A, McAllister AK. Synapse formation. In: Pfaff DW, Volkow ND, Rubenstein J, editors. Neuroscience in the 21st century: from basic to clinical. New York, NY: Springer; 2020. p. 1–18. https://doi.org/10.1007/978-1-4614-6434-1_190-1.

Moreno-Salinas AL, Holleran BJ, Ojeda-Muñiz EY, Correoso-Braña KG, Ribalta-Mena S, Ovando-Zambrano J-C, Leduc R, Boucard AA. Convergent selective signaling impairment exposes the pathogenicity of latrophilin-3 missense variants linked to inheritable ADHD susceptibility. Mol Psychiatry. 2022;27:2425–38. https://doi.org/10.1038/s41380-022-01537-3.

Mosca TJ, Hong W, Dani VS, Favaloro V, Luo L. Trans-synaptic Teneurin signalling in neuromuscular synapse organization and target choice. Nature. 2012;484:237–41. https://doi.org/10.1038/nature10923.

Müller A, Winkler J, Fiedler F, Sastradihardja T, Binder C, Schnabel R, Kungel J, Rothemund S, Hennig C, Schöneberg T, Prömel S. Oriented cell division in the C. elegans embryo is coordinated by G-protein signaling dependent on the adhesion GPCR LAT-1. PLoS Genet. 2015;11:e1005624. https://doi.org/10.1371/journal.pgen.1005624.

Muttathukunnel P, Frei P, Perry S, Dickman D, Müller M. Rapid homeostatic modulation of transsynaptic nanocolumn rings. Proc Natl Acad Sci USA. 2022;119:e2119044119. https://doi.org/10.1073/pnas.2119044119.

Nair D, Hosy E, Petersen JD, Constals A, Giannone G, Choquet D, Sibarita J-B. Super-resolution imaging reveals that AMPA receptors inside synapses are dynamically organized in nanodomains regulated by PSD95. J Neurosci. 2013;33:13204–24. https://doi.org/10.1523/JNEUROSCI.2381-12.2013.

Nair M, Bolyard C, Lee TJ, Kaur B, Yoo JY. Therapeutic application of brain-specific angiogenesis inhibitor 1 for cancer therapy. Cancers (Basel). 2021;13:3562. https://doi.org/10.3390/cancers13143562.

Najarro EH, Wong L, Zhen M, Carpio EP, Goncharov A, Garriga G, Lundquist EA, Jin Y, Ackley BD. Caenorhabditis elegans flamingo cadherin fmi-1 regulates GABAergic neuronal development. J Neurosci. 2012;32:4196–211. https://doi.org/10.1523/JNEUROSCI.3094-11.2012.

Nazarko O, Kibrom A, Winkler J, Leon K, Stoveken H, Salzman G, Merdas K, Lu Y, Narkhede P, Tall G, Prömel S, Araç D. A comprehensive mutagenesis screen of the adhesion GPCR Latrophilin-1/ADGRL1. iScience. 2018;3:264–78. https://doi.org/10.1016/j.isci.2018.04.019.

Nern A, Zhu Y, Zipursky SL. Local N-cadherin interactions mediate distinct steps in the targeting of lamina neurons. Neuron. 2008;58:34–41. https://doi.org/10.1016/j.neuron.2008.03.022.

Nimchinsky EA, Yasuda R, Oertner TG, Svoboda K. The number of glutamate receptors opened by synaptic stimulation in single hippocampal spines. J Neurosci. 2004;24:2054–64. https://doi.org/10.1523/JNEUROSCI.5066-03.2004.

Nishiguchi S, Kasai RS, Uchihashi T. Antiparallel dimer structure of CELSR cadherin in solution revealed by high-speed atomic force microscopy. Proc Natl Acad Sci USA. 2023;120:e2302047120. https://doi.org/10.1073/pnas.2302047120.

Nishimori H, Shiratsuchi T, Urano T, Kimura Y, Kiyono K, Tatsumi K, Yoshida S, Ono M, Kuwano M, Nakamura Y, Tokino T. A novel brain-specific p53-target gene, BAI1, containing thrombospondin type 1 repeats inhibits experimental angiogenesis. Oncogene. 1997;15:2145–50. https://doi.org/10.1038/sj.onc.1201542.

Nozawa K, Sogabe T, Hayashi A, Motohashi J, Miura E, Arai I, Yuzaki M. In vivo nanoscopic landscape of neurexin ligands underlying anterograde synapse specification. Neuron. 2022;110:3168–3185.e8. https://doi.org/10.1016/j.neuron.2022.07.027.

Nusser Z. Creating diverse synapses from the same molecules. Curr Opin Neurobiol. 2018;51:8–15. https://doi.org/10.1016/j.conb.2018.01.001.

O'Sullivan ML, de Wit J, Savas JN, Comoletti D, Otto-Hitt S, Yates JR, Ghosh A. FLRT proteins are endogenous latrophilin ligands and regulate excitatory synapse development. Neuron. 2012;73:903–10. https://doi.org/10.1016/j.neuron.2012.01.018.

Olah SS, Kareemo DJ, Buchta WC, Sinnen BL, Miller CN, Actor-Engel HS, Gookin SE, Winborn CS, Kleinjan MS, Crosby KC, Aoto J, Smith KR, Kennedy MJ. Acute reorganization of postsynaptic GABAA receptors reveals the functional impact of molecular nanoarchitecture at inhibitory synapses. Cell Rep. 2023;42:113331. https://doi.org/10.1016/j.celrep.2023.113331.

Opazo P, Choquet D. A three-step model for the synaptic recruitment of AMPA receptors. Mol Cell Neurosci. 2011;46:1–8. https://doi.org/10.1016/j.mcn.2010.08.014.

Ovando-Zambrano J-C, Arias-Montaño J-A, Boucard AA. Alternative splicing event modifying ADGRL1/latrophilin-1 cytoplasmic tail promotes both opposing and dual cAMP signaling pathways. Ann N Y Acad Sci. 2019;1456:168–85. https://doi.org/10.1111/nyas.14198.

Paavola KJ, Hall RA. Adhesion G protein-coupled receptors: signaling, pharmacology, and mechanisms of activation. Mol Pharmacol. 2012;82:777–83. https://doi.org/10.1124/mol.112.080309.

Park D, Bae S, Yoon TH, Ko J. Molecular mechanisms of synaptic specificity: spotlight on hippocampal and cerebellar synapse organizers. Mol Cells. 2018;41:373–80. https://doi.org/10.14348/molcells.2018.0081.

Pecot MY, Tadros W, Nern A, Bader M, Chen Y, Zipursky SL. Multiple interactions control synaptic layer specificity in the Drosophila visual system. Neuron. 2013;77:299–310. https://doi.org/10.1016/j.neuron.2012.11.007.

Pederick DT, Lui JH, Gingrich EC, Xu C, Wagner MJ, Liu Y, He Z, Quake SR, Luo L. Reciprocal repulsions instruct the precise assembly of parallel hippocampal networks. Science. 2021;372:1068–73. https://doi.org/10.1126/science.abg1774.

Pederick DT, Perry-Hauser NA, Meng H, He Z, Javitch JA, Luo L. Context-dependent requirement of G protein coupling for Latrophilin-2 in target selection of hippocampal axons. elife. 2023;12:e83529. https://doi.org/10.7554/eLife.83529.

Perez de Arce K, Schrod N, Metzbower SWR, Allgeyer E, Kong GK-W, Tang A-H, Krupp AJ, Stein V, Liu X, Bewersdorf J, Blanpied TA, Lucić V, Biederer T. Topographic mapping of the synaptic cleft into adhesive nanodomains. Neuron. 2015;88:1165–72. https://doi.org/10.1016/j.neuron.2015.11.011.

Purcell RH, Hall RA. Adhesion G protein-coupled receptors as drug targets. Annu Rev Pharmacol Toxicol. 2018;58:429–49. https://doi.org/10.1146/annurev-pharmtox-010617-052933.

Qian Y, Ma Z, Liu C, Li X, Zhu X, Wang N, Xu Z, Xia R, Liang J, Duan Y, Yin H. Structural insights into adhesion GPCR ADGRL3 activation and Gq, Gs, Gi, and G12 coupling. Molecular cell. 2022: 82(22) 4340–4352.e6. https://doi.org/10.1016/j.molcel.2022.10.009.

Qu Y, Glasco DM, Zhou L, Sawant A, Ravni A, Fritzsch B, Damrau C, Murdoch JN, Evans S, Pfaff SL, Formstone C, Goffinet AM, Chandrasekhar A, Tissir F. Atypical cadherins Celsr1-3 differentially regulate migration of facial branchiomotor neurons in mice. J Neurosci. 2010;30:9392–401. https://doi.org/10.1523/JNEUROSCI.0124-10.2010.

Racca C, Stephenson FA, Streit P, Roberts JDB, Somogyi P. NMDA receptor content of synapses in stratum radiatum of the hippocampal CA1 area. J Neurosci. 2000;20:2512–22. https://doi.org/10.1523/JNEUROSCI.20-07-02512.2000.

Raghavachari S, Lisman JE. Properties of quantal transmission at CA1 synapses. J Neurophysiol. 2004;92:2456–67. https://doi.org/10.1152/jn.00258.2004.

Ramsey AM, Tang A-H, LeGates TA, Gou X-Z, Carbone BE, Thompson SM, Biederer T, Blanpied TA. Subsynaptic positioning of AMPARs by LRRTM2 controls synaptic strength. Sci Adv. 2021;7:eabf3126. https://doi.org/10.1126/sciadv.abf3126.

Ranscht B. Cadherins: molecular codes for axon guidance and synapse formation. Int J Dev Neurosci. 2000;18:643–51. https://doi.org/10.1016/s0736-5748(00)00030-7.

Regan SL, Williams MT, Vorhees CV. Latrophilin-3 disruption: effects on brain and behavior. Neurosci Biobehav Rev. 2021;127:619–29. https://doi.org/10.1016/j.neubiorev.2021.04.030.

Roppongi RT, Karimi B, Siddiqui TJ. Role of LRRTMs in synapse development and plasticity. Neurosci Res. 2017;116:18–28. https://doi.org/10.1016/j.neures.2016.10.003.

Sachs S, Reinhard S, Eilts J, Sauer M, Werner C. Visualizing the trans-synaptic arrangement of synaptic proteins by expansion microscopy. Front Cell Neurosci. 2024;18:1328726. https://doi.org/10.3389/fncel.2024.1328726.

Saglietti L, Dequidt C, Kamieniarz K, Rousset M-C, Valnegri P, Thoumine O, Beretta F, Fagni L, Choquet D, Sala C, Sheng M, Passafaro M. Extracellular interactions between GluR2 and N-cadherin in spine regulation. Neuron. 2007;54:461–77. https://doi.org/10.1016/j.neuron.2007.04.012.

Sakamoto H, Ariyoshi T, Kimpara N, Sugao K, Taiko I, Takikawa K, Asanuma D, Namiki S, Hirose K. Synaptic weight set by Munc13-1 supramolecular assemblies. Nat Neurosci. 2018;21:41–9. https://doi.org/10.1038/s41593-017-0041-9.

Sando R, Südhof TC. Latrophilin GPCR signaling mediates synapse formation. eLife. 2021;10:e65717. https://doi.org/10.7554/eLife.65717.

Sando R, Bushong E, Zhu Y, Huang M, Considine C, Phan S, Ju S, Uytiepo M, Ellisman M, Maximov A. Assembly of excitatory synapses in the absence of glutamatergic neurotransmission. Neuron. 2017;94:312–321.e3. https://doi.org/10.1016/j.neuron.2017.03.047.

Sando R, Jiang X, Südhof TC. Latrophilin GPCRs direct synapse specificity by coincident binding of FLRTs and teneurins. Science. 2019;363:eaav7969. https://doi.org/10.1126/science.aav7969.

Sando R, Ho ML, Liu X, Südhof TC. Engineered synaptic tools reveal localized cAMP signaling in synapse assembly. J Cell Biol. 2022;221:e202109111. https://doi.org/10.1083/jcb.202109111.

Sanes JR, Yamagata M. Many paths to synaptic specificity. Annu Rev Cell Dev Biol. 2009;25:161–95. https://doi.org/10.1146/annurev.cellbio.24.110707.175402.

Sanes JR, Zipursky SL. Synaptic specificity, recognition molecules, and assembly of neural circuits. Cell. 2020;181:536–56. https://doi.org/10.1016/j.cell.2020.04.008.

Sara Y, Biederer T, Atasoy D, Chubykin A, Mozhayeva MG, Südhof TC, Kavalali ET. Selective capability of SynCAM and neuroligin for functional synapse assembly. J Neurosci. 2005;25:260–70. https://doi.org/10.1523/JNEUROSCI.3165-04.2005.

Scheiffele P, Fan J, Choih J, Fetter R, Serafini T. Neuroligin expressed in nonneuronal cells triggers presynaptic development in contacting axons. Cell. 2000;101:657–69. https://doi.org/10.1016/s0092-8674(00)80877-6.

Schoch S, Deák F, Königstorfer A, Mozhayeva M, Sara Y, Südhof TC, Kavalali ET. SNARE function analyzed in synaptobrevin/VAMP knockout mice. Science. 2001;294:1117–22. https://doi.org/10.1126/science.1064335.

Schoch S, Castillo PE, Jo T, Mukherjee K, Geppert M, Wang Y, Schmitz F, Malenka RC, Südhof TC. RIM1alpha forms a protein scaffold for regulating neurotransmitter release at the active zone. Nature. 2002;415:321–6. https://doi.org/10.1038/415321a.

Scholz N, Guan C, Nieberler M, Grotemeyer A, Maiellaro I, Gao S, Beck S, Pawlak M, Sauer M, Asan E, Rothemund S, Winkler J, Prömel S, Nagel G, Langenhan T, Kittel RJ. Mechano-dependent signaling by Latrophilin/CIRL quenches cAMP in proprioceptive neurons. Elife. 2017;6:e28360. https://doi.org/10.7554/eLife.28360.

Schroeder A, Vanderlinden J, Vints K, Ribeiro LF, Vennekens KM, Gounko NV, Wierda KD, de Wit J. A modular organization of LRR protein-mediated synaptic adhesion defines synapse identity. Neuron. 2018;99:329–344.e7. https://doi.org/10.1016/j.neuron.2018.06.026.

Senti K-A, Usui T, Boucke K, Greber U, Uemura T, Dickson BJ. Flamingo regulates R8 axon-axon and axon-target interactions in the Drosophila visual system. Curr Biol. 2003;13:828–32. https://doi.org/10.1016/s0960-9822(03)00291-4.

Shapiro L, Colman DR. The diversity of cadherins and implications for a synaptic adhesive code in the CNS. Neuron. 1999;23(3):427–30.

Shen K, Cowan CW. Guidance molecules in synapse formation and plasticity. Cold Spring Harb Perspect Biol. 2010;2:a001842. https://doi.org/10.1101/cshperspect.a001842.

Shen K, Fetter RD, Bargmann CI. Synaptic specificity is generated by the synaptic guidepost protein SYG-2 and its receptor, SYG-1. Cell. 2004;116:869–81. https://doi.org/10.1016/S0092-8674(04)00251-X.

Shi D-L. Planar cell polarity regulators in asymmetric organogenesis during development and disease. J Genet Genomics. 2023;50:63–76. https://doi.org/10.1016/j.jgg.2022.06.007.

Shiratsuchi T, Nishimori H, Ichise H, Nakamura Y, Tokino T. Cloning and characterization of BAI2 and BAI3, novel genes homologous to brain-specific angiogenesis inhibitor 1 (BAI1). Cytogenet Cell Genet. 1997;79:103–8. https://doi.org/10.1159/000134693.

Siddiqui TJ, Craig AM. Synaptic organizing complexes. Curr Opin Neurobiol. 2011;21:132–43. https://doi.org/10.1016/j.conb.2010.08.016.

Sigler A, Oh WC, Imig C, Altas B, Kawabe H, Cooper BH, Kwon H-B, Rhee J-S, Brose N. Formation and maintenance of functional spines in the absence of presynaptic glutamate release. Neuron. 2017;94:304–311.e4. https://doi.org/10.1016/j.neuron.2017.03.029.

Sigoillot SM, Iyer K, Binda F, González-Calvo I, Talleur M, Vodjdani G, Isope P, Selimi F. The secreted protein C1QL1 and its receptor BAI3 control the synaptic connectivity of excitatory inputs converging on cerebellar Purkinje cells. Cell Rep. 2015;10:820–32. https://doi.org/10.1016/j.celrep.2015.01.034.

Silva J-P, Lelianova VG, Ermolyuk YS, Vysokov N, Hitchen PG, Berninghausen O, Rahman MA, Zangrandi A, Fidalgo S, Tonevitsky AG, Dell A, Volynski KE, Ushkaryov YA. Latrophilin 1 and its endogenous ligand lasso/teneurin-2 form a high-affinity transsynaptic receptor pair with signaling capabilities. Proc Natl Acad Sci USA. 2011;108:12113–8. https://doi.org/10.1073/pnas.1019434108.

Soler-Llavina GJ, Fuccillo MV, Ko J, Südhof TC, Malenka RC. The neurexin ligands, neuroligins and leucine-rich repeat transmembrane proteins, perform convergent and divergent synaptic functions in vivo. Proc Natl Acad Sci USA. 2011;108:16502–9. https://doi.org/10.1073/pnas.1114028108.

Söllner C, Wright GJ. A cell surface interaction network of neural leucine-rich repeat receptors. Genome Biol. 2009;10:R99. https://doi.org/10.1186/gb-2009-10-9-r99.

Sperry RW. Regulative factors in the orderly growth of neural circuits. Growth. 1951:63–87.

Sperry RW. Chemoaffinity in the orderly growth of nerve fiber patterns and connections. Proc Natl Acad Sci USA. 1963;50:703–10. https://doi.org/10.1073/pnas.50.4.703.

Steimel A, Wong L, Najarro EH, Ackley BD, Garriga G, Hutter H. The flamingo ortholog FMI-1 controls pioneer-dependent navigation of follower axons in C. Elegans. Development. 2010;137:3663–73. https://doi.org/10.1242/dev.054320.

Südhof TC. Alpha-Latrotoxin and its receptors: neurexins and CIRL/latrophilins. Annu Rev Neurosci. 2001;24:933–62. https://doi.org/10.1146/annurev.neuro.24.1.933.

Südhof TC. Synaptic neurexin complexes: a molecular code for the logic of neural circuits. Cell. 2017;171:745–69. https://doi.org/10.1016/j.cell.2017.10.024.

Südhof TC. Towards an understanding of synapse formation. Neuron. 2018;100:276–93. https://doi.org/10.1016/j.neuron.2018.09.040.

Südhof TC. The cell biology of synapse formation. J Cell Biol. 2021;220:e202103052. https://doi.org/10.1083/jcb.202103052.

Sun S-Y, Li X-W, Cao R, Zhao Y, Sheng N, Tang A-H. Correlative assembly of subsynaptic nanoscale organizations during development. Front Synaptic Neurosci. 2022;14:748184. https://doi.org/10.3389/fnsyn.2022.748184.

Suzuki K, Elegheert J, Song I, Sasakura H, Senkov O, Matsuda K, Kakegawa W, Clayton AJ, Chang VT, Ferrer-Ferrer M, Miura E, Kaushik R, Ikeno M, Morioka Y, Takeuchi Y, Shimada T, Otsuka S, Stoyanov S, Watanabe M, Takeuchi K, Dityatev A, Aricescu AR, Yuzaki M. A synthetic synaptic organizer protein restores glutamatergic neuronal circuits. Science. 2020;369:eabb4853. https://doi.org/10.1126/science.abb4853.

Sylwestrak EL, Ghosh A. Elfn1 regulates target-specific release probability at CA1-interneuron synapses. Science. 2012;338:536–40. https://doi.org/10.1126/science.1222482.

Tamilselvan E, Sotomayor M. CELSR1, a core planar cell polarity protein, features a weakly adhesive and flexible cadherin ectodomain. Structure. 2024;32:476–491.e5. https://doi.org/10.1016/j.str.2024.01.003.

Tang A-H, Chen H, Li TP, Metzbower SR, MacGillavry HD, Blanpied TA. A trans-synaptic nanocolumn aligns neurotransmitter release to receptors. Nature. 2016;536:210–4. https://doi.org/10.1038/nature19058.

Thakar S, Wang L, Yu T, Ye M, Onishi K, Scott J, Qi J, Fernandes C, Han X, Yates JR, Berg DK, Zou Y. Evidence for opposing roles of Celsr3 and Vangl2 in glutamatergic synapse formation. Proc Natl Acad Sci USA. 2017;114:E610–8. https://doi.org/10.1073/pnas.1612062114.

Tian X, Li J, Valakh V, DiAntonio A, Wu C. Drosophila Rae1 controls the abundance of the ubiquitin ligase Highwire in post-mitotic neurons. Nat Neurosci. 2011;14:1267–75. https://doi.org/10.1038/nn.2922.

Tissir F, Goffinet AM. Atypical cadherins Celsr1-3 and planar cell polarity in vertebrates. Prog Mol Biol Transl Sci. 2013;116:193–214. https://doi.org/10.1016/B978-0-12-394311-8.00009-1.

Tissir F, Bar I, Jossin Y, De Backer O, Goffinet AM. Protocadherin Celsr3 is crucial in axonal tract development. Nat Neurosci. 2005;8:451–7. https://doi.org/10.1038/nn1428.

Tobaben S, Südhof TC, Stahl B. The G protein-coupled receptor CL1 interacts directly with proteins of the Shank family. J Biol Chem. 2000;275:36204–10. https://doi.org/10.1074/jbc.M006448200.

Trotter JH, Hao J, Maxeiner S, Tsetsenis T, Liu Z, Zhuang X, Südhof TC. Synaptic neurexin-1 assembles into dynamically regulated active zone nanoclusters. J Cell Biol. 2019;218:2677–98. https://doi.org/10.1083/jcb.201812076.

Uchida N, Honjo Y, Johnson KR, Wheelock MJ, Takeichi M. The catenin/cadherin adhesion system is localized in synaptic junctions bordering transmitter release zones. J Cell Biol. 1996;135:767–79. https://doi.org/10.1083/jcb.135.3.767.

Usui T, Shima Y, Shimada Y, Hirano S, Burgess RW, Schwarz TL, Takeichi M, Uemura T. Flamingo, a seven-pass transmembrane cadherin, regulates planar cell polarity under the control of frizzled. Cell. 1999;98:585–95. https://doi.org/10.1016/s0092-8674(00)80046-x.

Vanderhaeghen P, Polleux F. Developmental mechanisms patterning thalamocortical projections: intrinsic, extrinsic and in between. Trends Neurosci. 2004;27:384–91. https://doi.org/10.1016/j.tins.2004.05.009.

Varoqueaux F, Sigler A, Rhee J-S, Brose N, Enk C, Reim K, Rosenmund C. Total arrest of spontaneous and evoked synaptic transmission but normal synaptogenesis in the absence of Munc13-mediated vesicle priming. Proc Natl Acad Sci USA. 2002;99:9037–42. https://doi.org/10.1073/pnas.122623799.

Varoqueaux F, Aramuni G, Rawson RL, Mohrmann R, Missler M, Gottmann K, Zhang W, Südhof TC, Brose N. Neuroligins determine synapse maturation and function. Neuron. 2006;51:741–54. https://doi.org/10.1016/j.neuron.2006.09.003.

Verhage M, Maia AS, Plomp JJ, Brussaard AB, Heeroma JH, Vermeer H, Toonen RF, Hammer RE, van den Berg TK, Missler M, Geuze HJ, Südhof TC. Synaptic assembly of the brain in the absence of neurotransmitter secretion. Science. 2000;287:864–9. https://doi.org/10.1126/science.287.5454.864.

Vizurraga A, Adhikari R, Yeung J, Yu M, Tall GG. Mechanisms of adhesion G protein-coupled receptor activation. J Biol Chem. 2020;295:14065–83. https://doi.org/10.1074/jbc.REV120.007423.

Wang X-J, Zhang D-L, Xu Z-G, Ma M-L, Wang W-B, Li L-L, Han X-L, Huo Y, Yu X, Sun J-P. Understanding cadherin EGF LAG seven-pass G-type receptors. J Neurochem. 2014;131:699–711. https://doi.org/10.1111/jnc.12955.

Wang Y, Cao Y, Hays CL, Laboute T, Ray TA, Guerrero-Given D, Ahuja AS, Patil D, Rivero O, Kamasawa N, Kay JN, Thoreson WB, Martemyanov KA. Adhesion GPCR Latrophilin 3 regulates synaptic function of cone photoreceptors in a trans-synaptic manner. Proc Natl Acad Sci U S A. 2021a;118:e2106694118. https://doi.org/10.1073/pnas.2106694118.

Wang J, Miao Y, Wicklein R, Sun Z, Wang J, Jude KM, Fernandes RA, Merrill SA, Wernig M, Garcia KC, Südhof TC. RTN4/NoGo-receptor binding to BAI adhesion-GPCRs regulates neuronal development. Cell. 2021b;184:5869–5885.e25. https://doi.org/10.1016/j.cell.2021.10.016.

Williams ME, de Wit J, Ghosh A. Molecular mechanisms of synaptic specificity in developing neural circuits. Neuron. 2010;68:9–18. https://doi.org/10.1016/j.neuron.2010.09.007.

Woelfle R, D'Aquila AL, Lovejoy DA. Teneurins, TCAP, and latrophilins: roles in the etiology of mood disorders. Transl Neurosci. 2016;7:17–23. https://doi.org/10.1515/tnsci-2016-0004.

Wolterhoff N, Hiesinger PR. Synaptic promiscuity in brain development. Curr Biol. 2024;34:R102–16. https://doi.org/10.1016/j.cub.2023.12.037.

Wu Z, Sweeney LB, Ayoob JC, Chak K, Andreone BJ, Ohyama T, Kerr R, Luo L, Zlatic M, Kolodkin AL. A combinatorial semaphorin code instructs the initial steps of sensory circuit assembly in the drosophila CNS. Neuron. 2011;70:281–98. https://doi.org/10.1016/j.neuron.2011.02.050.

Xie Q, Li J, Li H, Udeshi ND, Svinkina T, Orlin D, Kohani S, Guajardo R, Mani DR, Xu C, Li T, Han S, Wei W, Shuster SA, Luginbuhl DJ, Quake SR, Murthy SE, Ting AY, Carr SA, Luo L. Transcription factor Acj6 controls dendrite targeting via a combinatorial cell-surface code. Neuron. 2022;110:2299–2314.e8. https://doi.org/10.1016/j.neuron.2022.04.026.

Xu N, Cao R, Chen S-Y, Gou X-Z, Wang B, Luo H-M, Gao F, Tang A-H. Structural and functional reorganization of inhibitory synapses by activity-dependent cleavage of neuroligin-2. Proc Natl Acad Sci USA. 2024;121:e2314541121. https://doi.org/10.1073/pnas.2314541121.

Yagi T, Takeichi M. Cadherin superfamily genes: functions, genomic organization, and neurologic diversity. Genes & development. 2000;15;14(10):1169–1180

Yamagata M, Herman JP, Sanes JR. Lamina-specific expression of adhesion molecules in developing chick optic tectum. J Neurosci. 1995;15:4556–71. https://doi.org/10.1523/JNEUROSCI.15-06-04556.1995.

Yamagishi S, Hampel F, Hata K, Del Toro D, Schwark M, Kvachnina E, Bastmeyer M, Yamashita T, Tarabykin V, Klein R, Egea J. FLRT2 and FLRT3 act as repulsive guidance cues for Unc5-positive neurons. EMBO J. 2011;30:2920–33. https://doi.org/10.1038/emboj.2011.189.

Yates LL, Papakrivopoulou J, Long DA, Goggolidou P, Connolly JO, Woolf AS, Dean CH. The planar cell polarity gene Vangl2 is required for mammalian kidney-branching morphogenesis and glomerular maturation. Hum Mol Genet. 2010;19:4663–76. https://doi.org/10.1093/hmg/ddq397.

Yogev S, Shen K. Cellular and molecular mechanisms of synaptic specificity. Annu Rev Cell Dev Biol. 2014;30:417–37. https://doi.org/10.1146/annurev-cellbio-100913-012953.

Yuzaki M. Two classes of secreted synaptic organizers in the central nervous system. Annu Rev Physiol. 2018;80:243–62. https://doi.org/10.1146/annurev-physiol-021317-121322.

Zhang LI, Poo MM. Electrical activity and development of neural circuits. Nat Neurosci. 2001;4(Suppl):1207–14. https://doi.org/10.1038/nn753.

Zhang RS, Liakath-Ali K, Südhof TC. Latrophilin-2 and latrophilin-3 are redundantly essential for parallel-fiber synapse function in cerebellum. eLife. 2020;9:e54443. https://doi.org/10.7554/eLife.54443.

Zhang X, Lin P-Y, Liakath-Ali K, Südhof TC. Teneurins assemble into presynaptic nanoclusters that promote synapse formation via postsynaptic non-teneurin ligands. Nat Commun. 2022;13:2297. https://doi.org/10.1038/s41467-022-29751-1.

Zhou L, Bar I, Achouri Y, Campbell K, De Backer O, Hebert JM, Jones K, Kessaris N, de Rouvroit CL, O'Leary D, Richardson WD, Goffinet AM, Tissir F. Early forebrain wiring: genetic dissection using conditional Celsr3 mutant mice. Science. 2008;320:946–9. https://doi.org/10.1126/science.1155244.

Zhu D, Li C, Swanson AM, Villalba RM, Guo J, Zhang Z, Matheny S, Murakami T, Stephenson JR, Daniel S, Fukata M. BAI1 regulates spatial learning and synaptic plasticity in the hippocampus. The Journal of clinical investigation. 2015: 1;125(4):1497–1508. https://doi.org/10.1172/JCI74603.

Zipursky SL, Sanes JR. Chemoaffinity revisited: dscams, protocadherins, and neural circuit assembly. Cell. 2010;143:343–53. https://doi.org/10.1016/j.cell.2010.10.009.

Zipursky SL, Wojtowicz WM, Hattori D. Got diversity? Wiring the fly brain with Dscam. Trends Biochem Sci. 2006;31:581–8. https://doi.org/10.1016/j.tibs.2006.08.003.

Chapter 6
Synaptic Cell Adhesion: A Structural Perspective

Sumit J. Bandekar, Szymon P. Kordon, and Demet Araç

Abstract Cell adhesion molecules (CAMs) play critical roles in mediating intercellular interactions in the context of the nervous system, such as guiding neuronal development, synapse formation and maturation, and synaptic plasticity. In addition to their extracellular adhesive roles, most CAMs induce intracellular signaling events and scaffold large protein complexes through intracellular domains. The molecular biology of how CAMs regulate synaptic development and function has been hugely advanced by decades of structural biology. These structures have illuminated multiple modes of CAM regulation, including how alternative splicing regulates CAM homotypic and heterotypic interactions. CAMs are diverse in size and contain a variety of adhesion domain classes such as immunoglobulin (Ig), leucine-rich repeats (LRR), and laminin G/neurexin/sex hormone (LNS). In this chapter, we detail structures of key synaptic adhesion complexes, including a mechanistic explanation of how these structures have informed functional work. Detailing the structural basis of synaptic adhesion provides a foundation for deciphering the complex interactions underlying neuronal connectivity and function in health and disease.

Keywords Cell adhesion molecule · Adhesion GPCR · Alternative splicing · Signaling · Synapse · Latrophilin

6.1 Introduction

Synapses represent the foundational units of communication within the human nervous system (NS), forming the labyrinthine network of connections between the 86 billion neurons that underlie all aspects of NS function (Scheiffele 2003; Südhof

S. J. Bandekar · S. P. Kordon · D. Araç (✉)
Department of Biochemistry and Molecular Biology, University of Chicago Neuroscience Institute, Institute for Biophysical Dynamics, and Center for Mechanical Excitability, University of Chicago, Chicago, IL, USA
e-mail: arac@uchicago.edu

E. T. Kavalali (ed.), *Nano-organization of the Synapse*, Advances in Neurobiology 48,
https://doi.org/10.1007/978-3-032-12594-1_6

and Malenka 2008; Missler et al. 2012; Südhof, 2018, 2021). Synapses allow neurons to transmit signals to one another, enabling the integration and processing of information, and this occurs in a controlled and reproducible manner (Scheiffele 2003; Südhof and Malenka 2008; Missler et al. 2012; Südhof, 2018, 2021). The dynamic properties of synapses, including their ability to undergo changes in strength and connectivity, underlie the brain's remarkable adaptability and capacity for learning and memory formation (Magee and Grienberger 2020). Moreover, synaptic dysfunction is implicated in a wide range of neurological and psychiatric disorders, highlighting the pivotal importance of synapses in maintaining nervous system health and function (Südhof, 2021). This chapter discusses the structural biology of cell adhesion molecules (CAMs) in synapse development and function. We detail representative structural studies of synaptic CAMs, and we review mechanistic details about synaptic adhesion and signaling events.

In the nervous system, CAMs play crucial roles in forming, organizing, maintaining, and pruning cell–cell interactions at various types of synapses (Yamagata et al. 2003; Missler et al. 2012; Südhof, 2018; Sanes and Zipursky 2020). In response to adhesive events between axons and dendrites, intracellular signaling events are instigated in both pre- and postsynaptic cells which trigger the assembly of the synaptic machinery, specialization, and other events (Fig. 6.1). These signaling events include second messenger generation, scaffolding of elaborate protein networks, changes in the cytoskeleton, and potentially others (Fig. 6.1). Furthermore, since synapses can diverge into many specializations, this requires a multitude of CAMs on both sides of the developing synapse (Südhof, 2018). Studies using proximity labeling experiments have identified the proteome at various synapse types (Ting et al. 2016; Cijsouw et al. 2018; van Oostrum et al. 2023). CAMs often function redundantly, which makes studying these systems difficult as perturbations can yield subtle phenotypes that are difficult to dissect (Südhof, 2018; Sanes and Zipursky 2020).

The focus of this review is on chemical rather than electrical synapses, and specifically excitatory synapses, as they have long been studied as model systems (Rollenhagen and Lübke 2006). Electron tomography analysis reveals that excitatory synapses typically have a synaptic cleft of 150–200 Å and can have a lateral size of ~500 nm (Fig. 6.1) (Burette et al. 2012; Tao et al. 2018; Cole and Reese 2023). The synaptic cleft is densely packed with many electron-dense components in a highly organized fashion (Burette et al. 2012; Biederer et al. 2017; Tao et al. 2018; Cole and Reese 2023); as early as 1959 a "band of extracellular material" was observed in the synaptic cleft (Gray 1959) and this was later shown to have even higher electron density than the cytoplasm, suggesting extremely high local concentrations of biological macromolecules (Zuber et al. 2005). Observed density in the synaptic cleft bridges across the synapse and connects to intracellular components on both sides, consistent with the idea of CAMs locked in a transsynaptic adhesive embrace and coordinating intracellular events on either side of the synaptic cleft (Fig. 6.1) (Burette et al. 2012; Biederer et al. 2017; Tao et al. 2018; Cole and Reese 2023). Studies using synaptosomes, which can be isolated from brain tissue and form a minimal functional synapse, or neurons grown on grids, have achieved

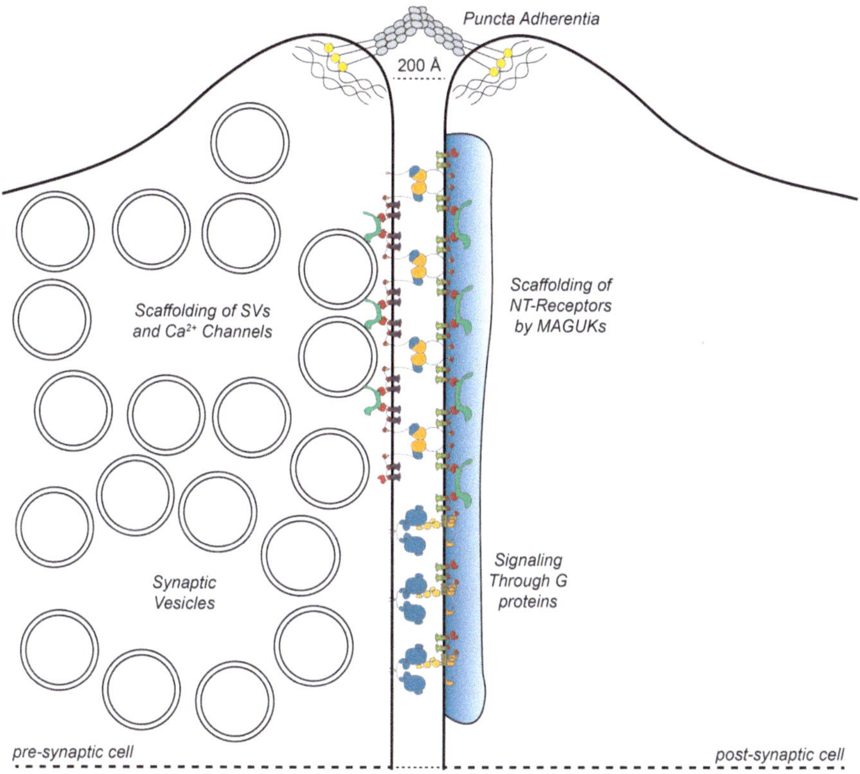

Fig. 6.1 *CAM complexes are key for synapse development.* An excitatory synapse is drawn roughly to scale, with the presynaptic terminal on the left and the postsynaptic terminal on the right. CAMs are shown in the synaptic cleft, and the events that they instigate are shown, such as presynpatic scaffolding (red and cyan scaffolding proteins) of synaptic vesicles (two concentric circles) and calcium channels (purple shapes). On the postsynaptic side, CAMs scaffold the tight clustering of neurotransmitter receptors such as NMDA or AMPA receptors (light green) and instigate intracellular signaling events such as G protein signaling. The postsynaptic density is shown as a blue gradient to indicate that the region is densely packed with molecules. Puncta adherentia toward the edge of the synapse are shown, including the cadherin molecules that mediate symmetric adhesion and their links to the actin cytoskeleton. The image displays only the top half of the synapse for clarity

higher resolution in cryogenic electron microscopy (cryo-EM) and may soon be able to resolve individual proteins at the synapse (Perez de Arce et al. 2015; Martinez-Sanchez et al. 2021; Sun et al. 2025; Held et al. 2024). Overall, the synaptic cleft represents a crowded environment full of a multitude of interactions between different macromolecules.

Both homophilic and heterophilic CAM interactions are required for synapses, and these coordinate to instigate a symphony of signaling events (Fig. 6.1) (Scheiffele 2003; Südhof and Malenka 2008; Togashi et al. 2009; Missler et al. 2012; Südhof, 2018, 2021). For instance, homophilic interactions of classical cadherins are present during early synapse development and contribute to cytoskeletal coordination through catenin signaling (Seong et al. 2015), and these interactions stabilize into *puncta adherentia*, which are found adjacent to the synaptic active zone (AZ) and help to maintain the structural integrity of the synapse (Togashi et al. 2009). The mature synapse is inherently asymmetric and requires heterophilic interactions between CAMs on axons and dendrites to develop (Südhof, 2021). On the presynaptic side, molecules such as neurexins (Nrxs) and receptor protein tyrosine phosphatases (RPTPs) connect to adaptor molecules such as calcium−/calmodulin-activated serine kinase (CASK) (Hata et al. 1996; Luo et al. 2020) and Liprinα (Marcó de la Cruz et al. 2024) to form the presynaptic AZ, which includes clustering of Ca^{2+} channels and formation of synaptic vesicles (SV) and the machinery for SV fusion. RPTP activity can also regulate synapse formation (Dunah et al. 2005) and this activity is regulated in response to RPTP adhesion and clustering (Xie et al. 2020). On the postsynaptic side, molecules such as the neuroligins and latrophilins have conserved C-terminal PSD-95/Discs-large/ZO-1 (PDZ) ligands that commonly interact with the PDZ domains of the membrane-associated guanylate kinase (MAGUK) or the SH3 and multiple ankyrin repeat (SHANK) families of scaffolding proteins (Irie et al. 1997; Tobaben et al. 2000; Zhu et al. 2016; Monteiro and Feng 2017; Südhof, 2021). MAGUK/SHANK proteins, such as postsynaptic density 95 (PSD-95), can multimerize and scaffold large protein complexes, which can induce the tight clustering of neurotransmitter receptors necessary for rapid response to neurotransmitter exocytosis (Chen et al. 2011b, 2015; Zhu et al. 2016). Recent work has begun to illustrate how MAGUKs and their scaffolded complexes can lead to phase-separated condensates (Zeng et al. 2019; Chen et al. 2020; Wu et al. 2021; Wang et al. 2024), consistent with the extremely high density of protein observed at the postsynaptic space in electron micrographs (Burette et al. 2012; Biederer et al. 2017; Tao et al. 2018; Cole and Reese 2023). Furthermore, the postsynaptic adhesion G protein-coupled receptors (aGPCRs), including latrophilins (ADGRLs) and brain angiogenesis inhibitors (ADGRBs), are present on the postsynaptic side and can signal through heterotrimeric G protein pathways (Stephenson et al. 2013; Lala and Hall 2022; Wang et al. 2024). ADGRBs can also coordinate cytoskeletal events directly through the RhoGEFs (Weng et al. 2019). Other signaling pathways, such as protein kinase A and protein kinase B, are important in synapse formation based on heterologous synapse formation assays (Jiang et al. 2021). To summarize, a plethora of signaling events are unleashed following the heterophilic interaction and tight clustering of CAMs across the synapse.

Our understanding of cell adhesion is guided in part by the differential adhesion hypothesis (DAH), which suggests that cells organize themselves to minimize their adhesive free energy, considering the strengths of homophilic and heterophilic interactions that they engage in (Honig and Shapiro, 2020). This is similar in principle to how oil and water separate when mixed. However, the DAH does not consider several variables such as heterogeneity of protein expression across the cell surface, cis interactions, downstream signaling, and the plethora of adhesion molecules present on the cell surface of neurites (Honig and Shapiro, 2020). Structural and molecular studies of the individual adhesion molecules that populate the neuron cell surface have helped further our understanding of how synaptic cell adhesion is guided (Fig. 6.2). Overall, CAM architecture manifests in a multitude of domain types and an extensive combination of these domains to make a variety of proteins (Honig and Shapiro, 2020). This system is further diversified by spatiotemporal regulation of alternative splicing, protein expression, and posttranslational modifications such as glycosylation (Missler et al. 2012; Südhof, 2018, 2021; Honig and Shapiro, 2020). These diversifications allow for the full complexity of the nervous system, including a full range of unique connections between neurons and specifications of a multitude of synapse types (Südhof, 2021). Atomic models, primarily derived from X-ray crystallography and accompanying structural data, such as electron microscopy and small angle X-ray scattering, have been instrumental in illuminating the mechanistic basis of CAM function at the synapse (Rudenko and Takahashi 2019). These data help us understand how CAMs operate within the tight distance constraints and high protein levels of the synaptic cleft, how they bind ligands, and how alternative splicing modulates CAM function.

6.2 Adhesion Domains Commonly Present in CAMs

CAMs consist of a variety of extracellular domains that are composed mostly of beta-strands (Fig. 6.3). Features of CAM domains include disulfide bonds, Ca^{2+}-binding sites, and glycosylation (Honig and Shapiro, 2020). CAMs are often composed of multiple smaller domains in series as opposed to a single large fold (Honig and Shapiro, 2020), potentially to allow for combinatorial shuffling of domains to increase protein diversity. Below, we outline several domain classes that are commonly found in CAMs and briefly discuss their overall features.

Immunoglobulin domains form the largest superfamily of domains present in CAMs and consist of a beta-sandwich fold often stabilized by disulfide bonds (Fig. 6.3) (Honig and Shapiro, 2020). Proteins such as RPTPs, MDGAs, SALMs, Slitrks, and more have Ig domains (Nam et al. 2011; Takahashi and Craig 2013; Kim et al. 2017; Won et al. 2019). Proteins with Ig domains often have fibronectin type III (FNIII) domains, which are also two-sheet beta-sandwich modules that are very similar in structure to Ig domains. Various surfaces of the Ig domain are used to interact with binding partners, although the longer lateral faces of the domain are typically used (Honig and Shapiro, 2020).

The laminin/neurexin/sex hormone (LNS) domain fold consists of a beta-sandwich domain with a hypervariable surface on the shorter side face composed of loops (Rudenko et al. 1999). The hypervariable surface often coordinates Ca^{2+} and can be used to bind a variety of proteins (Rudenko et al. 1999). This domain type is most famously found in neurexins in the context of the synapse but is present in other proteins as well.

Leucine-rich repeat (LRR) domains consist of the conserved amino acid sequence, LxxLxLxxN/CxL, and are found as a tandem repeat of individual motifs, which results in an overall horseshoe-like shape (Kobe and Kajava, 2001; Roppongi

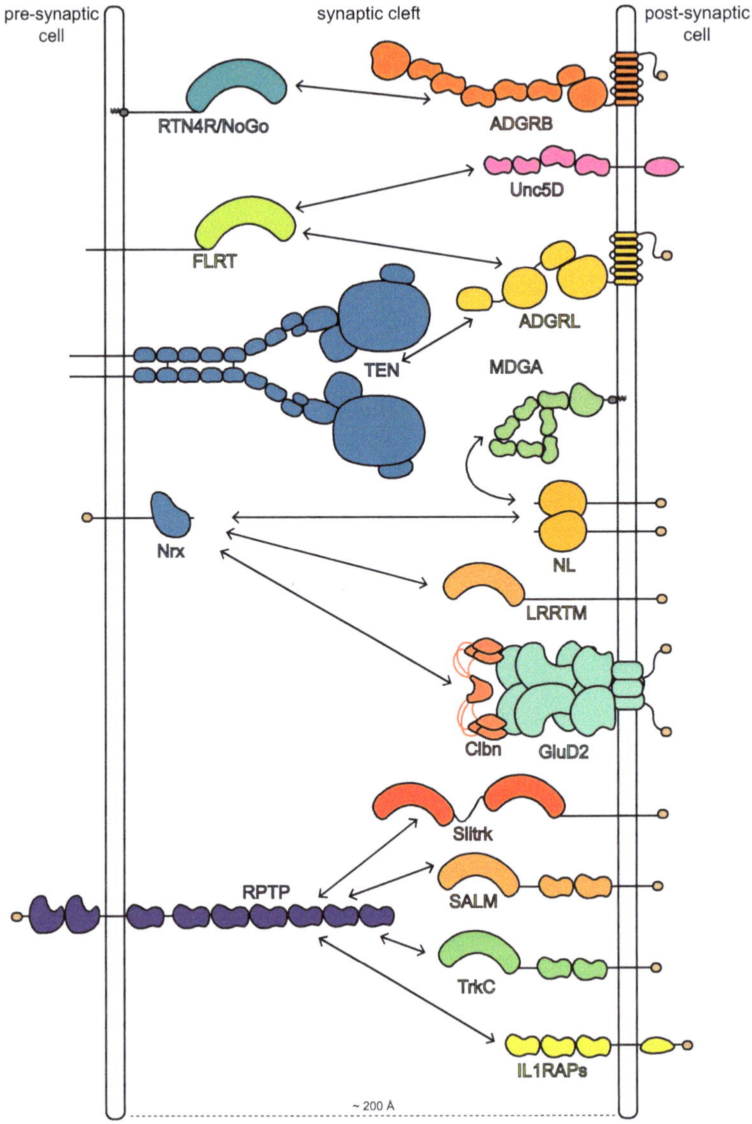

Fig. 6.2 *The Panoply of CAM interactions at the synapse leads to diverse outcomes***.** The CAMs covered in this chapter are presented here to convey the complexity of events that can occur during synaptic development. Presynaptic molecules are presented on the left side and postsynaptic molecules on the right side. Interactions validated by structural and mechanistic studies are shown using double-headed arrows.

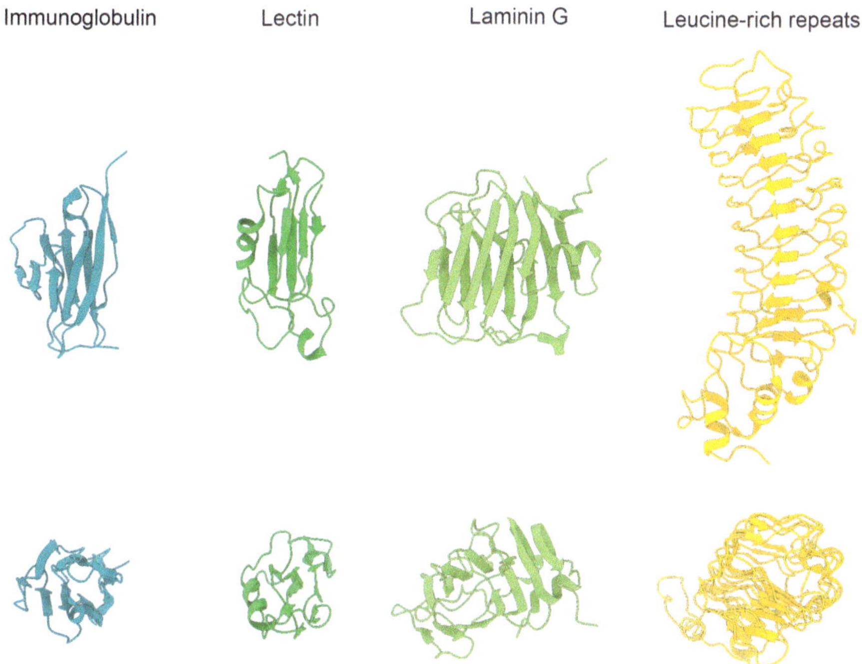

Fig. 6.3 *Common domain types observed in synaptic CAMs.* Common domains are shown using cartoon representations. Immunoglobulin domain: PDB 3 M45, lectin domain: PDB 6VHH, laminin G domain: PDB 2R1D, and leucine-rich repeat domain: PDB 5Z8X.

et al. 2017). The LRR is capped on the N- and C-terminus by different capping regions (N-Cap and C-Cap) (Kobe and Kajava, 2001; Roppongi et al. 2017). Various regions of the horseshoe can be used to interact with other proteins or, in some cases, proteoglycans. Synaptic proteins included in this class are the FLRT and LRRTM proteins, as well as the SALM molecules (Nam et al. 2011; Lu et al. 2015; Roppongi et al. 2017).

Other proteins have unique folds adapted from other protein classes. Teneurins are extremely large synaptic molecules that contain several domains organized around a large beta barrel and bear striking structural homology to bacterial Tc-toxins (Jackson et al. 2018; Li et al. 2018). Various surfaces of the beta barrel are used to interact with protein ligands. Neuroligin is another example, and it has an esterase domain similar to acetylcholinesterase, which is an α/β hydrolase fold (Araç et al. 2007). the neuroligin esterase domain is incompatible with catalysis (Araç et al. 2007). In the following sections, we detail representative structural studies of important synaptic CAMs and how they inform mechanistic details about synaptic adhesion and signaling events. Whenever possible, our structural renderings in the presented figures are to scale.

6.3 Neurexins and Their Interaction Partners

The presynaptic Neurexins (Nrxs) and their postsynaptic binding partners, most notably the neuroligins (NLs), are broadly known as some of the most famous synaptic adhesion molecules (Südhof 2017; Rudenko 2019; Gomez et al. 2021). They serve as a model system for how we understand synaptic CAMs. Neurexins were identified due to their role as the Ca^{2+}-dependent receptor for α-latrotoxin (αLTx) (Petrenko et al. 1990). These proteins have been studied for over 25 years and are involved in regulating synaptic specificity and maturation (Südhof 2017; Rudenko 2019; Gomez et al. 2021). Nrxs and NLs are tightly associated with autism spectrum disorder and other neurological conditions (Südhof 2017; Rudenko 2019; Gomez et al. 2021). In mammals, a vast array of Nrx protein products is generated and presented on the presynaptic surface through a combination of three genes (Nrx1-3), each with three alternative promoters (α, β, γ) and alternative splicing at six sites (SS1-6) (Schreiner et al. 2014; Treutlein et al. 2014; Fuccillo et al. 2015). Nrxs can interact with postsynaptic ligands depending on which version is produced and which ligands are present postsynaptically (Südhof 2017; Rudenko 2019). These ligands include various postsynaptic and soluble proteins, which are highlighted in the following sections (Südhof 2017; Rudenko 2019).

The longest Nrx version, Nrxα, has six LNS domains (LNS1-LNS6) with single EGF repeats between LNS1 and LNS2 (EGFA), LNS3 and LNS4 (EGFB), and LNS5 and LNS6 (EGFC), a heparan sulfate (HS) modification between LNS6 and the membrane, and a small transmembrane region (TMR) (Fig. 6.4a) (Südhof 2017; Gomez et al. 2021). Nrxβ contains only the LNS6 domain, a HS modification site, and the TMR (Südhof 2017; Gomez et al. 2021). Nrxγ is the shortest isoform and only has a short extracellular region that is HS-modified and the TMR (Südhof 2017; Gomez et al. 2021). At the far end of the intracellular C-terminus, Nrxs have a PDZ ligand, as many synaptic CAMs do (Fig. 6.4a) (Hata et al. 1996; Rudenko 2019; Gomez et al. 2021). Six alternatively spliced sites (SS1-6) generate further Nrx diversity. Sequencing studies detect thousands of neurexin isoforms within the mammalian brain (Schreiner et al. 2014; Treutlein et al. 2014; Fuccillo et al. 2015), and isolated neuron populations contain specific versions of neurexins rather than a broad array of neurexin isoforms (Fuccillo et al. 2015). Together, these data suggest that individual cells use a specific repertoire of neurexins to coordinate synaptic connections.

Nrx1β was the first to be structurally characterized (Fig. 6.4b) (Rudenko et al. 1999) and revealed the three-dimensional architecture of the LNS fold as two parallel beta sheets, each with seven strands. Importantly, this structure revealed that although SS2, SS3, and SS4 occur in different LNS domains, they all localize to the loop regions on the same surface of the LNS fold known as the hypervariable surface (Fig. 6.4b), suggesting that this surface forms a regulatory surface or a ligand-binding surface.

Over a decade later, the Nrx1α crystal structure was determined (Chen et al. 2011a), which consists of an L-shaped molecule with LNS2–5 making the long leg

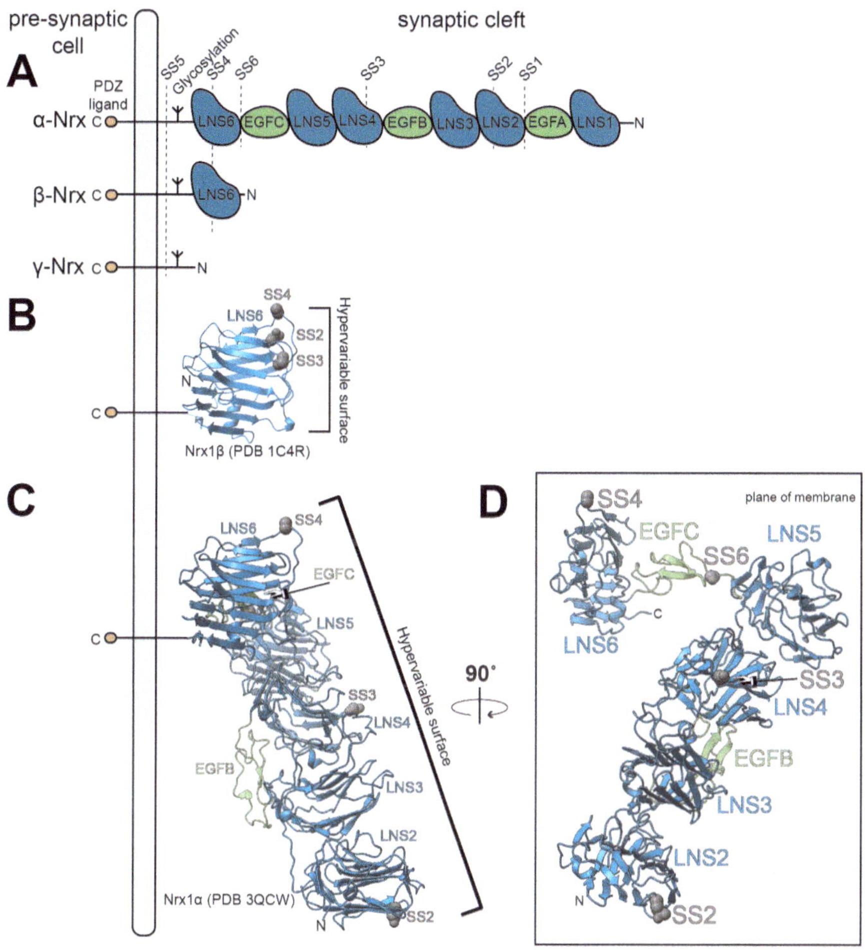

Fig. 6.4 *Features of the presynaptic hub molecule neurexin.* (**a**) Domain architectures of the neurexins. The sites of alternative splicing are shown as vertical dotted lines. SS: splice site. LNS: laminin G/neurexin/sex hormone. EGF: epidermal growth factor domain. (**b**) The structure of Nrx1β is shown in cartoon representation with splice sites highlighted using gray spheres. (**c**) The structure of Nrx1α is shown in cartoon representation with (**d**) an orthogonal view.

of the L and the LNS6 protruding ~90° from LNS2–5 (Fig. 6.4c, d). In relation to this L shape, the LNS1 domain and EGFA are not ordered in the crystal structures of Nrx1α. Additional work confirmed that there are two major points of flexibility in Nrx1α (Liu et al. 2018): first, between LNS1 and LNS2, and second, between LNS5 and LNS6. Based on their arrangement in Nrx1α (Chen et al. 2011a; Liu et al. 2018), the LNS domain hypervariable surfaces, some of which are ligand-binding sites, are present on the same side of the structure (Fig. 6.4c). This implies that multiple ligands could be recruited simultaneously if they bind to different LNS repeats (Liu et al. 2018). Furthermore, if protein partners need to interact with more than one LNS domain at the same time, it is possible in this arrangement (Liu et al. 2018). The orientation of the hypervariable surfaces also suggests the orientation of the Nrxα molecule relative to the membrane—the long leg of the L, including LNS1–5, likely sits parallel to the membrane surface (Chen et al. 2011a; Liu et al. 2018). This orients the hypervariable surfaces of the LNS domains toward the postsynaptic membrane for postsynaptic adhesion molecules to interact with (Liu et al. 2018). This Nrx1α structural data also revealed the three-dimensional positions of other splice inserts. SS1 is located between EGFA and LNS2, and SS6 is between LNS5 and EGFC (Fig. 6.4c), suggesting that these splice inserts could modulate the level of flexibility between domains and allow Nrx1α to access different binding partners in the synaptic cleft (Liu et al. 2018). Finally, the SS5 insert contains a putative protease cleavage site that can be cleaved by metalloproteases that are present in the synaptic cleft (Liu et al. 2018). Thus, the SS5 insert may be a mechanism to generate secreted Nrx modules or to turn over Nrx-mediated adhesion sites.

Nrxs can engage in several heterophilic protein–protein interactions, which lead to a multitude of adhesive outcomes (Südhof 2017; Rudenko 2019; Gomez et al. 2021). Furthermore, these heterophilic interactions are regulated by alternative splicing (Südhof 2017; Rudenko 2019).

Neuroligins (NLs) are the best-characterized Nrx binding partner and are membrane-tethered proteins with large extracellular domains with homology to α/β hydrolases such as acetylcholinesterases (Bemben et al. 2015). Following the extracellular domain (Fig. 6.5a), NLs have short TMRs, an intracellular region, and a PDZ ligand at the C-terminus (Fig. 6.5a) (Irie et al. 1997). There are four NLs in humans, and NLs are frequently mutated in autism spectrum disorder (ASD) (Bemben et al. 2015). Each NL is localized to different synaptic types, with NL1 and NL4 located at excitatory synapses, NL2 located at inhibitory synapses, and NL3 found at both types (Bemben et al. 2015; Marro et al. 2019). NLs are typically obligate homodimers; however, there is evidence for NL heterodimers (Budreck and Scheiffele 2007; Poulopoulos et al. 2012; Bemben et al. 2015). NLs are alternatively spliced at two sites known as splice site A and splice site B (SSA, SSB) (Fig. 6.5a), which regulate interactions with their binding partners (Chih et al. 2006; Koehnke et al. 2010; Lee et al. 2010).

The first structures of NL were determined in 2007; these were released coincident with Nrxβ–NL complex structures (Araç et al. 2007; Fabrichny et al. 2007; Chen et al. 2008). These structures revealed that NLs do have α/β hydrolase folds with a central twisted β-sheet that is surrounded by α-helices (Fig. 6.5a). The obligate NL dimer is mediated by two helices from each monomer, with several hydrophobic residues involved. The esterase catalytic site residues are not compatible with catalysis, and the entrance to the active site is not solvent-exposed as it would be in a catalytically competent esterase (Araç et al. 2007).

The Nrxβ–NL complex structures revealed the previously described Ca^{2+}-dependent interaction (Ichtchenko et al. 1996; Nguyen and Südhof 1997) between the hypervariable surface of the LNS6 domain of Nrxβ (Fig. 6.5b) and the side face of the esterase domain of NL (Fig. 6.5c, d) (Araç et al. 2007; Fabrichny et al. 2007; Chen et al. 2008). Strikingly, these structures revealed that the three-dimensional

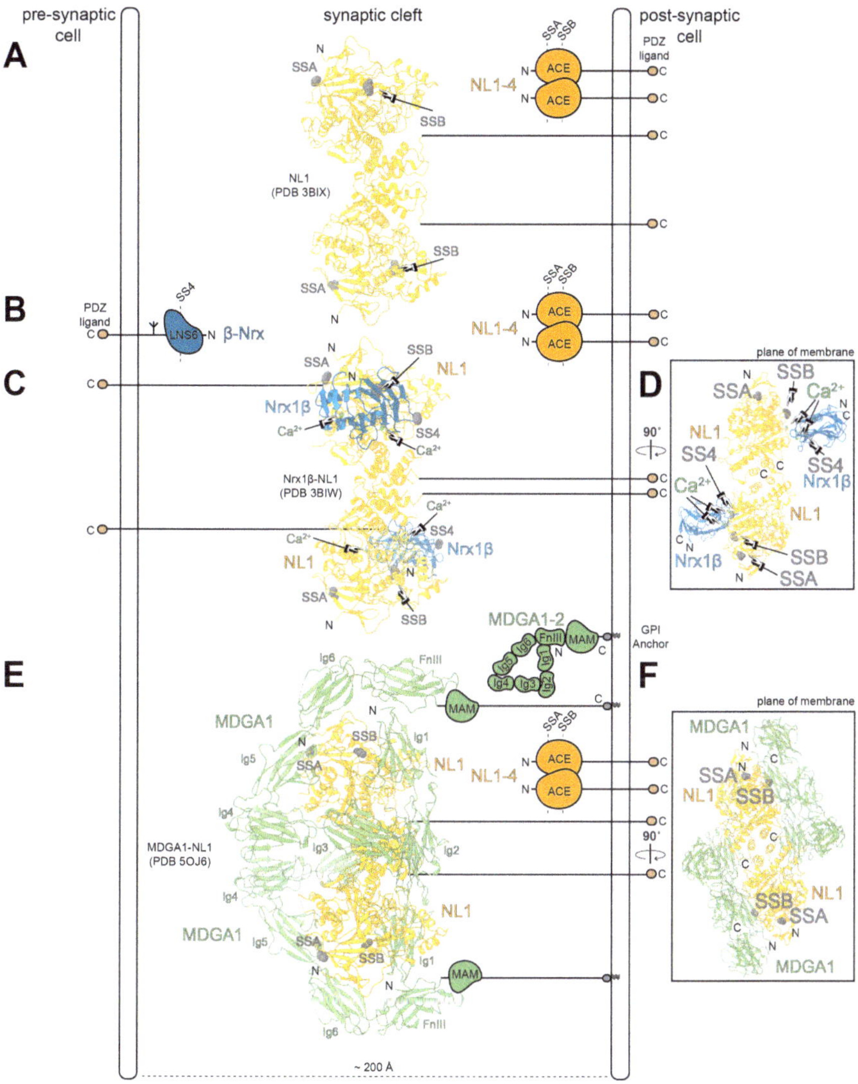

Fig. 6.5 *Features of Nrx/NL interaction and modulation by MDGA*. (**a**) Domain architectures and dimeric structure of the neuroligins. (**b**, **c**) Domain architectures and structures of the Nrx/NL interaction with orthogonal view shown in (**d**). (**e**) Domain layouts and structure of the MDGA/NL interaction with orthogonal view shown in (**f**).

location of both the Nrx SS4 and NL SSB splice inserts is directly at the Nrxβ–NL binding interface, providing an atomic explanation for previous observations of splice-site specific regulation of Nrxβ–NL binding (Fig. 6.5c, d) (Ichtchenko et al. 1996; Boucard et al. 2005; Comoletti et al. 2006). The presence of both Nrxβ SS4 and NL SSB reduces the affinity of the Nrxβ–NL interaction (Ichtchenko et al.

1996; Boucard et al. 2005; Comoletti et al. 2006), which follows as these inserts add extra residues directly at the Nrx–NL interface and could disrupt the tight network of interactions that form. Although there is no high-resolution structure of Nrxα in complex with any NL, biochemical studies have shown that these proteins do interact (Boucard et al. 2005).

The MAM-domain-containing glycosylphosphatidylinositol anchor proteins (MDGA) form a regulatory axis for the Nrx/NL interaction (Pettem et al. 2013; Connor et al. 2016). MDGA proteins are found on the postsynaptic side and can inhibit Nrx-/NL-dependent synapse development of both excitatory and inhibitory synapses (Pettem et al. 2013; Connor et al. 2016). MDGA proteins contain ECRs with six Ig domains (Ig1–6), one FnIII domain, and a MAM domain (Fig. 6.5e). At the C-terminus, these proteins are linked to the membrane by a glycosylphosphatidylinositol (GPI) anchor. Three groups determined crystal structures of the MDGA1 ECR in complex with NL (Elegheert et al. 2017; Gangwar et al. 2017; Kim et al. 2017). One structure used the full MDGA ECR, and this revealed a triangular architecture where the Ig1 domain contacts the FnIII domain (Fig. 6.5e, f) (Elegheert et al. 2017). Ig1/2 forms one face of the triangle, Ig3/4 the next, and Ig5/6 and FnIII the last face. Complexes of MDGA1 with NL1 revealed that the Ig1 domain forms extensive contacts with the side face of one NL protomer, and Ig2 contacts the other NL protomer in a Ca^{2+}-independent manner (Fig. 6.5e, f) (Elegheert et al. 2017; Gangwar et al. 2017; Kim et al. 2017). The binding surface of MDGA Ig1 to NL1 overlaps with LNS6 of Nrx binding to NL, which strongly suggests a mechanism for how MDGA proteins can regulate Nrx/NL interactions, through direct steric occlusion (Elegheert et al. 2017; Gangwar et al. 2017; Kim et al. 2017). The MDGA/NL interaction has a similar binding affinity as Nrx/NL, suggesting that these two interactions can compete depending on the expression levels of each protein present and which SS4 splice state Nrx is present in (Elegheert et al. 2017; Kim et al. 2017).

Nrxs also bind to another postsynaptic ligand, leucine-rich repeat transmembrane neuronal proteins (LRRTMs) (Ko et al. 2009; Siddiqui et al. 2010; Roppongi et al. 2017). LRRTMs are synaptogenic proteins implicated in neurological disorders such as schizophrenia, ASD, Alzheimer's disease, and others (Roppongi et al. 2017). There are four LRRTM proteins in mammals, and the LRRTM domain architecture consists of an extracellular LRR region with N- and C-caps, followed by a single-pass transmembrane region and an intracellular region (Fig. 6.6a). Additionally, there is a C-terminal PDZ-binding motif, which binds to MAGUKs, and this can induce clustering of excitatory neurotransmitter receptors (Roppongi et al. 2017). The structure of LRRTM2 in complex with Nrx1β revealed that LNS6 uses its hypervariable surface to interact with the C-cap of LRRTM2 in a Ca^{2+}-dependent manner, and that this binding interaction directly conflicts with the Nrx/NL interaction (Fig. 6.6b) (Yamagata et al. 2018). The binding affinity of Nrx/LRRTM2 is similar to Nrx/NL, but the binding affinity of Nrx to LRRTM2 is strikingly decreased in Nrx SS4+, whereas the shift due to SS4+ is more subtle for Nrx/NL (Yamagata et al. 2018). Thus, Nrx SS4+ will preferentially interact with NL over LRRTM.

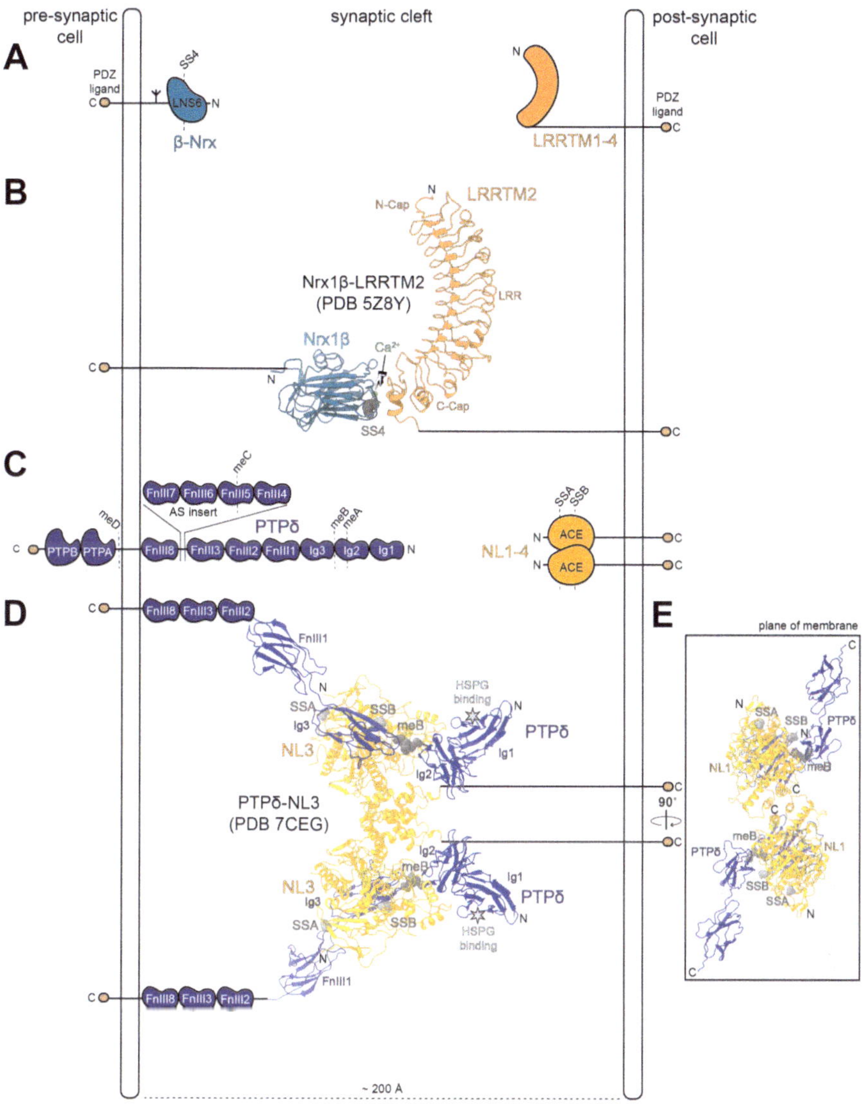

Fig. 6.6 *Features of the Nrx/LRRTM and RPTP/NL interactions*. (**a**) Domain architectures of neurexins and LRRTMs. (**b**) Structure of the Nrx/LRRTM interaction. (**c**) Domain layouts of the RPTP and NLs. (**d**) Structure of the RPTP/NL interaction with orthogonal view shown in (**e**).

The type IIA receptor protein tyrosine phosphatases (RPTPs) form another family of presynaptic adhesion molecules that interact with NLs in a splice-isoform-dependent manner (Takahashi and Craig 2013). RPTPs contribute to synaptic adhesion and synapse organization, and they are associated with ASD and schizophrenia (Takahashi and Craig 2013). RPTPs contain a series of three Ig domains (Ig1–3), four or eight FnIII domains depending on splicing (FnIII1–8), a short

TMR, and two protein tyrosine phosphatase domains (Fig. 6.6c). RPTP function is modulated by four alternatively spliced inserts termed mini-exons A–D (meA–meD), each encoding short amino acid sequences with less than ten residues (Takahashi and Craig 2013). The meA insert is located in Ig1, meB inserts are located between Ig2 and Ig3, meC is located in FnIII5, and meD is located in the region between the TMR and the first phosphatase domain. PTPδ is an RPTP that interacts with NL3, but only if the meB splice insert is absent (meB−) (Yoshida et al. 2021). Ig2 and Ig3 of PTPδ meB− interact with the esterase domain of NL3, where Ig3 sterically occludes LNS6 of Nrx from binding (Fig. 6.6d, e) (Yoshida et al. 2021). In contrast, PTPδ meB+ interacts with a separate set of ligands using Ig2 and Ig3; this is reviewed in a later section of this chapter (Takahashi and Craig 2013; Yoshida et al. 2021).

Recent work has shown the importance of soluble molecules in the trans-synaptic space, which can bridge pre- and postsynaptic CAMs together to mediate adhesion. Neurexophilins (Nxphs) are a family of secreted glycoproteins that are thought to play a role in GABAergic neurotransmission (Fig. 6.7a) (Born et al. 2014). Nxphs interact with the LNS2 domain of Nrx using a distinct surface, not the hypervariable surface employed for most Nrx interactions. This is done via a unique architecture where Nxph forms a contiguous β-sheet with Nrx LNS2, which extends the beta-sandwich fold of the LNS2 domain (Fig. 6.7b, c) (Wilson et al. 2019). Nxph also provides some side chains that interact with LNS2 residues (Wilson et al. 2019). This Nrx/Nxph interaction is regulated by Nrx splice site SS2, where SS2+ increases Nxph affinity for Nrx (Wilson et al. 2019). Two versions of SS2+ both increase the affinity of Nxph for Nrx (Wilson et al. 2019).

The cerebellin (Clbn) proteins are soluble molecules that link presynaptic Nrx to postsynaptic ionotropic Glutamate delta receptors (GluD1–2) to regulate levels of certain neurotransmitter receptors such as AMPA-type and NMDA-type receptors (Südhof 2023). This is a phenomenon that depends on the adhesive properties of the GluD receptor rather than ion transduction; the transmembrane domain of GluD can be removed without defects in synapse formation whereas the intracellular portion of GluD is necessary for synapse formation (Dai et al. 2021) There are four Clbns in humans, and they consist of cysteine-rich amino-terminal domains (NTDs) and C1q-like globular domains (Fig. 6.7d). Clbns form obligate trimers via their C1q domains, and these trimers can form dimers via their NTD to form a hexameric Clbn (Fig. 6.7d). Crystal structures and accompanying biophysical and biochemical data (Cheng et al. 2016; Elegheert et al. 2016) revealed that the amino-terminal domain (NTD) of GluD2 receptors can interact with a trimer of the Clbn1 C1q domain (Fig. 6.7e). Furthermore, a hexameric Clbn1 is necessary for interaction with Nrx1β, and the Clbn/Nrx1 interaction occurs only for Nrx1 SS4+ (Uemura et al. 2010; Elegheert et al. 2016). The interaction of Nrx1β with Clbn1 includes the hypervariable surface, but the high-resolution structure is not known (Elegheert et al. 2016).

For Nrxs, the SS4+ splice site serves as a master switch to regulate several interactions between LNS6 and postsynaptic adhesion partners. In the context of SS4+, interactions with NLs are slightly reduced in affinity, but the Nrx/NL pair still can

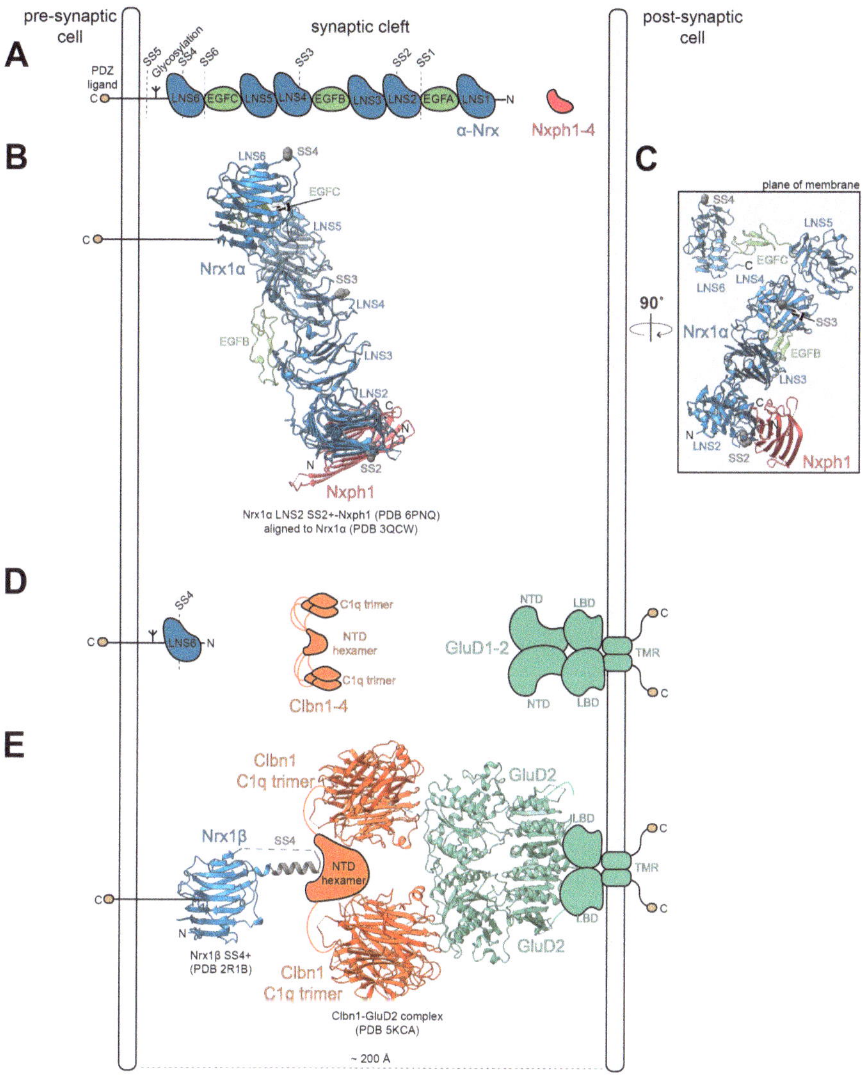

Fig. 6.7 *Features of the Nrx/Nxph and Nrx/Clbn/GluD2 interactions*. (**a**) Domain architectures of neurexins and neurexophilins. (**b**) Structure of the Nrx/Nxph interaction with orthogonal view in (**c**). (**d**) Domain layouts of the neurexins, cerebellins, and glutamate D2 receptors. (**e**) Structure of the Nrx/Clbn/GluD2 interaction.

interact (Ichtchenko et al. 1996; Boucard et al. 2005; Comoletti et al. 2006). With Nrx SS4+, NL is more likely to interact with MDGA than with Nrx, as the complex affinities are similar (Elegheert et al. 2017). Next, interactions between Nrxs and LRRTMs are strongly reduced (Yamagata et al. 2018). Finally, interactions between Nrxs and Clbns are strongly enhanced, which form a trans-synaptic bridge to GluD2 (Elegheert et al. 2016).

How these proteins mediate their effects at the synapse largely depends on their intracellular PDZ ligands interacting with PDZ domains, most commonly those of the MAGUK or SHANK families. Most of the abovementioned proteins have PDZ ligands in their C-termini, and the structures of neurexin bound to the MAGUKs calcium−/calmodulin-dependent serine protein kinase CASK (PDB 6NID) (Hata et al. 1996) and spinophilin (Muhammad et al. 2015) are known. Nrx interaction with CASK has been shown, and CASK can phosphorylate the intracellular tail of Nrx (Mukherjee et al. 2008) and nucleate filamentous actin from Nrx (Biederer and Südhof 2001). This Nrx/CASK interaction is crucial for the organization of the presynaptic active zone through the Liprinα proteins (Marcó de la Cruz et al. 2024); the structure of CASK bound to Liprinα is also known, providing a pathway from Nrx to Liprinα (Wei et al. 2011). On the postsynaptic side, neuroligin binds to the MAGUK PSD-95 (Irie et al. 1997) and several other PDZ domain proteins (Meyer et al. 2004), and the intracellular portion of GluD2 is necessary for its synaptic function through a PDZ interaction (Dai et al. 2021). NLs are shown to interact with a variety of MAGUKs (Irie et al. 1997; Meyer et al. 2004). MAGUKs have three PDZ domains as well as guanylate kinase-like domains that bind to phosphorylated peptides, and thus they link huge assemblies of molecules (Zhu et al. 2016) and likely play a master role in synapse organization.

6.4 Latrophilins and Their Binding Partners

Another CAM interaction hub is formed by the presynaptic teneurins (TENs) and fibronectin leucine-rich repeat transmembrane proteins (FLRTs) interacting with the postsynaptic latrophilins (ADGRLs or LPHNs). ADGRLs were originally identified due to their role as the Ca^{2+}-independent receptor of αLTx (Krasnoperov et al. 1997). ADGRLs are postsynaptic molecules that play key roles in synapse development through scaffolding of postsynaptic protein complexes as well as their ability to induce signaling through heterotrimeric G proteins (Sando and Südhof 2021; Wang et al. 2024). These key receptors integrate signals from various sources and output through several mechanisms (Sando et al. 2019), making them kingpins of the synapse (Fig. 6.8). ADGRL function can now be modulated due to the recent development of pioneering synthetic antigen binders (sABs) targeting ADGRLs (Kordon et al. 2023). ADGRLs are some of the most highly relevant synaptic proteins (Südhof, 2018), and through their interaction with the presynaptic TENs and FLRTs, the ADGRLs direct specific synapse development depending on the identity and splice isoform of each molecule present (Li et al. 2018; Sando et al. 2019). ADGRLs are associated with attention deficit hyperactivity disorder (ADHD), bipolar disorder, autism spectrum disorder (ASD), and cancers (Meza-Aguilar and Boucard 2014). Shortly after their discovery, ADGRLs were bioinformatically identified as members of the G protein-coupled receptor (GPCR) family (Lelianova et al. 1997; Sugita et al. 1998) and were later classified into the adhesion family of GPCRs (aGPCRs) (Fredriksson et al. 2003). There are three ADGRLs in

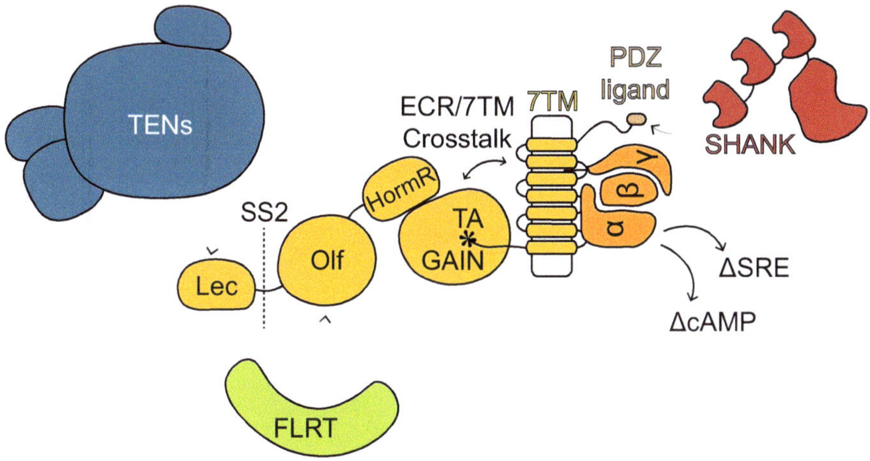

Fig. 6.8 *Latrophilins as a postsynaptic signal integrator***.** Various interactions and features of latrophilins are shown as cartoons, including binding partners, modes of signaling, and scaffolding interactions.

vertebrates, and the domain architecture consists of a lectin (Lec) domain, an olfactomedin (Olf) domain, a hormone receptor (HormR) domain, and the GPCR autoproteolysis-inducing (GAIN) domain (Araç et al. 2012) (Fig. 6.9a). Following this is a GPCR seven-transmembrane domain (7TM) as well as a C-terminal intracellular region (Krasnoperov et al. 1997) with a PDZ ligand. ADGRLs are alternatively spliced in at least five different sites (termed herein using nomenclature for *Mus musculus* ADGRL3 SS1-SS5 from Wang et al. 2024). SS1 occurs at the far N-terminus of the receptor before the Lec domain, SS2 occurs between the Lec and Olf domains (this was previously referred to as "SSA" (Boucard et al. 2014)), SS3 in the GAIN domain, SS4 in the third intracellular loop of the 7TM region, and SS5 at the far C-terminus (Wang et al. 2024). SS2 can regulate binding to teneurins (Boucard et al. 2014), and SS4/SS5 can modulate the synaptic function of ADGRLs (Wang et al. 2021b, 2024).

Biochemical studies in the early 2000s showed that ADGRLs consist of two non-covalently linked sequences, termed N-terminal fragment (NTF) and C-terminal fragment (CTF). The NTF and CTF are generated by posttranslational proteolysis at the GPCR proteolysis site (GPS) (Krasnoperov et al. 2002); the function of the GPS remained a mystery for over a decade. In 2012, the crystallographic structure of the ADGRL1 GAIN domain was revealed (Araç et al. 2012). This structure showed that the GAIN domain, the only ECR domain conserved throughout 32 of 33 aGPCRs (Prömel et al. 2013), houses its cleaved, yet associated, final β-strand, which is the N-terminus of the CTF (Araç et al. 2012), and this strand constitutes the GPS. The relevance of this was still unclear until work from two groups (Liebscher et al. 2014; Stoveken et al. 2015) revealed that the final β-strand of the GAIN domain acts as the orthosteric agonist of the 7TM, leading to its designation as the tethered agonist (TA). The GAIN domain shields the TA inside of it until the NTF is shed from the

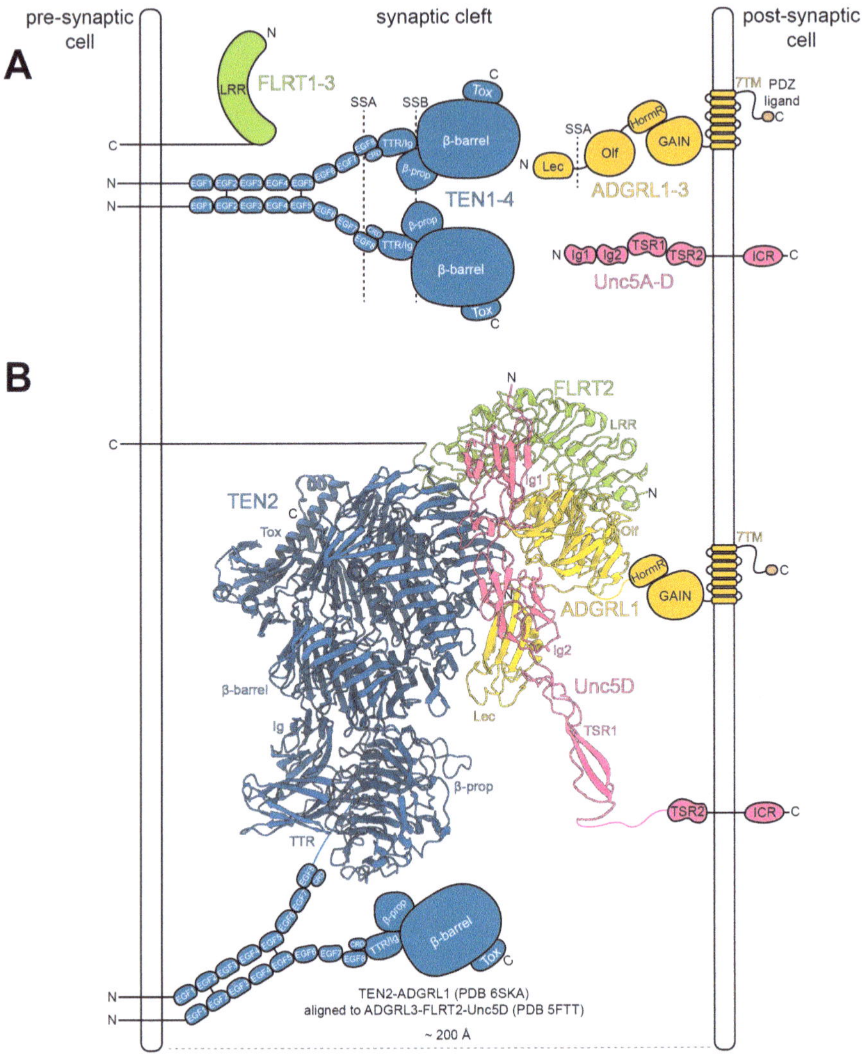

Fig. 6.9 *Network of Latrophilin interactions revealed by structural biology*. (**a**) Domain architectures of the proteins involved in ADGRL interaction. (**b**) Structural model of a super-complex between TEN/FLRT/ADGRL/UNC5D.

CTF; this is hypothesized to occur via ligand binding and/or force application (Vizurraga et al. 2020; Lala and Hall 2022). This TA-dependent mechanism for aGPCR activation now has been shown for many aGPCRs (Liebscher et al. 2014; Demberg et al. 2015, 2017; Schöneberg et al. 2015, 2016; Stoveken et al. 2015; Wilde et al. 2016; Perry-Hauser et al. 2022; Bernadyn et al. 2023) and the cryogenic electron microscopy (cryo-EM) structures of several aGPCRs (including ADGRL3) with the TA bound in the orthosteric site are available (Barros-Álvarez et al. 2022;

Ping et al. 2022; Qu et al. 2022; Xiao et al. 2022). Furthermore, cryo-EM structures of ADGRL3 in complex with the four major classes of heterotrimeric G proteins are available (Qian et al. 2022).

However, efficient cleavage in the GAIN domain is not required for several functions of ADGRLs, including synapse formation (Prömel et al. 2012; Sando et al. 2019), suggesting that ADGRLs can operate independently of TA-mediated activation. There is work that supports the idea that allosteric communication between the ECR and 7TM can regulate signaling in several aGPCRs (Kishore et al. 2016; Salzman et al. 2017; Kordon et al. 2023, 2024; Bandekar et al. 2025). Furthermore, available evidence suggests that some aGPCRs are not cleaved in the consensus GPS within the GAIN domain (Bui et al. 2023), and some do not respond to TA stimulation in assays for G protein signaling or downstream pathways (Favara et al. 2021; Dates et al. 2024). Emerging work has shown the relative ECR/7TM orientation of ADGRL3 using cryo-EM and demonstrated that changes in ADGRL3 signaling are linked to changes in ECR/7TM orientation (Kordon et al. 2024). Taken together, this evidence suggests that the ECRs of aGPCRs are not simply caps for the TA; they also communicate directly with the 7TM and thus merit intense study as the crown jewels of these enigmatic receptors.

The structure of the HormR domain is now known for several aGPCRs (Araç et al. 2012; Leon et al. 2020; Bandekar et al. 2025). Mysteriously, the function of the HormR domain remains unknown in any aGPCR, although HormR domains are present in several aGPCR subfamilies (Vizurraga et al. 2020; Lala and Hall 2022). Studies in the 2010s revealed the structural basis of the interaction of ADGRLs with their physiological ligands, the fibronectin leucine-rich transmembrane proteins (FLRTs) and the teneurins (TENs), which are detailed below.

FLRTs were identified as candidate ADGRL ligands in an affinity chromatography–mass spectrometry screen against the ADGRL ECR (O'Sullivan et al. 2012), and they were shown to interact directly with ADGRLs and to form trans-synaptic complexes. FLRTs were shown to regulate excitatory synapse function (O'Sullivan et al. 2012). There are three FLRTs in mammals (FLRT1–3) (Lacy et al. 1999), and they have an N-terminal LRR region with N- and C-terminal caps, a FnIII domain, and a short transmembrane region (Fig. 6.9a) (Lu et al. 2015). FLRTs form a weak dimer species through their LRR, which is thought to be a cis dimer (Seiradake et al. 2014; Lu et al. 2015).

Structural and biophysical studies performed by three competing groups revealed that the Olf domain of ADGRL interacts with the concave side of the FLRT LRR to make a tight complex (Jackson et al. 2015; Lu et al. 2015; Ranaivoson et al. 2015) (Fig. 6.9b) and that this interaction occurs in trans. Another postsynaptic FLRT ligand, uncoordinated 5 (UNC5), interacts with FLRT in a manner incompatible with FLRT dimerization (Fig. 6.9b) (Seiradake et al. 2014; Lu et al. 2015), although the relevance of this interaction to synaptic development is unclear.

The other major ligand of ADGRLs is the TEN family. TENs are a family of type II transmembrane proteins that play essential roles in neuronal development, synaptogenesis, and circuit formation (Young et al. 2013; Berns et al. 2018). TENs regulate synapse formation (Leamey and Sawatari 2014; Woelfle et al. 2016) and TEN perturbations impair synapse development (Silva et al. 2011; Hong et al. 2012; Mosca et al. 2012; Boucard et al. 2014; Mosca 2015). Mutations in TENs are linked to microphthalmia, congenital anosmia, essential tremors, and cancers (Aldahmesh et al. 2012; Ziegler et al. 2012; Hor et al. 2015; Alkelai et al. 2016; Chassaing et al. 2016; Graumann et al. 2017; Talamillo et al. 2017). There are four TENs in vertebrates (TEN1–TEN4), two in flies (TEN-m and TEN-a), and one in worms (Tucker et al. 2012; Tucker 2018; Wides 2019). TENs consist of an N-terminal intracellular region, a short TMR, and a large extracellular region with eight epidermal growth factor (EGF)-like repeats (EGF1–8) and a C-terminal region termed the TEN "superfold," which contains a cysteine-rich domain (CRD) that binds several Ca^{2+} ions, a transthyretin-like domain (TTR), two Ig-like domains (Ig), a β-propeller domain (β-prop), an extremely large β-barrel domain, and a toxin-like domain (Tox) (Fig. 6.9a) (Jackson et al. 2018; Li et al. 2018; Meijer et al. 2022). TENs form constitutive cis dimers linked by two intermolecular disulfide bonds in EGF2 and EGF5 (Li et al. 2018).

TENs were identified as binding partners of the ADGRLs using affinity chromatography and shown to interact trans-synaptically (Silva et al. 2011). The minimal TEN binding region of ADGRLs was identified as the Lec domain (Boucard et al. 2014). Furthermore, the presence of SS2 in ADGRL1 was shown to decrease binding affinity to TENs without affecting the affinity of the ADGRL/FLRT interaction (Boucard et al. 2014). Although TENs have long been studied biochemically (Feng et al. 2002), structural work proved elusive, as a large part of their sequences are not similar to any known eukaryotic protein. The first high-resolution structures of TENs were released in 2018 when two groups in parallel determined the structure of the monomeric TEN ECR (Jackson et al. 2018; Li et al. 2018). These structures revealed the striking architecture of the TEN ECR (Fig. 6.9b), termed the TEN "superfold" (TEN^{SF})—a module of domains organized around an extremely large β-barrel in an arrangement with strong structural resemblance to bacterial Tc toxins (Li et al. 2018). Within the TEN^{SF}, the Ig- and β-propeller cap the "bottom" of the β-barrel, and the Tox domain begins inside the β-barrel and protrudes out of the side, and is partially present outside of the β-barrel on its side surface. The TTR and CRD domains interact with the Ig domain as part of the same TEN^{SF} module (Meijer et al. 2022). Later work from the same two groups showed that the ADGRL Lec domain interacts with the TEN β-barrel on the side of the β-barrel roughly opposite to Tox (del Toro et al. 2020; Li et al. 2020) and the Olf domain also interacts with the top corner of the barrel (Fig. 6.9b). This interaction is compatible with FLRT-Olf binding, in agreement with previous reports that binding of both FLRT and TEN to ADGRL is required for excitatory synapse specification (Sando et al. 2019) and the triple complex of ADGRL/TEN/FLRT was shown using soluble ECR constructs (Li et al. 2020). The ADGRL/TEN complex structures suggested a mechanism for how the presence of SS2 decreases ADGRL/TEN affinity, as previously observed (Boucard et al. 2014). The SS2− structure (PDB 6SKA) (del Toro et al. 2020) (PDB 6SKA) has high-quality density for the Olf domain which also packs against TEN, whereas the SS2+ structure (PDB 6VHH) (Li et al. 2020) does not clearly resolve the Olf domain, suggesting that the splice insert, which adds five residues in between

Lec and Olf, increases the spacing between these two domains in order to modulate binding of Olf to TEN.

Based on the known binding modes of the ADGRL ligands, a super-complex of TEN/FLRT/ADGRL/UNC5D can be modeled, suggesting that ADGRLs form the center of a nexus of signals from several inputs (Fig. 6.9b) (Lu et al. 2015; Jackson et al. 2016; Li et al. 2018, 2020; Araç and Li 2019a, b).

Similar to the Nrx paradigm, alternative splicing of pre-synaptic TENs regulates their interactions with postsynaptic ADGRLs (Li et al. 2020). TENs are alternatively spliced at two sites in their ECR: SSA is found in EGF8, and SSB in the β-propeller domain (Berns et al. 2018). The presence of the SSB site (SSB+) is thought to mediate homophilic interaction of TENs, and this has been shown to disrupt the TEN/ADGRL interaction in an assay that used full-length proteins (Li et al. 2018, 2020). Structural work clarified that the SSB is present at a loop in the β-propeller domain and mediates contacts between two copies of the TEN^{SF}, suggesting it may mediate dimerization (Berns et al. 2018; Jackson et al. 2018). Mechanistic work agrees with the hypothesis that TEN dimerization via SSB disrupts the TEN/ADGRL interaction (Fig. 6.10) (Li et al. 2018, 2020). TEN −SSB has two freely rotating TEN^{SF}, which can productively interact with postsynaptic ADGRLs (Fig. 6.10a, b); TEN +SSB locks the two TEN^{SF} copies in a tight embrace via their β-propeller domains and limits the spatial extent of what they can explore (Fig. 6.10c, d). Thus, TEN +SSB is less likely to productively interact with ADGRLs. Furthermore, TEN2 SSB+ does not induce aggregation in trans with ADGRL3 (Li et al. 2018, 2020), and it shifts the synapse-forming properties of TEN2 from

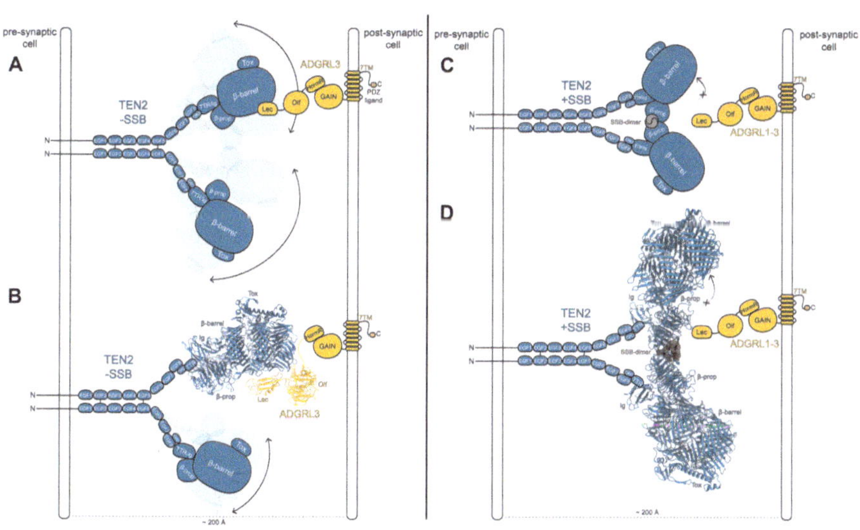

Fig. 6.10 *Structural models for regulation of the TEN/ADGRL interaction by alternative splicing.* (**a, b**) In the absence of the SSB insert, each TEN^{SF} can explore the synaptic cleft and find ADGRL molecules with which to interact (**c, d**) In the presence of SSB insert, the two TEN^{SF} are locked together and cannot explore the synaptic space to bind to postsynaptic ADGRLs.

excitatory (SSB−) to inhibitory (SSB+) (Li et al. 2018; Sando et al. 2019). Other dimer species of TENs are reported, including the mutagenesis-validated asymmetric dimer of the *D. melanogaster* TEN-m (Li et al. 2023). The compact dimer species of mammalian TEN3 has been reported, which is mediated by EGF8 and the Tox domains (Gogou et al. 2024), and this compact dimer only forms in TEN3 SSA+. This structure provides a hypothesis for how SSA could regulate access of TEN in the synaptic cleft, where SSA+ TEN3 would be locked in the compact dimer and sequestered from potential ligands. The compact TEN4 dimer is also reported; this dimer species is mediated by the CRD and the Tox domains (Meijer et al. 2022).

Downstream of receptor activation, ADGRLs stimulate the serum response element (SRE) pathway through the heterotrimeric G proteins $G\alpha_{12/13}$. This phenomenon is robust and consistently observed across several independent groups at different universities (Nazarko et al. 2018; Mathiasen et al. 2020; Barros-Álvarez et al. 2022; Perry-Hauser et al. 2022; Qian et al. 2022; Bui et al. 2023; Kordon et al. 2023; Dates et al. 2024; Wang et al. 2024). Additionally, ADGRLs have been shown to regulate the cyclic adenosine monophosphate (cAMP) pathway (Nazarko et al. 2018; Röthe et al. 2019; Sando and Südhof 2021; Qian et al. 2022; Kordon et al. 2023; Wang et al. 2024). Furthermore, the single splice variant of ADGRL3 which couples to $G\alpha_s$ is critical for ADGRL synaptic function (Wang et al. 2024), and the cAMP signaling pathway is known to be important for synapse formation (Jiang et al. 2021; Sando et al. 2022). Finally, ADGRLs have a PDZ ligand in their C-terminus, which can interact with the SHANK family of PDZ-domain-containing proteins (Kreienkamp et al. 2000, 2002; Tobaben et al. 2000). The structure of the C-terminal ADGRL PDZ ligand in complex with the SHANK PDZ domain is available (Ponna et al. 2018), showing that ADGRL interacts with the PDZ domain in a typical manner. Also, the intracellular C-terminus of ADGRL is necessary for synapse formation (Sando and Südhof 2021). SHANKs can scaffold a large array of proteins, including neurotransmitter receptors and the actin cytoskeleton, forming phase-separated macromolecular condensates, which may be important for clustering neurotransmitter receptors (Monteiro and Feng 2017; Wang et al. 2024). Thus, ADGRLs can output in two major modes: through postsynaptic signaling and through scaffolding of large protein complexes via the SHANK proteins.

The interaction of the Reticulon-4 receptors (RTN4R/Nogo) with the brain angiogenesis inhibitors (ADGRBs or BAIs) forms another interaction pair key for synapse development involving a postsynaptic aGPCR (Wang et al. 2021a). RTN4Rs are presynaptic proteins with two close homologs, RTN4L1 and RTN4L2 and the three form a group of neuronal receptors with the ADGRBs as bona fide ligands (Chong et al. 2018). RTN4Rs have an N-terminal LRR region, which is linked to the plasma membrane by a GPI anchor (Fig. 6.11a) (Wang et al. 2021a). ADGRBs are postsynaptic proteins (Stephenson et al. 2013) expressed in the nervous system and are important for synapse formation (Kakegawa et al. 2015; Sigoillot et al. 2015), and they are associated with schizophrenia (DeRosse et al. 2008; Liao et al. 2012). There are three ADGRBs in mammals (ADGRB1–3), and they have ECRs

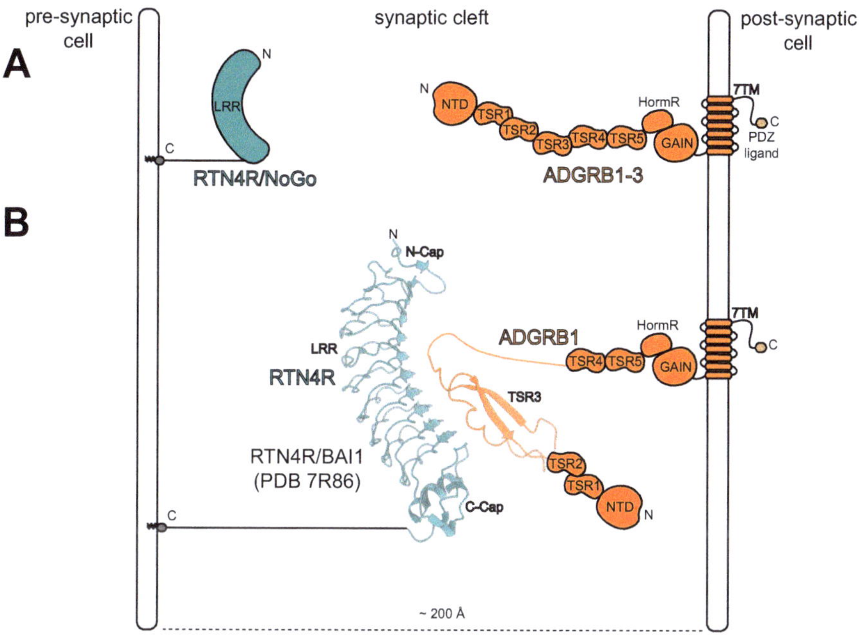

Fig. 6.11 *Structural basis for the RTN4R/ADGRB interaction.* (**a**) The domain layouts of RTN4R and ADGRBs. (**b**) The structure of the RTN4R/ADGRB interaction.

consisting of an amino-terminal domain (NTD), five thrombospondin repeats (TSR1–5), a HormR, and a GAIN domain (Fig. 6.11a) (Wang et al. 2021a). ADGRBs also have a GPCR 7TM as well as a C-terminal PDZ ligand (Stephenson et al. 2013).

The structural basis for the RTN4R/ADGRB1 interaction is known (Fig. 6.11b), the concave surface of the RTN4R LRR interacts with the TSR3 domain of ADGRB1 (Wang et al. 2021a).

ADGRBs can coordinate several events downstream of their activation, similar to ADGRLs. Signaling studies have provided evidence that ADGRBs signal to various pathways at least through $G_{12/13}$ and potentially through other means (Stephenson et al. 2013; Kishore et al. 2016). Additionally, the C-terminal ADGRB1 PDZ ligand can interact with a variety of PDZ domains, including PSD-95, in vitro (Stephenson et al. 2013). Other studies have shown that ADGRB1 can directly bind to the engulfment and cell motility protein (ELMO) via a short peptide sequence in the ICR. ELMO is an adaptor protein for the dedicator of cytokinesis protein (DOCK) (Weng et al. 2019), which is a Rho guanine nucleotide exchange factor (RhoGEF) that plays roles in coordinating cytoskeletal changes, and thus ADGRBs could regulate cytoskeletal activity through their ICR independently of G protein signaling.

Overall, the work described above strongly demonstrates how postsynaptic aGPCRs are integrators of adhesive signals and how they can disseminate these signals through multiple pathways, including G protein signaling, PDZ-ligand interactions to scaffold protein complexes, and RhoGEFs to modulate the actin cytoskeleton

(Lala and Hall 2022). These events are all required for the proper development of the postsynaptic density (Südhof, 2021).

6.5 Type IIa Receptor Protein Tyrosine Phosphatases and Their Interaction Partners

The type IIa receptor protein tyrosine phosphatases (referred to as RPTPs herein), which consist of PTPδ, PTPσ, and leukocyte-common antigen-related phosphatase (LAR), form another hub as presynaptic CAMs. They mediate synaptic adhesion and organization in a manner that is tightly regulated by alternative splicing (Takahashi and Craig 2013). RPTPs are implicated in neuropsychiatric conditions such as ASD, bipolar disorder, restless legs syndrome, and schizophrenia, and they have long been studied as molecules important for nervous system development (Takahashi and Craig 2013).

RPTPs have large ECRs that begin with three Ig domains (Ig1–3) and continue with either four or eight FnIII domains (FnIII1–8), depending on alternative splicing (Fig. 6.12a). A short TMR follows, and finally, the ICR has one functionally catalytic phosphatase domain (PTPA) and a second catalytically inert phosphatase domain (PTPB) (Takahashi and Craig 2013). This domain architecture is studded with four small alternatively spliced inserts termed mini-exons (meA–meD). meA is in Ig2, meB between Ig2 and Ig3, meC in FnIII5, and meD in the ICR between the TMR and PTPA (Takahashi and Craig 2013). As detailed in the previous Nrx hub section, RPTPs with meB− interact with NL3 (Yoshida et al. 2021). In the meB+ splice form, RPTPs have several ligands, which are the focus of this section (Takahashi and Craig 2013). RPTPs interact with proteoglycans, including heparan sulfate proteoglycans (HSPGs) and chondroitin sulfate proteoglycans (CSPGs), which affect the neuronal function of RPTPs (Coles et al. 2011). In a crystal structure of the LAR Ig1–Ig2 tandem, a highly conserved surface of positively charged residues on Ig1 binds to the small molecule mimic sucralose octasulfate (Coles et al. 2011). Heparan sulfate, as a mimic of HSPG binding, induced oligomerization of the PTPσ ECR but chondroitin sulfate did not, suggesting that these two glycans can differentially regulate RPTP clustering; the presence of HSPG may thus support the RPTP clustering needed to support molecular condensates and presynaptic specialization (Coles et al. 2011).

Several postsynaptic ligands of RPTPs have been confirmed by structural and functional work including synaptic adhesion-like molecules (SALM1–5) (Goto-Ito et al. 2018; Lin et al. 2018), Slit and Trk-like proteins (Slitrk1–6) (Yamagata et al. 2015a), neurotrophin receptor tropomyosin-related kinase C (TrkC) (Coles et al. 2014), interleukin-1-receptor accessory protein-like 1 (IL1RAPL1) (Yamagata et al. 2015b), and interleukin-1 receptor accessory protein (IL1RAcP) (Yamagata et al. 2015b). This interaction network forms a strong parallel to the Nrx hub in its organization, where one presynaptic molecule set (RPTPs) interacts with various

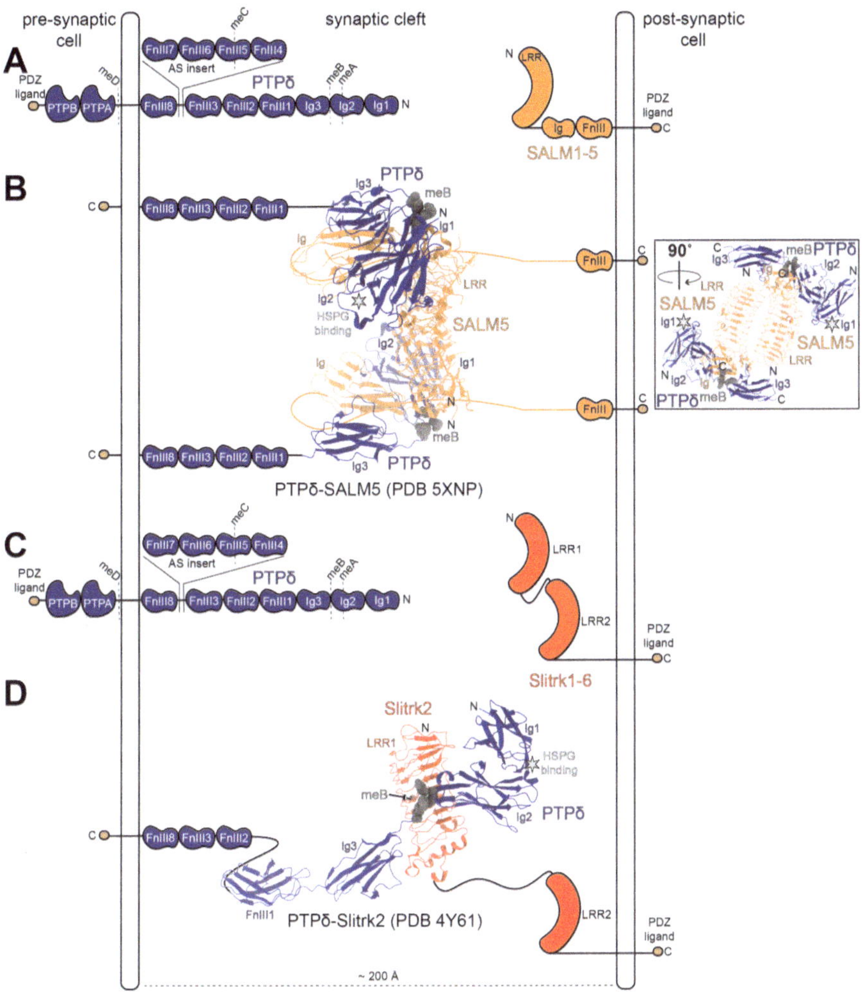

Fig. 6.12 *Structural basis for the RPTP/SALM and RPTP/Slitrk interactions*. (**a**) The domain layouts of PTPδ and SALMs. (**b**) The structure of the PTPδ/SALM5 interaction with orthogonal view on the right side. (**c**) The domain layouts of PTPδ and Slitrks. (**d**) The structure of the PTPδ/Slitrk2 interaction

postsynaptic partners, and the presynaptic molecule is regulated by alternative splicing (Takahashi and Craig 2013).

The interaction of RPTPs with SALMs (alternate name: leucine-rich fibronectin III-containing proteins—Lrfn) is key to synapse development and has been studied structurally (Nam et al. 2011). There are five SALMs in mammals (SALM1–5), and they are strongly associated with ASD and schizophrenia (Nam et al. 2011; Liu 2019). SALMs are postsynaptic and are involved in the development of both excitatory and inhibitory synapses, depending on the context (Nam et al. 2011; Liu

2019). SALMs have an LRR with N- and C-terminal cap regions, one Ig domain, one FnIII domain, a TMR, and a short ICR with a PDZ ligand (Fig. 6.12a) (Liu 2019). Crystallographic structures of SALM2 (Goto-Ito et al. 2018) and SALM5 (Goto-Ito et al. 2018; Lin et al. 2018) in complex with PTPδ are available. These structures revealed that PTPδ Ig2–Ig3 interacts with the SALM Ig domain and the C-cap of the SALM LRR using conserved residues (Fig. 6.12b). SALM dimerization occurs in an antiparallel manner and is critical for the synaptic functions of SALM as mutations in the dimerization interface reduce synaptic function (Goto-Ito et al. 2018; Lin et al. 2018). Additionally, PTPδ meB+ is present in all PTPδ-SALM structures, and although meB is close to the binding interface, it does not make any interactions with SALM (Fig. 6.12b). Interestingly, meB+ strongly enhances the binding affinity of the two proteins by >20-fold (Goto-Ito et al. 2018; Lin et al. 2018). One hypothesis to explain these results is that meB increases the spacing between Ig2 and Ig3 to allow Ig3 to productively bind to SALM (Goto-Ito et al. 2018).

Another key interaction for synapse development is that of the RPTPs with the Slitrks, which is necessary for both excitatory and inhibitory synapse development, depending on the context (Won et al. 2019). Slitrks are named due to their homology to the axon guidance molecules Slit and the Trk receptors, and they are highly expressed in the central nervous system. These key CAMs are linked to several neuropsychiatric disorders (Won et al. 2019). There are six Slitrks in mammals (Slitrk1–6), and they contain two LRR regions (LRR1 and LRR2), a short TMR, an ICR, which is suggested to signal through phosphorylation of key tyrosine residues (Fig. 6.12c) (Won et al. 2019), and a C-terminal PDZ ligand. The interaction of PTPδ with Slitrk2 (and LAR with Slitrk1) involves Ig2–Ig3 sandwiching the Slitrk LRR1 and a loop of the LRR1 C-cap (Fig. 6.12d) (Um et al. 2014; Yamagata et al. 2015a). In this case, meB directly makes several interactions with the inside of the concave surface of the LRR, and accordingly, meB-PTPδ does not exhibit any binding with Slitrk2 in surface plasmon resonance studies (Yamagata et al. 2015a). Furthermore, meB− LAR does not interact with Slitrk1 in cell-surface binding or cell aggregation experiments (Um et al. 2014). Interestingly, it was shown that the lateral interactions between different copies of the trans LAR/Slitrk1 heterodimer were important for heterologous synapse formation assays, suggesting that this species needs to cluster at high concentration to initiate synapse development (Um et al. 2014).

The RPTP–TrkC interaction is key for excitatory synapse formation (Takahashi et al. 2011). TrkCs have an LRR domain with N- and C-terminal cap regions and two Ig domains in the ECR (Fig. 6.13a) (Coles et al. 2014). This is followed by a TMR and a C-terminal PDZ ligand (Coles et al. 2014), and in some versions of TrkC, there is a tyrosine kinase domain at the C-terminus (Takahashi et al. 2011). Trks are tyrosine kinase molecules that signal in response to binding of the small protein neurotrophin, but TrkC has adapted to bind PTPσ instead, as TrkC neurotrophin binding is not required for synapse formation, whereas binding to PTPσ is (Takahashi et al. 2011). Furthermore, the variants of TrkC expressed in the synapse are versions lacking its kinase domain (Takahashi et al. 2011). The PTPσ/TrkC

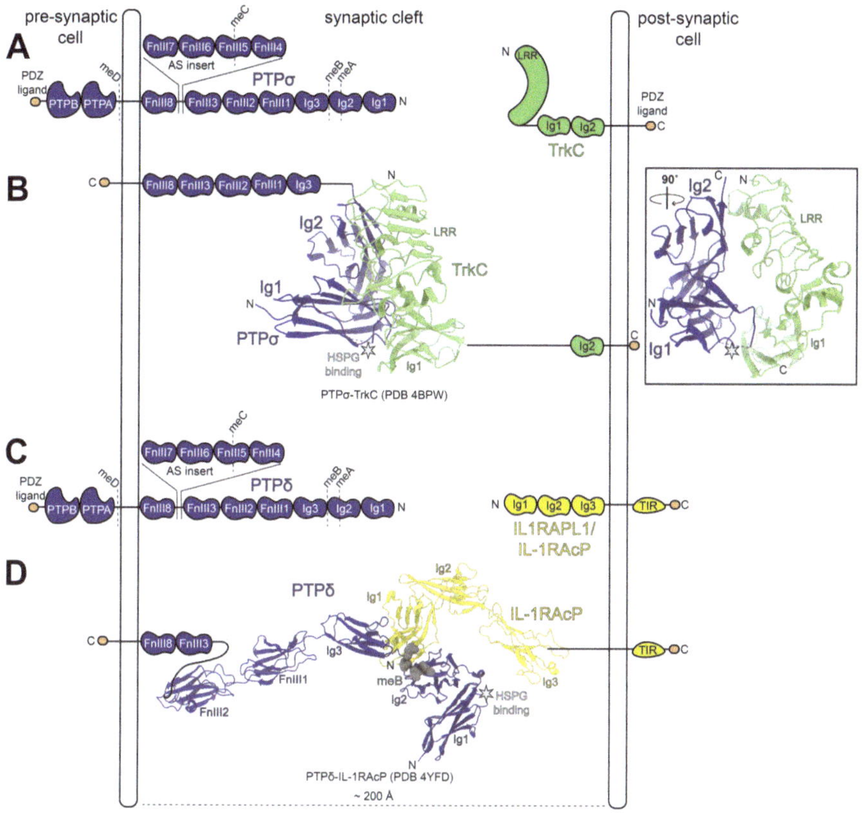

Fig. 6.13 *Structural basis for the RPTP/TrkC and RPTP/ILRAP interactions*. (**a**) The domain layouts of PTPσ and TrkC. (**b**) The structure of the PTPσ/TrkC interaction with orthogonal view on the right side. (**c**) The domain layouts of PTPδ and ILRAPs. (**d**) The structure of the PTPδ/ILRAP interaction

structure showed that Ig1/Ig2 of PTPσ interacts with the LRR and Ig1 of TrkC (Fig. 6.13b) (Coles et al. 2014). This sterically conflicts with the binding site for HSPG (Coles et al. 2011, 2014), and heparan sulfate can displace PTPσ from immobilized TrkC and disrupt artificial synapse formation through TrkC (Coles et al. 2014). Interestingly, this is the only RPTP ligand that binds to the Ig1/Ig2 tandem of any RPTP (Coles et al. 2014), in agreement with the hypothesis that TrkC competes with HSPG in a physiological context. All other characterized RPTP ligands bind to the Ig2/Ig3 tandem (Coles et al. 2014; Yamagata et al. 2015a, b; Lin et al. 2018).

IL-1RAcP and IL1RAPL1 (termed ILRAPs herein) are a group of postsynaptic molecules that partner with RPTPs for synapse formation (Yamagata et al. 2015b). ILRAPs are implicated in a range of disorders including ASD and X-linked intellectual disability (Pavlowsky et al. 2010). The two ILRAPs have three extracellular Ig domains (Ig1–Ig3), a short TMR, and an intracellular Toll/IL-1R homology domain (TIR) and C-terminal PDZ ligand (Pavlowsky et al. 2010) (Fig. 6.13c). The structures of both ILRAPs in complex with PTPδ are known (Yamagata et al. 2015b) and this work revealed that the ILRAP-Ig1 domain is cradled between PTPδ-Ig2 and PTPδ-Ig3, and furthermore ILRAP-Ig3 contacts PTPδ-Ig1 (Fig. 6.13d). Another structure of ILRAP in complex with meB− PTPδ was determined, and the ILRAP-Ig3/PTPδ-Ig1 were not observed, providing an explanation for meB+ regulation of the PTPδ/ILRAP interaction (Yamagata et al. 2015b). Similar to that proposed for SALM, meB is thought to increase the distance between PTPδ-Ig2 and PTPδ-Ig3 such that the ILRAP-Ig3/PTPδ-Ig1 can productively occur (Yamagata et al. 2015b). This agrees with binding affinity measurements where meB− PTPδ has lower affinity for ILRAPs (Yamagata et al. 2015b). Additionally, these structures revealed that meA, which is located on a loop of PTPδ-Ig2, is directly located in the interaction surface with ILRAP-Ig1. This provides an explanation for affinity measurements that showed the meA splice insert is a very precise switch that distinguishes between the two homologs IL-1RAcP and IL1RAPL1, where meA− PTPδ retains interactions with IL-1RAcP but loses measurable interaction with IL1RAPL1 (Yamagata et al. 2015b).

In the RPTP hub, splicing plays a master regulatory role. meA+ is necessary for the RPTP/IL1RAPL1 and RPTP/NL interactions, but dispensable for RPTP/IL–1RAcP (Yamagata et al. 2015b), RPTP/SALM5 (Goto-Ito et al. 2018), and RPTP/Slitrk (Yamagata et al. 2015a). The presence of RPTP meB+ eliminates the RPTP/NL interaction (Yoshida et al. 2021) and decreases the affinity of the RPTP/TrkC interaction (Coles et al. 2014), but enhances the affinity for RPTP/ILRAPs (Yamagata et al. 2015b), the RPTP/SALM interaction (Lin et al. 2018), and the RPTP/Slitrk interaction (Yamagata et al. 2015a).

Downstream of RPTPs on the presynaptic side, the scaffolding molecule Liprinα binds to the PTPB domain of all RPTPs (Marcó de la Cruz et al. 2024), and other proteins such as the RhoGEF Trio (Debant et al. 1996) also interact with PTPB; these proteins can organize and scaffold around the actin cytoskeleton. Liprinα proteins are master regulators of the presynaptic space; their deletion leads to empty boutons lacking presynaptic vesicles (Marcó de la Cruz et al. 2024). The structural basis for the interaction of Liprinα with PTPδ (Wakita et al. 2020) and LAR (Xie et al. 2020) is known and involves the sterile alpha motif (SAM) domains of Liprinα interacting with the PTPB domain of each RPTP. On the postsynaptic side, the Slitrks have PDZ ligands that interact with PSD-95 and SHANK proteins (Han et al. 2019), SALMs have a PDZ ligand and can interact with PSD-95 (Morimura et al. 2006; Wang et al. 2006), Slitrks interact with PSD-95 and SHANKs (Han et al. 2019), and ILRAPL1 can interact with PSD-95 (Pavlowsky et al. 2010). Thus, the RPTPs and their postsynaptic ligands can robustly scaffold the components required for productive synapse formation in a manner tightly regulated by alternative splicing.

6.6 Conclusions and Future Directions

This chapter summarizes several major points about the structural basis of synaptic adhesion, with a focus on the mechanisms by which alternative splicing modulates the function of synaptic CAMs. A variety of molecules are important for synaptic function, and they can be organized into several interaction hubs. The three most well-studied hubs presented here are the neurexin hub, the latrophilin hub, and the RPTP hub (Südhof, 2018). Alternative splicing, primarily of presynaptic molecules, affects the structure of CAMs, and this translates into functional effects (Südhof, 2018). Several possibilities are apparent, such as inserting or removing specific residue side chains into interfaces (Elegheert et al. 2016), affecting the spacing between domains (Lin et al. 2018), and affecting the conformational flexibility of CAMs in the synaptic cleft (Li et al. 2020), and other possibilities are likely. In response to productive CAM adhesion, scaffolding of dense phase-separated condensates occurs through interactions of intracellular sequences with proteins such as the Liprinαs and MAGUKs, and these are likely key for organizing the presynaptic active zone and the postsynaptic density (Südhof, 2021). How signaling events such as G protein and other pathways contribute to synapse formation is a hot topic (Jiang et al. 2021; Wang et al. 2024). The future of this field likely includes detailed structural and functional studies of more synaptic CAMs. Structural studies thus far have focused on the isolated ECRs of synaptic CAMs, and new exciting work, including both the TM and ECRs of CAMs, opens new doors to understanding CAM function (Kordon et al. 2024). New trans-synaptic interactions necessary for synapse formation continue to be structurally characterized (Patil et al. 2023). These studies will be expanded on by the cutting-edge technique of cryogenic electron tomography (cryo-ET), which has the potential to resolve individual proteins present in the context of real, vitrified synapses (Tao et al. 2018; Sun et al. 2025; Held et al. 2024). In conjunction with pioneering work on the molecular basis of CAM function (Marcó de la Cruz et al. 2024; Wang et al. 2024), future studies can characterize proteins at a level never seen before, and mechanistic work can use structural hypotheses to interrogate the details of how CAMs work and how mutations found in disease states can affect CAM functionality.

Ultimately, our work as a scientific community is oriented toward modulating these CAM interactions in disease states to serve human health needs, and the decades of structural and mechanistic work on CAMs have delivered many potential avenues for correcting altered synaptic function in disease states. The trans-synaptic adhesion of CAMs could be specifically blocked, or the lateral build-up of condensates could be blocked through interruption of intracellular scaffolding events, such as through MAGUKs or Liprinαs. The PDZ domain/ligand interaction is common in the function of many CAMs, and this interaction represents an excellent target for small-molecule intervention (Dev 2004; Christensen et al. 2019). Such discovery campaigns are underway for the treatment of neurological disorders (Dev 2004; Vogrig et al. 2013; Christensen et al. 2019). CAMs with direct signaling activity, such as aGPCRs, can be targeted by can be targeted by compounds to affect their signaling to affect their signaling (Kordon et al. 2023). Downstream signaling events can be modulated using traditional kinase, phosphatase, and other inhibitors (Jiang et al. 2021).

However, the most exciting possibilities illuminated by the studies outlined in this work include splice-isoform-specific modulators. For example, molecules that bind to the sequence of Nrx SS4 could disrupt Nrx/Clbn/GluD2 complex formation without affecting Nrx/NL binding, and molecules that bind to the TEN SSB sequence could disrupt TEN SSB dimer formation and promote ADGRL/TEN interaction in TEN SSB+. Antibody-like molecules, including nanobodies (Nbs) and synthetic antigen binders (sABs), are prime candidates for splice-isoform selective binders as they can target epitopes with exquisite specificity (Paduch et al. 2013). Antibodies can be made to target specific presynaptic and postsynaptic molecules, to target alternatively spliced sequences directly, or to stabilize or block the changed conformations that are directed by alternative splicing. Bi-specific antibodies that bind to the synaptic adhesion complex and promote adhesive events are also possible. Exciting studies have shown that sAB molecules can specifically disrupt the interaction of ADGRL with TEN, but not FLRT, and vice versa (Kordon et al. 2023), potentially providing a way to block excitatory synapse development. sABs are available that can modulate the signaling of ADGRL3 without affecting other ADGRLs (Kordon et al. 2023). The future of mechanistic molecular neurobiology is filled with challenges (Südhof, 2018), and structural biology, when combined with mechanistic studies using in vitro and in vivo systems, has the unique ability to provide us with atomic-level detail of the synapse and the broad, amazing array of proteins involved in its function. Protein structures have been a useful tool for decades and will continue to be extremely valuable in unraveling the mysteries of the human nervous system.

Acknowledgments Software used in the preparation of this chapter was installed and configured by SBGrid. The content is solely the responsibility of the authors and does not necessarily reflect the official views of the National Institute of General Medical Sciences or the National Institutes of Health. This research was supported by the following grants: National Institutes of Health Grant R35GM148412 to D.A. and National Institutes of Health Grant F32GM142266 and K99GM157487 to S.J.B. Molecular graphics and analyses were performed with UCSF ChimeraX, developed by the Resource for Biocomputing, Visualization, and Informatics at the University of California, San Francisco, with support from National Institutes of Health R01-GM129325 and the Office of Cyber Infrastructure and Computational Biology, National Institute of Allergy and Infectious Diseases.

Methods Structural analysis and figure preparation were performed using UCSF ChimeraX (Goddard et al. 2018; Pettersen et al. 2021; Meng et al. 2023). Sequence conservation analysis was performed using PROMALS3D (Pei et al. 2008).

Bibliography

Aldahmesh MA, et al. Homozygous null mutation in ODZ3 causes microphthalmia in humans. Genet Med. 2012;14(11):900–4. https://doi.org/10.1038/gim.2012.71.

Alkelai A, et al. A role for TENM1 mutations in congenital general anosmia. Clin Genet. 2016;90(3):211–9. https://doi.org/10.1111/cge.12782.

Araç D, Li J. Teneurin structure: splice variants of a bacterial toxin homolog specifies synaptic connections. Front Neurosci. 2019a;13(JUL):1–5. https://doi.org/10.3389/fnins.2019.00838.

Araç D, Li J. Teneurins and latrophilins: two giants meet at the synapse. Curr Opin Struct Biol. 2019b;54:141–51. https://doi.org/10.1016/j.sbi.2019.01.028.

Araç D, et al. Structures of Neuroligin-1 and the Neuroligin-1/Neurexin-1β complex reveal specific protein-protein and protein-Ca2+ interactions. Neuron. 2007;56(6):992–1003. https://doi.org/10.1016/j.neuron.2007.12.002.

Araç D, et al. A novel evolutionarily conserved domain of cell-adhesion GPCRs mediates autoproteolysis. EMBO J. 2012;31(6):1364–78. https://doi.org/10.1038/emboj.2012.26.

Bandekar, S.J., et al. (2025) Structural basis for regulation of CELSR1 by a compact module in its extracellular region. Nat Commun 16, 3972. https://doi.org/10.1038/s41467-025-59319-8

Barros-Álvarez X, et al. The tethered peptide activation mechanism of adhesion GPCRs. Nature. 2022;604(7907):757–62. https://doi.org/10.1038/s41586-022-04575-7.

Bemben MA, et al. The cellular and molecular landscape of neuroligins. Trends Neurosci. 2015:496–505. https://doi.org/10.1016/j.tins.2015.06.004.

Bernadyn TF, et al. GPR114/ADGRG5 is activated by its tethered peptide agonist because it is a cleaved adhesion GPCR. J Biol Chem. 2023;299(10) https://doi.org/10.1016/j.jbc.2023.105223.

Berns DS, et al. Teneurin-3 controls topographic circuit assembly in the hippocampus. Nature. 2018;554(7692):328–33. https://doi.org/10.1038/nature25463.

Biederer T, Südhof TC. CASK and protein 4.1 support F-actin nucleation on neurexins. J Biol Chem. 2001;276(51):47869–76. https://doi.org/10.1074/jbc.M105287200.

Biederer T, Kaeser PS, Blanpied TA. Transcellular nanoalignment of synaptic function. Neuron. 2017;96(3):680–96. https://doi.org/10.1016/j.neuron.2017.10.006.

Born G, et al. Modulation of synaptic function through the a-neurexin-specific ligand neurexophilin-1. Proc Natl Acad Sci U S A. 2014;111(13) https://doi.org/10.1073/pnas.1312112111.

Boucard AA, et al. A splice code for trans-synaptic cell adhesion mediated by binding of neuroligin 1 to α- and β-neurexins. Neuron. 2005;48(2):229–36. https://doi.org/10.1016/j.neuron.2005.08.026.

Boucard AA, Maxeiner S, Südhof TC. Latrophilins function as heterophilic cell-adhesion molecules by binding to teneurins: regulation by alternative splicing. J Biol Chem. 2014;289(1):387–402. https://doi.org/10.1074/jbc.M113.504779.

Budreck EC, Scheiffele P. Neuroligin-3 is a neuronal adhesion protein at GABAergic and glutamatergic synapses. Eur J Neurosci. 2007;26(7):1738–48. https://doi.org/10.1111/j.1460-9568.2007.05842.x.

Bui DLH, et al. The adhesion GPCRs CELSR1-3 and LPHN3 engage G proteins via distinct activation mechanisms. Cell Rep. 2023;42(6):112552. https://doi.org/10.1016/j.celrep.2023.112552.

Burette AC, et al. Electron tomographic analysis of synaptic ultrastructure. J Comp Neurol. 2012;520(12):2697–711. https://doi.org/10.1002/cne.23067.

Chassaing N, et al. Confirmation of TENM3 involvement in autosomal recessive colobomatous microphthalmia. Am J Med Genet A. 2016;170(7):1895–8. https://doi.org/10.1002/ajmg.a.37667.

Chen X, et al. Structural basis for synaptic adhesion mediated by neuroligin-neurexin interactions. Nat Struct Mol Biol. 2008;15(1):50–6. https://doi.org/10.1038/nsmb1350.

Chen F, et al. The structure of neurexin 1α reveals features promoting a role as synaptic organizer. Structure. 2011a;19(6):779–89. https://doi.org/10.1016/j.str.2011.03.012.

Chen X, et al. PSD-95 is required to sustain the molecular organization of the postsynaptic density. J Neurosci. 2011b;31(17):6329–38. https://doi.org/10.1523/JNEUROSCI.5968-10.2011.

Chen X, et al. PSD-95 family MAGUKs are essential for anchoring AMPA and NMDA receptor complexes at the postsynaptic density. Proc Natl Acad Sci U S A. 2015;112(50):E6983–92. https://doi.org/10.1073/pnas.1517045112.

Chen X, et al. Phase separation at the synapse. Nat Neurosci. 2020;23:301–10. https://doi.org/10.1038/s41593-019-0579-9.

Cheng S, et al. Conformational plasticity in the Transsynaptic Neurexin-Cerebellin-glutamate receptor adhesion complex. Structure. 2016;24(12):2163–73. https://doi.org/10.1016/j.str.2016.11.004.

Chih B, Gollan L, Scheiffele P. Alternative splicing controls selective trans-synaptic interactions of the Neuroligin-Neurexin complex. Neuron. 2006;51(2):171–8. https://doi.org/10.1016/j.neuron.2006.06.005.

Chong ZS, et al. Pooled extracellular receptor-ligand interaction screening using CRISPR activation. Genome Biol. 2018;19(1) https://doi.org/10.1186/s13059-018-1581-3.

Christensen NR, et al. PDZ domains as drug targets. Adv Ther. 2019;2(7):1800143. https://doi.org/10.1002/adtp.201800143.

Cijsouw T, et al. Mapping the proteome of the synaptic cleft through proximity labeling reveals new cleft proteins. Proteomes. 2018;6(4) https://doi.org/10.3390/proteomes6040048.

Cole AA, Reese TS. Transsynaptic assemblies link domains of presynaptic and postsynaptic intracellular structures across the synaptic cleft. J Neurosci. 2023;43(33):5883–92. https://doi.org/10.1523/JNEUROSCI.2195-22.2023.

Coles CH, et al. Proteoglycan-specific molecular switch for RPTPσ clustering and neuronal extension. Science. 2011;332(6028):484–8. https://doi.org/10.1126/science.1200840.

Coles CH, et al. Structural basis for extracellular cis and trans RPTPσ signal competition in synaptogenesis. Nat Commun. 2014;5 https://doi.org/10.1038/ncomms6209.

Comoletti D, et al. Gene selection, alternative splicing, and post-translational processing regulate neuroligin selectivity for β-neurexins. Biochemistry. 2006;45(42):12816–27. https://doi.org/10.1021/bi0614131.

Connor SA, et al. Altered cortical dynamics and cognitive function upon Haploinsufficiency of the autism-linked excitatory synaptic suppressor MDGA2. Neuron. 2016;91(5):1052–68. https://doi.org/10.1016/j.neuron.2016.08.016.

Dai J, et al. GluD1 is a signal transduction device disguised as an ionotropic receptor. Nature. 2021;595(7866):261–5. https://doi.org/10.1038/s41586-021-03661-6.

Dates AN, et al. Heterogeneity of tethered agonist signaling in adhesion G protein-coupled receptors. Cell Chem Biol. 2024;31(8):1542–53. https://doi.org/10.1016/j.chembiol.2024.03.004.

Debant, A. et al. (1996) The multidomain protein trio binds the LAR transmembrane tyrosine phosphatase, contains a protein kinase domain, and has separate rac-specific and rho-specific guanine nucleotide exchange factor domains (GTPase/immunoglobulin/pleckstrin homology/protein tyrosine phosphorylation/spectrin-like repeats). Available at: https://www.pnas.org.

del Toro D, et al. Structural basis of Teneurin-Latrophilin interaction in repulsive guidance of migrating neurons. Cell. 2020;180(2):323–339.e19. https://doi.org/10.1016/j.cell.2019.12.014.

Demberg LM, et al. Identification of the tethered peptide agonist of the adhesion G protein-coupled receptor GPR64/ADGRG2. Biochem Biophys Res Commun. 2015;464(3):743–7. https://doi.org/10.1016/j.bbrc.2015.07.020.

Demberg LM, et al. Activation of adhesion G protein-coupled receptors: agonist specificity of Stachel sequence-derived peptides. J Biol Chem. 2017;292(11):4383–94. https://doi.org/10.1074/jbc.M116.763656.

DeRosse P, et al. The genetics of symptom-based phenotypes: toward a molecular classification of schizophrenia. Schizophr Bull. 2008;34(6):1047–53. https://doi.org/10.1093/schbul/sbn076.

Dev KK. Making protein interactions druggable: targeting PDZ domains. Nat Rev Drug Discov. 2004:1047–56. https://doi.org/10.1038/nrd1578.

Dunah AW, et al. LAR receptor protein tyrosine phosphatases in the development and maintenance of excitatory synapses. Nat Neurosci. 2005;8(4):458–67. Available at:. https://doi.org/10.1038/nn1416.

Elegheert J, et al. Structural basis for integration of GluD receptors within synaptic organizer complexes. Science. 2016;353(6296):295–9. https://doi.org/10.1126/science.aae0104.

Elegheert J, et al. Structural mechanism for modulation of synaptic Neuroligin-Neurexin signaling by MDGA proteins. Neuron. 2017;95(4):896–913.e10. https://doi.org/10.1016/j.neuron.2017.07.040.

Fabrichny IP, et al. Structural analysis of the synaptic protein Neuroligin and its β-Neurexin complex: determinants for folding and cell adhesion. Neuron. 2007;56(6):979–91. https://doi.org/10.1016/j.neuron.2007.11.013.

Favara DM, et al. Elevated expression of the adhesion GPCR ADGRL4/ELTD1 promotes endothelial sprouting angiogenesis without activating canonical GPCR signalling. Sci Rep. 2021;11(1) https://doi.org/10.1038/s41598-021-85408-x.

Feng K, et al. All four members of the ten-m/Odz family of transmembrane proteins form dimers. J Biol Chem. 2002;277(29):26128–35. https://doi.org/10.1074/jbc.M203722200.

Fredriksson R, et al. The G-protein-coupled receptors in the human genome form five main families. Phylogenetic analysis, paralogon groups, and fingerprints. Mol Pharmacol. 2003;63(6):1256–72. https://doi.org/10.1124/mol.63.6.1256.

Fuccillo MV, et al. Single-cell mRNA profiling reveals cell-type-specific expression of Neurexin isoforms. Neuron. 2015;87(2):326–40. https://doi.org/10.1016/j.neuron.2015.06.028.

Gangwar SP, et al. Molecular Mechanism of MDGA1: regulation of Neuroligin 2:Neurexin Trans-synaptic Bridges. Neuron. 2017;94(6):1132–1141.e4. https://doi.org/10.1016/j.neuron.2017.06.009.

Goddard TD, et al. UCSF ChimeraX: meeting modern challenges in visualization and analysis. Protein Sci. 2018;27(1):14–25. https://doi.org/10.1002/pro.3235.

Gogou C, et al. Alternative splicing controls teneurin-3 compact dimer formation for neuronal recognition. Nat Commun. 2024;15(1) https://doi.org/10.1038/s41467-024-47763-x.

Gomez AM, Traunmüller L, Scheiffele P. Neurexins: molecular codes for shaping neuronal synapses. Nat Rev Neurosci. 2021;22:137–51. https://doi.org/10.1038/s41583-020-00415-7.

Goto-Ito S, et al. Structural basis of trans-synaptic interactions between PTPδ and SALMs for inducing synapse formation. Nat Commun. 2018;9(1) https://doi.org/10.1038/s41467-017-02417-z.

Graumann R, et al. Expression of teneurins is associated with tumor differentiation and patient survival in ovarian cancer. PLoS One. 2017;12(5):1–24. https://doi.org/10.1371/journal.pone.0177244.

Gray EG. AXO-somatic and AXO-dendritic synapses of the cerebral cortex: an electron microscope study. J Anat. 1959;93:420–33.

Han KA, et al. Slitrk2 controls excitatory synapse development via PDZ-mediated protein interactions. Sci Rep. 2019;9(1) https://doi.org/10.1038/s41598-019-53519-1.

Hata Y, Butz S, Südhof TC. CASK: a novel dlg/PSD95 homolog with an N-terminal Calmodulin-dependent protein kinase domain identified by interaction with neurexins. J Neurosci. 1996;16:2488.

Held RG, Liang J, Brunger AT. Nanoscale architecture of synaptic vesicles and scaffolding complexes revealed by cryo-electron tomography. Proc Natl Acad Sci. 2024;121(27) https://doi.org/10.1073/pnas.2403136121.

Hong W, Mosca TJ, Luo L. Teneurins instruct synaptic partner matching in an olfactory map. Nature. 2012;484(7393):201–7. https://doi.org/10.1038/nature10926.

Honig B, Shapiro L. Adhesion protein structure, molecular affinities, and principles of cell-cell recognition. Cell. 2020;181(3):520–35. https://doi.org/10.1016/j.cell.2020.04.010.

Hor H, et al. Missense mutations in TENM4, a regulator of axon guidance and central myelination, cause essential tremor. Hum Mol Genet. 2015;24(20):5677–86. https://doi.org/10.1093/hmg/ddv281.

Ichtchenko K, Nguyen T, Südhof TC. Structures, alternative splicing, and neurexin binding of multiple neuroligins. J Biol Chem. 1996;271(5):2676–82. https://doi.org/10.1074/jbc.271.5.2676.

Irie M, et al. Binding of Neuroligins to PSD-95. Available at: https://www.science.org. Science. 1997;277:1511.

Jackson VA, et al. Structural basis of latrophilin-FLRT interaction. Structure. 2015;23(4):774–81. https://doi.org/10.1016/j.str.2015.01.013.

Jackson VA, et al. Super-complexes of adhesion GPCRs and neural guidance receptors. Nat Commun. 2016;7 https://doi.org/10.1038/ncomms11184.

Jackson VA, et al. Structures of Teneurin adhesion receptors reveal an ancient fold for cell-cell interaction. Nat Commun. 2018;9(1):1–9. https://doi.org/10.1038/s41467-018-03460-0.

Jiang X, Sando R, Südhof TC. Multiple signaling pathways are essential for synapse formation induced by synaptic adhesion molecules. Proc Natl Acad Sci. 2021;118(3) https://doi.org/10.1073/pnas.2000173118.

Kakegawa W, et al. Anterograde C1ql1 signaling is required in order to determine and maintain a single-winner climbing fiber in the mouse cerebellum. Neuron. 2015;85(2):316–29. https://doi.org/10.1016/j.neuron.2014.12.020.

Kim JA, et al. Structural insights into modulation of Neurexin-Neuroligin trans-synaptic adhesion by MDGA1/Neuroligin-2 complex. Neuron. 2017;94(6):1121–1131.e6. https://doi.org/10.1016/j.neuron.2017.05.034.

Kishore A, et al. Stalk-dependent and stalk-independent signaling by the adhesion G protein-coupled receptors GPR56 (ADGRG1) and BAI1 (ADGRB1). J Biol Chem. 2016;291(7):3385–94. https://doi.org/10.1074/jbc.M115.689349.

Ko J, et al. LRRTM2 functions as a Neurexin ligand in promoting excitatory synapse formation. Neuron. 2009;64(6):791–8. https://doi.org/10.1016/j.neuron.2009.12.012.

Kobe B, Kajava AV. The leucine-rich repeat as a protein recognition motif. Curr Opin Struct Biol. 2001;11(6):725–32. https://doi.org/10.1016/S0959-440X(01)00266-4.

Koehnke J, et al. Splice form dependence of β-neurexin/neuroligin binding interactions. Neuron. 2010;67(1):61–74. https://doi.org/10.1016/j.neuron.2010.06.001.

Kordon SP, et al. Isoform- and ligand-specific modulation of the adhesion GPCR ADGRL3/Latrophilin3 by a synthetic binder. Nat Commun. 2023;14(1) https://doi.org/10.1038/s41467-023-36312-7.

Kordon, S.P., et al. Conformational coupling between extracellular and transmembrane domains modulates holo-adhesion GPCR function. Nat Commun 15, 10545 (2024). https://doi.org/10.1038/s41467-024-54836-4

Krasnoperov VG, et al. α-Latrotoxin stimulates exocytosis by the interaction with a neuronal G-protein-coupled receptor. Neuron. 1997;18(6):925–37. https://doi.org/10.1016/S0896-6273(00)80332-3.

Krasnoperov V, et al. Post-translational proteolytic processing of the calcium-independent receptor of α-latrotoxin (CIRL), a natural chimera of the cell adhesion protein and the G protein-coupled receptor: role of the G protein-coupled receptor proteolysis site (GPS) motif. J Biol Chem. 2002;277(48):46518–26. https://doi.org/10.1074/jbc.M206415200.

Kreienkamp HJ, et al. The calcium-independent receptor for α-latrotoxin from human and rodent brains interacts with members of the ProSAP/SSTRIP/shank family of multidomain proteins. J Biol Chem. 2000;275(42):32387–90. https://doi.org/10.1074/jbc.C000490200.

Kreienkamp H, et al. Interaction of G-protein-coupled receptors with synaptic scaffolding proteins. Biochern. Biophys. Res. Cornrnun. 2002; Available at: http://portlandpress.com/biochemsoctrans/article-pdf/30/4/464/535450/bst0300464.pdf

Lacy SE, et al. Identification of FLRT1, FLRT2, and FLRT3: a novel family of transmembrane leucine-rich repeat proteins. Genomics. 1999;62(3):417–26. https://doi.org/10.1006/geno.1999.6033.

Lala T, Hall RA. Adhesion G protein-coupled receptors: structure, signaling, physiology, and pathophysiology. Physiol Rev. 2022;102(4):1587–624. https://doi.org/10.1152/physrev.00027.2021.

Leamey CA, Sawatari A. The teneurins: new players in the generation of visual topography. Semin Cell Dev Biol. 2014;35:173–9. https://doi.org/10.1016/j.semcdb.2014.08.007.

Lee H, Dean C, Isacoff E. Alternative splicing of neuroligin regulates the rate of presynaptic differentiation. J Neurosci. 2010;30(34):11435–46. https://doi.org/10.1523/JNEUROSCI.2946-10.2010.

Lelianova VG, et al. α-Latrotoxin receptor, latrophilin, is a novel member of the secretin family of G protein-coupled receptors. J Biol Chem. 1997;272(34):21504–8. https://doi.org/10.1074/jbc.272.34.21504.
Leon K, et al. Structural basis for adhesion G protein-coupled receptor Gpr126 function. Nat Commun. 2020;11(1) https://doi.org/10.1038/s41467-019-14040-1.
Li J, et al. Structural basis for Teneurin function in circuit-wiring: a toxin motif at the synapse. Cell. 2018;173(3):735–748.e15. https://doi.org/10.1016/j.cell.2018.03.036.
Li J, et al. Alternative splicing controls teneurin-latrophilin interaction and synapse specificity by a shape-shifting mechanism. Nat Commun. 2020;11(1):2140. https://doi.org/10.1038/s41467-020-16029-7.
Li J, Bandekar SJ, Araç D. The structure of fly Teneurin-m reveals an asymmetric self-assembly that allows expansion into zippers. EMBO Rep. 2023;24(6):1–15. https://doi.org/10.15252/embr.202256728.
Liao HM, et al. Identification and characterization of three inherited genomic copy number variations associated with familial schizophrenia. Schizophr Res. 2012;139(1–3):229–36. https://doi.org/10.1016/j.schres.2012.05.015.
Liebscher I, et al. A tethered agonist within the Ectodomain activates the adhesion G protein-coupled receptors GPR126 and GPR133. Cell Rep. 2014;9(6):2018–26. https://doi.org/10.1016/j.celrep.2014.11.036.
Lin Z, et al. Structural basis of SALM5-induced PTPδ dimerization for synaptic differentiation. Nat Commun. 2018;9(1) https://doi.org/10.1038/s41467-017-02414-2.
Liu H. Synaptic organizers: synaptic adhesion-like molecules (SALMs). Curr Opin Struct Biol. 2019;54:59–67. https://doi.org/10.1016/j.sbi.2019.01.002.
Liu J, et al. Structural plasticity of Neurexin 1α: implications for its role as synaptic organizer. J Mol Biol. 2018;430(21):4325–43. https://doi.org/10.1016/j.jmb.2018.08.026.
Lu YC, et al. Structural basis of Latrophilin-FLRT-UNC5 interaction in cell adhesion. Structure. 2015;23(9):1678–91. https://doi.org/10.1016/j.str.2015.06.024.
Luo F, et al. Neurexins cluster ca 2+ channels within the presynaptic active zone. EMBO J. 2020;39(7) https://doi.org/10.15252/embj.2019103208.
Magee JC, Grienberger C. Synaptic plasticity forms and functions. Annu Rev Neurosci. 2020;43:95–117. https://doi.org/10.1146/annurev-neuro-090919-022842.
Marcó de la Cruz B, et al. Liprin-α proteins are master regulators of human presynapse assembly. Nat Neurosci. 2024;27(4):629–42. https://doi.org/10.1038/s41593-024-01592-9.
Marro SG, et al. Neuroligin-4 regulates excitatory synaptic transmission in human neurons. Neuron. 2019;103(4):617–626.e6. https://doi.org/10.1016/j.neuron.2019.05.043.
Martinez-Sanchez A, et al. Trans-synaptic assemblies link synaptic vesicles and neuroreceptors. Sci Adv. Available at: https://www.science.org. 2021;7
Mathiasen, S., et al. G12/13 is activated by acute tethered agonist exposure in the adhesion GPCR ADGRL3. Nat Chem Biol 16, 1343–1350 (2020). https://doi.org/10.1038/s41589-020-0617-7
Meijer DH, et al. Teneurin4 dimer structures reveal a calcium-stabilized compact conformation supporting homomeric trans-interactions. EMBO J. 2022;41(9):e107505. https://doi.org/10.15252/embj.2020107505.
Meng EC, et al. UCSF ChimeraX: tools for structure building and analysis. Protein Sci. 2023;32(11) https://doi.org/10.1002/pro.4792.
Meyer G, et al. The complexity of PDZ domain-mediated interactions at glutamatergic synapses: a case study on neuroligin. Neuropharmacology. 2004;47(5):724–33. https://doi.org/10.1016/j.neuropharm.2004.06.023.
Meza-Aguilar DG, Boucard AA. Latrophilins updated. Biomol Concepts. 2014;5(6):457–78. https://doi.org/10.1515/bmc-2014-0032.
Missler M, Südhof TC, Biederer T. Synaptic cell adhesion. Cold Spring Harb Perspect Biol. 2012;4(4) https://doi.org/10.1101/cshperspect.a005694.
Monteiro P, Feng G. SHANK proteins: roles at the synapse and in autism spectrum disorder. Nat Rev Neurosci. 2017;18:147–57. https://doi.org/10.1038/nrn.2016.183.

Morimura N, et al. Comparative analysis of structure, expression and PSD95-binding capacity of Lrfn, a novel family of neuronal transmembrane proteins. Gene. 2006;380(2):72–83. https://doi.org/10.1016/j.gene.2006.05.014.

Mosca TJ. On the Teneurin track: a new synaptic organization molecule emerges. Front Cell Neurosci. 2015;9(MAY):1–14. https://doi.org/10.3389/fncel.2015.00204.

Mosca TJ, et al. Trans-synaptic Teneurin signalling in neuromuscular synapse organization and target choice. Nature. 2012;484(7393):237–41. https://doi.org/10.1038/nature10923.

Muhammad K, et al. Presynaptic spinophilin tunes neurexin signalling to control active zone architecture and function. Nat Commun. 2015;6 https://doi.org/10.1038/ncomms9362.

Mukherjee K, et al. CASK functions as a Mg2+-independent Neurexin kinase. Cell. 2008;133(2):328–39. https://doi.org/10.1016/j.cell.2008.02.036.

Nam J, Mah W, Kim E. The SALM/Lrfn family of leucine-rich repeat-containing cell adhesion molecules. Semin Cell Dev Biol. 2011;22(5):492–8. https://doi.org/10.1016/j.semcdb.2011.06.005.

Nazarko O, et al. A Comprehensive Mutagenesis Screen of the Adhesion GPCR Latrophilin-1/ADGRL1. iScience. 2018;3:264–78. https://doi.org/10.1016/j.isci.2018.04.019.

Nguyen T, Südhof TC. Binding properties of neuroligin 1 and neurexin 1β reveal function as heterophilic cell adhesion molecules. J Biol Chem. 1997;272(41):26032–9. https://doi.org/10.1074/jbc.272.41.26032.

O'Sullivan ML, et al. FLRT proteins are endogenous Latrophilin ligands and regulate excitatory synapse development. Neuron. 2012;73(5):903–10. https://doi.org/10.1016/j.neuron.2012.01.018.

Paduch M, et al. Generating conformation-specific synthetic antibodies to trap proteins in selected functional states. Methods. 2013;60(1):3–14. https://doi.org/10.1016/j.ymeth.2012.12.010.

Patil DN, et al. Structure of the photoreceptor synaptic assembly of the extracellular matrix protein pikachurin with the orphan receptor GPR179. Sci Signal. 2023;16(795) https://doi.org/10.1126/scisignal.add9539.

Pavlowsky A, et al. A postsynaptic signaling pathway that may account for the cognitive defect due to IL1RAPL1 mutation. Curr Biol. 2010;20(2):103–15. https://doi.org/10.1016/j.cub.2009.12.030.

Pei J, Kim BH, Grishin NV. PROMALS3D: a tool for multiple protein sequence and structure alignments. Nucleic Acids Res. 2008;36(7):2295–300. https://doi.org/10.1093/nar/gkn072.

Perez de Arce K, et al. Topographic mapping of the synaptic cleft into adhesive Nanodomains. Neuron. 2015;88(6):1165–72. https://doi.org/10.1016/j.neuron.2015.11.011.

Perry-Hauser NA, et al. Disentangling autoproteolytic cleavage from tethered agonist–dependent activation of the adhesion receptor ADGRL3. J Biol Chem. 2022;298(12) https://doi.org/10.1016/j.jbc.2022.102594.

Petrenko, A.G. et al. (1990) Isolation and properties of the a-latrotoxin receptor, EMBO J.

Pettem KL, et al. Interaction between autism-linked MDGAs and neuroligins suppresses inhibitory synapse development. J Cell Biol. 2013;200(3):321–36. https://doi.org/10.1083/jcb.201206028.

Pettersen EF, et al. UCSF ChimeraX: structure visualization for researchers, educators, and developers. Protein Sci. 2021;30(1):70–82. https://doi.org/10.1002/pro.3943.

Ping YQ, et al. Structural basis for the tethered peptide activation of adhesion GPCRs. Nature. 2022;604(7907):763–70. https://doi.org/10.1038/s41586-022-04619-y.

Ponna SK, et al. Structural basis for PDZ domain interactions in the post-synaptic density scaffolding protein Shank3. J Neurochem. 2018;145(6):449–63. https://doi.org/10.1111/jnc.14322.

Poulopoulos A, et al. Homodimerization and isoform-specific heterodimerization of neuroligins. Biochem J. 2012;446(2):321–30. https://doi.org/10.1042/BJ20120808.

Prömel S, et al. The GPS motif is a molecular switch for bimodal activities of adhesion class G protein-coupled receptors. Cell Rep. 2012;2(2):321–31. https://doi.org/10.1016/j.celrep.2012.06.015.

Prömel S, Langenhan T, Araç D. Matching structure with function: the GAIN domain of adhesion-GPCR and PKD1-like proteins. Trends Pharmacol Sci. 2013;34(8):470–8. https://doi.org/10.1016/j.tips.2013.06.002.

Qian Y, et al. Structural insights into adhesion GPCR ADGRL3 activation and Gq, Gs, Gi, and G12 coupling. Mol Cell. 2022;82(22):4340–4352.e6. https://doi.org/10.1016/j.molcel.2022.10.009.
Qu X, et al. Structural basis of tethered agonism of the adhesion GPCRs ADGRD1 and ADGRF1. Nature. 2022;604(7907):779–85. https://doi.org/10.1038/s41586-022-04580-w.
Ranaivoson FM, et al. Structural and mechanistic insights into the Latrophilin3-FLRT3 complex that mediates glutamatergic synapse development. Structure. 2015;23(9):1665–77. https://doi.org/10.1016/j.str.2015.06.022.
Rollenhagen A, Lübke JHR. The morphology of excitatory central synapses: from structure to function. Cell Tissue Res. 2006:221–37. https://doi.org/10.1007/s00441-006-0288-z.
Roppongi RT, Karimi B, Siddiqui TJ. Role of LRRTMs in synapse development and plasticity. Neurosci Res. 2017;116:18–28. https://doi.org/10.1016/j.neures.2016.10.003.
Röthe J, et al. Involvement of the adhesion GPCRs Latrophilins in the regulation of insulin release. Cell Rep. 2019;26(6):1573–1584.e5. https://doi.org/10.1016/j.celrep.2019.01.040.
Rudenko G. Neurexins — versatile molecular platforms in the synaptic cleft. Curr Opin Struct Biol. 2019;54:112–21. https://doi.org/10.1016/j.sbi.2019.01.009.
Rudenko G, Takahashi H. Editorial overview: synapses: bridging the gap between their structure and function. Curr Opin Struct Biol. 2019;54:iii–vii. https://doi.org/10.1016/j.sbi.2019.05.020.
Rudenko G, et al. The structure of the ligand-binding domain of Neurexin I: regulation of LNS domain function by alternative splicing. Cell. 1999; Available at: http://www3.ebi.ac.uk/tops/
Salzman GS, et al. Stachel-independent modulation of GPR56/ADGRG1 signaling by synthetic ligands directed to its extracellular region. Proc Natl Acad Sci U S A. 2017;114(38):10095–100. https://doi.org/10.1073/pnas.1708810114.
Sando R, Südhof TC. Latrophilin gpcr signaling mediates synapse formation. eLife. 2021;10 https://doi.org/10.7554/eLife.65717.
Sando R, Jiang X, Südhof TC. Latrophilin GPCRs direct synapse specificity by coincident binding of FLRTs and teneurins. Science. 2019;363(6429):eaav7969. https://doi.org/10.1126/science.aav7969.
Sando R, et al. Engineered synaptic tools reveal localized cAMP signaling in synapse assembly. J Cell Biol. 2022;221(2) https://doi.org/10.1083/jcb.202109111.
Sanes JR, Zipursky SL. Synaptic specificity, recognition molecules, and assembly of neural circuits. Cell. 2020;181(3):536–56. https://doi.org/10.1016/j.cell.2020.04.008.
Scheiffele P. Cell-cell signaling during synapse formation in the CNS. Annu Rev Neurosci. 2003;26:485–508. https://doi.org/10.1146/annurev.neuro.26.043002.094940.
Schöneberg T, et al. Tethered agonists: a new mechanism underlying adhesion G protein-coupled receptor activation. J Recept Signal Transduct. 2015;35(3):220–3. https://doi.org/10.3109/10799893.2015.1072978.
Schöneberg T, Kleinau G, Brüser A. What are they waiting for? – tethered agonism in G protein-coupled receptors. Pharmacol Res. 2016;108:9–15. https://doi.org/10.1016/j.phrs.2016.03.027
Schreiner D, et al. Targeted combinatorial alternative splicing generates brain region-specific repertoires of Neurexins. Neuron. 2014;84(2):386–98. https://doi.org/10.1016/j.neuron.2014.09.011.
Seiradake E, et al. FLRT structure: balancing repulsion and cell adhesion in cortical and vascular development. Neuron. 2014;84(2):370–85. https://doi.org/10.1016/j.neuron.2014.10.008.
Seong E, Yuan L, Arikkath J. Cadherins and catenins in dendrite and synapse morphogenesis. Cell Adhes Migr. 2015;9(3):202–13. https://doi.org/10.4161/19336918.2014.994919.
Siddiqui TJ, et al. LRRTMs and neuroligins bind neurexins with a differential code to cooperate in glutamate synapse development. J Neurosci. 2010;30(22):7495–506. https://doi.org/10.1523/JNEUROSCI.0470-10.2010.
Sigoillot SM, et al. The secreted protein C1QL1 and its receptor BAI3 control the synaptic connectivity of excitatory inputs converging on cerebellar purkinje cells. Cell Rep. 2015;10(5):820–32. https://doi.org/10.1016/j.celrep.2015.01.034.
Silva JP, et al. Latrophilin 1 and its endogenous ligand lasso/teneurin-2 form a high-affinity transsynaptic receptor pair with signaling capabilities. Proc Natl Acad Sci U S A. 2011;108(29):12113–8. https://doi.org/10.1073/pnas.1019434108.

Stephenson JR, et al. Brain-specific angiogenesis inhibitor-1 signaling, regulation, and enrichment in the postsynaptic density. J Biol Chem. 2013;288(31):22248–56. https://doi.org/10.1074/jbc.M113.489757.

Stoveken HM, et al. Adhesion G protein-coupled receptors are activated by exposure of a cryptic tethered agonist. Proc Natl Acad Sci U S A. 2015;112(19):6194–9. https://doi.org/10.1073/pnas.1421785112.

Südhof TC. Synaptic Neurexin complexes: a molecular code for the logic of neural circuits. Cell. 2017;171(4):745–69. https://doi.org/10.1016/j.cell.2017.10.024.

Südhof TC. Towards an understanding of synapse formation. Neuron. 2018;100(2):276–93. https://doi.org/10.1016/j.neuron.2018.09.040.

Südhof TC. The cell biology of synapse formation. J Cell Biol. 2021;220(7):1–18. https://doi.org/10.1083/jcb.202103052.

Südhof TC. Cerebellin–neurexin complexes instructing synapse properties. Curr Opin Neurobiol. 2023;81:102727. https://doi.org/10.1016/j.conb.2023.102727.

Südhof TC, Malenka RC. Understanding synapses: past, present, and future. Neuron. 2008:469–76. https://doi.org/10.1016/j.neuron.2008.10.011.

Shuzo Sugita, et al. α-Latrotoxin Receptor CIRL/Latrophilin 1 (CL1) Defines an Unusual Family of Ubiquitous G-protein-linked Receptors: G-PROTEIN COUPLING NOT REQUIRED FOR TRIGGERING EXOCYTOSIS, Journal of Biological Chemistry, Volume 273, Issue 49, 1998, Pages 32715-32724, https://doi.org/10.1074/jbc.273.49.32715.

Rong Sun, et al. (2025) The postsynaptic density in excitatory synapses is composed of clustered, heterogeneous nanoblocks. J Cell Biol 2 224 (6): e202406133. doi: https://doi.org/10.1083/jcb.202406133

Takahashi H, Craig AM. Protein tyrosine phosphatases PTPδ, PTPς, and LAR: presynaptic hubs for synapse organization. Trends Neurosci. 2013:522–34. https://doi.org/10.1016/j.tins.2013.06.002.

Takahashi H, et al. (2011) Postsynaptic TrkC and presynaptic PTPσ function as a bidirectional excitatory synaptic organizing complex. Neuron. 69(2):287–303. https://doi.org/10.1016/j.neuron.2010.12.024.

Talamillo A, et al. ODZ1 allows glioblastoma to sustain invasiveness through a Myc-dependent transcriptional upregulation of RhoA. Oncogene. 2017;36(12):1733–44. https://doi.org/10.1038/onc.2016.341.

Tao CL, et al. Differentiation and characterization of excitatory and inhibitory synapses by cryo-electron tomography and correlative microscopy. J Neurosci. 2018;38(6):1493–510. https://doi.org/10.1523/JNEUROSCI.1548-17.2017.

Ting AY, et al. Proteomic analysis of unbounded cellular compartments: synaptic clefts. Cell. 2016;166(5):1295–1307.e21. https://doi.org/10.1016/j.cell.2016.07.041.

Tobaben S, Sudhof TC, Stahl B. The G protein-coupled receptor CL1 interacts directly with proteins of the shank family. J Biol Chem. 2000;275(46):36204–10. https://doi.org/10.1074/jbc.M006448200.

Togashi H, Sakisaka T, Takai Y. Cell adhesion molecules in the central nervous system. Cell Adhes Migr. 2009;3(1):29–35. https://doi.org/10.4161/cam.3.1.6773.

Treutlein B, et al. Cartography of neurexin alternative splicing mapped by single-molecule long-read mRNA sequencing. Proc Natl Acad Sci U S A. 2014;111(13) https://doi.org/10.1073/pnas.1403244111.

Tucker RP. Teneurins: domain architecture, evolutionary origins, and patterns of expression. Front Neurosci. 2018;12(December):1–14. https://doi.org/10.3389/fnins.2018.00938.

Tucker RP, et al. Phylogenetic analysis of the teneurins: conserved features and premetazoan ancestry. Mol Biol Evol. 2012;29(3):1019–29. https://doi.org/10.1093/molbev/msr271.

Uemura T, et al. Trans-synaptic interaction of GluRδ2 and neurexin through Cbln1 mediates synapse formation in the cerebellum. Cell. 2010;141(6):1068–79. https://doi.org/10.1016/j.cell.2010.04.035.

Um JW, et al. Structural basis for LAR-RPTP/Slitrk complex-mediated synaptic adhesion. Nat Commun. 2014;5 https://doi.org/10.1038/ncomms6423.

van Oostrum M, et al. The proteomic landscape of synaptic diversity across brain regions and cell types. Cell. 2023;186(24):5411–5427.e23. https://doi.org/10.1016/j.cell.2023.09.028.

Vizurraga A, et al. Mechanisms of adhesion G protein coupled receptor activation. J Biol Chem. 2020:jbc.REV120.007423. https://doi.org/10.1074/jbc.rev120.007423.

Vogrig A, et al. Structure-based design of PDZ ligands as inhibitors of 5-HT2A receptor/PSD-95 PDZ1 domain interaction possessing anti-hyperalgesic activity. ACS Chem Biol. 2013;8(10):2209–16. https://doi.org/10.1021/cb400308u.

Wakita M, et al. Structural insights into selective interaction between type IIa receptor protein tyrosine phosphatases and Liprin-α. Nat Commun. 2020;11(1) https://doi.org/10.1038/s41467-020-14516-5.

Wang CY, et al. A novel family of adhesion-like molecules that interacts with the NMDA receptor. J Neurosci. 2006;26(8):2174–83. https://doi.org/10.1523/JNEUROSCI.3799-05.2006.

Wang J, et al. RTN4/NoGo-receptor binding to BAI adhesion-GPCRs regulates neuronal development. Cell. 2021a;184(24):5869–5885.e25. https://doi.org/10.1016/j.cell.2021.10.016.

Wang Y, et al. Adhesion GPCR Latrophilin 3 regulates synaptic function of cone photoreceptors in a trans-synaptic manner. Proc Natl Acad Sci. 2021b;118(45) https://doi.org/10.1073/pnas.2106694118.

Wang S, et al. Alternative splicing of latrophilin-3 controls synapse formation. Nature. 2024;118(45):128–35. https://doi.org/10.1038/s41586-023-06913-9.

Wei Z, et al. Liprin-mediated large signaling complex organization revealed by the Liprin-α/CASK and Liprin-α/Liprin-β complex structures. Mol Cell. 2011;43(4):586–98. https://doi.org/10.1016/j.molcel.2011.07.021.

Weng Z, et al. Structure of BAI1/ELMO2 complex reveals an action mechanism of adhesion GPCRs via ELMO family scaffolds. Nat Commun. 2019;10(1) https://doi.org/10.1038/s41467-018-07938-9.

Wides R. The natural history of Teneurins: a billion years of evolution in three key steps. Front Neurosci. 2019;13(March):1–15. https://doi.org/10.3389/fnins.2019.00109.

Wilde C, et al. The constitutive activity of the adhesion GPCR GPR114/ADGRG5 is mediated by its tethered agonist. FASEB J. 2016;30(2):666–73. https://doi.org/10.1096/fj.15-276220.

Wilson SC, et al. Structures of neurexophilin–neurexin complexes reveal a regulatory mechanism of alternative splicing. EMBO J. 2019;38(22):1–16. https://doi.org/10.15252/embj.2019101603.

Woelfle R, D'Aquila AL, Lovejoy DA. Teneurins, TCAP, and Latrophilins: roles in the etiology of mood disorders. Transl Neurosci. 2016;7(1):17–23. https://doi.org/10.1515/tnsci-2016-0004.

Won SY, Lee P, Kim HM. Synaptic organizer: slitrks and type IIa receptor protein tyrosine phosphatasess. Curr Opin Struct Biol. 2019;54:95–103. https://doi.org/10.1016/j.sbi.2019.01.010.

Wu X, et al. Vesicle tethering on the surface of phase-separated active zone condensates. Mol Cell. 2021;81(1):13–24.e7. https://doi.org/10.1016/j.molcel.2020.10.029.

Xiao P, et al. Tethered peptide activation mechanism of the adhesion GPCRs ADGRG2 and ADGRG4. Nature. 2022;604(7907):771–8. https://doi.org/10.1038/s41586-022-04590-8.

Xie X, et al. Structural basis of liprin-α-promoted LAR-RPTP clustering for modulation of phosphatase activity. Nat Commun. 2020;11(1):1. https://doi.org/10.1038/s41467-019-13949-x.

Yamagata M, Sanes JR, Weiner JA. Synaptic adhesion molecules. Curr Opin Cell Biol. 2003;15(5):621–32. https://doi.org/10.1016/S0955-0674(03)00107-8.

Yamagata A, Sato Y, et al. Structure of Slitrk2-PTP complex reveals mechanisms for splicing-dependent trans-synaptic adhesion. Sci Rep. 2015a;5 https://doi.org/10.1038/srep09686.

Yamagata A, Yoshida T, et al. Mechanisms of splicing-dependent trans-synaptic adhesion by PTPδ -IL1RAPL1/IL-1RAcP for synaptic differentiation. Nat Commun. 2015b;6 https://doi.org/10.1038/ncomms7926.

Yamagata A, et al. Structural insights into modulation and selectivity of transsynaptic neurexin–LRRTM interaction. Nat Commun. 2018;9(1):1–11. https://doi.org/10.1038/s41467-018-06333-8.

Yoshida T, et al. Canonical versus non-canonical transsynaptic signaling of neuroligin 3 tunes development of sociality in mice. Nat Commun. 2021;12(1) https://doi.org/10.1038/s41467-021-22059-6.

Young TR, et al. Ten-m2 is required for the generation of binocular visual circuits. J Neurosci. 2013;33(30):12490–509. https://doi.org/10.1523/JNEUROSCI.4708-12.2013.

Zeng M, et al. Phase separation-mediated TARP/MAGUK complex condensation and AMPA receptor synaptic transmission. Neuron. 2019;104(3):529–543.e6. https://doi.org/10.1016/j.neuron.2019.08.001.

Zhu J, Shang Y, Zhang M. Mechanistic basis of MAGUK-organized complexes in synaptic development and signalling. Nat Rev Neurosci. 2016;17(4):209–23. https://doi.org/10.1038/nrn.2016.18.

Ziegler A, et al. Teneurin protein family: an emerging role in human tumorigenesis and drug resistance. Cancer Lett. 2012;326(1):1–7. https://doi.org/10.1016/j.canlet.2012.07.021.

Zuber B, et al. The mammalian central nervous synaptic cleft contains a high density of periodically organized complexes. Proc Natl Acad Sci U S A. 2005;102:19192. Available at: www.pnas.orgcgidoi, https://doi.org/10.1073pnas.0509527102

Chapter 7
Scales of Postsynaptic Nanostructure: Molecules, Nanoclusters, and Nanodomains

Martina Damenti, Aaron D. Levy, and Thomas A. Blanpied

Abstract Dissecting the mechanisms of synaptic transmission touches on nearly all fields of neuroscience. Of particular recent importance is the discovery that protein distribution within single synapses is highly organized across multiple spatial scales, ranging from the nanoscale accumulation of just a few protein molecules to larger domains with unique multiprotein compositions. Here, we address recent data regarding postsynaptic molecular organization. We argue that the complexity of synaptic nanostructure generates functional capabilities that can fine-tune synaptic strength and that far exceed the classical limits of quantal synaptic transmission. We focus first on the critical scaffold protein PSD-95 as a case study for how to approach the emergent problem of describing and classifying forms of protein organization, including trans-synaptic "nanocolumn" relationships. Then, we discuss recent work identifying new features of NMDA receptor subsynaptic organization that appear likely to regulate the patterns of neural activity that can induce synaptic plasticity. Overall, we assert that these mechanisms of molecular coordination at scales of 20–150 nm enhance the synapse's ability to tune synaptic transmission, carry out detailed biochemical signaling, and allow more complex impacts on the cell.

Keywords Nanocluster · Nanodomain · Nanocolumn · Postsynaptic scaffold proteins · Localization microscopy · Ionotropic glutamate receptors

M. Damenti · A. D. Levy · T. A. Blanpied (✉)
Department of Pharmacology and Physiology, Program in Neuroscience, and University of Maryland-Medicine Institute for Neuroscience Discovery (UM-MIND), University of Maryland School of Medicine, Baltimore, MD, USA
e-mail: TBlanpied@som.umaryland.edu

E. T. Kavalali (ed.), *Nano-organization of the Synapse*, Advances in Neurobiology 48,
https://doi.org/10.1007/978-3-032-12594-1_7

7.1 Synaptic Strength Is Graded

Synapse structure shapes information flow through neural circuits, and fine-tuning synaptic function through various forms of plasticity is critical for most human cognition and behavior. This chapter focuses on our current understanding of the molecular organization creating glutamatergic synaptic structure, particularly postsynaptic structure, with specific attention to how various elements of this nanoscale organization may regulate and diversify synaptic function.

The function of a synapse is often summarized in the seemingly simple value of its "strength"—its contribution to postsynaptic excitation when the synapse is activated. Synaptic strength varies widely between synapses, but at any particular synapse, it is typically thought of as a specific, singular value, subject to modulation over time by various mechanisms of neural plasticity. Indeed, students of neuroscience from an early point in their education are raised on the classic notions of quantal transmission (Purves et al. 2018; Kandel et al. 2021). In this model, derived from peripheral and nerve-muscle synapses, the strength of individual synaptic responses follows discrete steps depending on the number of vesicles (i.e., quanta) released and receptors activated. Responses can thus be mathematically described by the quantal parameters N, P, and Q, which have traditionally been interpreted as reflecting, respectively, the number of presynaptic release sites, the probability of release at each one following an action potential, and the number of postsynaptic receptors that produce a given response following release.

While this classic model has driven enormous insight and great practical knowledge, in an important sense, the typical glutamatergic contacts that make up the majority of the brain's synapses do not use quantal transmission. This is because the simplifying assumptions of the model break down under conditions that typify brain synapses: a somewhat small and variable amount of neurotransmitter is released per vesicle, a low concentration of glutamate can fail to activate all available receptors, and receptors can be spaced widely from the site of neurotransmitter release and are thus activated with a variety of probabilities and to a variety of conductance states (Biederer et al. 2017). Thus, transmission at single brain synapses most typically is highly variable with a skewed or log-normal amplitude distribution (Zhang and Trussell 1994; Buzsáki and Mizuseki 2014; Greger and Watson 2024), and individual responses following fusion of single vesicles thus commonly appear to follow a graded, rather than quantal, distribution. To be clear, most predictions of quantal theory stemming from presynaptic behavior are routinely upheld (Sahara and Takahashi 2001; Malagon et al. 2016), but a nonquantal postsynaptic response is a prominent feature in neocortical synapses and elsewhere (Bekkers and Stevens 1995; Ishikawa et al. 2002), and synaptic mechanisms should be analyzed with this in mind.

7.2 Nanocolumn Alignment of Release Sites and Receptors Contributes to Graded Synaptic Transmission

Many elements contribute to the graded amplitude of postsynaptic responses, and we leave most to previous excellent reviews (Lisman et al. 2007; Nicoll 2017; Guzikowski and Kavalali 2021; Gou et al. 2022; Droogers and MacGillavry 2023; Stockwell et al. 2024). Here, we focus on the relatively new and specific field of synapse nanostructure, or the organization of proteins on nanometer scales. The central observation from recent super-resolution imaging is that proteins frequently occur in groupings at particular locations *within the border of the synapse,* termed nanoclusters (NCs) (Biederer et al. 2017; Zieger and Choquet 2021). Protein NCs can occur outside synapses as well, but they are prominent features of synaptic protein organization, and identifying the mechanisms that produce them will likely offer deep insights into physiological and pathological processes. Considerable evidence now indicates that the functional state or weight distribution of a synapse is established not by the total number of each molecular type in the synapse, but by the number of molecules within these NCs and the position of NCs within the synapse, particularly with respect to one another (Sinnen et al. 2017; Haas et al. 2018; Ramsey et al. 2021).

Arguably, the most prominent example of how nanostructure impacts synaptic transmission is the nanometer-scale trans-synaptic alignment of AMPA receptor NCs with synaptic vesicle release sites, as indicated by NCs of the vesicle priming proteins RIM or Munc13. This alignment, termed the "nanocolumn" (Tang et al. 2016), is well supported by a diverse set of observations in several synapse types of mammalian brain and other organisms (Biederer et al. 2017; Sarkar et al. 2020; Martinez-Sanchez et al. 2021; Muttathukunnel et al. 2021; Sachs et al. 2024; Unterauer et al. 2024; Sun et al. 2025). NCs and the nanocolumn model appear to offer a straightforward means to maximize receptor activation and suggest a platform for previously unexpected mechanisms of plasticity based on nanostructural changes. The strongest evidence for this comes from observations that a variety of trans-synaptic adhesion complexes exert control over nanocolumn alignment (Haas et al. 2018; Fukata et al. 2021; Ramsey et al. 2021; Han et al. 2022; Kim et al. 2022; Lloyd et al. 2023) and can shape synapse strength even on rapid time scales by modulating the proportion of synaptic AMPARs that are aligned versus nonaligned with release sites (Ramsey et al. 2021). Key aspects of the nanocolumn are structurally unique in different synapse types (Sarkar et al. 2020; Dharmasri et al. 2024), consistent with the idea that nanocolumn features may contribute substantially to synapse plasticity and functional diversity. The expanding evidence for nanocolumn alignment, mechanisms that create it, its role in mediating synaptic strength and plasticity, and its role in disease mechanisms have all been the subject of recent reviews (Chen et al. 2018; Choquet 2018; Guzikowski and Kavalali 2021; Gou et al. 2022; Fukata et al. 2024; Xu et al. 2025); we refer readers to those excellent resources rather than revisit them here.

Here, we will address two topics. First, we will discuss the multiple levels of molecular organization within single postsynaptic densities (PSDs). As part of this, we will attempt to clarify confusing nomenclature that we think has been problematic for progress in the field. For the most part, we will present our point of view based on our own work, and we hope others will find this framework useful. Second, we will illustrate how NCs reveal a parcellation of organization and function within single synapses, greatly increasing the computational capacity of a synapse. For both these points, we will focus on the postsynaptic scaffold molecules, particularly PSD-95, that accumulate, retain, and position receptors in synapses. Due to their importance and the intense analysis they have received, they provide an instructive case study for synapse molecular architecture overall. We then expand on both points with reference to the recent examination of NMDA receptor subsynaptic distribution.

7.3 Scaffold Architecture on a Nested Set of Organizational Scales

The postsynaptic scaffolding protein PSD-95 and other MAGUK family members are a core pillar of excitatory postsynaptic nanostructure, given their role in retaining proteins, including glutamate receptors, at the synapse. Thus, understanding their organization provides insight into key characteristics of a synaptic organization. We know far more about the organization of PSD-95 as opposed to the other members of the MAGUK family or any of the other scaffolding proteins found in the synapse, and the characteristics of its organization provide a helpful guide for understanding some rules of thumb as well as some ambiguity persisting in the field.

Such biological discoveries go hand in hand with the technology that made them possible, and it remains likely that some aspects of our understanding of PSD-95 organization are constrained by the technology used to evaluate it. Electron microscopy has remained a central method for examining synapse structure, and recent developments present opportunities for identifying protein distributions in correlation with EM (Held et al. 2024). Nevertheless, this remains challenging, and much of what we know now comes from leveraging the rapid evolution of fluorescence microscopy methods, which more easily enable the identification of multiple proteins.

Unfortunately, as the power of super-resolution imaging methods to disclose postsynaptic protein organization has increased and the techniques have proliferated, so has the terminology used to describe the findings. This has resulted in a profusion of nano-related terms describing observations that sometimes or often overlap. Nanodomain, nanomodule, nanocolumn, subsynaptic domain, subdomain, cluster, subcluster, nanocluster, nanoblock, and probably many other terms are deployed with sometimes unclear rationale and unstated relationships. Thus, in summarizing key observations, one of our goals is to be clear on the nomenclature we use, hopefully helping address this problematic surplus of confusing terminology. To be clear, it is difficult at this point to put forward with high confidence a single, comprehensive model of synaptic molecular architecture, but we will assemble our observations as to how different scales of molecular assembly may relate to one another. Our apologies to those whose excellent data and preferred terms are not well represented here.

7.3.1 Structure of the Overall Postsynaptic Density

At the level of the synapse as a whole, the PSD is generally typified as a contiguous region of high protein density, though adopting a wide variety of shapes and sizes. The outline of the overall PSDs seen by EM or fluorescence super-resolution imaging is complex and varied, but morphological types have been classified as simple (continuous and generally round or oval), segmented, horseshoe-shaped, and perforated (containing holes without high postsynaptic protein density and also lacking apposed docked vesicles) (Harris and Weinberg 2012). Some ambiguity has arisen about the degree to which the absolute density of protein in the PSD has been overstated by reliance on chemical fixation and contrast methods in EM, but a variety of live-cell imaging observations, ranging from dissociated cultures to living brain, show high densities of PSD-95 and other proteins stably accumulated at synapses (MacGillavry et al. 2013; Fortin et al. 2014; Masch et al. 2018; Rimbault et al. 2021).

PSD-95 is often taken as an indicator of the lateral extent of the PSD as a whole. In EM tomograms, "vertical filaments" consistent with the structure of PSD-95 occur throughout and all the way to the edge of the PSD (Chen et al. 2008). The density of PSD-95 by immunolabeling EM when averaged across synapses is nearly constant throughout the interior (Degiorgis et al. 2006) and rolls off extremely quickly at the edge of the synapse (Degiorgis et al. 2006; Perez de Arce et al. 2015). It is therefore an attractive notion that PSD-95 fills out the entirety of the PSD, but this has not been systematically tested at single synapses rather than averaged populations, and other scaffolds might preferentially occupy interior areas or the edges of individual PSDs. In an analysis of PSD-95 and SAP102 in single PSDs of cultured neurons, these two MAGUKs neither exclusively avoided nor overlapped each other (Metzbower et al. 2024), indicating that some parts of the synapse may be populated by just one or the other, but that PSD-95 did provide an excellent estimate of PSD size in most cases. Individual PSD-95 molecules within the synapse are highly immobile (Blanpied et al. 2008; Chazeau et al. 2014), and turnover of synaptic PSD-95 molecules is extremely slow, probably exchanging only a few molecules per minute in an activity-regulated fashion (Blanpied et al. 2008; Xu et al. 2008; Fortin et al. 2014; Gürth et al. 2023). Nevertheless, overall PSD structure shows morphological dynamics on the order of minutes, driven in part by cytoskeletal treadmilling (Blanpied et al. 2008; Kerr and Blanpied 2012), suggesting that MAGUK positioning within the PSD may be conceived of as a flexible matrix.

7.3.2 Building Nanostructural Motifs from Single Molecules

PSD-95 organization *within* this overall structure can be described at multiple length scales (Fig. 7.1). Beginning at the molecular level, reports of the mean intermolecular spacing of PSD-95 within the PSD have varied from ~7 nm (Gürth et al. 2023; Shaib et al. 2024) to ~18 nm (Chen et al. 2008). This is a wide range that could stem from technical factors (variability in labeling density, imaging resolution, and detection reliability) or real biological differences between preparations, such as unknown developmental or activity-regulated states. It is tempting to think of these molecules as elements in a regular matrix, and some evidence for this has been reported (Gürth et al. 2023). However, it is not clear what mechanism might establish a high repeatability of spacing. One possibility is that PSD-95 spacing could be regularized via interactions with further cytosolic elements within the PSD, most notably including the GKAP/SAPAP family that has been postulated to form an interlinked meshwork

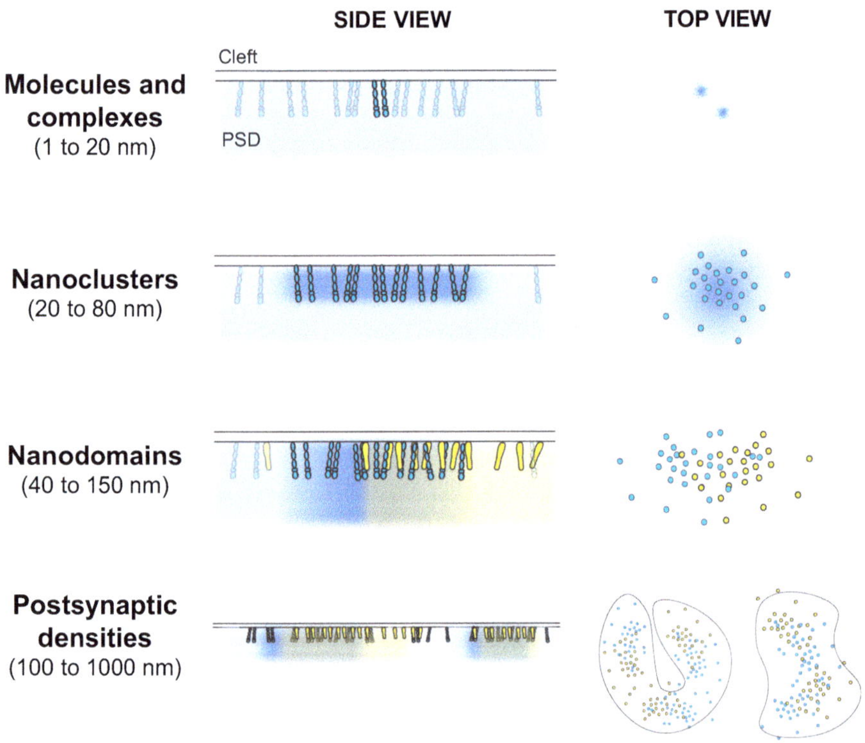

Fig. 7.1 *Scales of organization within synapses.* Individual molecules may be found in isolation or in complex with binding partners, as in traditional biochemical analyses. Groupings or subsets of a single molecule are termed nanoclusters. Regions of the synapse within which multiple proteins concentrate are termed nanodomains. Entire synapses (only PSDs are shown here, outlined as they might be defined based on EM images) can comprise multiple nanodomains of complex sizes, shapes, and protein makeup

with Shank proteins (Chen et al. 2008; Hayashi et al. 2009), though direct evidence for a patterning influence of GKAP/Shank on PSD-95 is lacking. In addition, several factors would seem to instead promote variable intermolecular distances. For example, many of the plentiful interaction partners of PSD-95 suggest that spacing would occur at distances established by the receptor complex structure. Indeed, mice carrying genetically tagged PSD-95 analyzed by imaging (Morris et al. 2024) or biochemical assays followed by mass spectrometry (Frank et al. 2016, 2017) show that the scaffold appears frequently in distinguishable dimers. The PDZ-binding tails of abundant AMPAR/TARP complexes and NMDA receptors in synapses establish dimeric or tetrameric PSD-95 binding sites and likely establish the spacing of MAGUK proteins at this scale. This assembly of a receptor with MAGUK proteins has been referred to as a prototypical "super-complex" building block of synaptic architecture (Frank et al. 2017). An attractive summary model at this point is that MAGUK proteins with their many interaction partners form a semiregular molecular mesh over the area of the PSD, though also, as discussed next, establish the basis for emergent, larger-scale organization.

At larger scales within the synapse, the spatial arrangement of PSD-95 and most other molecules that have been measured is neither random nor regular. Clear groupings of PSD-95 over larger dimensions than single receptor complexes are apparent. Early observations were made possible by the application of SMLM techniques PALM and dSTORM (MacGillavry et al. 2013; Nair et al. 2013) as well as STED (Fukata et al. 2013). These techniques facilitated nanoscale analysis of high-density labeling in single synapses (Dani et al. 2010), thus avoiding the problem of needing to average over many synapses after low-stoichiometry labeling. MacGillavry et al. (2013) first reported PSD-95 groupings of about 80 nm in diameter with PALM. We call these structures "nanoclusters" (NCs) of PSD-95 in order to indicate nanoscale groups of a single protein type. NCs of PSD-95 appeared as higher-density, ~80–130 nm diameter structures located within the larger, ~300 nm diameter synaptic region of lower-density PSD-95 (but note that overall synaptic PSD-95 density was still much higher than nonsynaptic areas). Estimates of the total number of PSD-95 molecules being imaged corresponded well to the expected number of molecules present in PSDs as deduced from biochemistry, raising confidence that the NCs represented an organizational substructure within single synapses. Importantly, PSD-95 NCs are observed in live super-resolution imaging both in vitro and in vivo (MacGillavry et al. 2013; Masch et al. 2018), indicating that they do not arise from fixation artifacts or cell culture.

One might expect that NCs of PSD-95 would be visible by EM. However, in transmission EM (which lacks the ability to detect specific protein types), the total density of proteins across the face of the PSD is roughly constant, not revealing strong peaks and valleys (Burette et al. 2012). Nevertheless, at higher 3D resolution afforded by cryo-EM tomography (cryo-ET), automated analysis of isolated PSDs has shown substructures roughly 120 nm in diameter within the PSD that span its entire thickness from membrane to cytosol (Jung et al. 2023), potentially consistent with nanodomains that include PSD-95. Cryo-ET of synaptosomes and brain slices also suggests the presence of somewhat smaller (~20–60 nm) membrane-associated "nanoblocks" of higher density (Sun et al. 2025). Note that in our terminology, we refer to both these as nanodomains, not nanoclusters because they presumably are regions achieving high density through accumulation of more than one protein. These observations provide further support for the idea that NCs and nanodomains are physical structures within single, continuous PSDs.

While NCs of other proteins within single synapses are sometimes quite discrete with clear borders and little "interstitial" protein between NCs, PSD-95 NCs are more reminiscent of a "hill" of elevated density in the synapse surrounded by lower-density PSD-95 instead of PSD-95-free regions. The enrichment of PSD-95 density within the NC is variable and generally modest, ~1.5-fold, meaning that identifying an NC and defining its size and border can be challenging and quite sensitive to the imaging resolution and labeling density of the protein. Nevertheless, recent observations using DNA-Exchange PAINT with nearly single-protein resolution have reiterated these characteristics (Anderson et al. 2023; Unterauer et al. 2024), and such structures, including the nested scale of PSD-95 NCs within larger features of PSDs (Zhang et al. 2025a, b), have been observed with a large range of techniques in brain and cultured neurons (Farley et al. 2015; Hruska et al. 2018; Masch et al. 2018; Helm et al. 2021; Shaib et al. 2022; Zhang et al. 2025b).

Organization on these multimolecular distance scales is important to investigate not just because it helps to understand synaptic function, but also since it demonstrates biological order on scales not well represented by longstanding structural biology, biochemical, or traditional confocal microscopy analyses. As the techniques to map the location of a high proportion of single molecules at near-molecular resolution proliferate and are optimized (Reinhardt et al. 2023; Sahl et al. 2024), our understanding of the organization of PSD-95 and other molecules on many spatial scales should continually improve.

7.3.3 Nanoscale Assembly of Other Scaffold Molecules Within the PSD

PSD-95 is the most abundant and well-studied postsynaptic scaffold, and information lags on others, but it is clear that PSD-95 is not the only scaffold relevant to establishing synapse nanostructure. Thus, a key question is whether regions of the PSD with a relatively low density of PSD-95 are "filled in" with other MAGUK scaffolds. As an alternative, the multiple scaffold proteins known to be present might instead be spatially redundant and overlapping within single synapses. A particularly interesting case is the MAGUK family member SAP102, which is expressed in most excitatory synapses (Cizeron et al. 2020) but thought to be especially important at early developmental stages when PSD-95 expression is still rising (Petralia et al. 2005; Sheng and Hoogenraad 2007; Elias et al. 2008). Though known to be widespread throughout the synapse (Degiorgis et al. 2006), it also organizes in NCs (Metzbower et al. 2024). SAP102 NCs are relatively smaller

(~60 nm in diameter) and denser than those of PSD-95 across several weeks of development in culture. Surprisingly, though, these NCs of SAP102 appear positioned nearly independently of PSD-95: within various NCs of either protein, there is a wide range of abundances of the other, including some where both are dense but some where the other is relatively sparse. About 40% of the total synaptic area contains both proteins, with PSD-95-only areas making up the majority of the single-MAGUK area of the PSD. Within the areas where both are present, the NCs of *both* proteins are larger than those outside of it. Thus, it appears that the internal structure of a PSD is comprised of different nanoscale subregions, and the characteristic features of a protein's organization are different depending on whether it is in a parcel that contains other specific molecules.

7.3.4 PSD Function Defined by Subsynaptic Nanodomains

This example of multiscaffold imaging emphasizes what we feel is an important principle of synaptic nanoarchitecture: within even the same synapse, nanodomains can exist that contain unique sets or ratios of proteins. Indeed, this conception of subsynaptic organization is not limited to the MAGUK family but extends to critical cytosolic scaffold proteins of the PSD, such as Shank, GKAP, and Homer (MacGillavry et al. 2013), and is evident in the frequent segregation of AMPARs and NMDA receptors (NMDARs) within single synapses (Goncalves et al. 2020; Hruska et al. 2022), suggesting that formation of molecularly unique nanodomains is a key and conserved property of the postsynaptic compartment. It is straightforward to speculate that scaffold-defined nanodomains are capable of numerous distinct functional roles (Fig. 7.2). Capturing diffusing glutamate receptors and organizing them with respect to presynaptic release sites is arguably the most straightforward role of the MAGUKs and certainly the aspect that is best understood. Indeed, the nanocolumn typifies a synaptic nanodomain. In addition, the parcellation of MAGUK types itself is critical in many ways, since these scaffolds are thought to organize not just receptors, but other transmembrane proteins such as ion channels, adhesion proteins, and signaling or regulatory molecules within the cytosol. Thus, most generally, the presence of heterogeneous nanodomains within synapses appears poised to add scale, complexity, and nuance to transmission at even single contact sites. They not only modulate postsynaptic responsiveness but also establish biochemically distinctive regions within synapses that enable multiple concurrent forms of information transfer.

The mechanisms by which synaptic nanodomains arise remain unclear. An influential recent model of PSD assembly, and indeed synaptic assembly overall, proposes that this synaptic compartment forms as a condensate through liquid–liquid phase separation of the constituent proteins (Zeng et al. 2016; Emperador-Melero and Kaeser 2019; Feng et al. 2019). An intrinsic feature of phase-separated droplets is their lack of internal structure and the ability for proteins to diffuse freely within them, features obviously not representative of native synapses. Thus, in this model,

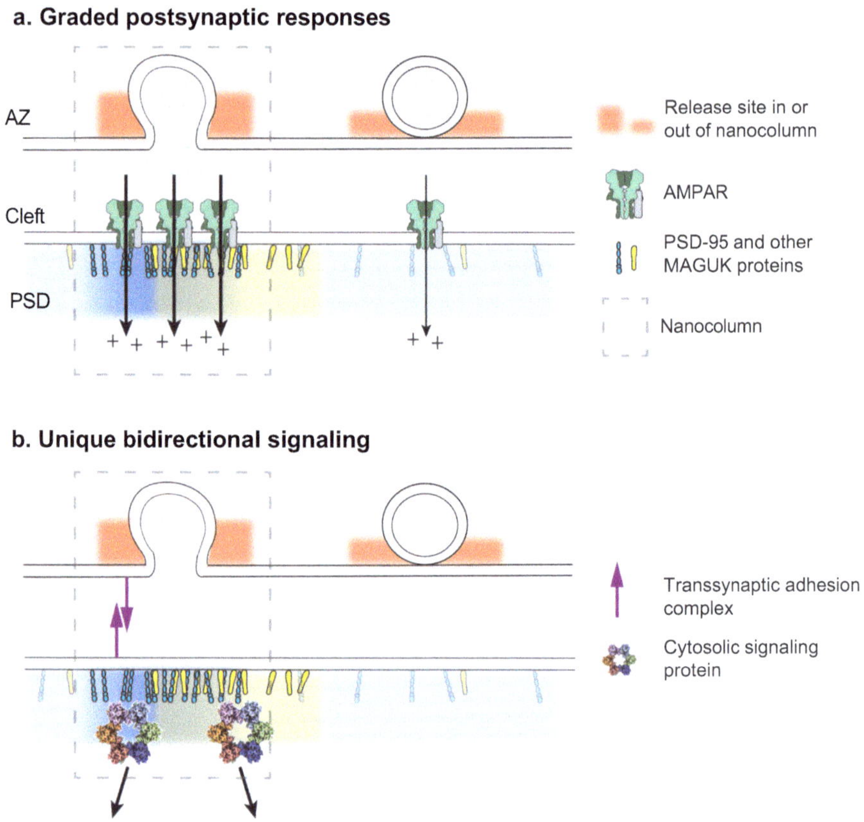

Fig. 7.2 *Nanodomain functions elaborate synaptic communication*. **(a)** Nanodomains of scaffold proteins concentrate receptors, and the probability of receptor activation is enhanced when these receptor-containing nanodomains are trans-synaptically aligned with presynaptic release sites in the nanocolumn. Release at sites outside the nanocolumn activates fewer receptors. **(b)** Nanodomains of scaffold proteins and receptors can accumulate unique signaling constituents. These may include cytosolic elements to produce intracellular signaling (as illustrated by CaMKII, PDB8T15), or bidirectional cell-to-cell communication mediated by a unique complement of cleft-resident adhesion molecules. The time scale and information content of such communication differ dramatically from excitatory postsynaptic currents, expanding the computational complexity of synaptic transmission

creating the extensive and complex nanodomain structure observed in synapses would presumably arise through a set of subsegregation processes (Hosokawa et al. 2021), wherein the numerous synaptic proteins having multiple weak interactions with potential neighbors self-sort into various “subphases” in the overall synaptic compartment. The behavior of such subphases on the nanoscale remains unclear, particularly in cells, but such segregation would also need to explain how certain proteins, including PSD-95, show their characteristically limited mobility. This could potentially be established through a stabilizing influence of interactions with proteins of the synaptic cleft. Though initially developed primarily as a hypothesis based on in vitro protein behavior, accumulating evidence suggests that aspects of phase separation behavior seen in vitro are likely at play in native synapses. Indeed, disrupting liquid–liquid phase separation via aliphatic alcohol has remarkably specific effects on action potential-evoked neurotransmitter release with little effect on spontaneous transmission (Guzikowski and Kavalali 2024), consistent with the idea that release sites in the nanocolumn preferentially engage in evoked transmission (Tang et al. 2016). This is an exciting area of active research, and it will be important to carefully evaluate which aspects of in vitro biochemical behavior are influential in the full cellular context.

7.3.5 Alternative Nomenclature

Considering terminology, note that some authors have used "subsynaptic domain" or SSD similarly to our definition of nanocluster. We prefer to think of a domain as a space in which different components can be assembled. That is, we use the term "nanodomain" to indicate a *subregion of the synapse* where multiple proteins gather in some fashion, as discussed further below. Thus, in our terminology, PSD-95 molecules can assemble in a NC, which itself can be a key element within a nanodomain enriched for both PSD-95 and other proteins such as receptors or other cytosolic proteins (Fig. 7.1). Notably, some authors have conversely argued to use SSD more or less synonymously with our use of "nanodomain" (Yang and Specht 2019), which appears sensible so long as one differentiates accumulations of one protein from accumulations of many, and keeps in mind that limited imaging resolution can blur not only the image but also the ability to discriminate substructures with different names.

Another alternative nomenclature for describing PSD nanostructure was proposed initially based on STED imaging in live cell cultures and in fixed brain slices (Hruska et al. 2018, 2022). Similar to Fukata et al. (2013), this study discriminated distinct structures of PSD-95 in single spines, separated by enough space to be individually resolved. These assemblies ranged in size from ~90 to 200 nm in diameter, and their number correlated with spine size. Because during chemical potentiation the number of such structures per spine varied even more prominently than did their size, they were termed "nanomodules" (NMs). This observation has thus proven particularly attractive for the study of synaptic plasticity, shining light on mechanisms in spines during LTP (Fukata et al. 2024).

However, the relationship between NMs and NCs remains unclear. For example, even after potentiation, more than 60% of observed spines were reported to have single NMs, but separation between NMs in single spines that have more than one was often larger than the typical diameter of synapses, suggesting NMs might represent entire PSDs, rather than PSD substructure like NCs. Thus, at some level, the NM terminology seems incompatible with the notion of NCs found within a larger continuous PSD, and clarifying the relationship between a PSD, an NM, and NCs remains a work in progress. Are some NMs essentially just entire PSDs? If so, do they contain or create NCs within them? A recent study combining diffraction-limited spine segmentation with dSTORM imaging of PSD-95 in brain slices reported a hierarchical organization of PSD-95 in spines, with one or more assemblies per spine that each contained smaller structures of comparable size to what we call here a NC (Zhang et al. 2025a, b). The number of larger assemblies per spine correlated with spine size, consistent with the NM terminology (Hruska et al. 2018, 2022), and the larger assemblies contained more NCs. It remains to be determined whether the common synapses that would be characterized as "simple" PSDs by EM have widely separated "modules" of PSD-95, and if so, what is the intervening molecular structure. Are single synapses built from a set of NMs, or are NMs best thought of as components for building functional complexity at the level of the spine rather than the level of single synaptic junctions? In spines that add NMs during plasticity, do those NMs maintain their distinct individual molecular identities, or do they share protein among themselves? We suspect that answering these central questions will require further advances in the resolution of live-cell imaging and in more systematic molecular-scale imaging in tissues under a variety of fixation conditions.

7.3.6 *Diversity of Nanostructure*

Given this nested set of architectural elements, key questions are what aspects of scaffold nanostructure are conserved across synapse types, and what features are adapted within the local synaptic environment to mediate unique organization and functionality in different synapses. Systematic, high-resolution comparison between synapse types is not yet a rich database, but the findings so far are intriguing. The most common report is that PSDs generally contain just one or two PSD-95 NCs. In the mouse hippocampus (Broadhead et al. 2016) and neocortex (Jones et al. 2024), synapses can contain multiple PSD-95 NCs, but most synapses (~70%) are constituted by just one (of an overall median diameter ~ 130 nm). A larger proportion of synapses in the mouse neocortex was reported to contain multiple NCs of PSD-95 using a version of Expansion Microscopy (Sarkar et al. 2022), potentially owing to the higher imaging resolution compared to gSTED or PALM, but the study did not discriminate the cell type of the presynaptic contact, which contributes substantially to postsynaptic nanostructural diversity (Dharmasri et al. 2024; Jones et al. 2024). Interestingly, single nanoassemblies within a PSD have slightly larger size and denser localizations (Broadhead et al. 2016), suggesting a potential subcategorization of NCs.

Synaptic characteristics also vary substantially at contacts between different cell types in a circuit, but nanocolumn features have not been systematically evaluated across cell types. To begin this, a direct comparison of two major excitatory synapse types, Dharmasri et al. (2024) analyzed the protein content and nanostructure of excitatory synapses onto either excitatory neurons or parvalbumin-expressing interneurons in culture. At a broad level, PSD-95 nanostructure was quite similar between these two synapse types, showing a fairly continuous expanse of PSD-95 containing about two NCs per synapse. This is a particularly interesting indication that this aspect of nanostructure is robust in the face of the differing geometry of axo-spinous and axo-dendritic synapses, as well as presumably divergent sets of receptor and adhesion molecule PSD-95 binding partners expressed by these two types of postsynaptic neurons. Nevertheless, several features of PSD-95 organization differed between the two cell types, including most notably the size of the NCs, which were nearly twice as large in the synapses onto parvalbumin interneurons (Fig. 7.3). Characteristics of Munc13–1 NCs differed between these synapse types even though the presynaptic cell type was presumably identical, and it will be important to determine whether postsynaptic cell organizing molecules drive this presynaptic differentiation to regulate afferent synaptic efficacy. Critically, overall nanocolumn structure was conserved, consistent with widespread identification in many synapses in vivo, but specific nanocolumn characteristics differed between these synapse types, including the size of aligned nanodomains, and unanticipated features such as the spatial extent over which PSD-95 concentrated near Munc13–1 NCs.

The overall conservation of nanocolumn alignment between different cell types highlights its key role in governing neurotransmission. At the same time, the distinct

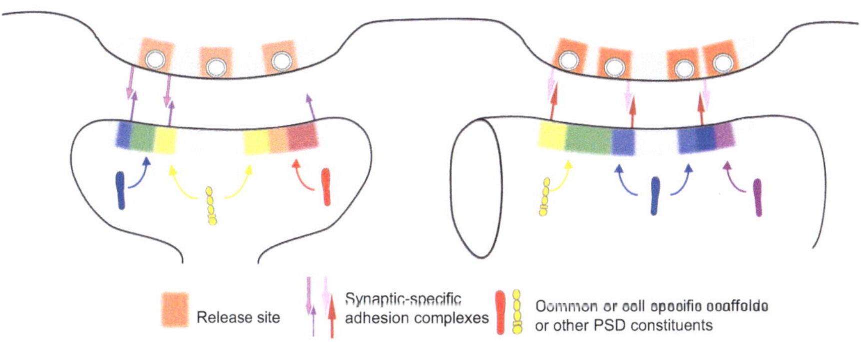

Fig. 7.3 *Nanodomain assembly contributes to synaptic function diversity*. Assembly of unique nanodomains in each cell arises from the unique expression pattern of MAGUKs and other proteins. Nanocolumn structure is conserved at synapses onto diverse cell types. However, specifics of the organization within the active zone and PSD differ, even for molecules like PSD-95 and Munc13–1 that are expressed in common. Trans-synaptic spatial relationships are in turn substantially different. In this figure, we compare two synapses onto different cell types, based on our findings (Dharmasri et al. 2024). A common set of PSD scaffold building blocks can each form nanoclusters (colored density gradients) that assemble into both unique and shared nanodomains (combinations of density gradients). These are likely aided in assembly by different adhesion proteins that govern their trans-synaptic alignment with presynaptic nanoclusters of proteins such as Munc13–1, which themselves may be uniquely governed by cell type. These tight spatial relationships are expected to define and tune the functional and biochemical properties of the synapse needed for its specific circuit role. In sum, the complement of nanodomain types and their assembly and positioning add functional diversity to the computational ability of the synapse

variations on a theme we have observed between cell type-specific connections indicate nanodomain organization is likely a key player in tuning the responses of those connections in neural circuits. In our discussion here, we have focused primarily on postsynaptic mechanisms, but presynaptic protein organization independently influences synaptic strength as well. Critically, trans-synaptic alignment of these NCs into the nanocolumn integrates the influence of nanoclusters and nanodomain architecture in each cell, impacting both the magnitude of single responses as well as more complex frequency-dependent responses. Thus, defining in many cell types the nanostructural characteristics of key proteins over multiple spatial scales, along with the trans-synaptic relationships among proteins, will be essential for relating cellular expression profiles to functional diversity across the brain.

7.4 Synapse Nanoarchitecture Regulates NMDAR Activation

An important case study in the functional role of nanodomain organization is our growing understanding of NMDA receptor (NMDAR) positioning. Nanocolumn reorganization likely contributes to the expression of synaptic plasticity by modifying AMPAR activation during behavioral learning. However, the induction of the most-studied forms of synaptic plasticity relies on NMDARs, and thus it is critical to elucidate the mechanisms regulating NMDAR activation. Classically, it has been recognized that the coincidence of depolarization after agonist binding is the major mechanism of NMDAR functional regulation, endowing a stimulus frequency-dependence to receptor activation and its Ca^{2+} flux. Here, we suggest that nanodomain organization can be a strong cellular point of control governing how NMDARs at individual synapses respond, and that can influence how the intensity and pattern of presynaptic activity may trigger plasticity. The data are considerably less abundant than for AMPARs, but the findings illustrate and expand on the themes of this review.

Biophysical characteristics of NMDARs indicate that their activation may be sensitive to their distance from release sites in a similar fashion to AMPARs. Importantly, the steady-state affinity of NMDARs for glutamate, which is generally moderately high for NMDARs and fairly low for AMPARs, is not directly relevant during the rapidly changing conditions of synaptic transmission. Most important are the kinetics of glutamate binding and the transitions to open and desensitized states after this. Modeling based on the single-channel rate constants of these transitions after glutamate release from a single synaptic vesicle shows, as a rule of thumb, that the GluN2B-containing NMDARs open 25–50% less often if they are displaced from the release site by ~100 nm, quite similar to the drop-off experienced by AMPARs (Santucci and Raghavachari 2008). GluN2A-containing receptors are less sensitive, decreasing their responsiveness by about 25% from one side to the other of a 300-nm synapse. Thus, modifying receptor proximity to release sites on its face appears to be an effective way to change the postsynaptic response to an unchanged release event. Unfortunately, the numerical models of receptor activation contain numerous parameters that are difficult to constrain with high confidence, and so quantitative predictions can be variable, but in models, it would be wise to account for the stark differences between the kinetics of GluN2A and GluN2B rather than simply predicting the behavior of a generic NMDAR of no particular subtype.

7.4.1 Subtype-Specific Characteristics of NMDAR NCs

A popular concept is that NMDARs tend to be located more toward the center of the PSD than toward the edge, an observation which has been supported with a number of papers and techniques (Pérez-Otaño et al. 2006; Chen et al. 2008; Goncalves et al. 2020; Li et al. 2021). However, this is typically measured as quite a mild effect and may stem from the low labeling density typical of classic immunogold EM, which necessitates averaging the relative positions of receptors across many synapses, preventing discrimination of subsynaptic patterns within individual synapses. Super-resolution microscopy has shown NMDAR organization is in fact quite individualized in different synapses (Dani et al. 2010) and typified by the presence of NMDAR subtype-specific NCs (Kellermayer et al. 2018). Indeed, NCs of NMDARs are routinely found throughout single synapses, including near the edge of the PSD (Tarusawa et al. 2009; Dani et al. 2010; Anderson et al. 2023) (and conversely, AMPARs are commonly found throughout PSDs, including near the center). It is our sense that many of the mechanisms and functional impacts of NMDAR subsynaptic positioning are more likely to be explained by the robust presence and positioning of receptor NCs, rather than the mild overall tendency for NMDARs to be located on average toward the center of synapses.

Some differences between the number and size of GluN2A and GluN2B NCs have been reported in cultured neurons (Kellermayer et al. 2018; Anderson et al. 2023), though they are generally similar to one another: 2–10 per synapse, about 60 nm in diameter. These colocalize with PSD-95 NCs (MacGillavry et al. 2013), consistent with their robust association with the scaffold (Frank and Grant 2017), and GluN2B in particular was seen to be enriched within PSD-95 NCs identified by dSTORM as we describe below. Many NMDARs, probably about half (Gray et al. 2011; Tovar et al. 2013; Zhang et al. 2025a, b), are "triheteromeric" receptors that contain two different types of GluN2 or GluN3 subunit in addition to the mandatory GluN1. All existing imaging data about endogenous NMDARs, including their synaptic organization, subcellular trafficking, and biosynthesis and degradation, have targeted single subunits due to the lack of reagents that distinguish specific triheteromeric subtypes. Thus, the triheteromeric receptors have been extremely difficult to detect. At high imaging resolutions, though, a population of triheteromeric NMDARs is suggested by closely colocalized GluN2A and GluN2B labels (Kellermayer et al. 2018; Anderson et al. 2023), indicating the presence of synaptic nanodomains in which either separate diheteromeric receptor types (GluN1/2A and GluN1/2B) closely intermix or triheteromeric GluN1/2A/2B receptors reside. The population of suspected triheteromers seen in these data does not appear to be as large as the electrophysiological and cryo data indicate (Gray et al. 2011; Tovar et al. 2013; Zhang et al. 2025a, b), suggesting that other detection methods may be required either at higher resolution or with different probes that detect the presence of more than one subunit type (Anderson et al. 2023).

7.4.2 Nanodomain Organization at a Subset of Release Sites Impacts NMDAR Activation

Given the potential sensitivity of NMDARs to the glutamate concentration-time profile following vesicle fusion, a key question is whether NMDA NCs align with presynaptic release sites. Surprisingly, the NCs on average show (Anderson et al. 2023) a de-enrichment from PSD areas closest to release sites, as labeled with Munc13 immunostaining (Böhme et al. 2016; Reddy-Alla et al. 2017; Sakamoto et al. 2018), quite different than the nanocolumn enrichment of AMPARs. One could imagine that this could be due to a lack of binding sites because, for instance, PSD-95 directly across from Munc13 sites is already occupied by AMPARs and other residents. Alternatively, receptor segregation could be explained in terms of liquid–liquid phase separation, in that NMDARs and AMPARs separate differentially through phase unmixing. Regardless, though NMDAR and AMPAR NCs each are found widely through the synapse, they appear to define distinguishable nanodomains, potentially including differential abundance of PSD-95 and SAP102 that shows some preferential ability to control trafficking of NMDARs rather than AMPARs (Chen et al. 2012).

Despite their lack of enrichment to release sites on average, the most illuminating aspect of NMDAR positioning is that the receptors are in fact relatively *well* accumulated around just a *subset* of release sites (Anderson et al. 2023), highlighting the specificity of nanodomain construction as illustrated in Fig. 7.4. While NMDARs are generally positioned away from release sites, they are significantly more associated with release sites that are in the nanocolumn—that is, those release sites across from a high accumulation of postsynaptic PSD-95. Further, the NCs of NMDARs that were in these nanocolumn-positive positions were denser than NCs elsewhere in the synapse, containing either more receptors or more closely spaced receptors. And interestingly, the two NMDAR subunits were more co-enriched with one another at the nanocolumn, potentially indicating more abundant triheteromeric receptors. Together, this reveals a preferential positioning of enlarged, heterogeneous NMDAR nanodomains near release sites aligned with PSD-95. Note that, unlike AMPAR alignment, little is known about adhesion mechanisms controlling NMDAR positioning, though the LAR-RPTP family modulates NMDAR-mediated transmission (Sclip and Sudhof 2020; Kim et al. 2022). Critically, though, simulations based on the measured protein locations within individual synapses showed that overall NMDA open probability is higher than predicted for synapses in which the locations of release or receptors were randomized, and indicated that the average

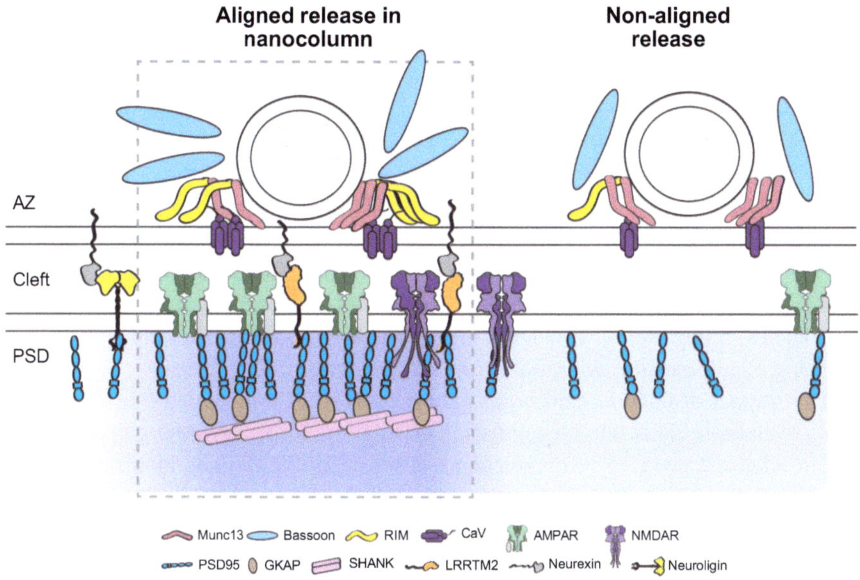

Fig. 7.4 *AMPAR and NMDAR signaling via postsynaptic nanodomains associated with unique presynaptic release sites***.** As illustrated in this summary model, subsets of nanocolumn-resident release sites are enriched postsynaptically with AMPARs as well as NMDARs that are laterally displaced compared to AMPARs. Active zones contain molecularly diverse release sites likely with differing vesicle priming and release properties, and whose functional impact depends in part on differential trans-synaptic alignment. Some release sites accumulating less Munc13 are not associated with postsynaptic nanoclusters of PSD-95 or NMDARs. PSD-95 nanocluster density in the nanocolumn is also denser, potentially aiding or aided by phase separation. Cell adhesion molecules mediate coordinated alignment of pre- and postsynaptic proteins. Within the PSD, enriched receptors and other scaffolding proteins accumulate a unique variety of signaling proteins that establish biochemically unique signaling capacity in distinct nanodomains of the synapse. Receptor nanodomains near nanocolumn release sites contain GluN2A and GluN2B subunits; NMDARs accumulated near release sites will be activated more efficiently and potentially transduce unique signaling typified by CaMKII, due to the presence of a mixed population of intracellular C-termini. Model adapted from Anderson et al. (2023)

NMDAR open probability trends higher when glutamate is released at sites in the nanocolumn versus outside the nanocolumn (Anderson et al. 2023). Thus, receptor organizational characteristics are dependent on the residents of their nanodomain within the synapse, including the trans-synaptic context.

How this detailed positioning of NMDARs within synapses corresponds to *functionally characterized* fusion sites will be critical to determine. Vesicle fusion events are categorized as synchronous with action potentials, briefly delayed (asynchronous), or independent of action potentials (spontaneous), and substantial evidence indicates that these events are mediated by at least partially nonoverlapping molecular mechanisms in the terminal (Kaeser and Regehr 2014; Guzikowski and Kavalali 2021). All these types of release events are known to activate NMDARs to mediate a variety of physiological outcomes (Sutton et al. 2006; Kavalali 2015; Wang et al. 2020; Li et al. 2021). Though the above discussion highlights our advancing understanding of NMDAR molecular context, it is not clear yet what molecular organization at distinct active zone sites supports specific forms of release (Kaeser and Regehr 2014). Nevertheless, in an elegant study using rapid freezing following single action potentials in neuron culture, Li et al. visualized labeled NMDARs along with the sites of exocytic events that could be categorized as synchronous or asynchronous. Whereas AMPARs clustered near exocytic sites of synchronous release as expected in the nanocolumn model, NMDARs preferentially clustered near sites of asynchronous release (Li et al. 2021). To reconcile these findings and approaches, a key technical need is the means to combine functional subsynaptic mapping of release sites with multiprotein nanostructural imaging.

The impact of nanodomain organization on NMDAR-mediated transmission is likely even more refined in the regulation of "nonionotropic" signaling that does not require channel opening or conductance (Park et al. 2022). The binding of either agonist or co-agonist induces conformational changes within the long NMDAR cytosolic C-terminal tails (Dore et al. 2015; Ferreira et al. 2017) thought to trigger second messenger cascades leading to, among other things, actin filament destabilization and dendritic spine shrinkage (Stein et al. 2015, 2021; Dore et al. 2017). Because binding of just one molecule of glutamate appears sufficient to induce at least some GluN1 structural changes (Vyklicky et al. 2021), the nonionotropic activation of NMDARs likely is much less sensitive to distance from the site of release and thus has the potential of being differentially influenced by synaptic nanostructure in comparison to ionotropic activation. To evaluate the impact of the nanostructure on ionotropic versus nonionotropic signaling, it would be critical to be able to spatially discriminate activated receptors directly in situ, which may become feasible with advances in super-resolution microscopy with techniques able to reach the molecular scale, such as MINFLUX (Sahl et al. 2024) or DNA-PAINT with its RESI iteration (Reinhardt et al. 2023).

7.5 Conclusions and Future Directions

Many aspects of the characteristics, mechanisms, and impact of synaptic nanostructure remain to be elucidated. Abundant new options for super-resolution imaging in brain tissue (Narayanasamy et al. 2021; Delling et al. 2024; Hümpfer et al. 2024; Pérez-Garza et al. 2024) should speed our determination of how nanodomains are built differentially by cells expressing different subsets of molecules, how gradients of protein content and synapse size along dendrites may help create cellular subregions of nanostructurally similar synapses, and how activity may shape nanostructure in these unique molecular contexts. Mechanisms of molecular coordination at scales of 50–150 nm, including the roles and specific mechanisms of phase separation (Hayashi et al. 2021), will be particularly challenging and informative to unravel, including how nanodomain composition regulates protein mobility and exchange on highly local levels. Functionally, modeling based on increasingly detailed molecular maps of synapses (Unterauer et al. 2024) along with deep optical physiology at single synapses (Barlow et al. 2024) offers hope for deciphering how nanostructure shapes postsynaptic responses to patterned activity, or whether and how biochemical signaling is compartmentalized by PSD nanodomains. We have argued that nanostructural organization elaborates the ability of synapses to carry out complex biochemical calculations, tuning the performance of the synapse and mediating more complex impacts on the cell. To determine how these features contribute to information transfer (Samavat et al. 2024), memory formation, and the broader operation of neural circuits will require new molecular tools that permit specific control of synaptic nanostructure during tailored physiological and behavioral assays.

We end with the most obvious point of all, but one that apparently still needs to be made explicit. The information in this chapter, as in probably all of this book, was obtained through federal funding of basic biomedical research. That research is expensive, but the payoff is enormous. Drugs or therapies that modify synapses and synaptic transmission are among the most common in psychiatry and neurology, and deepening our fundamental understanding of synapse organization and operation continually offers abundant new ideas for how to control brain function. Thus, new and better treatments will only come about via long-term and politically indestructible funding of basic science. And maybe even more self-evident, though the work summarized in this chapter is a nanometer-scale step forward in the long journey of human achievement, the discoveries represented here from the past 10 or so years of work form the foundation for the unimaginable successes that will come in the next 10 years and the years after that. It used to go without saying, but now we should all take every opportunity to say it: cutting off research funding now will mean the delay and loss of knowledge for generations.

Acknowledgments We thank Poorna A. Dharmasri, Ph.D., for initial consultations and detailed discussions of this work, and members of the Blanpied lab for many helpful conversations. Support was provided by the NIH (R37 MH080046 and R21 MH141559).

References

Anderson MC, Nigam A, Johnson JW, Blanpied, Thomas A. Direct visualization of triheteromeric NMDA receptor trafficking and organization. Society for Neuroscience Abstracts PST123.05. 2023.

Anderson MC, Dharmasri PA, Damenti M, Metzbower SR, Laghaei R, Blanpied TA, Levy AD. Trans-synaptic molecular context of NMDA receptor nanodomains. 2025; https://doi.org/10.1101/2023.12.22.573055. version 2

Barlow ST, Levy AD, Contreras M, Anderson MC, Blanpied TA. Dissecting the functional heterogeneity of glutamatergic synapses with high-throughput optical physiology. bioRxiv. 2024;

Bekkers JM, Stevens CF. Quantal analysis of EPSCs from small numbers of synapses in hippocampal cultures. J Neurophysiol. 1995;73:1145.

Biederer T, Kaeser PS, Blanpied TA. Transcellular nanoalignment of synaptic function. Neuron. 2017;96:680–96.
Blanpied TA, Kerr JM, Ehlers MD. Structural plasticity with preserved topology in the postsynaptic protein network. Proc Natl Acad Sci USA. 2008;105:12587–92.
Böhme MA, et al. Active zone scaffolds differentially accumulate Unc13 isoforms to tune Ca2+ channel–vesicle coupling. Nat Neurosci. 2016;19:1311–20.
Broadhead MJ, Horrocks MH, Zhu F, Muresan L, Benavides-Piccione R, DeFelipe J, Fricker D, Kopanitsa MV, Duncan RR, Klenerman D, Komiyama NH, Lee SF, Grant SGN. PSD95 nanoclusters are postsynaptic building blocks in hippocampus circuits. Sci Rep. 2016;6:24626.
Burette AC, Lesperance T, Crum J, Martone ME, Volkmann N, Ellisman MH, Weinberg RJ. Electron tomographic analysis of synaptic ultrastructure. J Comp Neurol. 2012;520:2697–711.
Buzsáki G, Mizuseki K. The log-dynamic brain: how skewed distributions affect network operations. Nat Rev Neurosci. 2014;15:264–78.
Chazeau A, Mehidi A, Nair D, Gautier JJ, Leduc C, Chamma I, Kage F, Kechkar A, Thoumine O, Rottner K, Choquet D, Gautreau A, Sibarita JB, Giannone G. Nanoscale segregation of actin nucleation and elongation factors determines dendritic spine protrusion Anaël. EMBO J. 2014;33:2745–64.
Chen X, Winters CA, Azzam R, Li X, Galbraith JA, Leapman RD, Reese TS. Organization of the core structure of the postsynaptic density. Proc Natl Acad Sci USA. 2008;105:4453–8.
Chen BS, Gray JA, Sanz-Clemente A, Wei Z, Thomas EV, Nicoll RA, Roche KW. SAP102 mediates synaptic clearance of NMDA receptors. Cell Rep. 2012;2:1120–8.
Chen H, Tang A, Blanpied TA. Subsynaptic spatial organization as a regulator of synaptic strength and plasticity. Curr Opin Neurobiol. 2018;51:147–53.
Choquet D. Linking nanoscale dynamics of AMPA receptor organization to plasticity of excitatory synapses and learning. J Neurosci. 2018;38:9318–29.
Cizeron M, Qiu Z, Koniaris B, Gokhale R, Komiyama NH, Fransén E, Grant SGN. A brainwide atlas of synapses across the mouse life span. Science. 2020;369:270–5.
Dani A, Huang B, Bergan J, Dulac C, Zhuang X. Superresolution imaging of chemical synapses in the brain. Neuron. 2010;68:843–56.
Degiorgis JA, Galbraith JA, Dosemeci A, Chen X, Reese TS. Distribution of the scaffolding proteins PSD-95, PSD-93, and SAP97 in isolated PSDs. Brain Cell Biol. 2006;35:239–50.
Delling JP, Bauer HF, Gerlach-Arbeiter S, Schön M, Jacob C, Wagner J, Pedro MT, Knöll B, Boeckers TM. Combined expansion and STED microscopy reveals altered fingerprints of postsynaptic nanostructure across brain regions in ASD-related SHANK3-deficiency. Mol Psychiatry. 2024;29(10):2997–3009.
Dharmasri PA, Levy AD, Blanpied TA. Differential nanoscale organization of excitatory synapses onto excitatory vs. inhibitory neurons. Proc Natl Acad Sci USA. 2024;121:e2315379121.
Dore K, Aow J, Malinow R. Agonist binding to the NMDA receptor drives movement of its cytoplasmic domain without ion flow. Proc Natl Acad Sci. 2015;112:14705–10.
Dore K, Stein IS, Brock JA, Castillo PE, Zito K, Sjöström PJ. Unconventional NMDA receptor signaling. J Neurosci. 2017;37:10800–7.
Droogers WJ, MacGillavry HD. Plasticity of postsynaptic nanostructure. Mol Cell Neurosci. 2023;124:103819.
Elias GM, Elias LAB, Apostolides PF, Kriegstein AR, Nicoll RA. Differential trafficking of AMPA and NMDA receptors by SAP102 and PSD-95 underlies synapse development. Proc Natl Acad Sci. 2008;105:20953–8.
Emperador-Melero J, Kaeser PS. Liquid active zones for controlling the phases of synaptic transmission. Mol Cell. 2019;73:859–60.
Farley MM, Swulius MT, Waxham MN. Electron tomographic structure and protein composition of isolated rat cerebellar, hippocampal and cortical postsynaptic densities. Neuroscience. 2015;304:286–301.

Feng Z, Chen X, Zeng M, Zhang M. Phase separation as a mechanism for assembling dynamic postsynaptic density signalling complexes. Curr Opin Neurobiol Mol Neurosci. 2019;57:1–8.
Ferreira JS, Papouin T, Ladépêche L, Yao A, Langlais VC, Bouchet D, Dulong J, Mothet JP, Sacchi S, Pollegioni L, Paoletti P, Richard Oliet SH, Groc L. Co-agonists differentially tune GluN2B-NMDA receptor trafficking at hippocampal synapses. elife. 2017;6:1–22.
Fortin DA, Tillo SE, Yang G, Rah J-C, Melander JB, Bai S, Soler-Cedeno O, Qin M, Zemelman BV, Guo C, Mao T, Zhong H. Live imaging of endogenous PSD-95 using ENABLED: a conditional strategy to fluorescently label endogenous proteins. J Neurosci. 2014;34:16698–712.
Frank RAW, Grant SGN. Supramolecular organization of NMDA receptors and the postsynaptic density. Curr Opin Neurobiol. 2017;45:139–47.
Frank RAW, Komiyama NH, Ryan TJ, Zhu F, O'Dell TJ, Grant SGN. NMDA receptors are selectively partitioned into complexes and supercomplexes during synapse maturation. Nat Commun. 2016;7(17):1–13.
Frank RAW, Zhu F, Komiyama NH, Grant SGN. Hierarchical organization and genetically separable subfamilies of PSD95 postsynaptic supercomplexes. J Neurochem. 2017;142:504–11.
Fukata Y, Dimitrov a, Boncompain G, Vielemeyer O, Perez F, Fukata M. Local palmitoylation cycles define activity-regulated postsynaptic subdomains. J Cell Biol. 2013;202:145–61.
Fukata Y, Chen X, Chiken S, Hirano Y, Yamagata A, Inahashi H, Sanbo M, Sano H, Goto T, Hirabayashi M, Kornau H-C, Prüss H, Nambu A, Fukai S, Nicoll RA, Fukata M. LGI1-ADAM22-MAGUK configures transsynaptic nanoalignment for synaptic transmission and epilepsy prevention. Proc Natl Acad Sci USA. 2021;118:e2022580118.
Fukata Y, Fukata M, MacGillavry HD, Nair D, Hosy E. Celebrating the birthday of AMPA receptor nanodomains: illuminating the nanoscale organization of Excitatory Synapses with 10 nanocandles. J Neurosci. 2024;44:e2104232024.
Goncalves J, Bartol TM Jr, Camus C, Levet F, Menegolla AP, Sejnowski TJ, Sibarita JB, Vivaudou M, Choquet D, Hosy E. Nanoscale co-organization and coactivation of AMPAR, NMDAR, and mGluR at excitatory synapses. Proc Natl Acad Sci USA. 2020;117:14503–11.
Gou X-Z, Ramsey AM, Tang A-H. Re-examination of the determinants of synaptic strength from the perspective of superresolution imaging. Curr Opin Neurobiol. 2022;74:102540.
Gray JA, Shi Y, Usui H, During MJ, Sakimura K, Nicoll RA. Distinct modes of AMPA receptor suppression at developing synapses by GluN2A and GluN2B: single-cell NMDA receptor subunit deletion in vivo. Neuron. 2011;71:1085–101.
Greger IH, Watson JF. 'Mini analysis' is an unreliable reporter of synaptic changes. 2024; https://doi.org/10.1101/2024.10.26.620084.
Gürth C-M, Lima MA d RBF, Palacios VM, Hubrich J, Delgado ARC, Mougios N, Opazo F, D'Este E. Neuronal activity modulates the incorporation of newly translated PSD-95 into a robust structure as revealed by STED and MINFLUX. 2023; https://doi.org/10.1101/2023.10.18.562700.
Guzikowski NJ, Kavalali ET. Nano-Organization at the synapse: segregation of distinct forms of neurotransmission. Front Synaptic Neurosci. 2021;13:796498.
Guzikowski NJ, Kavalali ET. Functional specificity of liquid-liquid phase separation at the synapse. Nat Commun. 2024;15:10103.
Haas KT, Compans B, Letellier M, Bartol TM, Grillo-Bosch D, Sejnowski TJ, Sainlos M, Choquet D, Thoumine O, Hosy E. Pre-post synaptic alignment through neuroligin-1 tunes synaptic transmission efficiency. Elife. 2018;7:e31755.
Han Y, Cao R, Qin L, Chen LY, Tang A-H, Südhof TC, Zhang B. Neuroligin-3 confines AMPA receptors into nanoclusters, thereby controlling synaptic strength at the calyx of Held synapses. Sci Adv. 2022;8:eabo4173.
Harris KM, Weinberg RJ. Ultrastructure of synapses in the mammalian brain. Cold Spring Harb Perspect Biol. 2012;4(5):1–30.
Hayashi MK, Tang C, Verpelli C, Narayanan R, Stearns MH, Xu R-M, Li H, Sala C, Hayashi Y. The postsynaptic density proteins Homer and Shank form a polymeric network structure. Cell. 2009;137:159–71.

Hayashi Y, Ford LK, Fioriti L, McGurk L, Zhang M. Liquid-liquid phase separation in physiology and pathophysiology of the nervous system. J Neurosci. 2021;41:834–44.

Held RG, Liang J, Brunger AT. Nanoscale architecture of synaptic vesicles and scaffolding complexes revealed by cryo-electron tomography. Proc Natl Acad Sci U S A. 2024;121:e2403136121.

Helm MS, Dankovich TM, Mandad S, Rammner B, Jähne S, Salimi V, Koerbs C, Leibrandt R, Urlaub H, Schikorski T, Rizzoli SO. A large-scale nanoscopy and biochemistry analysis of postsynaptic dendritic spines. Nat Neurosci. 2021;24:1151.

Hosokawa T, Liu PW, Cai Q, Ferreira JS, Levet F, Butler C, Sibarita JB, Choquet D, Groc L, Hosy E, Zhang M, Hayashi Y. CaMKII activation persistently segregates postsynaptic proteins via liquid phase separation. Nat Neurosci. 2021;24:777–85.

Hruska M, Henderson N, Le Marchand SJ, Jafri H, Dalva MB, Marchand SJL, Jafri H, Dalva MB, Le Marchand SJ, Jafri H, Dalva MB. Synaptic nanomodules underlie the organization and plasticity of spine synapses. Nat Neurosci. 2018;21:671–82.

Hruska M, Cain RE, Dalva MB. Nanoscale rules governing the organization of glutamate receptors in spine synapses are subunit specific. Nat Commun. 2022;13(1):920.

Hümpfer N, Thielhorn R, Ewers H. Expanding boundaries – a cell biologist's guide to expansion microscopy. J Cell Sci. 2024;137:jcs260765.

Ishikawa T, Sahara Y, Takahashi T. A single packet of transmitter does not saturate postsynaptic glutamate receptors. Neuron. 2002;34:613–21.

Jones G, Akter Y, Shifflett V, Hruska M. Nanoscale analysis of functionally diverse glutamatergic synapses in the neocortex reveals input and layer-specific organization. bioRxiv. 2024:2024.05.01.592008.

Jung JH, Chen X, Reese TS. Cryo-EM tomography and automatic segmentation delineate modular structures in the postsynaptic density. Front Synaptic Neurosci. 2023;15:1123564.

Kaeser PS, Regehr WG. Molecular mechanisms for synchronous, asynchronous, and spontaneous neurotransmitter release. Annu Rev Physiol. 2014;76:333–63.

Kandel E, Koester J, Mack S, Hudspeth A. Principles of neural science. New York: McGraw Hill/Medical; 2021.

Kavalali ET. The mechanisms and functions of spontaneous neurotransmitter release. Nat Rev Neurosci. 2015;16:5–16.

Kellermayer B, Ferreira JS, Dupuis JP, Levet F, Grillo-Bosch D, Bard L, Linarès-Loyez J, Bouchet D, Choquet D, Rusakov DA, Bon P, Sibarita JB, Cognet L, Sainlos M, Carvalho AL, Groc L. Differential nanoscale topography and functional role of GluN2-NMDA receptor subtypes at glutamatergic synapses. Neuron. 2018;100:106–119.e7.

Kerr JM, Blanpied TA. Subsynaptic AMPA receptor distribution is acutely regulated by actin-driven reorganization of the postsynaptic density. J Neurosci. 2012;32:658–73.

Kim J, Wulschner LEG, Oh WC, Ko J. Trans-synaptic mechanisms orchestrated by mammalian synaptic cell adhesion molecules. BioEssays. 2022;44:1–11.

Li S, Raychaudhuri S, Lee SA, Brockmann MM, Wang J, Kusick G, Prater C, Syed S, Falahati H, Ramos R, Bartol TM Jr, Hosy E, Watanabe S. Asynchronous release sites align with NMDA receptors in mouse hippocampal synapses. Nat Commun. 2021;12:1–13.

Lisman JE, Raghavachari S, Tsien RW. The sequence of events that underlie quantal transmission at central glutamatergic synapses. Nat Rev Neurosci. 2007;8:597–609.

Lloyd BA, Han Y, Roth R, Zhang B, Aoto J. Neurexin-3 subsynaptic densities are spatially distinct from Neurexin-1 and essential for excitatory synapse nanoscale organization in the hippocampus. Nat Commun. 2023;14:4706.

MacGillavry HD, Song Y, Raghavachari S, Blanpied TA. Nanoscale scaffolding domains within the postsynaptic density concentrate synaptic AMPA receptors. Neuron. 2013;78:615–22.

Malagon G, Miki T, Llano I, Neher E, Marty A. Counting vesicular release events reveals binomial release statistics at single glutamatergic synapses. J Neurosci. 2016;36:4010–25.

Martinez-Sanchez A, Laugks U, Kochovski Z, Papantoniou C, Zinzula L, Baumeister W, Lučić V. Trans-synaptic assemblies link synaptic vesicles and neuroreceptors. Sci Adv. 2021;7:eabe6204.

Masch JM, Steffens H, Fischer J, Engelhardt J, Hubrich J, Keller-Findeisen J, D'Este E, Urban NT, Grant SGN, Sahl SJ, Kamin D, Hell SW. Robust nanoscopy of a synaptic protein in living mice by organic-fluorophore labeling. Proc Natl Acad Sci USA. 2018;115:E8047–56.

Metzbower SR, Levy AD, Dharmasri PA, Anderson MC, Blanpied TA. Distinct SAP102 and PSD-95 Nano-organization defines multiple types of synaptic scaffold protein domains at single synapses. J Neurosci. 2024;44:e1715232024.

Morris K, Bulovaite E, Kaizuka T, Schnorrenberg S, Adams C, Komiyama N, Mendive-Tapia L, Grant SGN, Horrocks MH. Sequential replacement of PSD95 subunits in postsynaptic supercomplexes is slowest in the cortex. eLife. 2024;13:RP99303.

Muttathukunnel P, Frei P, Perry S, Dickman D, Müller M. Rapid homeostatic modulation of transsynaptic nanocolumn rings. bioRxiv. 2021:2021.06.15.448550.

Nair D, Hosy E, Petersen J, Constals A, Giannone G, Choquet D, Sibarita JB. Super-resolution imaging reveals that AMPA receptors inside synapses are dynamically organized in Nanodomains regulated by PSD95. J Neurosci. 2013;33:13204–24.

Narayanasamy KK, Stojic A, Li Y, Sass S, Hesse MR, Deussner-Helfmann NS, Dietz MS, Kuner T, Klevanski M, Heilemann M. Visualizing synaptic multi-protein patterns of neuronal tissue with DNA-assisted single-molecule localization microscopy. Front Synaptic Neurosci. 2021;13:671288.

Nicoll RA. A brief history of long-term potentiation. Neuron. 2017;93:281–90.

Park DK, Stein IS, Zito K. Ion flux-independent NMDA receptor signaling. Neuropharmacology. 2022;210:109019.

Perez de Arce K, Schrod N, Metzbower SR, Allgeyer E, Kong GK-W, Tang A-H, Krupp AJAJ, Stein V, Liu X, Bewersdorf J, Blanpied TA, Lucic V, Biederer T. Topographic mapping of the synaptic cleft into adhesive Nanodomains. Neuron. 2015;88:1165–72.

Pérez-Garza J, Orea J, Deane Z, Raimondi G, Tripp R, Charles I, Ostroff L. Ultraplex microscopy: versatile highly-multiplexed molecular labeling and imaging across scale and resolution. bioRxiv. 2024:2024.08.17.605585.

Pérez-Otaño I, Luján R, Tavalin SJ, Plomann M, Modregger J, Liu X-B, Jones EG, Heinemann SF, Lo DC, Ehlers MD. Endocytosis and synaptic removal of NR3A-containing NMDA receptors by PACSIN1/syndapin1. Nat Neurosci. 2006;9:611–21.

Petralia RS, Sans N, Wang Y-X, Wenthold RJ. Ontogeny of postsynaptic density proteins at glutamatergic synapses. Mol Cell Neurosci. 2005;29:436–52.

Purves D, Augustine G, Fitzpatrick D, Hall WC, LaMantia A, Mooney R, White LE. Neuroscience. Sinauer; 2018.

Ramsey AM, Tang A-H, LeGates TA, Gou X-Z, Carbone BE, Thompson SM, Biederer T, Blanpied TA. Subsynaptic positioning of AMPARs by LRRTM2 controls synaptic strength. Sci Adv. 2021;7:eabf3126.

Reddy-Alla S, et al. Stable positioning of Unc13 restricts synaptic vesicle fusion to defined release sites to promote synchronous neurotransmission. Neuron. 2017;95:1350–1364.e12.

Reinhardt SCM, Masullo LA, Baudrexel I, Steen PR, Kowalewski R, Eklund AS, Strauss S, Unterauer EM, Schlichthaerle T, Strauss MT, Klein C, Jungmann R. Ångström-resolution fluorescence microscopy. Nature. 2023;617:711–6.

Rimbault C, Breillat C, Compans B, Toulmé E, Vicente FN, Fernandez M. Engineering paralog-specific PSD-95 recombinant binders as minimally interfering multimodal probes for advanced imaging techniques. 2021; 1–32.

Sachs S, Reinhard S, Eilts J, Sauer M, Werner C. Visualizing the trans-synaptic arrangement of synaptic proteins by expansion microscopy. Front Cell Neurosci. 2024;18:1328726.

Sahara Y, Takahashi T. Quantal components of the excitatory postsynaptic currents at a rat central auditory synapse. J Physiol. 2001;536:189–97.

Sahl SJ, Matthias J, Inamdar K, Weber M, Khan TA, Brüser C, Jakobs S, Becker S, Griesinger C, Broichhagen J, Hell SW. Direct optical measurement of intramolecular distances with angstrom precision. Science (New York, NY). 2024;386:180–7.

Sakamoto H, Ariyoshi T, Kimpara N, Sugao K, Taiko I, Takikawa K, Asanuma D, Namiki S, Hirose K. Synaptic weight set by Munc13-1 supramolecular assemblies. Nat Neurosci. 2018;21:41–55.

Samavat M, Bartol TM, Bromer C, Hubbard DD, Dakota C, Kuwajima M, Mendenhall JM, Parker PH, Bowden JB, Abraham WC, Sejnowski TJ, Harris KM, Diego S, Jolla L, Health B, Zealand N. Long-term potentiation produces a sustained expansion of synaptic information storage capacity in adult rat hippocampus. BioRXiv. 2024;

Santucci DM, Raghavachari S. The effects of NR2 subunit-dependent NMDA receptor kinetics on synaptic transmission and CaMKII activation. PLoS Comput Biol. 2008;4:e1000208.

Sarkar D, Kang J, Wassie AT, Schroeder ME, Peng5 Z, Tarr TB, Tang I-H, Niederst E, Young JZ, Tsai L-H, Blanpied TA, Boyden ES. Expansion revealing: decrowding proteins to unmask invisible brain nanostructures. BioRXiv. 2020;

Sarkar D, Kang J, Wassie AT, Schroeder ME, Peng Z, Tarr TB, Tang A, Niederst ED, Young JZ, Su H, Park D, Yin P, Tsai L, Blanpied TA, Boyden ES. Revealing nanostructures in brain tissue via protein decrowding by iterative expansion microscopy. Nat Biomed Eng. 2022;6:1057–73.

Sclip A, Sudhof TC. LAR receptor phospho-tyrosine phosphatases regulate NMDA-receptor responses. elife. 2020;9:1–21.

Shaib AH, et al. Expansion microscopy at one nanometer resolution. bioRxiv. 2022:2022.08.03.502284.

Shaib AH, et al. One-step nanoscale expansion microscopy reveals individual protein shapes. Nat Biotechnol. 2024;43:1539.

Sheng M, Hoogenraad CC. The postsynaptic architecture of excitatory synapses: a more quantitative view. Annu Rev Biochem. 2007;76:823–47.

Sinnen BL, Bowen AB, Forte JS, Gibson ES, Acqua MLD, Kennedy MJ. Optogenetic control of synaptic composition and function. Neuron. 2017;93:646–660.e5.

Stein IS, Gray JA, Zito K. Non-ionotropic NMDA receptor signaling drives activity-induced dendritic spine shrinkage. J Neurosci. 2015;35:12303–8.

Stein IS, Park DK, Claiborne N, Zito K. Non-ionotropic NMDA receptor signaling gates bidirectional structural plasticity of dendritic spines. Cell Rep. 2021;34:108664.

Stockwell I, Watson JF, Greger IH. Tuning synaptic strength by regulation of AMPA glutamate receptor localization. Bioessays. 2024;46(7):2400006.

Sun R, Allen JP, Mao Z, Wilson L, Haider M, Alten B, Zhou Z, Wang X, Zhou Q. The postsynaptic density in excitatory synapses is composed of clustered, heterogeneous nanoblocks. J Cell Biol. 2025;224:e202406133.

Sutton MA, Ito HT, Cressy P, Kempf C, Woo JC, Schuman EM. Miniature neurotransmission stabilizes synaptic function via tonic suppression of local dendritic protein synthesis. Cell. 2006;125:785–99.

Tang A-H, Chen H, MacGillavry HD, Metzbower SR, Blanpied TA, Li TP. A trans-synaptic nanocolumn aligns neurotransmitter release to receptors. Nature. 2016;536:210.

Tarusawa E, Matsui K, Budisantoso T, Molnár E, Watanabe M, Matsui M, Fukazawa Y, Shigemoto R. Input-specific intrasynaptic arrangements of ionotropic glutamate receptors and their impact on postsynaptic responses. J Neurosci. 2009;29:12896–908.

Tovar KR, McGinley MJ, Westbrook GL. Triheteromeric NMDA receptors at hippocampal synapses. J Neurosci. 2013;33:9150–60.

Unterauer EM, Shetab Boushehri S, Jevdokimenko K, Masullo LA, Ganji M, Sograte-Idrissi S, Kowalewski R, Strauss S, Reinhardt SCM, Perovic A, Marr C, Opazo F, Fornasiero EF, Jungmann R. Spatial proteomics in neurons at single-protein resolution. Cell. 2024;187:1785–1800.e16.

Vyklicky V, Stanley C, Habrian C, Isacoff EY. Conformational rearrangement of the NMDA receptor amino-terminal domain during activation and allosteric modulation. Nat Commun. 2021;12:2694.

Wang: https://www.pnas.org/cgi/doi/10.1073/pnas.2016416117. PNAS. 2020:117(49): 31438–31447.

Xu W, Schlüter OM, Steiner P, Czervionke BL, Sabatini BL, Malenka RC. Molecular dissociation of the role of PSD-95 in regulating synaptic strength and LTD. Neuron. 2008;57:248–62.

Xu N, Chen S-Y, Tang A-H. Tuning synapse strength by nanocolumn plasticity. Trends Neurosci. 2025;48:200–12.

Yang X, Specht CG. Subsynaptic domains in super-resolution microscopy: the treachery of images. Front Mol Neurosci. 2019;12:1–8.

Zeng M, Shang Y, Araki Y, Guo T, Huganir RL, Zhang M. Phase transition in postsynaptic densities underlies formation of synaptic complexes and synaptic plasticity. Cell. 2016;166:1163–1175.e12.

Zhang S, Trussell LO. Voltage clamp analysis of excitatory synaptic transmission in the avian nucleus magnocellularis. J Physiol. 1994;480:123–36.

Zhang M, Feng J, Xie C, Song N, Jin C, Wang J, Zhao Q, Zhang L, Wang B, Sun Y, Guo F, Li Y, Zhu S. Assembly and architecture of endogenous NMDA receptors in adult cerebral cortex and hippocampus. Cell. 2025a;188:1198–207.

Zhang J, Vaidya R, Pendharkar R, Chung HJ, Selvin PR. Revealing synaptic nanostructure distribution through automatic dendritic spine segmentation and single-molecule localization microscopy. 2025b; https://doi.org/10.1101/2024.07.15.603616. version 2

Zieger HL, Choquet D. Nanoscale synapse organization and dysfunction in neurodevelopmental disorders. Neurobiol Dis. 2021;158:105453.

Chapter 8
Functional Nano-segregation of Distinct Forms of Neurotransmission

Natalie J. Guzikowski, Camille S. Wang, and Ege T. Kavalali

Abstract Within the single micron of the synapse, three distinct modes of neurotransmission, driven by synchronous, asynchronous, and spontaneous neurotransmitter release, occur concurrently. In this chapter, we discuss the synaptic nano-organization comprised of neurotransmitter release machinery, molecular platforms, scaffolding proteins, and liquid complexes that support the discrete signaling of these three modes of neurotransmission. This robust nano-organization supports unique functional roles for each discrete mode at both excitatory glutamatergic and inhibitory GABAergic synapses. Modular nanocolumn organization of excitatory synapses and largely single-domain organization of inhibitory synapses maintain homeostatic plasticity within neural circuits. These recent findings support a basic design principle where the single synapse is a highly ordered and compartmentalized unit that by the functional nano-segregation of distinct forms of neurotransmission shapes synaptic efficacy, determines neurotransmission reliability, and tunes plasticity. The development of novel tools will be instrumental in further elucidating the nano-environment of the synapse, essential to both uncovering mechanisms underlying neurological disorders as well as their treatment.

Keywords Synapse nano-organization · Spontaneous neurotransmitter release · Synchronous neurotransmitter release · Asynchronous neurotransmitter release · Super-resolution imaging · Photobleaching

8.1 Three Modes of Neurotransmission

In the 1960s, the canonical synaptic transmission pathway was established. It posited that neurotransmission is initiated by an action potential arriving at the presynaptic terminal, presynaptic calcium influx is triggered, synaptic vesicles fuse with

N. J. Guzikowski · C. S. Wang · E. T. Kavalali (✉)
Vanderbilt Brain Institute, Vanderbilt University, Nashville, TN, USA

Department of Pharmacology, Vanderbilt University, Nashville, TN, USA
e-mail: ege.kavalali@vanderbilt.edu

E. T. Kavalali (ed.), *Nano-organization of the Synapse*, Advances in Neurobiology 48,
https://doi.org/10.1007/978-3-032-12594-1_8

the presynaptic membrane, and ultimately release neurotransmitter into the synaptic space to bind to postsynaptic receptors (Katz 1969; Südhof 2013). This seminal work set the stage for the investigation of neurotransmission at numerous synapse types across the nervous system in diverse animal models. This original view of synaptic transmission still holds true today, albeit fortified with knowledge of the specificity and diversity of synaptic signaling and the molecular mechanisms that mediate and regulate neurotransmission.

Today, three different modes of neurotransmission are categorized by their timescale relative to a stimulus and their calcium dependence. Action potential-dependent release is composed of synchronous and asynchronous phases, where synchronous release strictly adheres to the timing of incoming presynaptic action potentials, and asynchronous release is only loosely coupled to stimulation. In contrast, spontaneous release happens independent of action potentials, in a quantal manner, where single synaptic vesicles fuse and release neurotransmitter in a quasi-random fashion (Kavalali 2015). Originally, spontaneous release events were viewed as random errors at the synaptic terminal, deviating from canonical action potential-dependent release, and described as biological noise (Fatt and Katz 1950). However, defined signaling pathways and functional roles of this mode of synaptic signaling have now been explicitly demonstrated to be fundamental in neuronal function (Guzikowski and Kavalali 2021).

All modes of neurotransmission are quantal, where neurotransmitters are packaged and released in discrete units (synaptic vesicles) or multiples of that unit, called quantum (Katz 1969). Due to electrophysiological readouts, it is easily misconstrued that evoked release is larger than spontaneous release. However, evoked and spontaneous fusion are both quantal in nature, making the location and timing of their release the drivers of their function. Studies conducted at the *Drosophila* neuromuscular junction, as well as hippocampal synapses, have demonstrated that a single active zone is capable of synchronous, asynchronous, and spontaneous release, while some synapses exclusively execute spontaneous or evoked neurotransmission (Atasoy et al. 2008; Melom et al. 2013; Peled et al. 2014; Reese and Kavalali 2016).

Each of the three modes of neurotransmission relies on distinct molecular frameworks to accomplish complex information processing (Kononenko and Haucke 2015; Chanaday and Kavalali 2017). The expansion of our synaptic "worldview" to include the nuances of different modes of neurotransmission relays the importance of the molecular pathways and synaptic structures that support release. Parallel signaling by these three release modes necessitates release segregation and underscores the diversity of function within a single synapse. To elucidate how one neuron influences its targets, we must consider how multiple molecular pathways and modes of release regulate synaptic efficacy. This influence can encompass a diverse group of signals, such as large depolarizations triggering an action potential, a decrease in spontaneous calcium currents through NMDA receptors, or a change in tonic inhibition. How this symphony of signaling is mediated by specialized separable structures that regulate distinct functional pathways at the nanometer scale will be the focus of this chapter.

8.2 Distinct Molecular Machinery of Different Release Modes

The current proposed structure of the synapse incorporates the nano-organization of proteins throughout the synapse, including synaptic vesicle molecular markers, protein density gradients in the active zone, synaptic cleft, receptor localization, and postsynaptic density (Guzikowski and Kavalali 2021). When simplified, the synapse can be viewed as a directional network vital for information flow, where diverse SNARE proteins and calcium sensors define synaptic vesicle pools supporting segregated signaling pathways.

Historically, the organization of synaptic vesicles within the presynaptic terminal was defined by their propensity to fuse in response to stimulation and their ultrastructural localization. The classical view on the pool organization of synaptic vesicles comprises the readily releasable pool (vesicles that fuse first in response to stimulation) and the reserve pool (the pool that replenishes the readily releasable pool, which shows limited synaptic vesicle trafficking during physiological activity) (Alabi and Tsien 2012). However, this pool organization cannot fully account for the dynamics of release with regard to different modes of neurotransmission. Elucidation of the heterogeneity of neurotransmitter release machinery and how it dictates release kinetics and vesicle trafficking forges a more accurate classification of synaptic vesicle organization in the presynaptic terminal. The diversity in synaptic vesicle molecular composition of v-SNAREs (vesicle-soluble *N*-ethylmaleimide-sensitive factor (NSF) attachment protein receptor), membrane-localized machinery (t-SNAREs), and calcium sensors distinguishes synaptic vesicle populations, thus indicating that unique molecular compositions may designate synaptic vesicles for different forms of release (Crawford and Kavalali 2015). Furthermore, there is increasing evidence that different calcium sources mediate distinct modes of release, thus factor into the synaptic vesicle pool organization (Kavalali 2020; Wang et al. 2023).

8.2.1 Synaptic Vesicle-Associated SNAREs, v-SNAREs

The canonical SNARE complex includes synaptic vesicle-associated protein Synaptobrevin 2 (Syb2, also called VAMP2) that forms a complex with the plasma membrane SNARE synaptosomal-associated protein 25 (SNAP25) and syntaxin 1 to drive rapid action potential-evoked synaptic vesicle fusion (Südhof and Rothman 2009). Early experiments investigating the selective loss of canonical SNAREs revealed that this loss induced large effects on evoked release, while spontaneous release was still present. The genetic deletion of Syb2 in mice caused the almost absolute loss of synchronous calcium-dependent evoked release, while residual spontaneous release, and to some extent, asynchronous release, were still present (Schoch et al. 2001; Deák et al. 2004). This initial work stimulated further research into which SNAREs are mediating different modes of neurotransmission, as molecular perturbations selectively affected spontaneous and evoked release in distinguishable ways.

Fluorescence imaging and electrophysiology experiments have shown that Vps10p-tail-interactor-1a (vti1a) containing vesicles traffic in the absence of activity, and vesicle-associated membrane protein 7 (VAMP7) containing vesicles also preferentially traffic in response to resting calcium signals; thus, both are v-SNAREs specifically regulating spontaneous release (Fig. 8.1) (Hua et al. 2011; Ramirez et al. 2012; Bal et al. 2013) (for a full list of SNAREs associated with the synaptic vesicle proteome, see Takamori et al. 2006). Furthermore, despite both synchronous and asynchronous release being action potential dependent, vesicle-associated membrane protein 4 (VAMP4) selectively maintains asynchronous release, as demonstrated by differential trafficking with minimal Syb2 trafficking overlap (Raingo et al. 2012). In addition to these alternative v-SNAREs maintaining distinct trafficking pathways, biochemical evidence is consistent with their distinct functions. For instance, VAMP4 forms stable SNARE complexes independent of Syb2, and these complexes do not interact with synaptotagmin-1 and complexins, two proteins essential for rapid synchronous evoked release (Fig. 8.1) (Raingo et al. 2012).

The elucidation of specific v-SNAREs that selectively maintain different modes of release reveals a synaptic terminal composed of subpopulations of vesicles defined by their heterogeneous distribution of molecular components (Ramirez et al. 2012; Revelo et al. 2014; Walter et al. 2014). Alternative v-SNAREs act as synaptic vesicle molecular tags for the organization of synaptic vesicles into pools, conferring distinct release properties, downstream functional roles, and the selective regulation of neurotransmitter release (Mori et al. 2021).

8.2.2 Target Membrane SNAREs, t-SNAREs

In SNAP25 knockout mice, there is the complete loss of calcium-dependent evoked release, while ~10–13% of spontaneous release remains (Washbourne et al. 2002; Bronk et al. 2007). These studies demonstrate a divergence in the molecular machinery utilized by two different modes of release. Furthermore, SNAP23, a homolog of SNAP25, has a suggested role in asynchronous release due to its preferred association with Syt-7 (asynchronous calcium sensor) (Chieregatti et al. 2004; Weber et al. 2014). While the aforementioned studies describe how release can be supported by alternative SNAREs, the majority of release is supported by the canonical SNARE proteins (Syb2, SNAP25, Synataxin1). Quantitative mass spectrometry has provided great insight into the mean protein copy number per synapse, demonstrating high copy numbers of the canonical SNAREs compared to alternative v-SNAREs (VAMP2: 26448.00 ± 661.62, VAMP4: 100.59 ± 10.03, vti1a: 50.55 ± 2.51, VAMP7: N/A), SNAPs (SNAP25: 26686.08 ± 5287.39, SNAP23: 265.61 ± 17.75) (Wilhelm et al. 2014).

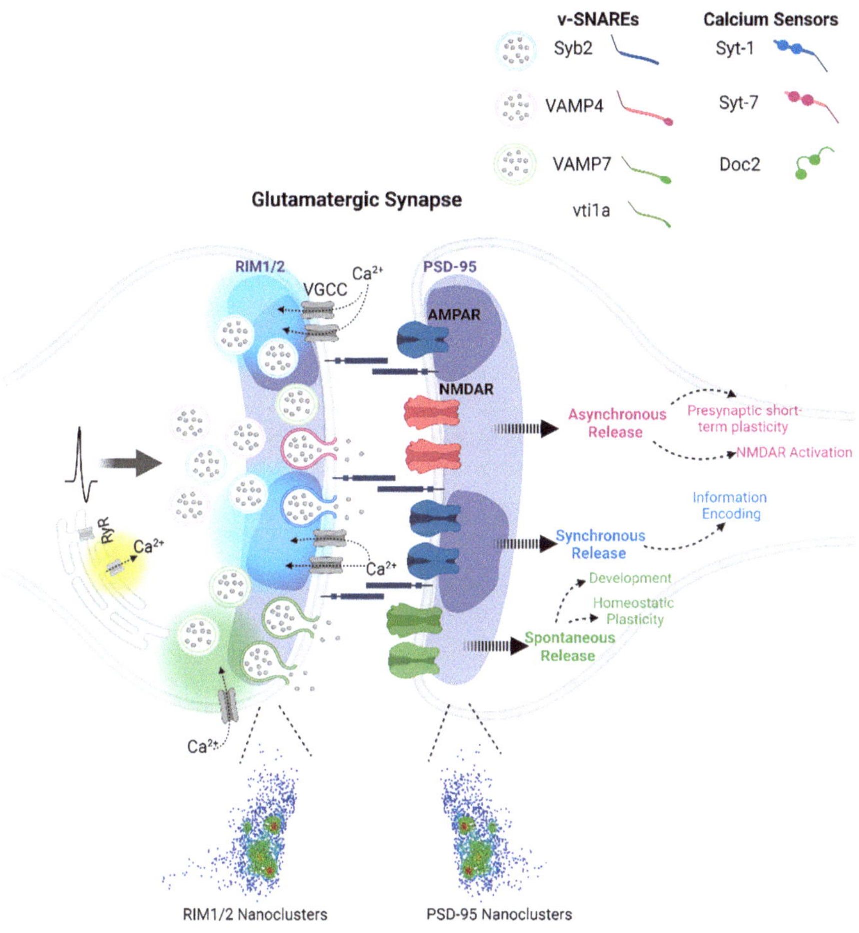

Fig. 8.1 Schematic representation of a glutamatergic synapse with molecular machinery, location of release, nanodomains of calcium sparks, and downstream signaling pathways highlighted in blue for synchronous evoked release, pink for asynchronous evoked release, and green for spontaneous neurotransmission. Canonical SNAREs and calcium sensors are utilized by all modes of release, but alternative SNAREs are highlighted for asynchronous and spontaneous release. Heat map schematics of RIM1/2 and PSD-95 are examples of super-resolution single-molecule localizations emphasizing the alignment of increased RIM1/2 and PSD-95 molecule clustering facilitating evoked release (purple color gradient within neuron)

8.2.3 SNAREopathies

A recent focus on the electrophysiological phenotype of "SNAREopathies" and genetic disorders of neurotransmitter release machinery has provided interesting insight into not only disease phenotypes, but the basal function of canonical SNARE machinery (Baker et al. 2015, 2018; Engel et al. 2016; Lipstein et al. 2017; Bradberry et al. 2020; Lammertse et al. 2020; Simmons et al. 2020; Vardar et al. 2020; Alten et al. 2021; Uzay and Kavalali 2023). For example, the electrophysiological characterization of 10 de novo SNAP25 patient variants showcases the unique synaptic impairments that result from different mutations within the same protein. Strikingly, the SNAP25 L50S variant results in the specific dysregulation of only spontaneous release and is sufficient to cause developmental epileptic encephalopathy. Furthermore, SNAP25 V48F and D166Y variants have more extreme alterations in spontaneous relative to evoked neurotransmission and also cause developmental epileptic encephalopathies. These recent findings on the role of aberrant spontaneous neurotransmission in neurological disease reflect the utility of different modes of release at the synapse and the specific function spontaneous neurotransmission has in larger-scale pathways (Alten et al. 2021). All in all, the utility of alternative SNARE proteins in distinct modes of release, in combination with data demonstrating specific regions of SNAP25 imparting release mode specificity, suggests that not one molecule decides whether a vesicle fuses spontaneously or in response to stimuli, but a combination of molecular tags, location, and molecular platforms.

8.2.4 Distinct Vesicle Recycling Pools

Pharmacological disruption of the recycling pathway revealed that the synaptic vesicle pools that maintain inhibitory spontaneous and evoked release are separable (Chung et al. 2010). The inhibition of a primary recycling GTPase, dynamin, immediately caused the depression of evoked release; yet, spontaneous release persisted despite dysregulated vesicle trafficking. This result suggested that the GABAergic vesicles for synchronous and asynchronous release come from one pool and spontaneous release from another (Chung et al. 2010). Interestingly, early stages of development are characterized by high levels of spontaneous release and resting synaptic vesicle cycling as compared to minimal evoked release (Molnár et al. 2002; Mozhayeva et al. 2002; Andreae et al. 2012). Previously, the importance of spontaneous activity in synapse formation has been speculative due to the lack of a proposed function of developmentally elevated spontaneous release (Andreae and Burrone 2018). However, key studies at the *Drosophila* neuromuscular junction during development have demonstrated how the frequency of spontaneous release events is directly correlated with synaptic structure. Experimentally facilitating mPSCs leads to the subsequent increase in presynaptic boutons, and the blockade of mPSCs leads to abnormal neuromuscular junctions characterized by reduced growth and surface area (Choi et al. 2014). In addition, spontaneous glutamate release events are thought to regulate dendritic arbors by acting as cues for dendritic outgrowth, a process not modulated by evoked release (Andreae and Burrone 2015). Spontaneous release maintains these developmental pathways independent of evoked release, delineating spontaneous release-specific biochemical pathways (Andreae and Burrone 2018).

8.2.5 Calcium Sensitivity and Sensors

Variations in the molecular machinery to maintain synchronous, asynchronous, and spontaneous release extend beyond SNARE machinery to include calcium-sensing proteins. The differences in calcium sensitivity between the modalities of release are complex, with degrees of absolute dependence, discrete sources of calcium mediating release, as well as divergent calcium dependences between excitatory and inhibitory spontaneous release (Courtney et al. 2018; Williams and Smith 2018; Lin et al. 2020). Synchronous release relies on nanodomain calcium increases upon action potential arrival, facilitated by a tight spatial relationship between VGCCs and vesicles, creating a fast temporal gradient with an incoming action potential (Dittman and Ryan 2019). In contrast, asynchronous release is not explicitly linked to a nanodomain, but a less localized increase in calcium (Eggermann et al. 2011; Mendonça et al. 2022).

Synaptotagmins are a family of calcium-sensing proteins implicated in synaptic vesicle exocytosis. Synaptotagmin-1 (Syt-1) is required for synchronous release by mediating the calcium sensitivity of fast evoked neurotransmitter release. While the loss of Syt-1 impairs synchronous release, it unclamps spontaneous release and augments asynchronous release, thereby regulating release bidirectionally (Maximov and Südhof 2005; Liu et al. 2009). Syt-1 is a low-affinity calcium isoform, making it only a reliable calcium sensor following voltage-gated calcium channel (VGCC) opening with subsequent nanodomain elevation of calcium. In contrast, synaptotagmin-7 (Syt-7), the proposed asynchronous calcium sensor, has a tenfold higher calcium affinity than Syt-1, rendering it more integral on a longer timescale post-VGCC opening. Syt-7 loss-of-function mutations drastically reduce asynchronous release and associated synaptic vesicle trafficking, providing the mechanism for the differential timing of these two types of action potential-dependent releases (Fig. 8.1) (Geppert et al. 1994; Sugita et al. 2002; Bacaj et al. 2013; Li et al. 2017).

Fast synchronous release is thought to be the main mode of information transfer due to its tight temporal coupling with presynaptic action potentials. Nevertheless, loss of synchronous neurotransmission by in vivo knockdown of Syt-1 in the hippocampus still allowed for the formation of fear memories, albeit with impairments, suggesting that asynchronous release is sufficient for the majority of the memory. This result indicates that action potential bursts and subsequent asynchronous release can be critical drivers for information encoding (Xu et al. 2012).

During asynchronous release, vesicle fusion occurs with a 10–100 ms delay after an action potential. However, neuron- and synapse-specific diversity in the degree of asynchronous release makes the generalization of the physiological significance of asynchronous release difficult. Syt-7, a putative asynchronous calcium sensor, is also required for short-term presynaptic facilitation at some synapses (Jackman et al. 2016), while at fast synapses, with minimal asynchronous release, Syt-7 is not required for short-term plasticity, but the fidelity of synchronous release instead, allowing for prolonged synaptic signaling (Luo and Südhof 2017).

Although the precise nature of calcium sensitivity of spontaneous release remains controversial, soluble calcium sensors of the Doc2-like protein family are thought to regulate spontaneous release in addition to synaptotagmin (Groffen et al. 2010; Pang et al. 2011). A quadruple knockdown strategy to eliminate members of the Doc2 family caused a reduction in spontaneous neurotransmission, while action potential-evoked neurotransmission remained relatively unaffected (Fig. 8.1). This protein loss also caused a subsequent increase in synaptic strength, suggesting that spontaneous neurotransmission is able to communicate independently with the postsynaptic neuron and trigger downstream signaling cascades that regulate the synaptic state (Ramirez et al. 2017). This regulation of release via calcium dynamics shows even more complexity, with excitatory spontaneous signaling preferentially regulated by Doc2α and inhibitory spontaneous signaling by both Doc2β and Syt-1 (Courtney et al. 2018). Quantitative mass spectrometry has revealed the mean calcium sensor copy number per synapse, providing insight into the relationship of sensor utility and abundance (Syt-1: 10332.00 ± 1079.20, Syt-7: 182.64 ± 3.54, Doc2α/β: 3696.50 ± 164.19) (Wilhelm et al. 2014).

8.2.6 Calcium Sources for Different Modes of Release

Calcium is a critical ion in neuronal physiology and is highly regulated within the pre-synapse. While the calcium dependence of evoked release is undisputed (Südhof 2013), its dependency and source(s) for spontaneous release remain controversial (Vyleta and Smith 2011; Ermolyuk et al. 2013; Dai et al. 2015; Tsintsadze et al. 2017; Wang et al. 2023). While calcium appears to be a mobile and freely moving ion within synapses, multiple studies have found that it is organized within domains. Presynaptic neuronal terminals express predominantly three subtypes of the Ca_v2 family, which contribute to neurotransmission and evoked calcium influx: $Ca_v2.1$ (or P/Q-), $Ca_v2.2$ (or N-), and $Ca_v2.3$ (or R-type) channels. Entry through these channels generates a local micro- or nanodomain of high calcium concentration (Catterall 2011; Nanou and Catterall 2018). A nanodomain is defined as a region of tens to hundreds of nanometers within which molecules are densely accumulated. A calcium nanodomain within the presynaptic terminal is caused by the opening of a few tightly coupled VGCCs to induce vesicle fusion (Yang et al. 2010; Heine and Holcman 2020).

In 1984, Chad and Eckert published the first model of calcium "domains" within a synapse, proposing that a calcium channel opening could lead to a local increase in calcium concentration that drops off steeply with distance from the channel (Chad and Eckert 1984). This work was further expounded upon a year later, when Simon and Llinás concluded that these domains do not overlap in squid presynaptic terminals (Simon and Llinás 1985). An important experiment in squid giant synapses compared the effects of buffer 1,2-bis(2-aminophenoxy)ethane-N,N,N′,N′-tetraacetic acid (BAPTA), a fast calcium chelator, with ethylene glycol tetraacetic acid (EGTA), a slower calcium chelator. It was found that neurotransmitter release was completely inhibited by only BAPTA (Adler et al. 1991). Furthermore, using an indicator protein n-Aequorin-J, which only fluoresces at calcium concentrations in the micromolar range or higher, squid giant presynaptic terminals indicated highly localized fluorescent regions (Llinás et al. 1992). These results support that evoked calcium release occurs within discrete nanodomains at the squid giant synapse.

At mammalian presynaptic terminals, evidence for nanodomain structuring of calcium signaling also exists. Over the course of development, the spatial coupling between open calcium channels and release sites tightens from "microdomains" to "nanodomains," where the opening of fewer VGCCs that are tightly coupled to synaptic vesicles is sufficient to induce release (Fedchyshyn and Wang 2005). This is accompanied by a decrease in calcium sensor sensitivity (Kochubey et al. 2009). These results suggest that with increased maturation, the spatial coupling between calcium channels and calcium sensors tightens to nanodomain signaling, which facilitates synchronized fusion events with minimal synaptic delay (Wang and Augustine 2014). Calcium signals are spatially organized at both pre- and postsynaptic terminals, which serve to regulate precise synaptic signaling mechanisms (McCarthy and Kavalali 2024).

To understand the spatial organization of distinct calcium signals within the presynapse, Wang et al. used photobleaching to examine whether presynaptic Ca^{2+} signals are spatially segregated within an individual axonal terminal (Wang et al. 2023). GCaMP8s was tagged to synaptobrevin-2, a presynaptic vesicle protein, in order to measure presynaptic calcium signals. Using this new sensor, three sources of presynaptic Ca^{2+} were detected: evoked PreCTs, spontaneous PreCTs, and baseline Ca^{2+} signals. Photobleaching at rest did not affect sPreCTs or ePreCTs, but it did decrease baseline Ca^{2+} signals. Photobleaching with stimulation significantly decreased the detection of ePreCTs; however, baseline Ca^{2+} signals were not further decreased. These differential effects of photobleaching on each Ca^{2+} signal reveal that they occur in spatially nonoverlapping nanodomains within the synapse at rest. In states of elevated activity, however, such as with stimulation, these domains may intermix and overlap. Interestingly, baseline calcium from an asynchronous/stochastic opening of VGCCs drives the majority of excitatory spontaneous glutamate release (Fig. 8.1). Other studies support the transient nature of calcium domains—a separate study estimated that ~9 VGCCs could be transiently confined to 1–2 nanodomains and dwell there for ~100 ms. VGCCs further from vesicle fusion sites can have a greater impact on asynchronous and spontaneous neurotransmitter release (Heine and Holcman 2020). The mobility of VGCCs enables dynamic regulation of evoked release probability, as well as the modulation of other forms of neurotransmission such as asynchronous and spontaneous release (Ermolyuk et al. 2013; Heck et al. 2019). How these domains are maintained, however, remains an open question.

In comparison to basal presynaptic calcium signaling, cell stress pathways also bolster the premise that, within an individual bouton, different sources of calcium have release mode specificity. During chronic ER stress, store-operated calcium entry (SOCE) specifically increases excitatory spontaneous neurotransmission, driven by Syt-7, creating excess glutamate release implicated in disease pathways.

ER stress triggered by altered Ca^{2+} homeostasis is a hallmark of numerous neurodegenerative diseases, suggesting that release mode-specific synaptic abnormalities contribute to neurodegeneration (Chanaday et al. 2021). Ultimately, different calcium sensors and calcium sources create a dynamic signaling network specifically tied to distinct modes of neurotransmission.

8.3 Synaptic Cleft Neurotransmitter Nano-gradient

The traditional picture of the synapse assumes a homogeneous structure whereby synaptic strength is governed by the number of receptors in the postsynaptic density, and neurotransmitter is released at any place within the active zone (Lisman et al. 2007). Under previous assumptions, neurotransmitter release activated the same subset of receptors regardless of its mode or timing; however, the dynamics of neurotransmitter diffusion and receptor activation challenge this notion.

Neurotransmitters at high enough concentrations for receptor activation in the synaptic cleft are transient and localized. Following quantal release, only ~25% of AMPARs and ~ 10% of NMDARs are activated on a single spine (Liu et al. 1999; Mainen et al. 1999; McAllister and Stevens 2000; Nimchinsky et al. 2004; Goncalves et al. 2020). The probability of AMPAR activation declines with distance from glutamate release due to both the low affinity of AMPARs for glutamate and the rapid diffusion of neurotransmitters following release (Franks et al. 2003; Raghavachari and Lisman 2004). Modeling has predicted that the necessary concentration of glutamate to activate AMPARs following synaptic vesicle fusion is extremely brief and localized, whereby presynaptic release most likely creates a "hotspot" of neurotransmitter, activating only a subset of receptors (Biederer et al. 2017). The implication of the neurotransmitter nano-gradient is that the nanoscale organization of release sites relative to receptors is fundamental in the efficacy of neurotransmission.

Synaptic plasticity involves both pre- and postsynaptic mechanisms that harness numerous molecules and signaling cascades. However, if we focus on receptor-mediated postsynaptic potentiation, nano-organization at the synapse can be vital, where not just the *number* of postsynaptic receptors, but also their location is a critical determinant of synaptic strength and efficacy (Haas et al. 2018). For instance, during LTP induction, preventing AMPAR surface diffusion markedly impairs potentiation in vitro, in addition to inhibiting behavioral aspects of contextual learning (Penn et al. 2017). Modeling predicts that increasing receptor density is more efficient than merely increasing synaptic area, as the same degree of AMPAR current potentiation can be achieved by reducing inter-receptor distances by 30–35% or by increasing AMPAR number by 100–200% (Franks et al. 2003; Savtchenko and Rusakov 2014). However, not until the advent of super-resolution microscopy has the nanoscale organization of a synapse been explicitly linked to synaptic strength.

With this tool, these earlier proposals were validated with the visualization of intra-synapse AMPAR clusters. Chemical LTP induction leads to nanodomain alignment alterations of increased PSD-95 density, followed by transsynaptic realignment within nanoclusters (Tang et al. 2016). In addition, structural plasticity and changes in spine and bouton size are related to synaptic molecular architecture in that synaptic reorganization is modular, whereby increases in spine size are accompanied by the addition of individual nano-modules (Hruska et al. 2018). These studies collectively posit that the redistribution of receptors alone can alter synaptic strength, and this reorganization can be a more resource-effective means for postsynaptic potentiation.

8.4 Excitatory Synapse—Segregation of Synchronous, Asynchronous, and Spontaneous Neurotransmission

Extensive work at excitatory synapses via the application of use-dependent drugs and optical imaging has demonstrated the segregation of spontaneous and evoked neurotransmission (Atasoy et al. 2008; Sara et al. 2011; Melom et al. 2013; Peled et al. 2014; Reese and Kavalali 2016). The distinct functional roles of evoked and spontaneous release at excitatory synapses provide a rationale for robust segregation maintained by prominent nano-organization within single active zones that link pre- and postsynaptic compartments (Tang et al. 2016; Biederer et al. 2017; Maschi and Klyachko 2017; Guzikowski and Kavalali 2021).

8.4.1 Segregation of Postsynaptic Receptors

The application of use-dependent drugs and optical imaging has demonstrated that different modes of neurotransmission target distinct postsynaptic receptors (Atasoy et al. 2008; Melom et al. 2013; Peled et al. 2014). Initial studies at excitatory synapses took advantage of use-dependent receptor blockers to probe postsynaptic receptor segregation. The application of the use-dependent NMDAR blocker, MK-801, demonstrated a near-complete segregation of NMDAR-mediated response to spontaneous and evoked glutamate release (Fig. 8.1) (Atasoy et al. 2008; Reese and Kavalali 2016). The use of philanthotoxin, a use-dependent blocker of GluR2 lacking AMPARs, extended this notion to evoked and spontaneous glutamate release-dependent activation of distinct sets of AMPARs (Sara et al. 2011; Peled et al. 2014). These results demonstrate the independence of evoked and spontaneous release probabilities, as well as the spatial segregation of the two forms of release.

8.4.2 Nanocolumn Organization of the Excitatory Synapse

The advancement of super-resolution light microscopy techniques in conjunction with cryogenic electron microscopy (cryo-EM) approaches has uncovered prominent nano-organization at single synapses supported by molecular nano-scaffolds that link pre- and postsynaptic compartments. This nano-organization appears to privilege discrete regions of the synapse for evoked release. Synaptic nanocolumns may play a role in facilitating evoked release differentially, providing the framework by which the synapse is able to maintain different modes of release via parallel pathways.

Early optical studies hinted at synaptic nano-organization while examining clustering dynamics of receptor subpopulations. Relative to the extrasynaptic space, AMPARs are organized in nanodomains where the number of clusters increases with synapse size (MacGillavry et al. 2013; Biederer et al. 2017). These clusters are mobile and dynamic, as AMPARs can diffuse in and out, creating a dynamic environment of AMPAR localization (Nair et al. 2013). Therefore, the distribution of receptors in the excitatory postsynaptic density is not homogeneous, as AMPARs and NMDARs are both arranged in clusters (Goncalves et al. 2020). Moreover, these receptor clusters are found in areas enriched in postsynaptic proteins, providing a proteomic network supporting nanoscale receptor organization (MacGillavry et al. 2013). A principal protein of interest is PSD-95 due to both its role in anchoring AMPARs via TARP binding and its co-enrichment with AMPAR clusters at the PSD (Chen et al. 2011; MacGillavry et al. 2013; Nair et al. 2013; Tang et al. 2016). Consequently, PSD-95 is thought to be the master organizer of excitatory postsynaptic organization, facilitating the heterogeneous enrichment of receptors and other scaffolding proteins such as GKAP, Shank, and Homer, and thereby creating a range of protein expression within the postsynapse (Tang et al. 2016; Biederer et al. 2017).

An influential study demonstrated that presynaptic loci with a high density of proteins vital for vesicle priming and fusion (such as RIM1/2, Munc13, and bassoon) were associated with parallel postsynaptic gradients of PSD-95 and AMPARs (Tang et al. 2016). Here, presynaptic RIM1/2 clusters show the largest degree of correspondence with postsynaptic PSD-95, whereas presynaptic Munc13 and Bassoon showed weaker association with their postsynaptic scaffold counterparts. These findings established a transsynaptic nanocolumn model via the co-alignment and enrichment of both pre- and postsynaptic proteomes, facilitating release at designated points along the active zone that are aligned with specific receptor subpopulations.

The organization of the synapse in nanocolumns deconstructs the classical homogeneous view of single active zone synapses and proposes a heterogeneous molecular platform that favors different modes of release at specific regions, governing distinct downstream biochemical signaling pathways. This nanocolumn substructure was defined by the co-enrichment of pre- and postsynaptic proteins such as high local densities of RIM1, PSD-95, and GluA2. The mapping of evoked and spontaneous synaptic vesicle fusion in relation to the molecular layout of these key proteins demonstrated that evoked fusion is privileged at areas of dense postsynaptic ensembles, whereas spontaneous release happens within a greater area of the active zone. Active zone regions with the highest likelihood of release were aligned with the densest AMPAR areas, optimizing the potency of neurotransmission (Tang et al. 2016). This work demonstrated that AMPARs are enriched within 80 nm nanodomains aligned with presynaptic RIM nanodomains; however, how AMPAR clusters are maintained juxtaposed to release sites across the synapse was originally unclear.

The extracellular cleavage of leucine-rich repeat transmembrane 2 (LRRTM2), a cell adhesion molecule that binds postsynaptic PSD-95 and presynaptic neurexin, has recently been demonstrated to be vital in regulating this pre- and postsynaptic alignment. The acute proteolytic cleavage of LRRTM2, which disrupts extracellular neurexin binding, led to the rapid repositioning of AMPARs away from RIM nanodomains, specifically reducing the amplitude of evoked PSCs, while spontaneous release was unaffected. The enrichment of LRRTM2 in the nanocolumn upholds release and receptor nanoclustering by facilitating the alignment of AMPARs for evoked release (Ramsey et al. 2021). Taken together, this result demonstrates that LRRTM2 acts as a structural link mediating the transsynaptic alignment necessary for evoked neurotransmitter release.

The bias of these defined nanocolumns toward one mode of release bolsters the premise that different modes of release activate distinct postsynaptic receptor subpopulations and provides a substrate for plasticity. This nanocolumn organization posits that active zone areas with the highest likelihood of release are aligned with dense receptor regions (Tang et al. 2016), creating a conduit for synchronous evoked release required for fast synaptic communication (Fig. 8.1). However, multiple questions remain. First, synaptic nanostructure has been shown to facilitate evoked release; however, it is unknown if there are molecular platforms in place to support spontaneous release. Furthermore, are there other transsynaptic molecules enriched in nanocolumns? If so, what is their role in development? In addition, are there other mechanisms besides direct interactions supporting pre- and postsynaptic domains, i.e., steric hindrance (Biederer et al. 2017), that impact nano-organization? Lastly, how does this synaptic nanostructure affect synaptic vesicle pools? As discussed above, synaptic vesicle pools are defined by their molecular identity, but how do nanodomains traffic and facilitate release at distinct fusion sites of molecularly diverse synaptic vesicles?

8.4.3 Novel Method to Probe Excitatory Subsynaptic Organization

In the process of developing optical sensors, scientists and engineers aim to make their probes increasingly "photostable," or as resistant to being photobleached as possible, to maximize the observation time of such probes. Interestingly, some experimental approaches take advantage of photobleaching's properties to investigate physiological processes. Photobleaching is the irreversible process by which an activated fluorophore becomes damaged and can no longer fluoresce. An interesting property of photobleaching is that it only affects active fluorescent probes. During imaging, a fluorophore repeatedly goes through the excitation-emission cycle, as excitation light activates electrons into their excited state, and an emission light is emitted as the electrons decay back to their ground state. In their excited state, electrons can enter a triplet excited state during which they are susceptible to being photobleached. Photobleaching is the cumulative effect of fluorophore loss from the excitation-emission cycle, and the rate of photobleaching correlates with the excitation intensity. Once photobleached, a fluorophore can no longer participate in the excitation-emission cycle (Song et al. 1995; McQuarrie and Simon 1997) (Fig. 8.2a, b). As such, photobleaching not only blocks fluorescence detection, but it does so in a use-dependent and permanent manner.

The use dependence of photobleaching could be harnessed to understand cellular physiology, and indeed, it has been intentionally employed in a multitude of ways. It is used in kymographs to measure protein transport, to study the presence of liquid–liquid phase separation. One of its most well-known uses is a method known as fluorescence recovery after photobleaching (FRAP), used to determine the dynamics and diffusional properties of molecules. In FRAP, the initial signal of a fluorescence-tagged molecule is measured with an optical microscope. Then a smaller region is chosen, and the laser power is intensely and quickly increased within that region such that the active fluorophores in that region are photobleached. Then the laser power returns to its initial levels, and the rate of fluorescence recovery in the photobleached region is measured. The fluorescence recovery occurs due to intermixing of unbleached probes with photobleached probes, and the rate of recovery can reveal properties such as the mobility of proteins within a membrane, protein trafficking, and organelle continuity. For instance, fluorescence recovery has been used to reveal that PSD95 has less turnover compared to other postsynaptic scaffolding proteins, such as Shank (Kuriu et al. 2006), and STORM microscopy confirmed a laminar organization of the postsynaptic density (Dani et al. 2010).

The principle of FRAP can not only be employed using a laser-based microscope, but also with an epifluorescence microscope over the entire field of view. Using this approach to examine the spatial organization of fluorescently tagged proteins can reveal the nano-physiology of synapses (Fig. 8.2c, d). For instance, in 2022, Wang et al. employed the use-dependent effects of photobleaching to probe the subsynaptic organization of postsynaptic glutamate sensors (Wang et al. 2022). Prior studies have used STORM microscopy to characterize a nanocolumn of

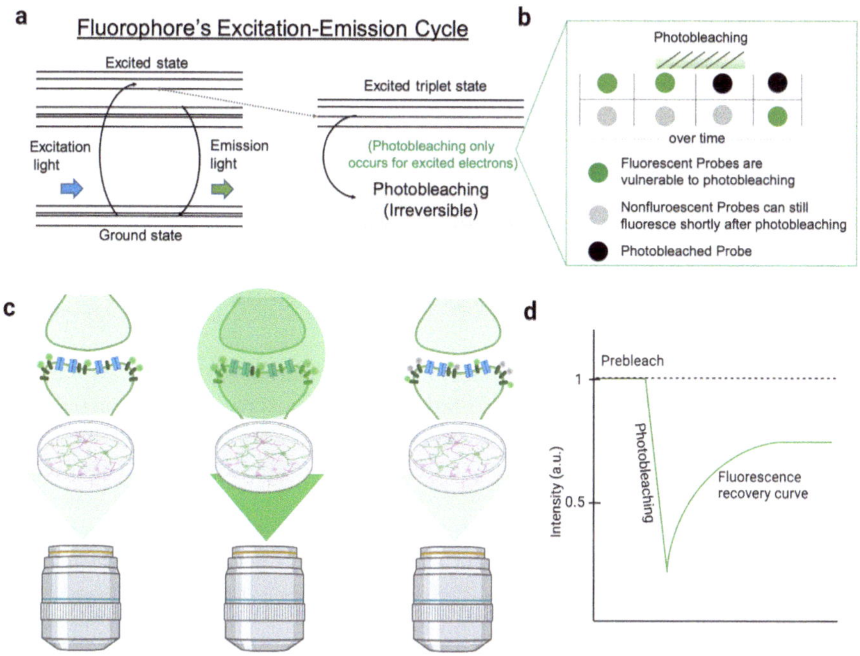

Fig. 8.2 Photobleaching can be used as a tool to probe subsynaptic organization. (**a**) Fluorophores absorb energy via an excitation light and move from the ground state to the excited singlet state. They then return to the ground state, a process during which they may emit fluorescence (known as the emission light). While in the excited singlet state, fluorophores can alternatively enter an excited triplet state, where they are susceptible to being photobleached. (**b**) Photobleaching is a process in which fluorophores are permanently altered such that they cannot fluoresce anymore. Only fluorescent fluorophores are susceptible to photobleaching, while nonfluorescent fluorophores are not. In this way, photobleaching is a use-dependent process. Once a fluorophore is photobleached, they are irreversibly damaged and can no longer fluoresce. (**c**) Photobleaching experiment schematic, where the baseline activity of neurons is recorded, followed by photobleaching, then recording of fluorescent activity after photobleaching (green fluorophores = fluorescing, gray fluorophores = photobleached). (**d**) A graph demonstrating normalized fluorescence at baseline, the drop in fluorescence after photobleaching, and the fluorescence recovery curve, which reflects the diffusion of non-photobleached fluorescence molecules into the field of view, and the diffusion of photobleached fluorescence molecules out of the field of view

pre- and postsynaptic proteins that aligns with evoked neurotransmitter release (Tang et al. 2016). However, the structure of spontaneous release remains not well understood. Using iGluSnFR, a fluorescent glutamate indicator, they were able to make single synapse level measurements and, furthermore, reliably observe both evoked and spontaneous quantal events. Evoked and spontaneous glutamate release are differentially photobleached, where evoked release is photobleached at a much faster rate and spontaneous release is very resistant to photobleaching, even at 20–40 min. Furthermore, after photobleaching, the fluorescence recovery of spontaneous release occurs at a faster rate than evoked release fluorescence recovery

(Wang et al. 2022). These results suggest that evoked release likely occurs in clustered diffusion-restricted nanocolumns, while spontaneous glutamate events occur at more mobile diffuse event sites, ultimately strengthening the premise that evoked release is happening within specific hotspots at the synapse that are separable from spontaneous release sites.

8.4.4 Segregation of Release Modes in Glutamatergic Synapses Supports Homeostatic Plasticity

A key example of why excitatory nanocolumn organization segregating evoked and spontaneous release is vital in synaptic function is homeostatic plasticity. In response to global changes in neuronal activity, synaptic weights are multiplicatively up- or downscaled to maintain homeostatic set points of neurotransmission. This homeostatic synaptic scaling maintains the relative differences in synaptic weights between synapses on a neuron, vital for information processing, while still allowing the system to adapt to environmental levels of activity (Turrigiano and Nelson 2004; Kavalali and Monteggia 2020). Global changes in activity elicited by complete suppression of action potentials (tetrodotoxin treatment) or, conversely, disinhibition (bicuculine treatment) can regulate this scaling phenomenon by the up- or down-regulation of postsynaptic receptor density. In contrast to the synapse specificity of classical long-term potentiation, where input strength and synaptic strength scale in the same direction (strong stimulation leads to an increase in synaptic strength), with homeostatic plasticity, decreased activity leads to synaptic upscaling (and vice versa) throughout the neuron (Kavalali and Monteggia 2023).

While the maintenance of the synaptic dynamic range is regulated by global changes in network activity over days, spontaneous neurotransmission also acts as a signal for the system within hours. In this way, synaptic scaling can also be triggered by alterations in action potential-independent release, ultimately revealing a biochemical signaling pathway downstream of spontaneous release (Gonzalez-Islas et al. 2018). At the center of this intersection between spontaneous neurotransmission and homeostatic plasticity is brain-derived neurotrophic factor (BDNF), downstream of both excitatory glutamate release and inhibitory GABA release (Fig. 8.3).

Independent of network activity and evoked neurotransmission, spontaneous synaptic events suppress local dendritic protein translation, whereby basal levels of spontaneous release regulate receptor composition at synapses (Sutton et al. 2006). Therefore, a reduction in N-methyl-D-aspartate receptor (NMDAR)-mediated miniature excitatory postsynaptic currents (mEPSCs) via postsynaptic manipulations triggers eukaryotic elongation factor 2 kinase-mediated synaptic upscaling via local dendritic translation of BDNF (Fig. 8.3) (Sutton et al. 2006; Aoto et al. 2008; Reese and Kavalali 2015). In parallel, the loss of spontaneous specific release machinery, vti1a and VAMP7, triggers synaptic scaling of α-amino-3-hydroxy-5-methyl-4-isoxazolepropionic acid receptors (AMPARs), as reflected by increased mEPSC amplitudes (Crawford et al. 2017). Decreased spontaneous neurotransmission leads to synaptic upscaling. Not only is this NMDAR synaptic scaling pathway separable from the canonical NMDAR-mediated long-term potentiation (LTP) pathway, but it is also implicated as a substrate for disease intervention in major depressive disorder (Lin et al. 2018, 2021).

8.4.5 *Synaptic Loci of Excitatory Asynchronous Release Regulate Signaling*

Due to the tight temporal correlation between synchronous and asynchronous release, only recently have high-resolution optical techniques begun to reveal the synaptic loci of asynchronous release sites. The development of a system that allowed for neuronal stimulation followed by the immediate high-pressure freezing

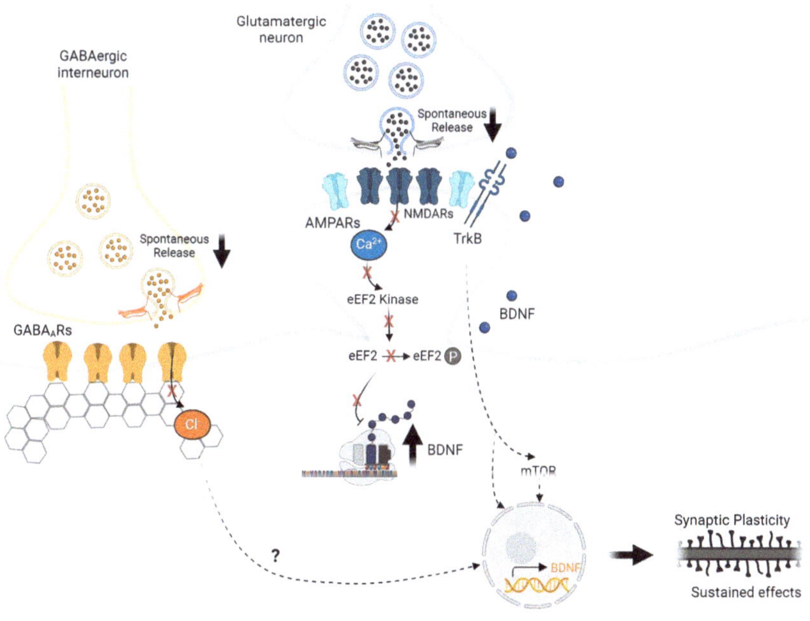

Fig. 8.3 Diagram of both inhibitory and excitatory input onto a pyramidal neuron. A reduction in spontaneous chloride currents mediated by GABA$_A$Rs leads to the subsequent upregulation of *Bdnf*, triggering subsequent excitatory downscaling by an unknown mechanism. At glutamatergic synapses, a reduction in spontaneous calcium currents mediated by NMDARs leads to a decrease in eukaryotic elongation factor 2 (eEF2) kinase activity, a decrease in basal levels of eEF2 phosphorylation, and subsequent de-suppression of local dendritic translation of BDNF. BDNF activation of tropomyosin receptor kinase B (TrkB) signaling is necessary for rapid (synaptic upscaling) and sustained homeostatic synaptic plasticity. While these two pathways share BDNF as a common signaling molecule, whether these pathways converge, act in parallel, or counterbalance each other is unknown

of the sample, "zap and freeze," allowed for the spatial and temporal analysis of both synchronous and asynchronous release at the active zone. A series of electron micrographs of synaptic ultrastructure revealed a distinct spatial organization of vesicle fusion following stimulation, with synchronous release happening throughout the active zone and asynchronous release, vesicles that fuse 11 ms after stimulation, concentrated at the center (Kusick et al. 2020). However, a study utilizing the fluorescent glutamate sensor, SF-iGluSnFR, in combination with deconvolution/2D Gaussian fitting, determined that the relative distribution of asynchronous release sites is more widely distributed relative to synchronous release sites within the presynaptic bouton (Mendonça et al. 2022). The use of super-resolution localization routines provided a 4 ms temporal and 75 nm spatial resolution. Despite their conflicting findings, each technique offers select advantages. "Zap and freeze" can localize release sites within the active zone. However, more than one release event cannot be captured per active zone, whereas live iGluSnFR imaging in combination with super-resolution localization routines can only define a broad release area but can compare the location of different release types within the same synapse. In addition, it remains a challenge to visualize spontaneous release using this approach ("zap and freeze"), as fusion events are not under the experimenter's control. Due to the differential calcium dependencies of synchronous and asynchronous release, future work elucidating the nanostructure of the synapse must clarify putative release sites relative to the nano-level clustering of VGCCs.

While the segregation of synchronous vs. asynchronous release is evident, only recently have groups begun to combine these high-resolution imaging methods to correlate structure to function. Excitatory synapses typically express three subtypes of glutamate receptors: NMDARs, AMPARs, and metabotropic glutamate receptors that display a centralized, peripheral, and dispersed organization within the PSD, respectively (Goncalves et al. 2020). The location of NMDARs is of particular interest due to the necessity of not only glutamate binding but also postsynaptic depolarization to relieve magnesium pore blockade for receptor activation (Seeburg et al. 1995). The ultrastructural analysis of synaptic fusion pits and postsynaptic receptors delineated AMPARs cluster at the periphery and NMDARs at the center of the PSD. Further computer modeling revealed the spatial organization of fusion sites with specific receptors, where asynchronous release is privileged adjacent to where NMDARs are localized (Fig. 8.1) (Li et al. 2021). The temporal difference between synchronous and asynchronous release is suspected to facilitate NMDAR activation due to the transsynaptic alignment of synchronous release sites with AMPARs. AMPAR activation leads to local depolarization, relieving NMDAR Mg^{2+} pore block, while asynchronous release sites provide glutamate binding with NMDARs, showcasing how the spatial segregation and synaptic alignment of two modes of release has functional implications.

8.4.6 *Asynchronous Release in Neurodegenerative/ Neurological Disorders*

Augmented asynchronous release is a disease phenotype in both animal models of neurodegenerative disease and human epileptic tissue. Furthermore, neurotransmitter release gradually desynchronizes in vitro throughout neuronal maturation and aging (Uzay et al. 2023). A mouse model of spinal muscular atrophy presented with dysregulated evoked neurotransmission with a large augmentation of asynchronous release (Ruiz et al. 2010). Similarly, amyloid precursor protein deficient neuromuscular synapses (Alzheimer's Disease model) have increased asynchronous release. Furthermore, in human epileptic tissue and a mouse model of epilepsy, there was increased asynchronous release from fast-spiking GABAergic interneurons (Jiang et al. 2012). These alterations in asynchronous release are not independent of other deficits in evoked neurotransmission and are implicated downstream of dysregulated calcium homeostasis. Nevertheless, the prominence of asynchronous release increases in pathological conditions (Kaeser and Regehr 2014).

8.4.7 *Nano-organization of Glutamate Receptor Subtypes*

These basic nano-architecture governing principles extend beyond receptor type to receptor subtype, where the rules governing AMPAR and NMDAR nano-location are subunit specific. The use of super-resolution microscopy revealed that GluA2 subunits are organized within subsynaptic nanodomains across from Syt-1 containing nano-modules, creating subsynaptic modules that follow scaling principles with spine size. However, GluA1-subunit containing AMPARs and GluN2B-subunit containing NMDARs are flexible and do not follow strict modular organization (Hruska et al. 2022). From the specificity of release-mode specific release site to receptor subtype organization, there is a precise and ordered architecture that determines the efficacy of neurotransmission.

8.5 Inhibitory Synapse—Segregation of Evoked and Spontaneous Neurotransmission

Historically, the main role of inhibitory signaling has been cast as modulating excitatory neurotransmission and regulating a neuron's propensity to fire action potentials, partially explaining the disproportionate emphasis on excitatory synapses. Synaptic neuroscience research has largely focused on the *Drosophila* neuromuscular junction and the mammalian excitatory synapse as model systems, facilitating the discovery of a molecularly heterogeneous excitatory synapse implicated in disease pathways and now a substrate for psychiatric disease intervention (Autry et al.

2011; Melom et al. 2013; Tang et al. 2016; Kavalali and Monteggia 2020; Ramsey et al. 2021; Suzuki et al. 2021). However, despite the fundamental importance of inhibitory neurotransmission, only in the last 5 years have studies begun to explore the correlation between GABAergic synaptic nanostructure, synaptic signaling, and downstream molecular pathways.

8.5.1 Center-Surround Organization of the Inhibitory Synapse

The GABAergic inhibitory postsynapse is often referred to as a thin sheet, as it is less densely populated with scaffolding proteins relative to the excitatory postsynaptic density. However, pharmacological and presynaptic evidence of release segregation suggests a postsynaptic mechanism may be present to mediate the downstream signaling of different modes of GABAergic neurotransmission. A recent study uncovered the partial segregation of evoked and spontaneous release at the inhibitory synapse. The utilization of picrotoxin, a use-dependent $GABA_AR$ channel blocker, demonstrated that approximately 40% of $GABA_ARs$ are exclusively activated by evoked release (Fig. 8.4) (Horvath et al. 2020).

Subsequently, Guzikowski and Kavalali (2022) took advantage of the molecularly specific interaction of antimalarial Artemisinins and the inhibitory synapse scaffold protein, gephyrin, to probe the functional organization of $GABA_AR$-mediated neurotransmission in central synapses. The intracellular postsynaptic binding site of Artemisinins allowed for the disruption of intrinsic synaptic structural elements to directly probe the cytosolic mechanisms that might mediate the segregation of evoked and spontaneous neurotransmission. With super-resolution microscopy and electrophysiology, the coordinated shrinkage of gephyrin synaptic cluster volume in concert with a decrease in mIPSC frequency was revealed. The reduced density of gephyrin molecules at the periphery of synaptic clusters made the edges of these clusters more vulnerable to Artemisinins, thereby causing the overall shrinkage of gephyrin clusters upon treatment. This allowed the use of Artemisinins as a tool to specifically disrupt peripheral gephyrin $GABA_AR$ binding and probe the spatial organization within the inhibitory synapse. These results suggest that the GABAergic postsynaptic apparatus possesses a concentric center-surround organization, where the periphery of gephyrin clusters selectively maintains spontaneous GABAergic neurotransmission, facilitating its autonomous function. Overall, this data demonstrates a contrasting organization to excitatory synapses with a uni-domain phenotype and an average of one gephyrin and $GABA_AR$ subsynaptic domain (Fig. 8.4) (Liu et al. 2020; Guzikowski and Kavalali 2022; Ramsay et al. 2023).

How this relates to the presynaptic organization of release sites, however, is unknown. Is there a concentric center-surround organization of scaffolding proteins, RIM1/2 and Munc-13, akin to excitatory nanocolumns? Does the uni-domain model of gephyrin organization extend beyond the postsynaptic density? Furthermore,

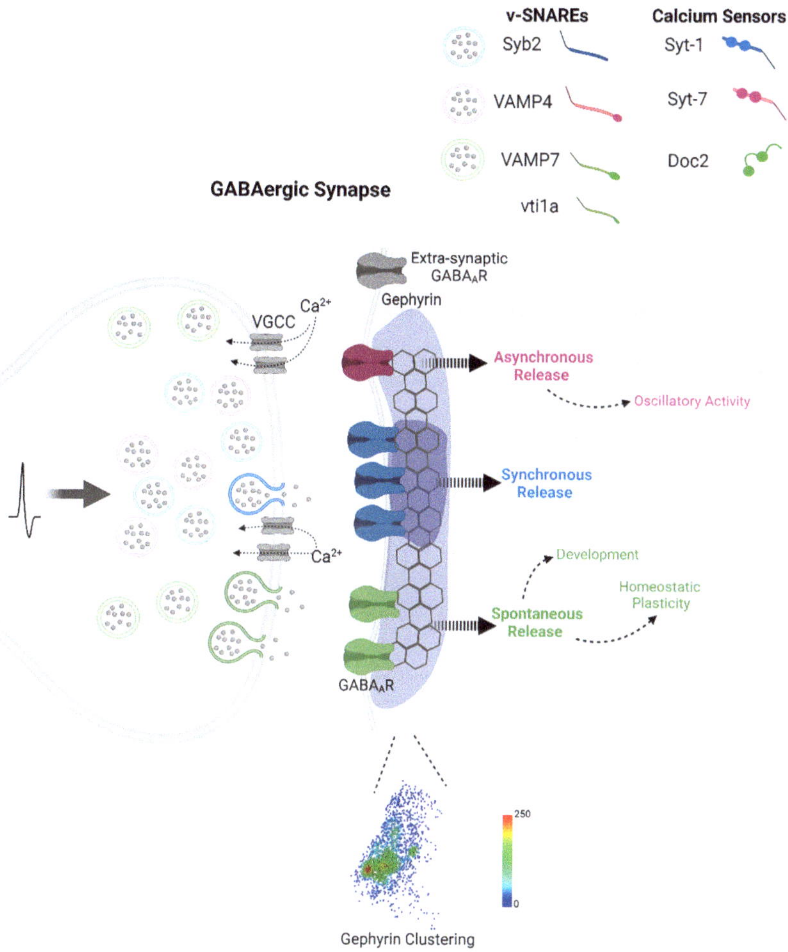

Fig. 8.4 Schematic representation of a GABAergic synapse with molecular machinery, location of release, and downstream signaling pathways highlighted in blue for synchronous evoked release, pink for asynchronous evoked release, and green for spontaneous neurotransmission (subsynaptic location of asynchronous release unknown). Canonical SNAREs and calcium sensors are utilized by all modes of release, but alternative SNAREs are highlighted for asynchronous and spontaneous release. Heat map schematic representation of gephyrin super-resolution single-molecule localizations emphasizing gephyrin uni-domain clustering facilitating evoked release with spontaneous release localized at the periphery

how do we conceptualize the fact that 40% of GABA$_A$Rs are exclusively activated by evoked release (Horvath et al. 2020) with the organization of gephyrin? Is there more fluidity in the nano-organization of the inhibitory synapses relative to excitatory and/or interneuron subtype specificity to this architecture?

8.5.2 *Bdnf* Is a Specific Target of Spontaneous GABAergic Release

Based on the findings from excitatory synapses, GABAergic release mode segregation may also indicate that there are specific signaling pathways downstream of each mode of release, potentially facilitated by the center-surround organization of evoked and spontaneous release at central GABAergic synapses (Horvath et al. 2020; Guzikowski and Kavalali 2022). The blockade of spontaneous GABAergic release alters calcium signaling and leads to multiplicative downscaling at excitatory synapses via *Bdnf* transcription (Horvath et al. 2021). Furthermore, this happens physiologically in the presence of activity, as the suppression of only spontaneous GABAergic neurotransmission in the presence of activity leads to an increase in *Bdnf* transcripts. However, whether this is due to the direct signaling of spontaneous chloride currents or alterations in the electrical gradient along the membrane, altering excitatory synapses is unknown. Together, these data suggest *Bdnf* is a regulated specific target of spontaneous GABAergic release, akin to excitatory synapses (Fig. 8.3).

Only recently has the biochemical signaling of inhibitory neurotransmission been addressed due to the few known targets of chloride. The discovery that with-no-lysine kinases (WNKs) function as chloride sensors mediating second messenger cascades downstream of GABAergic chloride current provides insight into the dynamic nature of the inhibitory synapse (Piala et al. 2014; Heubl et al. 2017; Chen et al. 2019). Further adding to the complexity of the inhibitory synapse, GABAergic signaling is excitatory during early development, influencing synapse plasticity and growth to generate functional circuits, but inhibitory during adulthood (Ben-Ari 2002). The majority of work investigating WNK signaling at synapses has been focused on the regulation of cation-chloride cotransporters (CCCs) in maintaining the internal Cl^- concentration (Piala et al. 2014; Huang et al. 2019). However, the link between spontaneous chloride currents and *Bdnf* transcription is an evolving area.

8.5.3 Asynchronous GABAergic Release

While recent work has uncovered a putative model for the spatial and therefore functional segregation of GABAergic evoked and spontaneous release, how the synchronous and asynchronous phases of evoked release fit are unclear. Unlike glutamatergic synapses, there is a large degree of interneuron diversity whereby interneuron subtypes should be considered in relation to synaptic phenotype. For example, cholecystokinin (CCK)–expressing GABAergic interneurons (innervate pyramidal cells at soma and proximal dendrites) are characterized by high asynchronous activity compared to other interneuron subtypes. Therefore, their prolonged asynchronous GABA release generates long-lasting inhibition that modulates circuit activity (Hefft and Jonas 2005). Based on CCK-expressing GABAergic interneurons' role in maintaining low-frequency oscillations of the hippocampus, their prominent asynchronous release is suspected to play a key role in generating hippocampal theta rhythms (Fig. 8.4) (Klausberger and Somogyi 2008; Rozov et al. 2019). This provides a key example of how release mode dynamics regulate brain circuit activity, creating the context to further study how synaptic architecture drives each mode of release.

The further development of tools to monitor inhibitory neurotransmission including the advancement of chloride sensors akin to GCaMP and the development of iGABASnFR to the same sensitivity and resolution as iGluSnFR is necessary. The development of tools that match the quality of calcium and glutamate indicators will greatly aid our understanding of the basic governing mechanisms of the synapse and further elucidate inhibition's functional role.

8.5.4 Technology Development to Study Inhibitory Subsynaptic Dynamics

Patch clamp electrophysiology, though the gold standard for studying ion channel activity, does not afford the level of spatial information needed to study single synapses. To further investigate single synapse level physiology, we need improved optical sensors. While iGluSnFR is a reliable sensor that has single synapse level resolution, its counterpart iGABASnFR (which measures GABAergic signaling) is typically used to measure bulk inhibitory transmission (Marvin et al. 2019) and does not have fast enough kinetics to measure events at individual synapses. Other approaches have used chloride-sensing probes to measure inhibitory transmission, with lesser success. For instance, SuperClomeleon and LSSmClopHensor, which use halide-sensitive YFP variants, have been useful to measure changes in chloride concentrations (Sulis Sato et al. 2017; Boffi et al. 2018). However, neither has fast enough kinetics to detect fast spiking inhibitory events and is primarily used to measure large shifts in Cl- concentration. This is in part due to the complicated pH sensitivity of this YFP variant, and that $GABA_A$Rs are permeable to HCO_3^- (Zajac et al. 2020). Another fluorescent probe, vesicular GABA transporter pHluorin (vGAT-pHluorin), which works similarly to vGluT-pHluorin, was developed to measure inhibitory exo- and endocytosis (Santos et al. 2013). The newer version of iGABASnFR2, which has faster kinetics and higher signal-to-noise ratio than the prior version, may hold promise as part of ongoing efforts to visualize individual GABAergic synapses (Marvin et al. 2019). The further improvement of optical sensors, such as for inhibitory neurotransmission, can open many doors to examine the spatial segregation of these forms of release at the single synapse level.

8.6 Dopaminergic Synaptic Organization

Dopamine is one of the most widely implicated molecules in reward and motivation, motor control, as well as disorders including, but not limited to, Parkinson's disease, substance use disorder, and schizophrenia. Despite its central role in normal and disease-related neuronal processes, the structural and functional organization of dopaminergic synapses remains to be characterized. Dopamine, like glutamate and GABA, is released from vesicles. Release can occur at synapses, as well as from cell bodies and dendrites (release sites more unique to dopamine), where they then act upon dopamine receptors, which are G-protein-coupled receptors (Liu and Kaeser 2019).

Despite not being a fast neurotransmitter, the precise, rapid, and efficient release of dopamine is evident, as opposed to purely volume transmission (Kershberg et al. 2022). Recent work investigating what synaptic architecture can support dopamine's fast influence on synapses and behavior has revealed distinct active zone-like structures, where ~30% of striatal dopamine axons contain active zone-like structures with RIM, necessary for action potential-dependent dopamine release (Howe and Dombeck 2016; Liu et al. 2018). These active zone-like complexes are analogous to conventional synapse release site scaffolding and are necessary for only action potential-dependent release, as spontaneous somatodendritic release persists despite the loss of RIM (Liu et al. 2018; Robinson et al. 2019; Banerjee et al. 2020, 2022; Tan et al. 2022). Consistent with these results, ~75% dopamine varicosities with vesicles do not have active zone proteins that are potentially "silent varicosities" in response to stimulation (Pereira et al. 2016; Liu et al. 2018; Banerjee et al. 2022; Özçete et al. 2024). This nascent investigation into dopaminergic synaptic organization is beginning to reveal different modalities of release segregation within the nervous system. This is an exciting and necessary area of future investigation with interesting questions including how these active zone-like complexes relate to postsynaptic target sites and receptors. How is release maintained at the other 75% of varicosities? How are these different release modalities related to function and ultimately behavior? Lastly, what are the implications of these recent results for other types of neuromodulators such as serotonin and acetylcholine?

8.6.1 Novel Tools to Study Other Types of Neurotransmission

In addition to GABA and glutamate, understanding the spatiotemporal organization of other neurotransmitters can shed light on normal and disease processes. New tools, such as the dLight and the $GRAB_{DA}$ family (Patriarchi et al. 2018; Sun et al. 2018), have already been used to make important discoveries at the neural circuit level (Kim and Schnitzer 2022). Photobleaching as a use-dependent tool, as well as other approaches, can be used to shed light on the spatial organization of different modes of dopamine release, from fast release to more tonic forms of release, and how they are spatially organized relative to each other within individual release sites.

Furthermore, acetylcholine (ACh) is a neurotransmitter that plays an important role in attention and wakefulness in dementia-related disorders such as Alzheimer's, and is implicated in the treatment of many neuropsychiatric drugs (Higley and Picciotto 2014; Moran et al. 2019). Yet, cholinergic transmission, especially at the synaptic level, remains poorly understood. A recently developed tool, called intensity-based acetylcholine sensing fluorescent reporter (iAchSnFR), has low micromolar affinity, high maximal fluorescence change, and millisecond level activation and decay kinetics. Other promising sensors have been developed for acetylcholine as well, such as GPCR-activation-based ACh ($GRAB_{ACh}$) (Jing et al. 2018, 2020).

Other tools have been developed to sense other molecules, such as serotonin (5HT) via iSeroSnFR, which was developed through introducing a number of mutations to a previously developed ACh sensor such that iSeroSnFR has a 15-fold higher response to 5HT. Another 5HT sensor, PsychLight, was developed to detect 5HT dynamics and study hallucinogenic drugs (Dong et al. 2021). Another neuromodulator important in not only normal physiology but also disease processes is endocannabinoids, which act as retrograde messengers to modulate presynaptic neurotransmitter release. While we have linked endocannabinoids to stress-related disorders, substance use disorders, neurodevelopment disorders, we still do not fully understand the organization of endocannabinoid release at synaptic sites. Newly developed tools such as $GRAB_{eCB}$ have the potential to help answer these questions.

There has also been the development of sensors for drugs, such as ketamine with iSketSnFR (Bera et al. 2019), which was also developed through introducing various mutations to an ACh sensor. iSketSnFR has the potential to visualize ketamine's target. New imaging tools can be applied to help us better understand the diffusional properties of ketamine and how it interacts with the spatial organization of synapses to enact its antidepressant and other clinical effects. Other variants have also developed sensors that can sense the presence of other drugs, such as nicotine (sensed by iNicSnFR) (Shivange et al. 2019) and opioids (detected by sensors such as M-SPOTIT) (Kroning and Wang 2021).

To advance our understanding of the physiology surrounding quantal events, we need sensors with improved kinetics, spatial resolution, increased stability, and greater fluorescence signal. The development of new techniques is certainly warranted, as well as innovative and creative applications of the tools we currently have to help us glean insight into synaptic physiology in novel ways.

8.7 Emerging Cellular Organization Principle: Liquid–Liquid Phase Separation

8.7.1 Fundamental Principles of LLPS

The neuron is a polarized cell, with numerous substructures that are bound by intercellular membranes (i.e., the nucleus, mitochondria, endoplasmic reticulum). However, a multitude of subcellular complexes do not possess an enclosing membrane to aid in compartmentalization, despite being autonomous functioning units.

Liquid–liquid phase separation (LLPS) is emerging as a fundamental chemical principle necessary for subcellular organization, whereby liquid condensates enriched in specific protein–protein or protein–nucleic acid complexes separate from a dilute phase, creating distinct phases within the cytosol (Li et al. 2012). LLPS is mediated by intrinsically disordered regions (IDRs) of proteins, which lack a stably folded secondary and tertiary structure, and therefore do not form site-specific stoichiometric interactions. IDRs of condensate-forming proteins instead interact via specific physicochemical principles mediated by multivalent interactions that create enriched pockets of proteins spatially distinct from the cytosol (Li et al. 2012; van der Lee et al. 2014; Milovanovic and De Camilli 2017; Mittag and Parker 2018; Berkeley and Debelouchina 2022; Pappu et al. 2023). A growing amount of research has revealed how LLPS mediates protein interactions throughout the cell and synapse (Brangwynne et al. 2009; Chen et al. 2020; Lyon et al. 2021; McDonald and Shen 2021; Wu et al. 2023).

8.7.2 Phase Separation at the Synapse

The synapse, a highly ordered and dynamic subcellular domain, maintains autonomy, organization, and high-fidelity neurotransmission without a bounding membrane. Since one-third of the eukaryotic genome contains proteins with IDRs, it follows that numerous synaptic proteins have liquid-like properties and phase separate in vitro (van der Lee et al. 2014). The synapse is a directional network vital for information flow, where LLPS is involved in protein organization at every level of neurotransmission: synaptic vesicle pool clustering (Milovanovic et al. 2018; Pechstein et al. 2020; Park et al. 2021, 2023; Wu et al. 2023), active zone scaffolding (Wu et al. 2019; Emperador-Melero et al. 2021), active zone assembly (McDonald et al. 2020), synaptic vesicle endocytosis (Day et al. 2021; Imoto et al. 2022, p. 20), and inhibitory and excitatory postsynaptic density scaffolding (Zeng et al. 2018, 2019; Bai et al. 2021).

In addition to basal synaptic organization, the premise that liquid complexes may aid in the functional segregation of distinct forms of neurotransmission is bolstered by the specific and exclusive nature of phase-separated groups of proteins. Reconstituted excitatory postsynaptic densities exclude inhibitory postsynaptic densities from the condensed phase, creating segregating assemblies (Zeng et al. 2018; Zhu et al. 2024). Furthermore, recent work suggests that separable synaptic vesicle pools are organized by distinct protein condensates, active zone RIM/RIM-BP droplets, and synapsin droplets (Wu et al. 2021), creating a synaptic topography composed of exclusive liquid complexes in addition to structured domain interactions that may mediate the nano-organization necessary for distinct signaling pathways within an individual synapse.

8.7.3 Nano-specificity of Liquid Condensates

Despite compelling evidence that synaptic proteins phase separate, it has been challenging to determine if the liquid condensate properties of these proteins are essential to the function of the synapse. Recently, Guzikowski and Kavalali (2024) took advantage of aliphatic alcohol, 1,6-hexanediol (1,6-HD) to differentiate between liquid condensates and solid-like complexes, specifically manipulating LLPS assemblies independent of structured domain interactions (Kroschwald et al. 2017). Upon acute LLPS disruption, calcium-mediated action potential-dependent release is dysregulated, while remarkably spontaneous neurotransmission is relatively unaffected (Guzikowski and Kavalali 2024). Given that proteins involved in every step of neurotransmission have phase separation properties, these results suggest that liquid condensates have spatial and functional specificity.

The characterized synaptic deficit in action potential-dependent neurotransmission is due to the phase separation of subsynaptic active zone nanoclusters, thus establishing that, within the single micron of the synapse, an immense degree of spatial and functional specificity is supported via the phase separation of RIM1/2 and Munc13 and subsequent tethering of VGCCs at the active zone, essential in regulating the efficacy and integrity of fast evoked calcium-dependent release. Importantly, despite their large micrometer-sized organization in vitro, synaptic condensates operate within the nanometer-level organization of the synapse to direct evoked release (Guzikowski and Kavalali 2024). The conservation of fundamental principles from liquid condensate biology to the nano-organization of the synapse demonstrates how a multitude of protein interactions work in synchrony to compartmentalize the synapse into distinct signaling zones. Thus, the organization of the presynaptic milieu, including both synaptic vesicle pools and active zone scaffolding complexes, is fluid in nature, where dynamic liquid condensates are essential for efficient and precise neurotransmission.

Liquid–liquid phase separation as a fundamental organizing principle within the neuron is an emerging idea of heightened interest. However, the correlation between in vitro phase separation studies to the functional output of the synapse is of paramount importance. Creative solutions to specifically manipulate LLPS-mediated complexes, independent of targeted structured protein domain interactions, protein trafficking, and genetic perturbations, are necessary to establish the nano-specificity and physiological relevance of LLPS. As researchers continue to investigate the molecular substrates that bestow unique functional roles to different modes of release, it will be essential to elucidate how structured domain interactions work in parallel with liquid condensates to organize the synapse into distinct signaling zones.

8.8 Conclusion

8.8.1 Future Directions—Technology and Tools

The largest challenge with understanding the ultrastructure of the synapse in relation to neurotransmission is the necessity of tools with both high spatial and temporal resolution. In order to fully elucidate the function of synaptic nanoscale organization, methods that provide high-resolution molecular information as well as the monitoring of synaptic vesicle release and retrieval must be used in parallel.

Photobleaching of fluorescent probes that can measure from single synapses has the potential to garner new insights that cannot be gleaned from other methods. Photobleaching as a tool sidesteps the requirement of developing new imaging tools to image molecules at the nanoscale level, reduces the amount of post-processing technology, and does not require nearly the level of data storage as is necessary in super-resolution microscopy. Because the same microscope is used for other live imaging experiments, this can increase access to study questions at subsynaptic levels. Nevertheless, photobleaching as a tool to probe subsynaptic organization has its limitations. For instance, while it can reveal the occurrence of spatial segregation or overlap of molecules at levels beyond the diffraction limit of light, the ability of photobleaching alone to identify specific molecules involved in forming and maintaining the spatial organization is limited. However, as more fluorescent probes continue to be developed with increasingly improved kinetics for live imaging purposes, this limitation could be overcome.

Using photobleaching in an intentional manner to probe subsynaptic organization can be applied to other types of imaging, such as two-photon imaging. This approach would be incredibly powerful, as two-photon imaging can penetrate deeper structures of the brain with a high level of resolution, such as ex vivo brain slices and in vivo brain tissue. While primary neuronal cultures are one of the best models to study synapses, studying synaptic physiology in a system with more of the complex brain architecture intact would reveal important insights. Studies using photobleaching to probe nano-physiology using other types of microscopes would first need to rigorously establish the foundation for using photobleaching in this system. To examine synaptic organization, one would first need to confirm that there is synapse-level resolution of their optical sensor in their imaging and model system. Then they would need to establish that their signal can be intentionally photobleached within a reasonable imaging time frame (such as before significant drift occurs, or before cells become unhealthy in an imaging chamber). If photobleaching at rest does not have a significant effect, one could increase the activation of the optical probe during the photobleaching period, such as via increased depolarization. After finding the settings with maximal photobleaching effects, one can then measure the rate of fluorescence recovery with time.

The novel use of photobleaching allows for the investigation of the subsynaptic organization during live imaging, and this approach could certainly be used to investigate the spatial organization of other aspects of synaptic physiology. Subsynaptic organization is difficult to resolve due to the diffraction limit of conventional light microscopy, which is ~200 nm, while the entire synapse itself is ~1 μm, and the active zone is ~200 nm. Furthermore, while super-resolution imaging can resolve molecules as small as 5–20 nm, there is an unavoidable temporal resolution tradeoff (Huang et al. 2010).

Computational methods have arisen that could be applied to resolve subsynaptic level organization during live imaging experiments (Newman et al. 2022). The recently developed method, QuaSOR, is a super-resolution method that can relate quantal transmission to synaptic architecture at intact glutamatergic synapses. Using QuaSOR, Newmann et al. were able to map locations of quantal transmission, quantify release probability and spontaneous frequency at individual synapses, and subsequently perform super-resolution imaging to link presynaptic proteins with synaptic neurotransmission properties (Newman et al. 2022).

Some groups have used denoising and deconvolution techniques to improve the detection of the timing and amplitude of release events using iGluSnFR. To perform the deconvolving technique, selected regions of interest (ROIs) were filtered using a low and high bandpass filter, which removed the baseline drift and high frequency noise, followed by deconvolution with the Fourier transform. Using this technique, they probed where synchronous and asynchronous release sites are located with respect to each other. The authors found that asynchronous exocytosis sites are enhanced in synapses with low release probability compared to synchronous sites and, furthermore, that the asynchronous sites are more scattered within the presynaptic release site (Mendonça et al. 2022). Another group developed an improved deep learning based denoising method for fluorescence imaging called DeepCAD-RT. Compared to prior methods, it has the advantage of performing real-time denoising of fluorescence imaging that can process fast neuronal dynamics using reduced memory consumption (Li et al. 2023).

To better probe single synapse dynamics, the development and improvement of optical sensors to measure activity at presynaptic terminals are necessary. As novel imaging techniques become available, our ability to interrogate the nano-physiology of synapses will grow, thereby bolstering our breadth of knowledge of synaptic physiology. Furthermore, heightening our understanding of neuropsychiatric disorders and consequently developing better treatments for them are desperately needed goals.

8.8.2 Concluding Thoughts

Different pathways that can drive, maintain, and regulate varying modes of neurotransmission imply distinct functional roles, with molecules mediating discrete biochemical pathways. There are dedicated functional roles that are unique and specific to each mode of release. Spontaneous neurotransmission is of specific note due to its stimulus-independent nature, contrasting canonical views of neural communication. Beyond spontaneous neurotransmission's electrical function in the regulation of action potential firing, spontaneous release also has a specialized biochemical role in both regulating synapse and circuit development, and in the maintenance of basal synapse dynamics in several organisms (Huntwork and Littleton 2007; Choi et al. 2014; Andreae and Burrone 2015; Banerjee et al. 2021). Furthermore, aberrant spontaneous neurotransmission in neurological disease as well as ketamine's rapid antidepressant action arising from spontaneous release modulation exemplifies the applicability of divergent modes of release at the synapse (Autry et al. 2011; Kavalali and Monteggia 2020; Alten et al. 2021).

The studies outlined herein present a synaptic nano-organization composed of neurotransmitter release machinery, molecular platforms, scaffolding proteins, and liquid complexes that support the discrete signaling of evoked and spontaneous neurotransmission at both excitatory glutamatergic and inhibitory GABAergic synapses. The excitatory modular nanocolumn and inhibitory uni-domain synaptic organization comprises a design principle that may ultimately mediate the segregation of release, shape synaptic efficacy, determine neurotransmission reliability, and tune plasticity.

In order to fully elucidate the function of synaptic nanoscale organization, methods that provide high-resolution molecular information as well as the monitoring of synaptic vesicle release and retrieval must be used in parallel. Future investigation should not be focused purely on excitatory neurotransmission but expanded to the diversity of GABAergic interneurons and neuromodulator signaling. How fundamental principles of structure and function converge across synapse types while displaying unique nuance will be instrumental in elucidating the nano-environment of the synapse, essential to uncovering mechanisms underlying both neurological diseases and their treatment.

References

Adler EM, Augustine GJ, Duffy SN, Charlton MP. Alien intracellular calcium chelators attenuate neurotransmitter release at the squid giant synapse. J Neurosci. 1991;11:1496–507. https://doi.org/10.1523/JNEUROSCI.11-06-01496.1991.

Alabi AA, Tsien RW. Synaptic vesicle pools and dynamics. Cold Spring Harb Perspect Biol. 2012;4:a013680. https://doi.org/10.1101/cshperspect.a013680.

Alten B, Zhou Q, Shin O-H, Esquivies L, Lin P-Y, White KI, et al. Role of aberrant spontaneous neurotransmission in SNAP25-associated encephalopathies. Neuron. 2021;109:59–72.e5. https://doi.org/10.1016/j.neuron.2020.10.012.

Andreae LC, Burrone J. Spontaneous neurotransmitter release shapes dendritic arbors via long-range activation of NMDA receptors. Cell Rep. 2015;10:873–82. https://doi.org/10.1016/j.celrep.2015.01.032.

Andreae LC, Burrone J. The role of spontaneous neurotransmission in synapse and circuit development. J Neurosci Res. 2018;96:354–9. https://doi.org/10.1002/jnr.24154.

Andreae LC, Fredj NB, Burrone J. Independent vesicle pools underlie different modes of release during neuronal development. J Neurosci. 2012;32:1867–74. https://doi.org/10.1523/JNEUROSCI.5181-11.2012.

Aoto J, Nam CI, Poon MM, Ting P, Chen L. Synaptic signaling by all-trans retinoic acid in homeostatic synaptic plasticity. Neuron. 2008;60:308–20. https://doi.org/10.1016/j.neuron.2008.08.012.

Atasoy D, Ertunc M, Moulder KL, Blackwell J, Chung C, Su J, et al. Spontaneous and evoked glutamate release activates two populations of NMDA receptors with limited overlap. J Neurosci. 2008;28:10151–66. https://doi.org/10.1523/JNEUROSCI.2432-08.2008.

Autry AE, Adachi M, Nosyreva E, Na ES, Los MF, Cheng P, et al. NMDA receptor blockade at rest triggers rapid behavioural antidepressant responses. Nature. 2011;475:91–5. https://doi.org/10.1038/nature10130.

Bacaj T, Wu D, Yang X, Morishita W, Zhou P, Xu W, et al. Synaptotagmin-1 and synaptotagmin-7 trigger synchronous and asynchronous phases of neurotransmitter release. Neuron. 2013;80:947–59. https://doi.org/10.1016/j.neuron.2013.10.026.

Bai G, Wang Y, Zhang M. Gephyrin-mediated formation of inhibitory postsynaptic density sheet via phase separation. Cell Res. 2021;31:312–25. https://doi.org/10.1038/s41422-020-00433-1.

Baker K, Gordon SL, Grozeva D, van Kogelenberg M, Roberts NY, Pike M, et al. Identification of a human synaptotagmin-1 mutation that perturbs synaptic vesicle cycling. J Clin Invest. 2015;125:1670–8. https://doi.org/10.1172/JCI79765.

Baker K, Gordon SL, Melland H, Bumbak F, Scott DJ, Jiang TJ, et al. SYT1-associated neurodevelopmental disorder: a case series. Brain. 2018;141:2576–91. https://doi.org/10.1093/brain/awy209.

Bal M, Leitz J, Reese AL, Ramirez DMO, Durakoglugil M, Herz J, et al. Reelin mobilizes a VAMP7-dependent synaptic vesicle pool and selectively augments spontaneous neurotransmission. Neuron. 2013;80:934–46. https://doi.org/10.1016/j.neuron.2013.08.024.

Banerjee A, Lee J, Nemcova P, Liu C, Kaeser PS. Synaptotagmin-1 is the Ca2+ sensor for fast striatal dopamine release. eLife. 2020;9:e58359. https://doi.org/10.7554/eLife.58359.

Banerjee S, Vernon S, Jiao W, Choi BJ, Ruchti E, Asadzadeh J, et al. Miniature neurotransmission is required to maintain drosophila synaptic structures during ageing. Nat Commun. 2021;12:4399. https://doi.org/10.1038/s41467-021-24490-1.

Banerjee A, Imig C, Balakrishnan K, Kershberg L, Lipstein N, Uronen R-L, et al. Molecular and functional architecture of striatal dopamine release sites. Neuron. 2022;110:248–265.e9. https://doi.org/10.1016/j.neuron.2021.10.028.

Ben-Ari Y. Excitatory actions of gaba during development: the nature of the nurture. Nat Rev Neurosci. 2002;3:728–39. https://doi.org/10.1038/nrn920.

Bera K, Kamajaya A, Shivange AV, Muthusamy AK, Nichols AL, Borden PM, et al. Biosensors show the pharmacokinetics of S-ketamine in the endoplasmic reticulum. Front Cell Neurosci. 2019;13:499. https://doi.org/10.3389/fncel.2019.00499.

Berkeley RF, Debelouchina GT. Chemical tools for study and modulation of biomolecular phase transitions. Chem Sci. 2022;13:14226–45. https://doi.org/10.1039/D2SC04907D.

Biederer T, Kaeser PS, Blanpied TA. Transcellular Nanoalignment of synaptic function. Neuron. 2017;96:680–96. https://doi.org/10.1016/j.neuron.2017.10.006.

Boffi JC, Knabbe J, Kaiser M, Kuner T. KCC2-dependent steady-state intracellular chloride concentration and pH in cortical layer 2/3 neurons of anesthetized and awake mice. Front Cell Neurosci. 2018;12:7. https://doi.org/10.3389/fncel.2018.00007.

Bradberry MM, Courtney NA, Dominguez MJ, Lofquist SM, Knox AT, Sutton RB, et al. Molecular basis for synaptotagmin-1-associated neurodevelopmental disorder. Neuron. 2020;107:52–64.e7. https://doi.org/10.1016/j.neuron.2020.04.003.

Brangwynne CP, Eckmann CR, Courson DS, Rybarska A, Hoege C, Gharakhani J, et al. Germline P granules are liquid droplets that localize by controlled dissolution/condensation. Science. 2009;324:1729–32. https://doi.org/10.1126/science.1172046.

Bronk P, Deák F, Wilson MC, Liu X, Südhof TC, Kavalali ET. Differential effects of SNAP-25 deletion on Ca2+ -dependent and Ca2+ -independent neurotransmission. J Neurophysiol. 2007;98:794–806. https://doi.org/10.1152/jn.00226.2007.

Catterall WA. Voltage-gated calcium channels. Cold Spring Harb Perspect Biol. 2011;3:a003947. https://doi.org/10.1101/cshperspect.a003947.

Chad JE, Eckert R. Calcium domains associated with individual channels can account for anomalous voltage relations of CA-dependent responses. Biophys J. 1984;45:993–9. https://doi.org/10.1016/S0006-3495(84)84244-7.

Chanaday NL, Kavalali ET. How do you recognize and reconstitute a synaptic vesicle after fusion? F1000Res. 2017;6:1734. https://doi.org/10.12688/f1000research.12072.1.

Chanaday NL, Nosyreva E, Shin O-H, Zhang H, Aklan I, Atasoy D, et al. Presynaptic store-operated Ca2+ entry drives excitatory spontaneous neurotransmission and augments endoplasmic reticulum stress. Neuron. 2021;109:1314–1332.e5. https://doi.org/10.1016/j.neuron.2021.02.023.

Chen X, Nelson CD, Li X, Winters CA, Azzam R, Sousa AA, et al. PSD-95 is required to sustain the molecular Organization of the Postsynaptic Density. J Neurosci. 2011;31:6329–38. https://doi.org/10.1523/JNEUROSCI.5968-10.2011.

Chen J-C, Lo Y-F, Lin Y-W, Lin S-H, Huang C-L, Cheng C-J. WNK4 kinase is a physiological intracellular chloride sensor. Proc Natl Acad Sci. 2019;116:4502–7. https://doi.org/10.1073/pnas.1817220116.

Chen X, Wu X, Wu H, Zhang M. Phase separation at the synapse. Nat Neurosci. 2020;23:301–10. https://doi.org/10.1038/s41593-019-0579-9.

Chieregatti E, Chicka MC, Chapman ER, Baldini G. SNAP-23 functions in docking/fusion of granules at low Ca2+. Mol Biol Cell. 2004;15:1918–30. https://doi.org/10.1091/mbc.e03-09-0684.

Choi BJ, Imlach WL, Jiao W, Wolfram V, Wu Y, Grbic M, et al. Miniature neurotransmission regulates drosophila synaptic structural maturation. Neuron. 2014;82:618–34. https://doi.org/10.1016/j.neuron.2014.03.012.

Chung C, Barylko B, Leitz J, Liu X, Kavalali ET. Acute dynamin inhibition dissects synaptic vesicle recycling pathways that drive spontaneous and evoked neurotransmission. J Neurosci. 2010;30:1363–76. https://doi.org/10.1523/JNEUROSCI.3427-09.2010.

Courtney NA, Briguglio JS, Bradberry MM, Greer C, Chapman ER. Excitatory and inhibitory neurons utilize different Ca2+ sensors and sources to regulate spontaneous release. Neuron. 2018;98:977–991.e5. https://doi.org/10.1016/j.neuron.2018.04.022.

Crawford DC, Kavalali ET. Molecular underpinnings of synaptic vesicle Pool heterogeneity. Traffic Cph Den. 2015;16:338–64. https://doi.org/10.1111/tra.12262.

Crawford DC, Ramirez DMO, Trauterman B, Monteggia LM, Kavalali ET. Selective molecular impairment of spontaneous neurotransmission modulates synaptic efficacy. Nat Commun. 2017;8:14436. https://doi.org/10.1038/ncomms14436.

Dai J, Chen P, Tian H, Sun J. Spontaneous vesicle release is not tightly coupled to voltage-gated calcium channel-mediated Ca2+ influx and is triggered by a Ca2+ sensor other than synaptotagmin-2 at the juvenile mice calyx of held synapses. J Neurosci. 2015;35:9632–7. https://doi.org/10.1523/JNEUROSCI.0457-15.2015.

Dani A, Huang B, Bergan J, Dulac C, Zhuang X. Superresolution imaging of chemical synapses in the brain. Neuron. 2010;68:843–56. https://doi.org/10.1016/j.neuron.2010.11.021.

Day KJ, Kago G, Wang L, Richter JB, Hayden CC, Lafer EM, et al. Liquid-like protein interactions catalyze assembly of endocytic vesicles. Nat Cell Biol. 2021;23:366–76. https://doi.org/10.1038/s41556-021-00646-5.

Deák F, Schoch S, Liu X, Südhof TC, Kavalali ET. Synaptobrevin is essential for fast synaptic-vesicle endocytosis. Nat Cell Biol. 2004;6:1102–8. https://doi.org/10.1038/ncb1185.

Dittman JS, Ryan TA. The control of release probability at nerve terminals. Nat Rev Neurosci. 2019;20:177–86. https://doi.org/10.1038/s41583-018-0111-3.

Dong C, Ly C, Dunlap LE, Vargas MV, Sun J, Hwang I-W, et al. Psychedelic-inspired drug discovery using an engineered biosensor. Cell. 2021;184:2779–2792.e18. https://doi.org/10.1016/j.cell.2021.03.043.

Eggermann E, Bucurenciu I, Goswami SP, Jonas P. Nanodomain coupling between Ca2+ channels and sensors of exocytosis at fast mammalian synapses. Nat Rev Neurosci. 2011;13:7–21. https://doi.org/10.1038/nrn3125.

Emperador-Melero J, Wong MY, Wang SSH, de Nola G, Nyitrai H, Kirchhausen T, et al. PKC-phosphorylation of Liprin-α3 triggers phase separation and controls presynaptic active zone structure. Nat Commun. 2021;12:3057. https://doi.org/10.1038/s41467-021-23116-w.

Engel AG, Selcen D, Shen X-M, Milone M, Harper CM. Loss of MUNC13-1 function causes microcephaly, cortical hyperexcitability, and fatal myasthenia. Neurol Genet. 2016;2:e105. https://doi.org/10.1212/NXG.0000000000000105.

Ermolyuk YS, Alder FG, Surges R, Pavlov IY, Timofeeva Y, Kullmann DM, et al. Differential triggering of spontaneous glutamate release by P/Q-, N- and R-type Ca2+ channels. Nat Neurosci. 2013;16:1754–63. https://doi.org/10.1038/nn.3563.

Fatt P, Katz B. Some observations on biological noise. Nature. 1950; 166:597–598

Fedchyshyn MJ, Wang L-Y. Developmental transformation of the release modality at the calyx of held synapse. J Neurosci. 2005;25:4131–40. https://doi.org/10.1523/JNEUROSCI.0350-05.2005.

Franks KM, Stevens CF, Sejnowski TJ. Independent sources of quantal variability at single glutamatergic synapses. J Neurosci. 2003;23:3186–95. https://doi.org/10.1523/JNEUROSCI.23-08-03186.2003.

Geppert M, Goda Y, Hammer RE, Li C, Rosahl TW, Stevens CF, et al. Synaptotagmin I: a major Ca2+ sensor for transmitter release at a central synapse. Cell. 1994;79:717–27. https://doi.org/10.1016/0092-8674(94)90556-8.

Goncalves J, Bartol TM, Camus C, Levet F, Menegolla AP, Sejnowski TJ, et al. Nanoscale co-organization and coactivation of AMPAR, NMDAR, and mGluR at excitatory synapses. Proc Natl Acad Sci USA. 2020;117:14503–11. https://doi.org/10.1073/pnas.1922563117.

Gonzalez-Islas C, Bülow P, Wenner P. Regulation of synaptic scaling by action potential-independent miniature neurotransmission. J Neurosci Res. 2018;96:348–53. https://doi.org/10.1002/jnr.24138.

Groffen AJ, Martens S, Arazola RD, Cornelisse LN, Lozovaya N, de Jong APH, et al. Doc2b is a high-affinity Ca2+ sensor for spontaneous neurotransmitter release. Science. 2010;327:1614–8. https://doi.org/10.1126/science.1183765.

Guzikowski NJ, Kavalali ET. Nano-Organization at the synapse: segregation of distinct forms of neurotransmission. Front. Synaptic Neurosci. 2021;13 https://doi.org/10.3389/fnsyn.2021.796498.

Guzikowski NJ, Kavalali ET. Nano-organization of spontaneous GABAergic transmission directs its autonomous function in neuronal signaling. Cell Rep. 2022;40 https://doi.org/10.1016/j.celrep.2022.111172.

Guzikowski NJ, Kavalali ET. Functional specificity of liquid-liquid phase separation at the synapse. Nat Commun. 2024;15:10103. https://doi.org/10.1038/s41467-024-54423-7.

Haas KT, Compans B, Letellier M, Bartol TM, Grillo-Bosch D, Sejnowski TJ, et al. Pre-post synaptic alignment through neuroligin-1 tunes synaptic transmission efficiency. elife. 2018; https://doi.org/10.7554/eLife.31755.

Heck J, Parutto P, Ciuraszkiewicz A, Bikbaev A, Freund R, Mitlöhner J, et al. Transient confinement of CaV2.1 Ca2+-channel splice variants shapes synaptic short-term plasticity. Neuron. 2019;103:66–79.e12. https://doi.org/10.1016/j.neuron.2019.04.030.

Hefft S, Jonas P. Asynchronous GABA release generates long-lasting inhibition at a hippocampal interneuron-principal neuron synapse. Nat Neurosci. 2005;8:1319–28. https://doi.org/10.1038/nn1542.

Heine M, Holcman D. Asymmetry between pre- and postsynaptic transient Nanodomains shapes neuronal communication. Trends Neurosci. 2020;43:182–96. https://doi.org/10.1016/j.tins.2020.01.005.

Heubl M, Zhang J, Pressey JC, Al Awabdh S, Renner M, Gomez-Castro F, et al. GABA a receptor dependent synaptic inhibition rapidly tunes KCC2 activity via the Cl − -sensitive WNK1 kinase. Nat Commun. 2017;8:1776. https://doi.org/10.1038/s41467-017-01749-0.

Higley MJ, Picciotto MR. Neuromodulation by acetylcholine: examples from schizophrenia and depression. Curr Opin Neurobiol. 2014;29:88–95. https://doi.org/10.1016/j.conb.2014.06.004.

Horvath PM, Piazza MK, Monteggia LM, Kavalali ET. Spontaneous and evoked neurotransmission are partially segregated at inhibitory synapses. eLife. 2020;9:e52852. https://doi.org/10.7554/eLife.52852.

Horvath PM, Chanaday NL, Alten B, Kavalali ET, Monteggia LM. A subthreshold synaptic mechanism regulating BDNF expression and resting synaptic strength. Cell Rep. 2021;36:109467. https://doi.org/10.1016/j.celrep.2021.109467.

Howe MW, Dombeck DA. Rapid signalling in distinct dopaminergic axons during locomotion and reward. Nature. 2016;535:505–10. https://doi.org/10.1038/nature18942.

Hruska M, Henderson N, Le Marchand SJ, Jafri H, Dalva MB. Synaptic nanomodules underlie the organization and plasticity of spine synapses. Nat Neurosci. 2018;21:671–82. https://doi.org/10.1038/s41593-018-0138-9.

Hruska M, Cain RE, Dalva MB. Nanoscale rules governing the organization of glutamate receptors in spine synapses are subunit specific. Nat Commun. 2022;13:920. https://doi.org/10.1038/s41467-022-28504-4.

Hua Z, Leal-Ortiz S, Foss SM, Waites CL, Garner CC, Voglmaier SM, et al. V-SNARE composition distinguishes synaptic vesicle pools. Neuron. 2011;71:474–87. https://doi.org/10.1016/j.neuron.2011.06.010.

Huang B, Babcock H, Zhuang X. Breaking the diffraction barrier: super-resolution imaging of cells. Cell. 2010;143:1047–58. https://doi.org/10.1016/j.cell.2010.12.002.

Huang H, Song S, Banerjee S, Jiang T, Zhang J, Kahle KT, et al. The WNK-SPAK/OSR1 kinases and the cation-chloride cotransporters as therapeutic targets for neurological diseases. Aging Dis. 2019;10:626–36. https://doi.org/10.14336/AD.2018.0928.

Huntwork S, Littleton JT. A complexin fusion clamp regulates spontaneous neurotransmitter release and synaptic growth. Nat Neurosci. 2007;10:1235–7. https://doi.org/10.1038/nn1980.

Imoto Y, Raychaudhuri S, Ma Y, Fenske P, Sandoval E, Itoh K, et al. Dynamin is primed at endocytic sites for ultrafast endocytosis. Neuron. 2022;110:2815–2835.e13. https://doi.org/10.1016/j.neuron.2022.06.010.

Jackman SL, Turecek J, Belinsky JE, Regehr WG. The calcium sensor synaptotagmin 7 is required for synaptic facilitation. Nature. 2016;529:88–91. https://doi.org/10.1038/nature16507.

Jiang M, Zhu J, Liu Y, Yang M, Tian C, Jiang S, et al. Enhancement of asynchronous release from fast-spiking interneuron in human and rat epileptic neocortex. PLoS Biol. 2012;10:e1001324. https://doi.org/10.1371/journal.pbio.1001324.

Jing M, Zhang P, Wang G, Feng J, Mesik L, Zeng J, et al. A genetically encoded fluorescent acetylcholine indicator for in vitro and in vivo studies. Nat Biotechnol. 2018;36:726–37. https://doi.org/10.1038/nbt.4184.

Jing M, Li Y, Zeng J, Huang P, Skirzewski M, Kljakic O, et al. An optimized acetylcholine sensor for monitoring in vivo cholinergic activity. Nat Methods. 2020;17:1139–46. https://doi.org/10.1038/s41592-020-0953-2.

Kaeser PS, Regehr WG. Molecular mechanisms for synchronous, asynchronous, and spontaneous neurotransmitter release. Annu Rev Physiol. 2014;76:333–63. https://doi.org/10.1146/annurev-physiol-021113-170338.

Katz B. The release of neural transmitter substances. Liverpool University Press; 1969.

Kavalali ET. The mechanisms and functions of spontaneous neurotransmitter release. Nature ReviewsNeuroscience. 2015; 16:5–16.

Kavalali ET. Neuronal Ca2+ signalling at rest and during spontaneous neurotransmission. J Physiol. 2020;598:1649–54. https://doi.org/10.1113/JP276541.

Kavalali ET, Monteggia LM. Targeting homeostatic synaptic plasticity for treatment of mood disorders. Neuron. 2020;106:715–26. https://doi.org/10.1016/j.neuron.2020.05.015.

Kavalali ET, Monteggia LM. Rapid homeostatic plasticity and neuropsychiatric therapeutics. Neuropsychopharmacology. 2023;48:54–60. https://doi.org/10.1038/s41386-022-01411-4.

Kershberg L, Banerjee A, Kaeser PS. Protein composition of axonal dopamine release sites in the striatum. eLife. 2022;11:e83018. https://doi.org/10.7554/eLife.83018.

Kim TH, Schnitzer MJ. Fluorescence imaging of large-scale neural ensemble dynamics. Cell. 2022;185:9–41. https://doi.org/10.1016/j.cell.2021.12.007.

Klausberger T, Somogyi P. Neuronal diversity and temporal dynamics: the Unity of hippocampal circuit operations. Science. 2008;321:53–7. https://doi.org/10.1126/science.1149381.

Kochubey O, Han Y, Schneggenburger R. Developmental regulation of the intracellular Ca2+ sensitivity of vesicle fusion and Ca2+-secretion coupling at the rat calyx of held. J Physiol. 2009;587:3009–23. https://doi.org/10.1113/jphysiol.2009.172387.

Kononenko NL, Haucke V. Molecular mechanisms of presynaptic membrane retrieval and synaptic vesicle reformation. Neuron. 2015;85:484–96. https://doi.org/10.1016/j.neuron.2014.12.016.

Kroning KE, Wang W. Designing a single protein-chain reporter for opioid detection at cellular resolution. Angew Chem Int Ed Engl. 2021;60:13358–65. https://doi.org/10.1002/anie.202101262.

Kroschwald S, Maharana S, Simon A. Hexanediol: a chemical probe to investigate the material properties of membrane-less compartments. Matters. 2017; https://doi.org/10.19185/matters.201702000010.

Kuriu T, Inoue A, Bito H, Sobue K, Okabe S. Differential control of postsynaptic density scaffolds via actin-dependent and -independent mechanisms. J Neurosci. 2006;26:7693–706. https://doi.org/10.1523/JNEUROSCI.0522-06.2006.

Kusick GF, Chin M, Raychaudhuri S, Lippmann K, Adula KP, Hujber EJ, et al. Synaptic vesicles transiently dock to refill release sites. Nat Neurosci. 2020;23:1329–38. https://doi.org/10.1038/s41593-020-00716-1.

Lammertse HCA, van Berkel AA, Iacomino M, Toonen RF, Striano P, Gambardella A, et al. Homozygous STXBP1 variant causes encephalopathy and gain-of-function in synaptic transmission. Brain. 2020;143:441–51. https://doi.org/10.1093/brain/awz391.

Li P, Banjade S, Cheng H-C, Kim S, Chen B, Guo L, et al. Phase transitions in the assembly of multivalent signaling proteins. Nature. 2012;483:336–40. https://doi.org/10.1038/nature10879.

Li YC, Chanaday NL, Xu W, Kavalali ET. Synaptotagmin-1- and Synaptotagmin-7-dependent fusion mechanisms target synaptic vesicles to kinetically distinct endocytic pathways. Neuron. 2017;93:616–631.e3. https://doi.org/10.1016/j.neuron.2016.12.010.

Li S, Raychaudhuri S, Lee SA, Brockmann MM, Wang J, Kusick G, et al. Asynchronous release sites align with NMDA receptors in mouse hippocampal synapses. Nat Commun. 2021;12:677. https://doi.org/10.1038/s41467-021-21004-x.

Li X, Li Y, Zhou Y, Wu J, Zhao Z, Fan J, et al. Real-time denoising enables high-sensitivity fluorescence time-lapse imaging beyond the shot-noise limit. Nat Biotechnol. 2023;41:282–92. https://doi.org/10.1038/s41587-022-01450-8.

Lin P-Y, Kavalali ET, Monteggia LM. Genetic dissection of presynaptic and postsynaptic BDNF-TrkB signaling in synaptic efficacy of CA3-CA1 synapses. Cell Rep. 2018;24:1550–61. https://doi.org/10.1016/j.celrep.2018.07.020.

Lin P-Y, Chanaday NL, Horvath PM, Ramirez DMO, Monteggia LM, Kavalali ET. VAMP4 maintains a Ca2+-sensitive pool of spontaneously recycling synaptic vesicles. J Neurosci. 2020;40:5389–401. https://doi.org/10.1523/JNEUROSCI.2386-19.2020.

Lin P-Y, Ma ZZ, Mahgoub M, Kavalali ET, Monteggia LM. A synaptic locus for TrkB signaling underlying ketamine rapid antidepressant action. Cell Rep. 2021;36:109513. https://doi.org/10.1016/j.celrep.2021.109513.

Lipstein N, Verhoeven-Duif NM, Michelassi FE, Calloway N, van Hasselt PM, Pienkowska K, et al. Synaptic UNC13A protein variant causes increased neurotransmission and dyskinetic movement disorder. J Clin Invest. 2017;127:1005–18. https://doi.org/10.1172/JCI90259.

Lisman JE, Raghavachari S, Tsien RW. The sequence of events that underlie quantal transmission at central glutamatergic synapses. Nat Rev Neurosci. 2007;8:597–609. https://doi.org/10.1038/nrn2191.

Liu C, Kaeser PS. Mechanisms and regulation of dopamine release. Curr Opin Neurobiol. 2019;57:46–53. https://doi.org/10.1016/j.conb.2019.01.001.

Liu G, Choi S, Tsien RW. Variability of neurotransmitter concentration and nonsaturation of postsynaptic AMPA receptors at synapses in hippocampal cultures and slices. Neuron. 1999;22:395–409. https://doi.org/10.1016/S0896-6273(00)81099-5.

Liu H, Dean C, Arthur CP, Dong M, Chapman ER. Autapses and networks of hippocampal neurons exhibit distinct synaptic transmission phenotypes in the absence of synaptotagmin I. J Neurosci. 2009;29:7395–403. https://doi.org/10.1523/JNEUROSCI.1341-09.2009.

Liu C, Kershberg L, Wang J, Schneeberger S, Kaeser PS. Dopamine secretion is mediated by sparse active zone-like release sites. Cell. 2018;172:706–718.e15. https://doi.org/10.1016/j.cell.2018.01.008.

Liu Y-T, Tao C-L, Zhang X, Xia W, Shi D-Q, Qi L, et al. Mesophasic organization of GABA a receptors in hippocampal inhibitory synapses. Nat Neurosci. 2020;23:1589–96. https://doi.org/10.1038/s41593-020-00729-w.

Llinás R, Sugimori M, Silver RB. Microdomains of high calcium concentration in a presynaptic terminal. Science. 1992;256:677–9. https://doi.org/10.1126/science.1350109.

Luo F, Südhof TC. Synaptotagmin-7-mediated asynchronous release boosts high-Fidelity synchronous transmission at a central synapse. Neuron. 2017;94:826–839.e3. https://doi.org/10.1016/j.neuron.2017.04.020.

Lyon AS, Peeples WB, Rosen MK. A framework for understanding functions of biomolecular condensates on molecular to cellular scales. Nat Rev Mol Cell Biol. 2021;22:215. https://doi.org/10.1038/s41580-020-00303-z.

MacGillavry HD, Song Y, Raghavachari S, Blanpied TA. Nanoscale scaffolding domains within the postsynaptic density concentrate synaptic AMPA receptors. Neuron. 2013;78:615–22. https://doi.org/10.1016/j.neuron.2013.03.009.

Mainen ZF, Malinow R, Svoboda K. Synaptic calcium transients in single spines indicate that NMDA receptors are not saturated. Nature. 1999;399:151–5. https://doi.org/10.1038/20187.

Marvin JS, Shimoda Y, Magloire V, Leite M, Kawashima T, Jensen TP, et al. A genetically encoded fluorescent sensor for in vivo imaging of GABA. Nat Methods. 2019;16:763–70. https://doi.org/10.1038/s41592-019-0471-2.

Maschi D, Klyachko VA. Spatiotemporal regulation of synaptic vesicle fusion sites in central synapses. Neuron. 2017;94:65–73.e3. https://doi.org/10.1016/j.neuron.2017.03.006.

Maximov A, Südhof TC. Autonomous function of Synaptotagmin 1 in triggering synchronous release independent of asynchronous release. Neuron. 2005;48:547–54. https://doi.org/10.1016/j.neuron.2005.09.006.

McAllister AK, Stevens CF. Nonsaturation of AMPA and NMDA receptors at hippocampal synapses. Proc Natl Acad Sci. 2000;97:6173–8. https://doi.org/10.1073/pnas.100126497.

McCarthy CI, Kavalali ET. Nano-organization of synaptic calcium signaling. Biochem Soc Trans. 2024;52:1459–71. https://doi.org/10.1042/BST20231385.

McDonald NA, Shen K. Finding functions of phase separation in the presynapse. Curr Opin Neurobiol. 2021;69:178–84. https://doi.org/10.1016/j.conb.2021.04.001.

McDonald NA, Fetter RD, Shen K. Assembly of synaptic active zones requires phase separation of scaffold molecules. Nature. 2020;588:454–8. https://doi.org/10.1038/s41586-020-2942-0.

McQuarrie D, Simon J. Physical chemistry: a molecular approach. 1st ed. University Science Books; 1997.

Melom JE, Akbergenova Y, Gavornik JP, Littleton JT. Spontaneous and evoked release are independently regulated at individual active zones. J Neurosci. 2013;33:17253–63. https://doi.org/10.1523/JNEUROSCI.3334-13.2013.

Mendonça PRF, Tagliatti E, Langley H, Kotzadimitriou D, Zamora-Chimal CG, Timofeeva Y, et al. Asynchronous glutamate release is enhanced in low release efficacy synapses and dispersed across the active zone. Nat Commun. 2022;13:3497. https://doi.org/10.1038/s41467-022-31070-4.

Milovanovic D, De Camilli P. Synaptic vesicle clusters at synapses: a distinct liquid phase? Neuron. 2017;93:995–1002. https://doi.org/10.1016/j.neuron.2017.02.013.

Milovanovic D, Wu Y, Bian X, De Camilli P. A liquid phase of synapsin and lipid vesicles. Science. 2018;361:604–7. https://doi.org/10.1126/science.aat5671.

Mittag T, Parker R. Multiple Modes of Protein–Protein Interactions Promote RNP Granule Assembly. J Mol Biol. 2018;430:4636–49. https://doi.org/10.1016/j.jmb.2018.08.005.

Molnár Z, López-Bendito G, Small J, Partridge LD, Blakemore C, Wilson MC. Normal development of embryonic Thalamocortical connectivity in the absence of evoked synaptic activity. J Neurosci. 2002;22:10313–23. https://doi.org/10.1523/JNEUROSCI.22-23-10313.2002.

Moran SP, Maksymetz J, Conn PJ. Targeting muscarinic acetylcholine receptors for the treatment of psychiatric and neurological disorders. Trends Pharmacol Sci. 2019;40:1006–20. https://doi.org/10.1016/j.tips.2019.10.007.

Mori Y, Takenaka K, Fukazawa Y, Takamori S. The endosomal Q-SNARE, Syntaxin 7, defines a rapidly replenishing synaptic vesicle recycling pool in hippocampal neurons. Commun Biol. 2021;4:1–13. https://doi.org/10.1038/s42003-021-02512-4.

Mozhayeva MG, Sara Y, Liu X, Kavalali ET. Development of vesicle pools during maturation of hippocampal synapses. J Neurosci. 2002;22:654–65. https://doi.org/10.1523/JNEUROSCI.22-03-00654.2002.

Nair D, Hosy E, Petersen JD, Constals A, Giannone G, Choquet D, et al. Super-resolution imaging reveals that AMPA receptors inside synapses are dynamically organized in Nanodomains regulated by PSD95. J Neurosci. 2013;33:13204–24. https://doi.org/10.1523/JNEUROSCI.2381-12.2013.

Nanou E, Catterall WA. Calcium channels, synaptic plasticity, and neuropsychiatric disease. Neuron. 2018;98:466–81. https://doi.org/10.1016/j.neuron.2018.03.017.

Newman ZL, Bakshinskaya D, Schultz R, Kenny SJ, Moon S, Aghi K, et al. Determinants of synapse diversity revealed by super-resolution quantal transmission and active zone imaging. Nat Commun. 2022;13:229. https://doi.org/10.1038/s41467-021-27815-2.

Nimchinsky EA, Yasuda R, Oertner TG, Svoboda K. The number of glutamate receptors opened by synaptic stimulation in single hippocampal spines. J Neurosci. 2004;24:2054–64. https://doi.org/10.1523/JNEUROSCI.5066-03.2004.

Özçete ÖD, Banerjee A, Kaeser PS. Mechanisms of neuromodulatory volume transmission. Mol Psychiatry. 2024:1–14. https://doi.org/10.1038/s41380-024-02608-3.

Pang ZP, Bacaj T, Yang X, Zhou P, Xu W, Südhof TC. Doc2 supports spontaneous synaptic transmission by a Ca2+-independent mechanism. Neuron. 2011;70:244–51. https://doi.org/10.1016/j.neuron.2011.03.011.

Pappu RV, Cohen SR, Dar F, Farag M, Kar M. Phase transitions of associative biomacromolecules. Chem Rev. 2023;123:8945–87. https://doi.org/10.1021/acs.chemrev.2c00814.

Park D, Wu Y, Lee S-E, Kim G, Jeong S, Milovanovic D, et al. Cooperative function of synaptophysin and synapsin in the generation of synaptic vesicle-like clusters in non-neuronal cells. Nat Commun. 2021;12:263. https://doi.org/10.1038/s41467-020-20462-z.

Park D, Wu Y, Wang X, Gowrishankar S, Baublis A, De Camilli P. Synaptic vesicle proteins and ATG9A self-organize in distinct vesicle phases within synapsin condensates. Nat Commun. 2023;14:455. https://doi.org/10.1038/s41467-023-36081-3.

Patriarchi T, Cho JR, Merten K, Howe MW, Marley A, Xiong W-H, et al. Ultrafast neuronal imaging of dopamine dynamics with designed genetically encoded sensors. Science. 2018;360:eaat4422. https://doi.org/10.1126/science.aat4422.

Pechstein A, Tomilin N, Fredrich K, Vorontsova O, Sopova E, Evergren E, et al. Vesicle clustering in a living synapse depends on a Synapsin region that mediates phase separation. Cell Rep. 2020;30:2594–2602.e3. https://doi.org/10.1016/j.celrep.2020.01.092.

Peled ES, Newman ZL, Isacoff EY. Evoked and spontaneous transmission favored by distinct sets of synapses. Curr Biol. 2014;24:484–93. https://doi.org/10.1016/j.cub.2014.01.022.

Penn AC, Zhang CL, Georges F, Royer L, Breillat C, Hosy E, et al. Hippocampal LTP and contextual learning require surface diffusion of AMPA receptors. Nature. 2017;549:384–8. https://doi.org/10.1038/nature23658.

Pereira DB, Schmitz Y, Mészáros J, Merchant P, Hu G, Li S, et al. Fluorescent false neurotransmitter reveals functionally silent dopamine vesicle clusters in the striatum. Nat Neurosci. 2016;19:578–86. https://doi.org/10.1038/nn.4252.

Piala AT, Moon TM, Akella R, He H, Cobb MH, Goldsmith EJ. Chloride sensing by WNK1 involves inhibition of autophosphorylation. Sci Signal. 2014;7:ra41. https://doi.org/10.1126/scisignal.2005050.

Raghavachari S, Lisman JE. Properties of quantal transmission at CA1 synapses. J Neurophysiol. 2004;92:2456–67. https://doi.org/10.1152/jn.00258.2004.

Raingo J, Khvotchev M, Liu P, Darios F, Li YC, Ramirez DMO, et al. VAMP4 directs synaptic vesicles to a pool that selectively maintains asynchronous neurotransmission. Nat Neurosci. 2012;15:738–45. https://doi.org/10.1038/nn.3067.

Ramirez DMO, Khvotchev M, Trauterman B, Kavalali ET. Vti1a identifies a vesicle pool that preferentially recycles at rest and maintains spontaneous neurotransmission. Neuron. 2012;73:121–34. https://doi.org/10.1016/j.neuron.2011.10.034.

Ramirez DMO, Crawford DC, Chanaday NL, Trauterman B, Monteggia LM, Kavalali ET. Loss of Doc2-dependent spontaneous neurotransmission augments glutamatergic synaptic strength. J Neurosci. 2017;37:6224–30. https://doi.org/10.1523/JNEUROSCI.0418-17.2017.

Ramsay HJ, Gookin SE, Ramsey AM, Kareemo DJ, Crosby KC, Stich DG, et al. AMPA and GABAA receptor nanodomains assemble in the absence of synaptic neurotransmitter release. Front Mol Neurosci. 2023;16. Available at: https://www.frontiersin.org/articles/10.3389/fnmol.2023.1232795. Accessed 23 Aug 2023

Ramsey AM, Tang A-H, LeGates TA, Gou X-Z, Carbone BE, Thompson SM, et al. Subsynaptic positioning of AMPARs by LRRTM2 controls synaptic strength. Sci Adv. 2021;7:eabf3126. https://doi.org/10.1126/sciadv.abf3126.

Reese AL, Kavalali ET. Spontaneous neurotransmission signals through store-driven Ca(2+) transients to maintain synaptic homeostasis. eLife. 2015;4 https://doi.org/10.7554/eLife.09262.

Reese AL, Kavalali ET. Single synapse evaluation of the postsynaptic NMDA receptors targeted by evoked and spontaneous neurotransmission. eLife. 2016;5 https://doi.org/10.7554/eLife.21170.

Revelo NH, Kamin D, Truckenbrodt S, Wong AB, Reuter-Jessen K, Reisinger E, et al. A new probe for super-resolution imaging of membranes elucidates trafficking pathways. J Cell Biol. 2014;205:591–606. https://doi.org/10.1083/jcb.201402066.

Robinson BG, Cai X, Wang J, Bunzow JR, Williams JT, Kaeser PS. RIM is essential for stimulated but not spontaneous somatodendritic dopamine release in the midbrain. eLife. 2019;8:e47972. https://doi.org/10.7554/eLife.47972.

Rozov A, Bolshakov AP, Valiullina-Rakhmatullina F. The ever-growing puzzle of asynchronous release. Front Cell Neurosci. 2019;13:28. https://doi.org/10.3389/fncel.2019.00028.

Ruiz R, Casañas JJ, Torres-Benito L, Cano R, Tabares L. Altered intracellular Ca2+ homeostasis in nerve terminals of severe spinal muscular atrophy mice. J Neurosci. 2010;30:849–57. https://doi.org/10.1523/JNEUROSCI.4496-09.2010.

Santos MS, Park CK, Foss SM, Li H, Voglmaier SM. Sorting of the vesicular GABA transporter to functional vesicle pools by an atypical dileucine-like motif. J Neurosci. 2013;33:10634–46. https://doi.org/10.1523/JNEUROSCI.0329-13.2013.

Sara Y, Bal M, Adachi M, Monteggia LM, Kavalali ET. Use-dependent AMPA receptor block reveals segregation of spontaneous and evoked glutamatergic neurotransmission. J Neurosci. 2011;31:5378–82. https://doi.org/10.1523/JNEUROSCI.5234-10.2011.

Savtchenko LP, Rusakov DA. Moderate AMPA receptor clustering on the nanoscale can efficiently potentiate synaptic current. Philos Trans R Soc Lond Ser B Biol Sci. 2014;369:20130167. https://doi.org/10.1098/rstb.2013.0167.

Schoch S, Deák F, Königstorfer A, Mozhayeva M, Sara Y, Südhof TC, et al. SNARE function analyzed in Synaptobrevin/VAMP knockout mice. Science. 2001;294:1117–22. https://doi.org/10.1126/science.1064335.

Seeburg PH, Burnashev N, Köhr G, Kuner T, Sprengel R, Monyer H. The NMDA receptor channel: molecular design of a coincidence detector. Recent Prog Horm Res. 1995;50:19–34. https://doi.org/10.1016/b978-0-12-571150-0.50006-8.

Shivange AV, Borden PM, Muthusamy AK, Nichols AL, Bera K, Bao H, et al. Determining the pharmacokinetics of nicotinic drugs in the endoplasmic reticulum using biosensors. J Gen Physiol. 2019;151:738–57. https://doi.org/10.1085/jgp.201812201.

Simmons RL, Li H, Alten B, Santos MS, Jiang R, Paul B, et al. Overcoming presynaptic effects of VAMP2 mutations with 4-aminopyridine treatment. Hum Mutat. 2020;41:1999–2011. https://doi.org/10.1002/humu.24109.

Simon SM, Llinás RR. Compartmentalization of the submembrane calcium activity during calcium influx and its significance in transmitter release. Biophys J. 1985;48:485–98. https://doi.org/10.1016/S0006-3495(85)83804-2.

Song L, Hennink EJ, Young IT, Tanke HJ. Photobleaching kinetics of fluorescein in quantitative fluorescence microscopy. Biophys J. 1995;68:2588–600. https://doi.org/10.1016/S0006-3495(95)80442-X.

Südhof TC. Neurotransmitter release: the last millisecond in the life of a synaptic vesicle. Neuron. 2013;80:675–90. https://doi.org/10.1016/j.neuron.2013.10.022.

Südhof TC, Rothman JE. Membrane fusion: grappling with SNARE and SM proteins. Science. 2009;323:474–7. https://doi.org/10.1126/science.1161748.

Sugita S, Shin O-H, Han W, Lao Y, Südhof TC. Synaptotagmins form a hierarchy of exocytotic Ca(2+) sensors with distinct Ca(2+) affinities. EMBO J. 2002;21:270–80. https://doi.org/10.1093/emboj/21.3.270.

Sulis Sato S, Artoni P, Landi S, Cozzolino O, Parra R, Pracucci E, et al. Simultaneous two-photon imaging of intracellular chloride concentration and pH in mouse pyramidal neurons in vivo. Proc Natl Acad Sci USA. 2017;114:E8770–9. https://doi.org/10.1073/pnas.1702861114.

Sun F, Zeng J, Jing M, Zhou J, Feng J, Owen SF, et al. A genetically encoded fluorescent sensor enables rapid and specific detection of dopamine in flies, fish, and mice. Cell. 2018;174:481–496.e19. https://doi.org/10.1016/j.cell.2018.06.042.

Sutton MA, Ito HT, Cressy P, Kempf C, Woo JC, Schuman EM. Miniature neurotransmission stabilizes synaptic function via tonic suppression of local dendritic protein synthesis. Cell. 2006;125:785–99. https://doi.org/10.1016/j.cell.2006.03.040.

Suzuki K, Kim J-W, Nosyreva E, Kavalali ET, Monteggia LM. Convergence of distinct signaling pathways on synaptic scaling to trigger rapid antidepressant action. Cell Rep. 2021;37:109918. https://doi.org/10.1016/j.celrep.2021.109918.

Takamori S, Holt M, Stenius K, Lemke EA, Grønborg M, Riedel D, et al. Molecular anatomy of a trafficking organelle. Cell. 2006;127:831–46. https://doi.org/10.1016/j.cell.2006.10.030.

Tan C, Wang SSH, de Nola G, Kaeser PS. Rebuilding essential active zone functions within a synapse. Neuron. 2022;110:1498–1515.e8. https://doi.org/10.1016/j.neuron.2022.01.026.

Tang A-H, Chen H, Li TP, Metzbower SR, MacGillavry HD, Blanpied TA. A trans-synaptic nanocolumn aligns neurotransmitter release to receptors. Nature. 2016;536:210–4. https://doi.org/10.1038/nature19058.

Tsintsadze T, Williams CL, Weingarten DJ, von Gersdorff H, Smith SM. Distinct actions of voltage-activated Ca2+ channel block on spontaneous release at excitatory and inhibitory central synapses. J Neurosci. 2017;37:4301–10. https://doi.org/10.1523/JNEUROSCI.3488-16.2017.

Turrigiano GG, Nelson SB. Homeostatic plasticity in the developing nervous system. Nat Rev Neurosci. 2004;5:97–107. https://doi.org/10.1038/nrn1327.

Uzay B, Kavalali ET. Genetic disorders of neurotransmitter release machinery. Front Synaptic Neurosci. 2023;15:1148957. https://doi.org/10.3389/fnsyn.2023.1148957.

Uzay B, Houcek A, Ma ZZ, Konradi C, Monteggia LM, Kavalali ET. Neurotransmitter release progressively desynchronizes in induced human neurons during synapse maturation and aging. Cell Rep. 2023;42:112042. https://doi.org/10.1016/j.celrep.2023.112042.

van der Lee R, Buljan M, Lang B, Weatheritt RJ, Daughdrill GW, Dunker AK, et al. Classification of intrinsically disordered regions and proteins. Chem Rev. 2014;114:6589–631. https://doi.org/10.1021/cr400525m.

Vardar G, Gerth F, Schmitt XJ, Rautenstrauch P, Trimbuch T, Schubert J, et al. Epilepsy-causing STX1B mutations translate altered protein functions into distinct phenotypes in mouse neurons. Brain. 2020;143:2119–38. https://doi.org/10.1093/brain/awaa151.

Vyleta NP, Smith SM. Spontaneous glutamate release is independent of calcium influx and tonically activated by the calcium-sensing receptor. J Neurosci. 2011;31:4593–606. https://doi.org/10.1523/JNEUROSCI.6398-10.2011.

Walter AM, Kurps J, de Wit H, Schöning S, Toft-Bertelsen TL, Lauks J, et al. The SNARE protein vti1a functions in dense-core vesicle biogenesis. EMBO J. 2014;33:1681–97. https://doi.org/10.15252/embj.201387549.

Wang L-Y, Augustine GJ. Presynaptic nanodomains: a tale of two synapses. Front Cell Neurosci. 2014;8:455. https://doi.org/10.3389/fncel.2014.00455.

Wang CS, Chanaday NL, Monteggia LM, Kavalali ET. Probing the segregation of evoked and spontaneous neurotransmission via photobleaching and recovery of a fluorescent glutamate sensor. eLife. 2022;11:e76008. https://doi.org/10.7554/eLife.76008.

Wang CS, Monteggia LM, Kavalali ET. Spatially non-overlapping Ca2+ signals drive distinct forms of neurotransmission. Cell Rep. 2023;42:113201. https://doi.org/10.1016/j.celrep.2023.113201.

Washbourne P, Thompson PM, Carta M, Costa ET, Mathews JR, Lopez-Benditó G, et al. Genetic ablation of the t-SNARE SNAP-25 distinguishes mechanisms of neuroexocytosis. Nat Neurosci. 2002;5:19–26. https://doi.org/10.1038/nn783.

Weber JP, Toft-Bertelsen TL, Mohrmann R, Delgado-Martinez I, Sørensen JB. Synaptotagmin-7 is an asynchronous calcium sensor for synaptic transmission in neurons expressing SNAP-23. PLoS One. 2014;9:e114033. https://doi.org/10.1371/journal.pone.0114033.

Wilhelm BG, Mandad S, Truckenbrodt S, Kröhnert K, Schäfer C, Rammner B, et al. Composition of isolated synaptic boutons reveals the amounts of vesicle trafficking proteins. Science. 2014;344:1023–8. https://doi.org/10.1126/science.1252884.

Williams CL, Smith SM. Calcium dependence of spontaneous neurotransmitter release. J Neurosci Res. 2018;96:335–47. https://doi.org/10.1002/jnr.24116.

Wu X, Cai Q, Shen Z, Chen X, Zeng M, Du S, et al. RIM and RIM-BP form presynaptic active-zone-like condensates via phase separation. Mol Cell. 2019;73:971–984.e5. https://doi.org/10.1016/j.molcel.2018.12.007.

Wu X, Ganzella M, Zhou J, Zhu S, Jahn R, Zhang M. Vesicle tethering on the surface of phase-separated active zone condensates. Mol Cell. 2021;81:13–24.e7. https://doi.org/10.1016/j.molcel.2020.10.029.

Wu X, Qiu H, Zhang M. Interactions between Membraneless condensates and membranous organelles at the Presynapse: a phase separation view of synaptic vesicle cycle. J Mol Biol. 2023;435:167629. https://doi.org/10.1016/j.jmb.2022.167629.

Xu W, Morishita W, Buckmaster PS, Pang ZP, Malenka RC, Südhof TC. Distinct neuronal coding schemes in memory revealed by selective erasure of fast synchronous synaptic transmission. Neuron. 2012;73:990–1001. https://doi.org/10.1016/j.neuron.2011.12.036.

Yang Y-M, Fedchyshyn MJ, Grande G, Aitoubah J, Tsang CW, Xie H, et al. Septins regulate developmental switching from microdomain to nanodomain coupling of ca(2+) influx to neurotransmitter release at a central synapse. Neuron. 2010;67:100–15. https://doi.org/10.1016/j.neuron.2010.06.003.

Zajac M, Chakraborty K, Saha S, Mahadevan V, Infield DT, Accardi A, et al. What biologists want from their chloride reporters—a conversation between chemists and biologists. J Cell Sci. 2020;133:jcs240390. https://doi.org/10.1242/jcs.240390.

Zeng M, Chen X, Guan D, Xu J, Wu H, Tong P, et al. Reconstituted postsynaptic density as a molecular platform for understanding synapse formation and plasticity. Cell. 2018;174:1172–1187.e16. https://doi.org/10.1016/j.cell.2018.06.047.

Zeng M, Díaz-Alonso J, Ye F, Chen X, Xu J, Ji Z, et al. Phase separation-mediated TARP/MAGUK complex condensation and AMPA receptor synaptic transmission. Neuron. 2019;104:529–543.e6. https://doi.org/10.1016/j.neuron.2019.08.001.

Zhu S, Shen Z, Wu X, Han W, Jia B, Lu W, et al. Demixing is a default process for biological condensates formed via phase separation. Science. 2024;384:920–8. https://doi.org/10.1126/science.adj7066.

Chapter 9
Optical Approaches to Dissect the Structure and Dynamics of the Synapse at Nanoscale Resolution

Jongyun Myeong and Vitaly A. Klyachko

Abstract The complex nanostructure and spatiotemporal dynamics of central synapses remain among the fundamental mysteries of neurobiology. The resolution of traditional microscopy techniques—constrained by the intrinsic limits of light diffraction—is largely insufficient to study central synapses effectively. Conventional imaging can resolve areas roughly the size of a synapse's active zone, that is, severalfold larger than the size of synaptic vesicles. Recent advances have generated several super-resolution imaging modalities that overcome or bypass the light diffraction limit to support studies of synaptic nanostructure. In this chapter, we present the principles, features, and limitations of the most common super-resolution imaging tools. Though these advancements have greatly improved our understanding of synaptic architecture and dynamics, significant challenges remain. Difficulties of translating the existing tools to in vivo applications, and the inherent trade-off between spatial and temporal resolution, continue to limit studies of the function of central synapses in native tissue.

Keywords Synaptic nanostructure · Super-resolution imaging · Structured illumination microscopy · Stochastic optical reconstruction microscopy · Stimulated emission depletion microscopy · Single-particle tracking

J. Myeong · V. A. Klyachko (✉)
Department of Cell Biology and Physiology, Washington University School of Medicine, St. Louis, MO, USA
e-mail: klyachko@wustl.edu

E. T. Kavalali (ed.), *Nano-organization of the Synapse*, Advances in Neurobiology 48,
https://doi.org/10.1007/978-3-032-12594-1_9

9.1 Introduction

9.1.1 *Challenges to Define Synaptic Nanostructure and Dynamics*

Quantal vesicle release represents a unitary event of information transmission at synapses. Yet, because synapses are so small (<1 μm for most central synapses, with an active zone typically <250 nm (Schikorski and Stevens 1997, 1999)), our understanding of the structure and organization of synapses in the brain is limited. Until recently, our ability to study synaptic architecture has been constrained by insufficient spatial resolution of most conventional experimental approaches; the techniques offering the highest spatial resolution can only resolve objects down to ~250 nm (Gramlich and Klyachko 2019; Schermelleh et al. 2019; Sigal et al. 2018; Toomre and Bewersdorf 2010). Moreover, we now know that synaptic organization is not static; proteins within both pre- and postsynaptic areas reorganize on various timescales, both over the course of normal activity and in response to many forms of synaptic plasticity (Choquet and Hosy 2020; Maschi and Klyachko 2015; Nair et al. 2013). However, techniques providing sufficient temporal resolution to reveal such dynamics generally sacrifice spatial resolution, ultimately limiting our ability to visualize the structure and function of synapses simultaneously. This is a critical gap in our knowledge, as dysfunctions of transmitter release mechanisms have been linked to numerous disease states, including epilepsy and neurodevelopmental disorders (Taoufik et al. 2018; Wang and Balice-Gordon 2022). The challenge is clear: we need to develop novel microscopy techniques that can visualize the smallest subsynaptic structures while tracking the fastest synaptic dynamics, with the overarching goal to perform these measurements in three dimensions and ultimately in vivo. Several novel technologies have begun to address these challenges in various ways, but some of the key limitations remain, including those posed by the heterogeneity of synaptic vesicle properties and release processes, and the inherent trade-off between spatial and temporal resolution.

9.1.2 *The Fundamental Limitations of Optical Microscopy Techniques*

The resolution of a microscope or an imaging system is defined by its ability to distinguish between two light sources emitted from the specimen. Over the three centuries since the development of the first single-lens microscope by Antonie van Leeuwenhoek (Cole 1939), innovation has generated increasingly complex compound microscopes and fluorescence reporters. However, while studies using these tools have deepened our knowledge of neuronal and synaptic architecture, the spatial resolution of optical imaging is fundamentally limited by the light diffraction limit. As an example, consider conventional wide-field microscopy in which all fluorophores are imaged at the same time. The captured image is an overlapping signal comprising myriad fluorescent molecules. The resolution of such images is defined by the Abbe's principle, which is based on the numerical aperture (NA) of the microscope objective and the wavelength of the emitted light. This principle postulates that light emitted by even the smallest fluorescently labeled object always results in a diffraction-limited spot represented by the point spread function (PSF) of the optical system, thus setting the resolution limit of an optical microscope to ~250–300 nm (Arizono et al. 2023; Betzig et al. 2006; Hell 2007; Sigal et al. 2018). This limit is roughly the size of a synaptic active zone (AZ; ~250 nm), while synaptic vesicles are much smaller (~40–50 nm). Thus, the fundamental resolution limit of conventional optical microscopy presents a major challenge to visualizing synaptic architecture.

Until the recent advances in live-imaging super-resolution microscopy tools, which we discuss below, only a family of near-field scanning probe microscopy techniques, such as scanning electron microscopy (SEM) or atomic force microscopy (AFM), were able to achieve nanometer spatial resolution (Siksou et al. 2009). These techniques provided the first fundamental observations of synaptic nanostructures. Since the pioneering work of Ramon y Cajal, studies using these tools have uncovered important details of synaptic ultrastructure in various model systems. For example, immunogold labeling EM studies were instrumental in identifying the location of transmitter receptors (Ottersen and Landsend 1997) and associated proteins (Valtschanoff and Weinberg 2001), including the precise location of AMPA and NMDA receptors (Takumi et al. 1999), and the proportion of NMDA versus non-NMDA glutamate receptors (Thomson and Deuchars 1994). Early EM analyses also revealed basic synaptic vesicle organization and number at presynaptic terminals (De Robertis and Bennett 1955; Palay 1956). Yet because only fixed samples can be used in applications of these near-field techniques, the temporal dynamics of synaptic architecture and the mechanistic details of synaptic vesicle release have not been amenable to these approaches.

9.1.3 Heterogeneity of Synaptic Vesicle Properties and Modes of Release Demands Techniques That Can Span a Wide Range of Spatial and Temporal Resolution

Adding to the challenge of visualizing functioning central synapses is the fact that the dynamics of synaptic vesicle release can vary considerably, even within a single synaptic terminal. Studies in the 1960s revealed that the probability of release is not the same across all vesicles; some are more easily released than others (Matthews 1966). In fact, we now recognize three major pools of vesicles that differ in their distribution within the terminal and readiness for release: (1) a small, readily releasable pool (RRP) of vesicles docked at the AZ and fully primed for release immediately upon stimulation; (2) a "recycling pool" of vesicles scattered throughout the nerve terminal that may undergo translocation and subsequent release upon moderate stimulation; and (3) a larger "reserved pool," which is only recruited upon strong stimulation after the recycling pool is exhausted (Denker and Rizzoli 2010). This heterogeneity poses additional challenges for developing technologies that can visualize single synaptic vesicle release events at nanoscale resolution and at a wide range of different timescales.

The process of vesicle release itself is also heterogeneous. Several distinct forms of neurotransmitter release take place within central synapses: (1) synchronous uni- and multivesicular release (UVR and MVR, respectively), either of which may occur within a few milliseconds of the arrival of an action potential (AP), (2) asynchronous release, that is, vesicle release that may persist for tens of milliseconds to several seconds after an AP, and (3) spontaneous release, which occurs in the absence of activity (Kavalali 2007). Each release type plays important roles in synaptic function and information processing (Kaeser and Regehr 2013). For example, MVR has been detected in many types of central synapses and has been implicated in a wide range of functions, including reliability of synaptic transmission, regulating synaptic integration, and inducing several forms of synaptic plasticity (Rudolph et al. 2015). As described above, spatiotemporal resolution constraints inherent to conventional microscopy techniques make it difficult to distinguish and study UVR and MVR in real-time at central synapses. As a result, we still do not know what determines the propensity for MVR and/or how common this form of release is in small central synapses. We also still do not fully understand to what extent these different forms of neurotransmitter release overlap spatially within the AZ, nor do we understand how release is regulated and/or structurally supported. Other important outstanding questions include the following: Do different forms of release share the same release sites, or are there specialized release sites that support specific forms of release? How do neural activity and/or cytoskeletal structure regulate the spatiotemporal properties of release? How is vesicle refilling regulated at the release sites? Answering these questions requires technology that can detect and track nanoscale events of vesicle release over a broad range of timescales (Figs. 9.3 and 9.4).

9.2 Super-Resolution Microscopy Techniques for the Study of Nanoscale Synaptic Architecture

To bypass the limitation of light diffraction, several so-called super-resolution approaches have been developed over the last two decades. These techniques, which can achieve nanoscale resolution, work around the diffraction barrier, rather than breaking it. In most cases, these techniques are based on the central premise that a single fluorescent light source can be localized with very high accuracy if enough photons are collected (Thompson et al. 2002). Examples of nanoscale resolution imaging techniques employing this concept include photoactivated localization microscopy (PALM) (Betzig et al. 2006), superresolution fluorescence localization microscopy (SRFLM) (Nelson and Hess 2014), stochastic optical reconstruction microscopy (STORM) (Rust et al. 2006), and various single-particle tracking (SPT) approaches (Alcor et al. 2009). These techniques have been described in detail in several excellent recent reviews (Alcor et al. 2009; Bates et al. 2008; Hell 2007; Huang et al. 2009; Maglione and Sigrist 2013; Nelson and Hess 2014; Sigrist and Sabatini 2011). In this chapter, we will focus on the subset of these approaches most commonly used in studies of synaptic nanostructure.

9.2.1 Structured Illumination Microscopy

The principle of structured illumination microscopy (SIM) relies on a concept that superposition of two periodic patterns with slightly different frequencies (the illumination pattern and the sample features) produce a so-called Moiré fringe pattern with a lower frequency than either of the original patterns, bringing out the finer (super-resolution) details of the sample to the resolvable range (Fig 9.1) (Gustafsson 1999, 2000; Gustafsson et al. 2008; Heintzmann and Huser 2017; Heintzmann et al. 2002). SIM utilizes a stripe-like illumination pattern which is systematically applied across the sample, enabling high-frequency (i.e., high-resolution) information that is normally not attainable with conventional illumination to be shifted into a resolvable range, and extracted through computational image processing (Heintzmann and Huser 2017; Kner et al. 2009). The concept of SIM was developed in the late 1960s by Lucosz (1966), who placed the structured illumination directly in front of the objective, then deconvolved the patterns by applying a matching structure. The initial applications of SIM improved spatial resolution by twofold relative to conventional confocal imaging (Gustafsson 2000). In later work, resolution was further enhanced by using nonlinear structured illumination patterns (Gustafsson 2005). Moreover, further advancements in both hardware and software have helped SIM achieve higher super-resolution (<100 nm), as well as ultrahigh imaging speeds (exceeding 500 frames per second (FPS)), a large field-of-view (>200 μm), long-term imaging (>1 h), and multicolor imaging (up to four colors) (Heintzmann and Huser 2017; Kner et al. 2009). This superior spatiotemporal resolution, combined with its compatibility with conventional fluorescent probes, makes it particularly well-suited for live-cell imaging, which is critical for the study of synaptic ultrastructure and function in native tissue. Nonlinear SIM can also be combined with light-sheet microscopy to observe the volume of a whole cell in 3D. This approach, called three-dimensional SIM (3D-SIM), has achieved axial resolution of 160 nm (Gustafsson et al. 2008) (Fig. 9.1).

Limitations of SIM Despite its superior performance in both spatial and temporal domains, some limitations remain. For example, modern SIM relies on sophisticated computational procedures for image reconstruction, but this can introduce artifacts. Trained users are needed to recognize and avoid these distortions. The

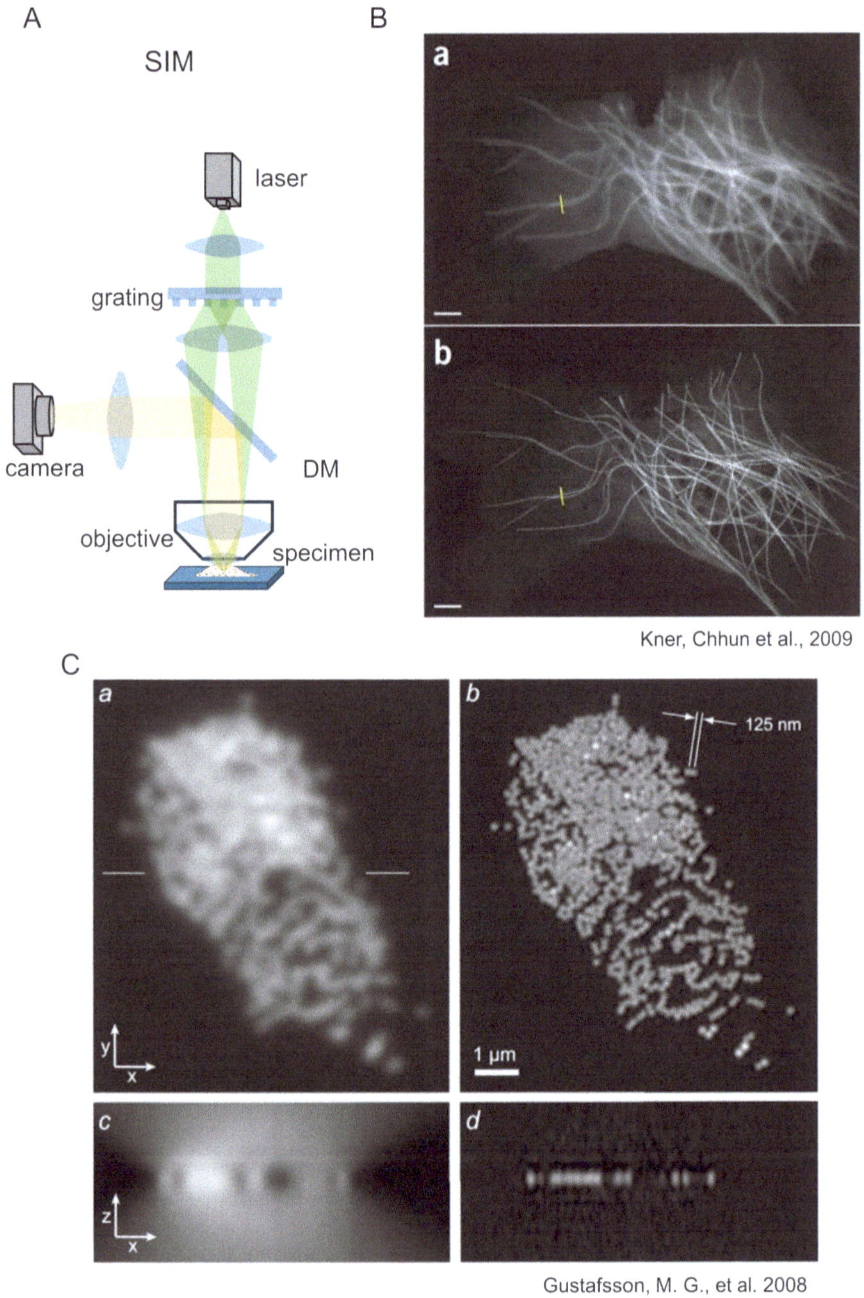

Fig. 9.1 SIM microscopy and its applications. (**a**) Schematic representation of SIM microscopy. (**b**) Comparison of TIRF microscopy (upper) and TIRF-SIM (bottom) images of the microtubule cytoskeleton in S2 cells. Scale bars, 2 µm. (**c**) Cluster of microspheres of nominal diameter 0.12 µm, imaged with conventional microscopy (left) and SIM (right). (**b**) and (**c**) are modified from Kner et al. (2009) and Gustafsson et al. (2008)

approach must also be resistant to photobleaching since it requires multiple exposures to light to capture the periodic interference pattern images.

Applications of SIM to date There are numerous permutations of SIM, each developed to meet specific imaging needs. Examples include two-dimensional SIM (2D-SIM), three-dimensional SIM (3D-SIM), total internal reflection fluorescence SIM (TIRF-SIM), lattice-SIM, and coherent SIM (Heintzmann and Huser 2017; Kner et al. 2009). Although SIM does not achieve the spatial resolution of some other super-resolution microscopy techniques, it has the advantage of requiring relatively minimal modifications to a conventional microscope or sample preparation protocols. Additionally, it can be readily applied to samples prepared for conventional wide-field fluorescence microscopy. As a result, this technique has been widely used in studies of the nanoscale structure of synapses (Hong et al. 2017; Kim et al. 2021; Paul et al. 2023). For instance, multicolor 3D-SIM has been applied to study the key components of inhibitory presynaptic terminals and postsynaptic domains, revealing that inhibitory synapses are organized into nanoscale domains composed of the AZ protein Rab3-interacting molecule (RIM), gephyrin scaffolds, and $GABA_A$Rs (Crosby et al. 2019). The number and volume of RIM clusters correlate with the number and volume of $GABA_A$ receptor clusters independently of synapse size, whereas presynaptic structures positive for the vesicular GABA transporter do not show such a correlation. A variation of SIM, known as Lattice-SIM, was used to study the fine nanoscale organization of a tripartite synapse in which Bestrophin-1 (Best1) is a GABA- and glutamate-permeable, Ca^{2+}-activated Cl^- channel, which is mainly expressed in astrocytes. Visualization of Best1 expression at the perisynaptic junctions of the tripartite synapses in the CA1 area of the mouse hippocampus using multicolor Lattice-SIM found that astrocytic Best-1 was distributed closer to the glutamatergic than GABAergic synapses (An et al. 2021). SIM has also been employed to examine the nanoscale organization of acetylcholine receptors (AChRs) at the neuromuscular junction (York and Zheng 2017), revealing that AChRs form stripes within the pretzel-shape NMJ in mice. SIM can also be used to reveal fine details of synaptic dysfunction in disease models. For example, in a mouse model of Alzheimer's disease—the APP/PS1 mice, Lattice SIM showed that Best1 distribution in tripartite synapses was significantly altered by moving away from the glutamatergic synapses while moving closer to the GABAergic synapses. These observations suggested that the localization of Best1-containing perisynapses is dynamic and is disregulated in the APP/PS1 mice (An et al. 2021). Recent developments using a Bayesian framework for SIM image reconstruction have made the technique more flexible in accommodating a wider range of illumination patterns (Lukes et al. 2014a, b; Paul et al. 2023; Pospisil et al. 2019). This has enabled imaging deeper into samples where scattering degrades high spatial frequency patterns typically used in SIM. For example, this approach has been applied to 150 μm thick slices of mouse brain tissue expressing a neuronal GFP marker, achieving a lateral resolution of 144 nm (Paul et al. 2023).

9.2.2 STORM and PALM

Stochastic optical reconstruction microscopy (STORM) and photoactivated localization microscopy (PALM) are a part of the single-molecule localization microscopy family. These approaches use stochastic switching of photoactivatable or switchable fluorophores to reduce their spatial overlap (Fig. 9.2) (Betzig et al. 1991, 2006; Rust et al. 2006; Shroff et al. 2008). Conceptually, this approach relies on a trade-off between temporal and spatial resolution; by exciting and emitting a small, sparsely distributed subpopulation of switchable fluorophores at any given time, the intensity profiles of single emitter molecules can be fitted with a Gaussian function to determine the precise localization of these molecules. In STORM, two fluorophores, called the reporter and activator, are typically conjugated to one antibody against the protein of interest. The reporter emits signals when excited by an imaging laser, while the activator is responsible for nonoverlapping activation of neighboring reporters. The efficiency of activation depends on the intensity of the activation laser and the ratio of reporters to activators. STORM achieves precise localization by temporally separating the activation of individual molecules in

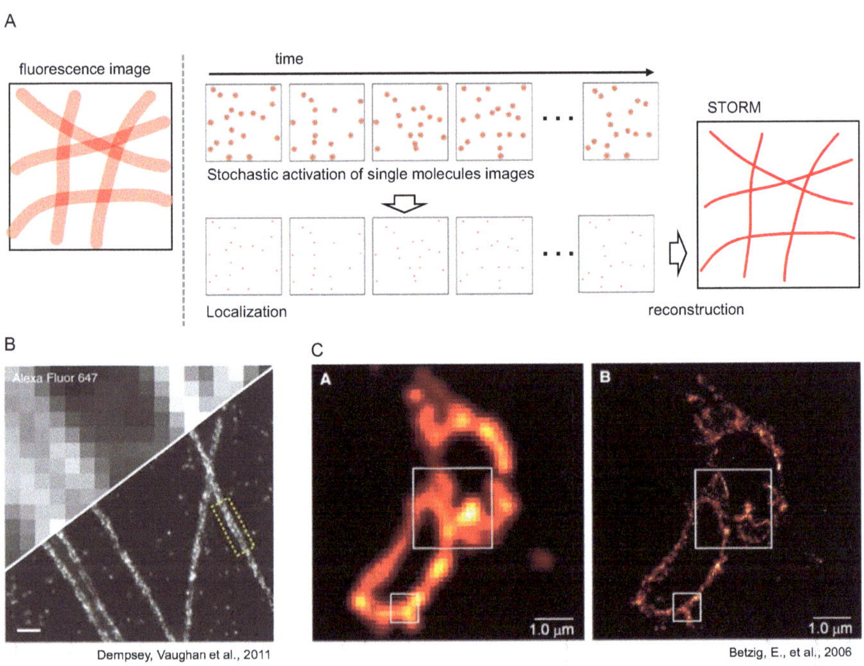

Fig. 9.2 STORM microscopy and its applications. (**a**) Schematic representation of STORM microscopy. (**b**) STORM images of microtubules immunostained and partially overlaid conventional fluorescence images (top left). Scale bars, 250 nm. (**c**) Comparative summed-molecule TIRF (left) and PALM (right) images of the same region within a cryo-prepared thin section from a COS-7 cell expressing the lysosomal transmembrane protein CD63. (**b**) and (**c**) are modified from Dempsey et al. (2011) and Betzig et al. (2006)

successive rounds of frame acquisition. In each frame, a small subset of fluorophores is localized; this information is accumulated over thousands of frames to generate a final super-resolution image. Notably, the resolution limits imposed by light diffraction on resolving multiple light sources observed in close proximity at the same time do not apply to the localization of individual emitter molecules. Thus, by reconstructing individual emitter locations over time, STORM can achieve a remarkable precision, on the order of ~10 nm (Henriques et al. 2011). Multicolor labeling is possible by using combinations of activator-reporter dye pairs (Dempsey et al. 2011; Oddone et al. 2014). The PALM technique is similar to STORM, with the key difference that STORM uses photoswitchable dyes while PALM uses photoactivatable fluorophores expressed in the cell through transfection or viral infection (Fig. 9.2) (Betzig et al. 1991, 2006; Henriques et al. 2011; Kamiyama and Huang 2012; Rust et al. 2006; Shroff et al. 2008).

Limitations of STORM and PALM Despite the versatility of STORM and PALM, limitations of these techniques stem from the possibility of reactivating the same fluorophore, leading to potential inaccuracies in quantitative measurements. STORM can also have artifacts due to numerous iterations to reconstruct a single image, requiring a complex post-acquisition reconstruction process. More critically, both techniques experience a trade-off between spatial and temporal resolution that initially limited their applicability to live-cell imaging. Finally, it is important to note that in these single-molecule localization approaches, there is a fundamental requirement to accumulate a sufficiently large number of single-molecule localizations to define the object of interest at the desired spatial resolution. This is often described by the Nyquist criterion, which defines the image resolution as a function of localization density and desired two- or three-dimensional imaging modality as $2/(\text{localization density})^{1/D}$ (Jones et al. 2011). For a typical desired spatial resolution in the 40–80 nm range, this Nyquist criterion effectively results in a temporal requirement for a 30–60 s per fully reconstructed STORM/PALM image, which initially limited the applicability of these tools to live-cell imaging. However, recent advances in STORM imaging have improved its temporal resolution to only a few seconds (Henriques et al. 2011; Jones et al. 2011).

Applications to Date STORM and PALM (and their numerous 2D and 3D implementations) have been widely used to study synaptic nanostructure. These studies have provided some of the most precise views of the synapse to date. For example, in a study using multicolor 3D-STORM, investigators showed that Piccolo and Bassoon, the two major AZ scaffold proteins, have a similar axial distribution and are organized in a highly oriented manner (Dani et al. 2010). In another study, 3D-STORM data revealed that Rab3-interacting molecule (RIM), which serves as a critical vesicle docking/priming factor, is organized in nanoclusters throughout the AZ with a preferential localization toward the AZ center (Tang et al. 2016). The data further reveal that the presynaptic RIM nanoclusters closely align with clusters of postsynaptic receptors and scaffolding proteins to form transsynaptic nanocolumns that align presynaptic release machinery with postsynaptic receptors. Supporting this finding, investigators used dual-color direct STORM (dSTORM; a variation of the STORM technique that does not require an additional activator dye to induce photoswitching) to reveal a correlation between neuroligin-1 and AMPA receptor nanodomains (Haas et al. 2018). Specifically, they found that a ~100 nm lateral shift between AMPAR nanoclusters and presynaptic nanodomains induced a significant reduction in AMPAR-mediated currents, suggesting that the high fidelity of postsynaptic responses relies on precise transsynaptic nanoalignment of pre- and postsynaptic machinery. Another study using 3D-STORM demonstrated that Munc13-1 is also a core component of the vesicle release sites at the AZ in hippocampal neurons, which organize the supramolecular complexes to recruit SNARE proteins essential for synaptic vesicle exocytosis (Sakamoto et al. 2018). Another study using dSTORM combined with TIRF revealed that voltage-gated calcium channels are recruited to the AZ through interactions with the scaffolding proteins RIM and RIM-BP. The study demonstrated that these proteins form self-assembled condensates via liquid–liquid phase separation, which plays a central role in organizing the functional components of the release sites at the AZ (McDonald et al. 2020). In summary, the broad range of recent implementations of STORM and PALM has brought important insights to our understanding of synaptic nano-organization.

9.3 Nanoscale Approaches Used to Study Dynamic Synaptic Processes

The nanostructure of central synapses is not static. In addition to vesicle release itself, which can happen within milliseconds to seconds of an AP or spontaneously at any time, vesicles also undergo a complex multistep recycling process in the interior of the synapse in preparation for the next round of release. In addition, vesicles are also transported and exchanged among neighboring synaptic boutons in an activity-dependent manner. Furthermore, many components of presynaptic docking and release machinery, for example, syntaxin-1 and Munc-18, have a large mobile fraction and translocate to/from nanoclusters at the AZ within synaptic terminals (Sieber et al. 2007; van den Bogaart et al. 2011; Willig et al. 2006). The development of pH-sensitive genetically encoded indicators has permitted analysis of individual synaptic vesicle fusion events at the plasma membrane (Kavalali and Jorgensen 2014). However, processes occurring within the interior of the nerve terminals, for example, vesicle recycling, are more difficult to access. Vesicle mobility during the recycling process is of particular interest because it may serve as a rate-limiting step in the recycling process (Shtrahman et al. 2005; Yeung et al. 2007). Extensive efforts to address these questions have inspired a new generation of super-resolution imaging techniques that also achieve high temporal resolution. We discuss a few of these techniques in this section, including stimulated emission depletion microscopy (STED) and a family of single-particle tracking (SPT) approaches, which are most widely used to study synaptic dynamics at the nanoscale.

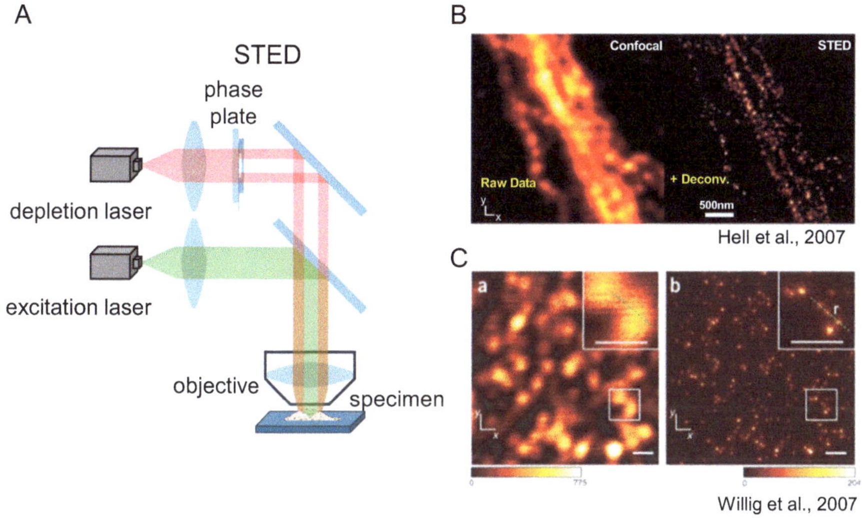

Fig. 9.3 STED microscopy and its applications. (**a**) Schematic representation of STED microscopy. (**b**) Neurofilaments in human neuroblastoma recorded in the confocal mode (left) and with STED after nonlinear deconvolution (right). (**c**) Raw data of confocal (left) and corresponding CW-STED (right) image of fluorescent 20-nm-diameter beads. Scale bars, 500 nm. (**b**) and (**c**) are modified from Hell (2007) and Willig et al. (2007)

9.3.1 Stimulated Emission Depletion (STED)

Stimulated emission depletion (STED) microscopy is a family of super-resolution microscopy techniques that leverage the reversible optical transitions between fluorescent and nonfluorescent states to inhibit fluorescence in a sample of interest in a controlled way (Fig. 9.3) (Hell 2003, 2007; Hell et al. 2004). In STED, the sample is illuminated by two laser beams: one is an excitation laser pulse with a Gaussian profile producing the typical diffraction-limited spot, and a second red-shifted depletion laser beam is focused in a donut-/toroid-like shape surrounding the first beam (Maschi and Klyachko 2015). A depletion beam is applied to return fluorophores to the ground state by stimulating emission at the periphery of the scanning focal spot. Thus, only the fluorophores in the center of the toroid continue to emit light. The size of the resulting nanometer-scale focal point scales inversely with the intensity of the depletion laser beam, and defines the resolution capacity of a STED microscope (20–50 nm in lateral dimensions in fixed and live preparations, respectively, in custom-designed STED microscopes (Hell 2003, 2007; Hell et al. 2004)). Rapid scanning of this focal spot across the sample provides a high temporal resolution (~25 frames/s). Extensive efforts to adapt and extend this approach have improved the temporal resolution of STED even beyond the video rate, and have validated its applications to 3D imaging in live cells with the axial resolution of ~100 nm (Osseforth et al. 2014; Westphal et al. 2008; Wildanger et al. 2009). Another extension of STED utilizes continuous wave beam excitation (CW STED) instead of pulse excitation. In this version, emission occurs continuously such that more photons can be captured per unit time, thus further increasing temporal and spatial resolution (Lauterbach et al. 2010; Willig et al. 2007).

Limitations of STED STED offers one of the best combinations of spatial and temporal resolution, with some limitations. One initial constraint of this imaging modality was a limited imaging area because the technique relies on scanning of the field of view. This constraint has limited the effective imaging area in early studies to a few square micrometers (Westphal et al. 2008; Willig et al. 2006). However, more recent efforts have adopted resonant scanning, which has expanded the imaging area to 50 × 50 μm^2 (Wu et al. 2015). Another limitation is photobleaching, which occurs as the sample is scanned by both the excitation beam and the high-intensity depletion beam. This effectively limits the imaging duration and quality. STED photobleaching is primarily caused by the depletion light acting upon the excited fluorophores; thus, there is a trade-off between spatial resolution and photobleaching as the power of the depletion laser increases. This limitation has been partially overcome by a modification to use resonant scanning, which extends fluorophore survival time by ~80% (Wu et al. 2015).

Applications of STED to Date Because this technique achieves not only nanoscale spatial resolution but also very high temporal resolution, STED has been used in numerous studies of synaptic nanoarchitecture and dynamics. For example, STED has been used to map the spatial organization of many AZ proteins, such as RIM, RIM-BP2, ELKS2, Basson, and Munc13-1 (Brockmann et al. 2020; Gaffield et al. 2006; Westphal et al. 2008; Wong et al. 2018), and their localization relative to the voltage-gated calcium channels (VGCCs). This is important because their relative localization determines the effective calcium diffusional spread and thus the timescale by which calcium reaches the synaptic vesicle release machinery (Bohme et al. 2019). In the NMJ synapse, dual-color STED has revealed an unexpected nonoverlapping spatial organization of scaffolding proteins Piccolo and Bassoon and their localization relative to the VGCCs, showing that clusters of Piccolo sandwiched clusters of Bassoon, which colocalized with P/Q-type VGCCs (Nishimune et al. 2016). STED also revealed nanoscale organization of the Liprin-α proteins relative to other AZ proteins, demonstrating that Liprin-α3, but not Liprin-α2, has a large component that colocalized with the AZ proteins Bassoon, RIM, Munc13, RIM-BP, and ELKS in excitatory synapses (Wong et al. 2018). This study further showed that synaptic vesicle tethering and docking were reduced in Liprin-α3-KO synapses and thus established important roles for Liprin-α3 in AZ assembly and function. STED studies further revealed the scaffolding role for RIM at the AZ, in which its Zn^{2+} finger domain binds to the C_2A domain of the priming factor Munc13, thereby inducing synaptic vesicle priming. Interestingly, this mechanism is not mediated by coupling Munc13 to other AZ proteins as previously thought, but by directly activating Munc13 (Deng et al. 2011). Multicolor time-gated modification of STED also showed an association between SNARE proteins anchored in synaptic vesicles and the scaffold protein intersectin-1 during exocytosis (Gerth et al. 2017). This association is important because intersectin-1 is thought to clear release sites following vesicle fusion by direct SH3 domain-mediated association with SNAREs and provides a mechanism for exo-endocytic coupling that allows synapses to sustain transmission during high-frequency activity. Another study using multicolor STED in the dendritic spines, both in vitro and in vivo, revealed that the structural plasticity of dendritic spines is mediated by coordinated increases in the number of precisely aligned modules of pre- and postsynaptic proteins (Hruska et al. 2018).

9.3.2 *Single-Particle Tracking (SPT)*

SPT is based on the principle that the localization of a single isolated emitter particle is not subject to the diffraction limit. The image of a single fluorescent particle is represented by the point spread function (PSF) of the optical system, thus its position can be localized with a precision severalfold better than the diffraction limit by fitting the intensity profile with a 2D-Gaussian function and determining its peak position (Fig. 9.4) (Park et al, 2022). However, Gaussian fitting is not the only way

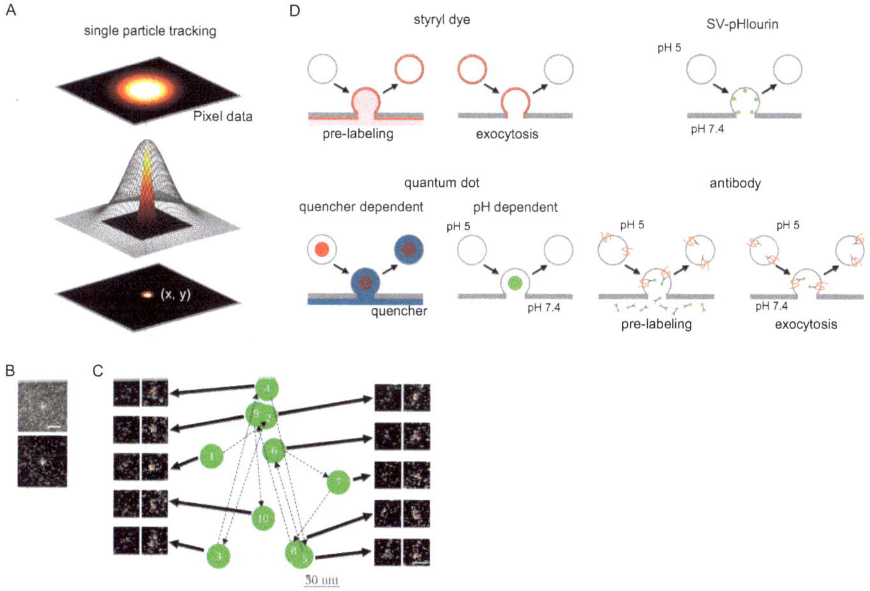

Fig. 9.4 SPT microscopy and its applications. (**a**) Schematic representation of SPT microscopy. A Gaussian fit of pixels is used to localize the fluorescence peak. (**b**) Raw images (top row) and background-subtracted images of released synaptic vesicles. Scale bars, 1 μm. (**c**) Sample map of ten subsequent vesicle fusion events within a single hippocampal bouton. For each fusion event, one frame preceding (left) and one frame following (right) the stimulus are shown. Numbers correspond to the temporal order of the events. Red dots show subpixel event localization in the corresponding frame. (Modified from Maschi and Klyachko (2017)). (**d**) Fluorescent probes for labeling and imaging secretory vesicles

to determine the single-particle location; other mathematical algorithms are even more efficient in terms of computational cost and speed. Some even make it possible to track position in real-time with nanometer accuracy (Small and Stahlheber 2014; Smith et al. 2010; Wolter et al. 2010). Typically, spatial precision offered by these techniques ranges from single nanometers (gold particles, polystyrene beads) to ~10–15 nm for quantum dots or ~20–50 nm for fluorescent molecules. In some cases, it has been possible to locate the fluorophore with 1-nanometer accuracy, an approach known as fluorescence imaging with one-nanometer accuracy (FIONA) (Park et al. 2007). In combination with additional postprocessing computational tools, such as multiple-hypothesis tracking, SPT can be used to simultaneously track multiple particles (tens to hundreds in some cases) in large and densely populated fields of view (Jaqaman et al. 2008). Another major advantage of SPT over other super-resolution approaches, particularly as they are applied to study the dynamics of synaptic processes, is its high temporal resolution. SPT temporal resolution is only limited by the signal-to-noise ratio; recent studies have achieved >100 Hz acquisition rates (Cognet et al. 2014; von Diezmann et al. 2017).

Limitations of SPT The main limitation of SPT is that it requires sparse particle labeling: the fluorescence reporters must be separated by distances greater than the diffraction limit or visualized at different time points. Therefore, only a subset of the biomolecules of interest or vesicles can be studied at a given time. Nevertheless, SPT imaging provides unprecedented access to single-molecule behavior within the natural context of living cells.

SPT Probes and Implications for Spatial and/or Temporal Resolution A wide range of nanolabels and fluorescent indicators are compatible with SPT. Initially, gold nanoparticles as small as 40 nm in size were used to visualize the dynamic processes in live cells (De Brabander et al. 1988; Geerts et al. 1987), for example, the movements of kinesin (Gelles et al. 1988). More recently, quantum dots—bright, largely nonphotobleaching reporters—have become popular (Dahan et al. 2003). Their popularity is primarily due to the fact that they improve the trade-off between spatial precision and temporal resolution because both are determined by the number of photons collected from the reporters. Utilization of SPT in applications to study synaptic processes has been greatly aided by the development of numerous fluorescent indicators. Many have been used to study vesicle trafficking modes, for example, exocytosis, endocytosis, and recycling, at the synapse (Alabi and Tsien 2012; Kavalali and Jorgensen 2014). Individual synaptic vesicles can be fluorescently labeled in nerve terminals via a number of approaches, most commonly via compensatory endocytosis and uptake of fluorescent lipophilic dyes (such as FM1-43 or SCG5) (Hoopmann et al. 2012; Peng et al. 2012; Wu et al. 2009), quantum dots (Q-dots) (Lee et al. 2012; Park et al. 2012; Zhang et al. 2009), or fluorescently labeled antibodies against specific vesicular proteins (Kavalali and Jorgensen 2014). Another type of probe can be localized in the lumen of the vesicle by genetically tagging it to a vesicular protein, such as a glutamate transporter or a vesicular SNARE protein. Sparse labeling, required for SPT, is achieved in this case via the use of pH-sensitive probes, most commonly one of the pHluorin family of indicators, which are quenched by the acidic lumen of the vesicle and become fluorescent only when exposed to neutral extracellular space, that is, upon vesicle exocytosis (Kavalali and Jorgensen 2014). In the approaches using vesicle labeling via compensatory endocytosis, the sparse labeling is typically achieved by controlling the number of release-evoking stimuli and, thus, compensatory vesicle uptake. At physiological temperatures, single AP stimulation leads to an uptake of at most one or two vesicles in each synaptic bouton in cultured hippocampal neurons (Peng et al. 2012). Adjusting the strength of the stimulus and the timing of dye application further allows specific labeling of early, late, or spontaneously endocytosed vesicles (Aravanis et al. 2003; Lemke and Klingauf 2005; Murthy and Stevens 1998; Peng et al. 2012; Ryan et al. 1997).

Applications of SPT to Date SPT has been widely used to study rapid synaptic processes, particularly vesicle exo- and endocytosis, vesicle transport, and dynamics during recycling, docking, and priming (Gormal and Meunier 2022; Gramlich and Klyachko 2019; Maschi and Klyachko 2015; Park et al. 2012). Several recent studies used SPT to quantify vesicle dynamics within and between synaptic boutons. SPT of individual vesicles labeled with lipophilic dyes (Forte et al. 2017; Gramlich and Klyachko 2017; Miki et al. 2020; Peng et al. 2012), quantum dots (Lee et al. 2012; Park et al. 2012), or fluorescently tagged nanobodies trapped in synaptic vesicles (Joensuu et al. 2016) showed that a large proportion of recycling vesicles undergo large-scale movements in central synapses during the recycling process and that their motion comprises a complex set of dynamic behaviors, including epochs of active, diffusive and stalled motion. These results confirm and extend earlier findings of extensive vesicle mobility in large ribbon-type synapses (Holt et al. 2004; Zenisek 2008; Zenisek et al. 2000). In other studies, analysis of SPT-derived vesicle trajectories using transient motion analysis algorithms found that vesicles undergo cycles of docking and undocking at the AZ (Maschi et al. 2018; Miki et al. 2020) and that vesicle-associated molecular motor myosin V controls vesicle retention at release sites in a calcium- and activity-dependent manner (Maschi et al. 2018). SPT studies using Q-dots also found a link between changes in vesicle mobility and synaptic plasticity, in which the induction of synaptic potentiation markedly increases vesicle mobility in synapses (Lee et al. 2012).

More recently, SPT has been used to detect and localize individual vesicle release events in central synapses (Maschi and Klyachko 2017; Miki et al. 2020; Tang et al. 2016). In combination with pH-sensitive genetically targeted fluorescent indicator pHluorin tagged to the vesicle lumen, SPT provided ~20 nm localization precision for single vesicle fusion events in hippocampal synapses (Maschi and Klyachko 2017, 2020). These studies revealed the presence of multiple distinct release sites within individual central synapses and demonstrated that release site properties are not uniform but form a gradient decreasing from the AZ center to its periphery. Combining SPT imaging with deep-learning computational tools has further permitted to distinguish different forms of vesicle release, including uni-vesicular

(UVR), multivesicular release (MVR), and asynchronous release (Malagon et al. 2023; Maschi et al. 2021; Maschi and Klyachko 2020; Mendonca et al. 2022). SPT approaches also uncovered the presence of two distinct populations of asynchronous events in central synapses, one of which occurs inside the AZ, and another one that occurs ectopically (Malagon et al. 2023; Mendonca et al. 2022). SPT applications thus have been instrumental in revealing some of the fundamental principles governing organization and spatiotemporal dynamics of different forms of release within individual AZs.

9.3.3 Total Internal Reflection Fluorescence (TIRF) Microscopy

Total internal reflection fluorescence (TIRF) microscopy, also known as evanescent wave microscopy, is a wide-field approach that exploits the evanescent wave resulting from a laser beam that is totally internally reflected at the interface between the coverslip and the sample (Axelrod 2001; Mattheyses et al. 2010; Schneckenburger 2005). The evanescent wave penetrates only about 100–150 nm into the sample above the coverslip and thus TIRF provides exquisite optical sectioning with very low background and minimal photobleaching (Axelrod 2001; Mattheyses et al. 2010). However, its advantages have a flip side: TIRF is only useful if the process of interest occurs very close to the coverslip; it cannot be used to image deeper into the sample. In addition, TIRF is not in itself a truly super-resolution approach. However, in combination with computational SPT tools, it can achieve some of the highest spatial and temporal resolution ever reported in studies of synaptic vesicle trafficking and release (Chen et al. 2013; Midorikawa and Sakaba 2015; Miki et al. 2024; Myeong and Klyachko 2022; Zenisek 2008; Zenisek et al. 2000).

How TIRF Microscopy Works According to Snell's law, when a light traveling through a high refractive index medium encounters an interface with a lower refractive index medium at a shallow angle, it undergoes total reflection—only a very thin layer of light, the evanescent field, propagates beyond the interface between two mediums. The TIRF microscope uses this evanescent field to illuminate fluorescent molecules very near the interface of the sample and the glass coverslip (Axelrod 2001; Mattheyses et al. 2010; Schneckenburger 2005). The depth of the evanescent field is typically on the order of 100 nm, and the evanescent wave vanishes exponentially with distance from the glass coverslip. Thus, even for cells placed directly on the coverslip, TIRF microscopy can excite fluorescent molecules only at the very thin layer beneath the plasma membrane; more remote cytosolic molecules remain in the dark. The ~100 nm depth of the illumination area is about two synaptic vesicle diameters and is comparable to the thickness of ultrathin sections used for electron microscopy. In terms of depth discrimination, this feature gives TIRF microscopy severalfold better resolution than confocal microscopy. Besides this depth resolution, minimum exposure of out-of-focus fluorophores to illumination light, and the ability to capture the entire field of view without scanning, provide TIRF microscopy with its key benefits of low background, minimal photobleaching, and higher time resolution (Schneckenburger 2005).

Applications of TIRF Microscopy to Date TIRF microscopy has been widely used to study synaptic vesicle dynamics and fusion in various model systems. In ribbon-type synaptic terminals of photoreceptors, TIRF imaging revealed high vesicle mobility and rapid docking/undocking dynamics at the plasma membrane (Chen et al. 2013; Zenisek 2008; Zenisek et al. 2000). In central mossy fiber terminals, TIRF data suggest an unexpected role of actin in restricting vesicle diffusion and determining the rate of vesicle recruitment (Midorikawa and Sakaba 2015; Miki et al. 2024). TIRF microscopy has also been applied in an artificial bilayer system to examine the role of α-Synuclein in stabilization of vesicle docking at the plasma membrane (Man et al. 2021), and in purified synaptic vesicles to monitor pH and electrical gradients (Farsi et al. 2016). The latter studies reveal distinct uptake mechanisms for glutamate and GABA in glutamatergic and GABAergic vesicles, respectively (Farsi et al. 2016).

Applying TIRF microscopy to the small central synapses has been more challenging. Small central synapses do not typically form directly onto glass surfaces, making it impossible to achieve single-vesicle resolution using conventional TIRF imaging of these structures. Recently, we reported the application of a near-TIRF modification to central synapses (Myeong and Klyachko 2022). In this modification, a small deviation from the critical angle allowed the evanescent wave to propagate deeper into the sample, ~500–800 nm, while retaining many advantages of regular TIRF, including low background, high signal-to-noise ratio, and high temporal resolution in the ~10 ms range. Under these conditions, we were able to perform robust single-vesicle release studies in hippocampal excitatory synapses, revealing that individual vesicle release events are not independent from each other, but can dynamically regulate the temporal and spatial properties of subsequent events via the engagement of a rapid astrocyte-dependent facilitation. This astrocyte-dependent form of synaptic facilitation operates at subsecond timescales in central neurons, dynamically amplifies multivesicular release, and induces release events with unique spatial organization (Myeong and Klyachko 2022). Near-TIRF was also used recently to study metabolic regulation of vesicle exo- and endocytosis in central synapses at a single-vesicle resolution (Myeong et al. 2024; Tiwari et al. 2024). This work revealed that glycolysis is dispensable for single-vesicle release, while mitochondrial inhibition shifts vesicle release closer to the AZ center and alters the efficiency of vesicle retrieval by increasing the occurrence of ultrafast endocytosis. Furthermore, measurements of vesicle release using near-TIRF revealed that hippocampal terminals can use oxidative fuels instead of glucose to maintain robust vesicle recycling and release evoked not only by single action potentials but also during high-frequency trains of activity.

9.4 3D and In Vivo Imaging

Obtaining 3D information of synaptic ultrastructure in native tissue is crucial for a deeper understanding of synaptic function. Yet numerous challenges associated with super-resolution imaging in complex brain tissue remain, which limit understanding of how functionally diverse synaptic populations in the brain organize their nanoarchitecture. Preserving native synaptic environment in living tissue and visualizing those functions in 3D requires thicker samples, which continue to pose challenges, including light scattering, increased fluorescence background, and large imaging volumes. Below, we discuss several imaging modalities that enable the acquisition of both lateral and axial information at nanoscale resolution, and recent applications of super-resolution imaging to studies in vivo.

9.4.1 Modalities Amenable to Super-Resolution 3D Imaging

Recent advancements have expanded the applicability of several super-resolution imaging approaches to 3D applications. Since the development of 3D STORM using optical astigmatism, which achieved a resolution of 20–30 nm in the lateral dimensions and 50–60 nm in the axial dimension (Huang et al. 2008a, b), it became one of the most widely used 3D super-resolution modalities. Multicolor 3D STORM was used to reveal the nanoscale organization of AZ scaffold proteins (Dani et al. 2010), to define supramolecular complexes comprising vesicle release sites (Sakamoto et al. 2018), and to identify the presence of transsynaptic nanocolumns aligning presynaptic release machinery with postsynaptic receptors (Tang et al. 2016). STED is another nanoscopy approach amenable to 3D applications. STED has been combined with single-domain nanobodies to perform 3D quantitative multiplex imaging of synaptic nano-organization in 6 μm brain cryosections. This approach revealed nanoscale diversity of functionally distinct glutamatergic synapses defined by afferent input and subcellular localization of individual synaptic connections (Jones et al. 2024). The data also show that spines receiving thalamocortical inputs are larger than spines receiving cortico-cortical inputs on both apical and basal dendrites, though the nano-organization of thalamocortical synapses appears to conform to different organizational principles and changes across the apical/basal axis. SIM can also be used for 3D imaging by generating structural illumination patterns along both the lateral and axial directions (Gustafsson et al. 2008). 3D-SIM has been used to reveal complex actin structures in neuron growth cones and observe the dynamics of cortical actin in hippocampal neurons and glia (Fiolka et al. 2012).

Another imaging modality that can aid in interrogation of the synapse ultrastructure in 3D is light sheet fluorescence microscopy (LSFM) (Bennett and Kim 2022; Gagliano et al. 2021). LSFM, also known as selective plane illumination microscopy (SPIM), is a wide-field method in which the sample is illuminated with a thin sheet of light introduced perpendicular to the detection axis. This method optically sections the sample and excites fluorophores within a thin slice around the image plane. This results in reduced fluorescence background and reduced photobleaching, which makes LSFM a useful approach for studies of 3D structures in thick samples (Hobson et al. 2022; Voie et al. 1993). Another 3D-capable imaging approach is highly inclined and laminated microscopy (HILO), which, as the name implies, utilizes a highly inclined, thin illumination beam. HILO increases image intensity and decreases background intensity, yielding a signal/background ratio about eightfold greater than epi-illumination in cultured mammalian cells (Tokunaga et al. 2008) or mouse brain tissue (Tang and Han 2018).

9.4.2 Modalities Amenable to Super-Resolution Imaging In Vivo

Key to the application of super-resolution imaging tools in vivo is the ability to acquire images within densely labeled tissue. This feat requires an optical sectioning capability, as well as techniques that label endogenous proteins without interfering with protein function and are bright, photostable, and have low phototoxicity (Willig 2022).

Thus far, the application of super-resolution imaging to in vivo systems has been limited. SIM has evolved to accommodate live cell in vivo imaging. Image acquisition is extremely fast (up to 400 frames per second) (Kner et al. 2009; Li et al. 2015), and fast reconstruction algorithms allow the real-time display of multicolor SIM data at a video rate (Markwirth et al. 2019). Recently, an in vivo SIM modality—optical-sectioning SIM (OS-SIM)—was described that improved upon the reconstruction algorithm and incorporated adaptive optics (Li et al. 2020; Turcotte et al. 2019). This method was successfully applied in vivo to image mouse cortex and zebrafish larval motor neurons (Li et al. 2020; Turcotte et al. 2019). In combination with two-photon imaging, investigators using SIM were also able to image the living mouse cortex to a depth of 120 μm, achieving a resolution of 119 nm and an acquisition rate of 3.5 images per second (Urban et al. 2018). Another promising approach for 3D super-resolution imaging is the 3D Airyscan microscope (Huff 2015; Sivaguru et al. 2018; Wang et al. 2020; Wu and Hammer 2021). The Airyscan microscope is similar to conventional confocal microscopes in that it uses a laser for single-point excitation. However, while a confocal has one detector and uses a pinhole to remove out-of-focus light, the Airyscan has 32 detectors arranged in a hexagonal array. Each detector acts as its own separate pinhole, and the entire detector array is used to calculate the origin of all emitted light. Functionally, this increases resolution up to 140 nm, nearly twice the resolution of conventional confocal imaging. Another imaging approach that has been used in in vivo settings combined concepts of STORM and SPT to develop the Quantal Synaptic Optical Reconstruction ("QuaSOR") approach (Newman et al. 2022). Investigators used QuaSOR to perform single-vesicle detection simultaneously in hundreds of synapses throughout the *Drosophila* larvae NMJ and were able to map locations of quantal vesicle release at individual synapses, including both action potential-evoked and spontaneous release (Newman et al. 2022).

9.5 New Directions

Although super-resolution imaging modalities have experienced major technological advances in the past two decades, significant challenges remain. A number of experimental and computational tools to improve both spatial and temporal resolution of several nanoscopy approaches are being developed and are briefly discussed below (Fig. 9.5).

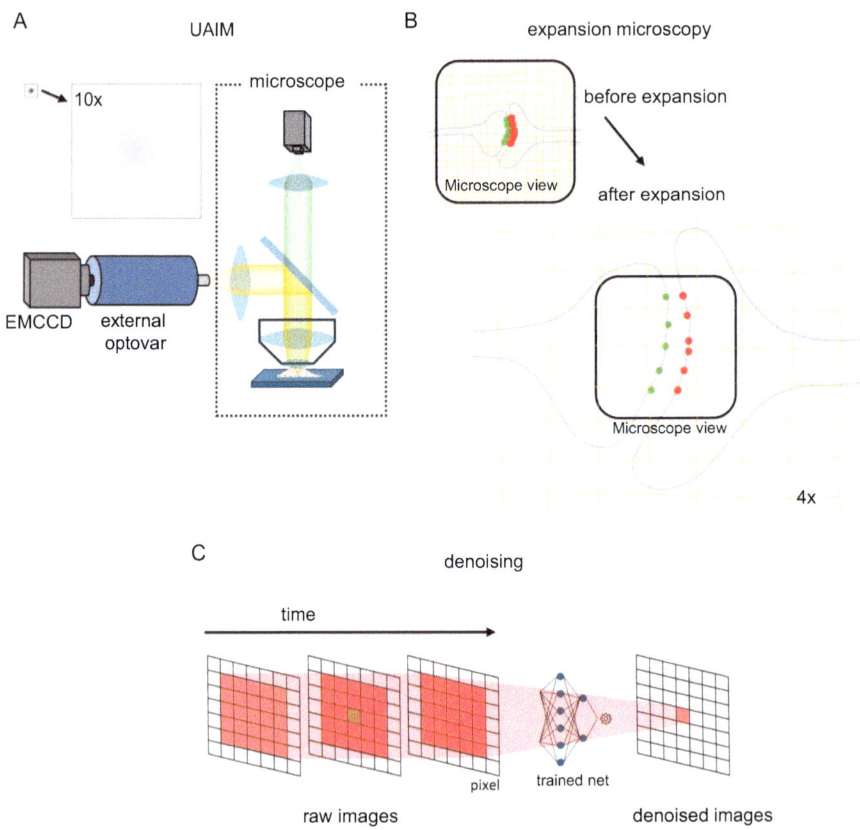

Fig. 9.5 New approaches to improve spatial and temporal resolution. Schematic representation of UAIM (**a**), expansion microscopy (**b**), and denoising approach (**c**)

9.5.1 Ultrahigh Accuracy Imaging Modality (UAIM)

Commonly used microscopy techniques use pixelated detectors to record fluorescence signals from the objects of interest collected by the microscope. These signals are then processed to produce images from which the precise localization of molecules is determined. Object localization data obtained using this approach are intrinsically limited by the pixel size of the detector array. Current detector technology places a limit on the minimal size of a pixel achievable in commercial cameras, resulting in only a small number of pixels forming a single-molecule image. UAIM seeks to address this limitation by utilizing additional optical magnification to further spread out the image across the detector array, which effectively reduces the area captured by an EMCCD's pixel (Fig. 9.5A). This, in turn, increases the number of pixels forming the image and reduces the number of photons reaching each pixel. Thus, when combined with single-photon counting approaches, UAIM improves localization accuracy by ~200% (Chao et al. 2013). Though promising for its potential to provide new insights into the nanostructure of very small subcellular spaces, to date, UAIM has not been used within the context of neurons or synapses.

9.5.2 Expansion Microscopy

Expansion microscopy employs a completely different approach to improving imaging resolution. This technique achieves super-high resolution by physically expanding the tissue sample using embedded swellable gels (Fig. 9.5B) (Chen et al. 2015; Karagiannis and Boyden 2018). This approach and its application to the study of synapses have recently been reviewed elsewhere (Karagiannis and Boyden 2018; Sneve and Piatkevich 2021). In several recent examples of applying expansion microscopy to study synaptic nanostructure, investigators were able to localize synaptic vesicle clusters, the presynaptic active zone scaffolding protein Bassoon, and the postsynaptic density identified by Homer 1, with ~25 nm resolution by expanding the samples tenfold (Chen et al. 2015; Park et al. 2019; Truckenbrodt et al. 2018, 2019). The resolution of expansion microscopy can be further enhanced by combining it with other super-resolution imaging modalities, such as 2D and 3D-STED (Gao et al. 2018). This combination of imaging approaches has been recently applied to the mouse brain, revealing a mosaic-like pattern formed by postsynaptic proteins (Delling et al. 2024; Karagiannis and Boyden 2018; Sneve and Piatkevich 2021).

9.5.3 Applications of Deep Learning to Image Denoising

Any super-resolution imaging approach faces the inherent trade-offs between resolution, imaging speed, and signal-to-noise ratio. Improving resolution in imaging data requires increasing the number of collected photons by increasing either the power of excitation or the sensitivity of the sensor. However, regardless of which approach is used to improve resolution, after a certain point, there is an inevitable increase in background noise. Denoising is a type of signal processing that attempts to extract underlying signals from noisy observations based on previous knowledge of the signal and noise (Fig. 9.5C). In other words, denoising is a statistical estimation of the most probable value based on previous statistical knowledge of the signal and the noise. Recent advancements in deep learning-based denoising algorithms have significantly reduced noise levels in live imaging data. For example, denoising algorithms were used to enhance both the spatial and temporal resolution of calcium imaging in neurons (Li et al. 2021), as well as membrane potential imaging (Eom et al. 2023; Wang et al. 2024). Applications of denoising algorithms to synaptic studies are a promising new direction, though only a few attempts have been reported to date. For example, a denoising process called forward-backward nonlinear denoising was recently used to improve resolution in single-vesicle imaging data in hippocampal synapses, permitting optical detection of three distinct modes of endocytosis (Chanaday and Kavalali 2018).

These exciting new directions, particularly if they can be applied in combination, have a strong potential to address many remaining unresolved questions in synaptic physiology by providing an unprecedented combination of spatial and temporal resolution. Yet further innovation is needed to address some of the key limitations remaining in the application of nanoscopy technology to studies of synaptic structure and function in vivo.

References

Alabi AA, Tsien RW. Synaptic vesicle pools and dynamics. Cold Spring Harb Perspect Biol. 2012;4:a013680.

Alcor D, Gouzer G, Triller A. Single-particle tracking methods for the study of membrane receptors dynamics. Eur J Neurosci. 2009;30:987–97.

An H, Koh W, Kang S, Nam MH, Lee CJ. Differential proximity of perisynaptic astrocytic Best1 at the excitatory and inhibitory tripartite synapses in APP/PS1 and MAOB-KO mice revealed by Lattice structured illumination microscopy. Exp Neurobiol. 2021;30:213–21.

Aravanis AM, Pyle JL, Tsien RW. Single synaptic vesicles fusing transiently and successively without loss of identity. Nature. 2003;423:643–7.

Arizono M, Idziak A, Quici F, Nagerl UV. Getting sharper: the brain under the spotlight of super-resolution microscopy. Trends Cell Biol. 2023;33:148–61.

Axelrod D. Total internal reflection fluorescence microscopy in cell biology. Traffic. 2001;2:764–74.

Bates M, Huang B, Zhuang X. Super-resolution microscopy by nanoscale localization of photo-switchable fluorescent probes. Curr Opin Chem Biol. 2008;12:505–14.

Bennett HC, Kim Y. Advances in studying whole mouse brain vasculature using high-resolution 3D light microscopy imaging. Neurophotonics. 2022;9:021902.

Betzig E, Trautman JK, Harris TD, Weiner JS, Kostelak RL. Breaking the diffraction barrier: optical microscopy on a nanometric scale. Science. 1991;251:1468–70.

Betzig E, Patterson GH, Sougrat R, Lindwasser OW, Olenych S, Bonifacino JS, Davidson MW, Lippincott-Schwartz J, Hess HF. Imaging intracellular fluorescent proteins at nanometer resolution. Science. 2006;313:1642–5.

Bohme MA, McCarthy AW, Grasskamp AT, Beuschel CB, Goel P, Jusyte M, Laber D, Huang S, Rey U, Petzoldt AG, et al. Rapid active zone remodeling consolidates presynaptic potentiation. Nat Commun. 2019;10:1085.

Brockmann MM, Zarebidaki F, Camacho M, Grauel MK, Trimbuch T, Sudhof TC, Rosenmund C. A trio of active zone proteins comprised of RIM-BPs, RIMs, and Munc13s Governs neurotransmitter release. Cell Rep. 2020;32:107960.

Chanaday NL, Kavalali ET. Optical detection of three modes of endocytosis at hippocampal synapses. eLife. 2018;7

Chao J, Ram S, Ward ES, Ober RJ. Ultrahigh accuracy imaging modality for super-localization microscopy. Nature methods. 2013;10:335–8.

Chen M, Van Hook MJ, Zenisek D, Thoreson WB. Properties of ribbon and non-ribbon release from rod photoreceptors revealed by visualizing individual synaptic vesicles. J Neurosci. 2013;33:2071–86.

Chen F, Tillberg PW, Boyden ES. Optical imaging. Expansion microscopy. Science. 2015;347:543–8.

Choquet D, Hosy E. AMPA receptor nanoscale dynamic organization and synaptic plasticities. Curr Opin Neurobiol. 2020;63:137–45.

Cognet L, Leduc C, Lounis B. Advances in live-cell single-particle tracking and dynamic super-resolution imaging. Curr Opin Chem Biol. 2014;20:78–85.

Cole FJ. Alle de brieven van Antoni van Leeuwenhoek. Nature. 1939;144:956–8.

Crosby KC, Gookin SE, Garcia JD, Hahm KM, Dell'Acqua ML, Smith KR. Nanoscale subsynaptic domains underlie the organization of the inhibitory synapse. Cell Rep. 2019;26:3284–3297.e3.

Dahan M, Levi S, Luccardini C, Rostaing P, Riveau B, Triller A. Diffusion dynamics of glycine receptors revealed by single-quantum dot tracking. Science. 2003;302:442–5.

Dani A, Huang B, Bergan J, Dulac C, Zhuang X. Superresolution imaging of chemical synapses in the brain. Neuron. 2010;68:843–56.

De Brabander M, Nuydens R, Geerts H, Hopkins CR. Dynamic behavior of the transferrin receptor followed in living epidermoid carcinoma (A431) cells with nanovid microscopy. Cell Motil Cytoskeleton. 1988;9:30–47.

De Robertis ED, Bennett HS. Some features of the submicroscopic morphology of synapses in frog and earthworm. J Biophys Biochem Cytol. 1955;1:47–58.

Delling JP, Bauer HF, Gerlach-Arbeiter S, Schon M, Jacob C, Wagner J, Pedro MT, Knoll B, Boeckers TM. Combined expansion and STED microscopy reveals altered fingerprints of postsynaptic nanostructure across brain regions in ASD-related SHANK3-deficiency. Mol Psychiatry. 2024;29:2997–3009.

Dempsey GT, Vaughan JC, Chen KH, Bates M, Zhuang X. Evaluation of fluorophores for optimal performance in localization-based super-resolution imaging. Nat Methods. 2011;8:1027–36.

Deng L, Kaeser PS, Xu W, Sudhof TC. RIM proteins activate vesicle priming by reversing autoinhibitory homodimerization of Munc13. Neuron. 2011;69:317–31.

Denker A, Rizzoli SO. Synaptic vesicle pools: an update. Front Synaptic Neurosci. 2010;2:135.

Eom M, Han S, Park P, Kim G, Cho ES, Sim J, Lee KH, Kim S, Tian H, Bohm UL, et al. Statistically unbiased prediction enables accurate denoising of voltage imaging data. Nat Methods. 2023;20:1581–92.

Farsi Z, Preobraschenski J, van den Bogaart G, Riedel D, Jahn R, Woehler A. Single-vesicle imaging reveals different transport mechanisms between glutamatergic and GABAergic vesicles. Science. 2016;351:981–4.

Fiolka R, Shao L, Rego EH, Davidson MW, Gustafsson MG. Time-lapse two-color 3D imaging of live cells with doubled resolution using structured illumination. Proc Natl Acad Sci U S A. 2012;109:5311–5.

Forte LA, Gramlich M, Klyachko VA. Activity-dependence of synaptic vesicle dynamics. J Neurosci. (in revision). 2017;37:10597.

Gaffield M, Rizzoli SO, Betz WJ. Mobility of synaptic vesicles in different pools in resting and stimulated frog motor nerve terminals. Neuron. 2006;51:317–25.

Gagliano G, Nelson T, Saliba N, Vargas-Hernandez S, Gustavsson AK. Light sheet illumination for 3D single-molecule super-resolution imaging of neuronal synapses. Front Synaptic Neurosci. 2021;13:761530.

Gao M, Maraspini R, Beutel O, Zehtabian A, Eickholt B, Honigmann A, Ewers H. Expansion Stimulated Emission Depletion Microscopy (ExSTED). ACS nano. 2018;12:4178–85.

Geerts H, De Brabander M, Nuydens R, Geuens S, Moeremans M, De Mey J, Hollenbeck P. Nanovid tracking: a new automatic method for the study of mobility in living cells based on colloidal gold and video microscopy. Biophys J. 1987;52:775–82.

Gelles J, Schnapp BJ, Sheetz MP. Tracking kinesin-driven movements with nanometre-scale precision. Nature. 1988;331:450–3.

Gerth F, Japel M, Pechstein A, Kochlamazashvili G, Lehmann M, Puchkov D, Onofri F, Benfenati F, Nikonenko AG, Fredrich K, et al. Intersectin associates with synapsin and regulates its nanoscale localization and function. Proc Natl Acad Sci U S A. 2017;114:12057–62.

Gormal RS, Meunier FA. Nanoscale organization of the pre-synapse: tracking the neurotransmitter release machinery. Curr Opin Neurobiol. 2022;75:102576.

Gramlich MW, Klyachko VA. Actin/Myosin-V- and activity-dependent inter-synaptic vesicle exchange in central neurons. Cell Rep. 2017;18:2096–104.

Gramlich MW, Klyachko VA. Nanoscale organization of vesicle release at central synapses. Trends Neurosci. 2019;42:425–37.

Gustafsson MG. Extended resolution fluorescence microscopy. Curr Opin Struct Biol. 1999;9:627–34.

Gustafsson MG. Surpassing the lateral resolution limit by a factor of two using structured illumination microscopy. J Microsc. 2000;198:82–7.

Gustafsson MG. Nonlinear structured-illumination microscopy: wide-field fluorescence imaging with theoretically unlimited resolution. Proc Natl Acad Sci U S A. 2005;102:13081–6.

Gustafsson MG, Shao L, Carlton PM, Wang CJ, Golubovskaya IN, Cande WZ, Agard DA, Sedat JW. Three-dimensional resolution doubling in wide-field fluorescence microscopy by structured illumination. Biophys J. 2008;94:4957–70.

Haas KT, Compans B, Letellier M, Bartol TM, Grillo-Bosch D, Sejnowski TJ, Sainlos M, Choquet D, Thoumine O, Hosy E. Pre-post synaptic alignment through neuroligin-1 tunes synaptic transmission efficiency. eLife. 2018;7

Heintzmann R, Huser T. Super-resolution structured illumination microscopy. Chem Rev. 2017;117:13890–908.

Heintzmann R, Jovin TM, Cremer C. Saturated patterned excitation microscopy–a concept for optical resolution improvement. J Opt Soc Am A Opt Image Sci Vis. 2002;19:1599–609.

Hell SW. Toward fluorescence nanoscopy. Nat Biotechnol. 2003;21:1347–55.

Hell SW. Far-field optical nanoscopy. Science. 2007;316:1153–8.

Hell SW, Dyba M, Jakobs S. Concepts for nanoscale resolution in fluorescence microscopy. Curr Opin Neurobiol. 2004;14:599–609.

Henriques R, Griffiths C, Hesper Rego E, Mhlanga MM. PALM and STORM: unlocking live-cell super-resolution. Biopolymers. 2011;95:322–31.

Hobson CM, Guo M, Vishwasrao HD, Wu Y, Shroff H, Chew TL. Practical considerations for quantitative light sheet fluorescence microscopy. Nat Methods. 2022;19:1538–49.

Holt M, Cooke A, Neef A, Lagnado L. High mobility of vesicles supports continuous exocytosis at a ribbon synapse. Curr Biol. 2004;14:173–83.

Hong S, Wilton DK, Stevens B, Richardson DS. Structured illumination microscopy for the investigation of synaptic structure and function. Methods Mol Biol. 2017;1538:155–67.

Hoopmann P, Rizzoli SO, Betz WJ. Imaging synaptic vesicle recycling by staining and destaining vesicles with FM dyes. Cold Spring Harb Protoc. 2012;2012:77–83.

Hruska M, Henderson N, Le Marchand SJ, Jafri H, Dalva MB. Synaptic nanomodules underlie the organization and plasticity of spine synapses. Nat Neurosci. 2018;21:671–82.

Huang B, Jones SA, Brandenburg B, Zhuang X. Whole-cell 3D STORM reveals interactions between cellular structures with nanometer-scale resolution. Nat Methods. 2008a;5:1047–52.

Huang B, Wang W, Bates M, Zhuang X. Three-dimensional super-resolution imaging by stochastic optical reconstruction microscopy. Science. 2008b;319:810–3.

Huang B, Bates M, Zhuang X. Super-resolution fluorescence microscopy. Annu Rev Biochem. 2009;78:993–1016.

Huff J. The Airyscan detector from ZEISS: confocal imaging with improved signal-to-noise ratio and super-resolution. Nat Methods. 2015;12:i–ii.

Jaqaman K, Loerke D, Mettlen M, Kuwata H, Grinstein S, Schmid SL, Danuser G. Robust single-particle tracking in live-cell time-lapse sequences. Nat Methods. 2008;5:695–702.

Joensuu M, Padmanabhan P, Durisic N, Bademosi AT, Cooper-Williams E, Morrow IC, Harper CB, Jung W, Parton RG, Goodhill GJ, et al. Subdiffractional tracking of internalized molecules reveals heterogeneous motion states of synaptic vesicles. J Cell Biol. 2016;215:277–92.

Jones SA, Shim SH, He J, Zhuang X. Fast, three-dimensional super-resolution imaging of live cells. Nat Methods. 2011;8:499–508.

Jones G, Akter Y, Shifflett V, Hruska M. Nanoscale analysis of functionally diverse glutamatergic synapses in the neocortex reveals input and layer-specific organization. 2024. bioRxiv: the preprint server for biology.

Kaeser PS, Regehr WG. Molecular mechanisms for synchronous, asynchronous, and spontaneous neurotransmitter release. Annu Rev Physiol. 2013;76:333–63.

Kamiyama D, Huang B. Development in the STORM. Dev Cell. 2012;23:1103–10.

Karagiannis ED, Boyden ES. Expansion microscopy: development and neuroscience applications. Curr Opin Neurobiol. 2018;50:56–63.

Kavalali ET. Multiple vesicle recycling pathways in central synapses and their impact on neurotransmission. J Physiol. 2007;585:669–79.

Kavalali ET, Jorgensen EM. Visualizing presynaptic function. Nat Neurosci. 2014;17:10–6.

Kim GT, Bahn S, Kim N, Choi JH, Kim JS, Rah JC. Efficient and accurate synapse detection with selective structured illumination microscopy on the putative regions of interest of ultrathin serial sections. Front Neuroanat. 2021;15:759816.

Kner P, Chhun BB, Griffis ER, Winoto L, Gustafsson MG. Super-resolution video microscopy of live cells by structured illumination. Nat Methods. 2009;6:339–42.

Lauterbach MA, Keller J, Schonle A, Kamin D, Westphal V, Rizzoli SO, Hell SW. Comparing video-rate STED nanoscopy and confocal microscopy of living neurons. J Biophotonics. 2010;3:417–24.

Lee S, Jung KJ, Jung HS, Chang S. Dynamics of multiple trafficking behaviors of individual synaptic vesicles revealed by quantum-dot based presynaptic probe. PloS One. 2012;7:e38045.

Lemke EA, Klingauf J. Single synaptic vesicle tracking in individual hippocampal boutons at rest and during synaptic activity. J Neurosci. 2005;25:11034–44.

Li D, Shao L, Chen BC, Zhang X, Zhang M, Moses B, Milkie DE, Beach JR, Hammer JA 3rd, Pasham M, et al. ADVANCED IMAGING. Extended-resolution structured illumination imaging of endocytic and cytoskeletal dynamics. Science. 2015;349:aab3500.

Li Z, Zhang Q, Chou S-W, Newman Z, Turcotte R, Natan R, Dai Q, Isacoff EY, Ji N. Fast wide-field imaging of neuronal structure and function with optical sectioning in vivo. Sci Adv. 2020;6:eaaz3870.

Li X, Zhang G, Wu J, Zhang Y, Zhao Z, Lin X, Qiao H, Xie H, Wang H, Fang L, et al. Reinforcing neuron extraction and spike inference in calcium imaging using deep self-supervised denoising. Nat Methods. 2021;18:1395–400.

Lucosz W. Optical systems with resolving powers exceeding the classical limit. J Opt Soc Am. 1966;56:1463–71.

Lukes T, Hagen GM, Křížek P, Švindrych Z, Fliegel K, Klima M. Comparison of image reconstruction methods for structured illumination microscopy. SPIE Proc. 2014a;9129:607–19.

Lukes T, Krizek P, Svindrych Z, Benda J, Ovesny M, Fliegel K, Klima M, Hagen GM. Three-dimensional super-resolution structured illumination microscopy with maximum a posteriori probability image estimation. Opt Express. 2014b;22:29805–17.

Maglione M, Sigrist SJ. Seeing the forest tree by tree: super-resolution light microscopy meets the neurosciences. Nat Neurosci. 2013;16:790–7.

Malagon G, Myeong J, Klyachko VA. Two forms of asynchronous release with distinctive spatiotemporal dynamics in central synapses. eLife. 2023;12

Man WK, Tahirbegi B, Vrettas MD, Preet S, Ying L, Vendruscolo M, De Simone A, Fusco G. The docking of synaptic vesicles on the presynaptic membrane induced by alpha-synuclein is modulated by lipid composition. Nat Commun. 2021;12:927.

Markwirth A, Lachetta M, Monkemoller V, Heintzmann R, Hubner W, Huser T, Muller M. Video-rate multi-color structured illumination microscopy with simultaneous real-time reconstruction. Nat Commun. 2019;10:4315.

Maschi D, Klyachko VA. A nanoscale resolution view on synaptic vesicle dynamics. Synapse. 2015;69:256–67.

Maschi D, Klyachko VA. Spatiotemporal regulation of synaptic vesicle fusion sites in central synapses. Neuron. 2017;94:65–73 e63.

Maschi D, Klyachko VA. Spatiotemporal dynamics of multi-vesicular release is determined by heterogeneity of release sites within central synapses. elife. 2020;9

Maschi D, Gramlich MW, Klyachko VA. Myosin V functions as a vesicle tether at the plasma membrane to control neurotransmitter release in central synapses. eLife. 2018;7

Maschi D, Gramlich MW, Klyachko VA. Myosin V regulates spatial localization of different forms of neurotransmitter release in central synapses. Front Synaptic Neurosci. 2021;13:650334.

Matthews EK. The presynaptic effects of quaternary ammonium compounds on the acetylcholine metabolism of a sympathetic ganglion. Br J Pharmacol Chemother. 1966;26:552–66.

Mattheyses AL, Simon SM, Rappoport JZ. Imaging with total internal reflection fluorescence microscopy for the cell biologist. J Cell Sci. 2010;123:3621–8.

McDonald NA, Fetter RD, Shen K. Assembly of synaptic active zones requires phase separation of scaffold molecules. Nature. 2020;588:454–8.

Mendonca PRF, Tagliatti E, Langley H, Kotzadimitriou D, Zamora-Chimal CG, Timofeeva Y, Volynski KE. Asynchronous glutamate release is enhanced in low release efficacy synapses and dispersed across the active zone. Nat Commun. 2022;13:3497.

Midorikawa M, Sakaba T. Imaging Exocytosis of Single Synaptic Vesicles at a Fast CNS Presynaptic Terminal. Neuron. 2015;88:492–8.

Miki T, Midorikawa M, Sakaba T. Direct imaging of rapid tethering of synaptic vesicles accompanying exocytosis at a fast central synapse. Proc Natl Acad Sci U S A. 2020;117:14493–502.

Miki T, Okamoto Y, Ueno-Umegai M, Toyofuku R, Hattori S, Sakaba T. Single-vesicle imaging reveals actin-dependent spatial restriction of vesicles at the active zone, essential for sustained transmission. Proc Natl Acad Sci U S A. 2024;121:e2402152121.

Murthy VN, Stevens CF. Synaptic vesicles retain their identity through the endocytic cycle. Nature. 1998;392:497–501.

Myeong J, Klyachko VA. Rapid astrocyte-dependent facilitation amplifies multi-vesicular release in hippocampal synapses. Cell Rep. 2022;41:111820.

Myeong J, Stunault MI, Klyachko VA, Ashrafi G. Metabolic regulation of single synaptic vesicle exo- and endocytosis in hippocampal synapses. Cell Rep. 2024;43:114218.

Nair D, Hosy E, Petersen JD, Constals A, Giannone G, Choquet D, Sibarita JB. Super-resolution imaging reveals that AMPA receptors inside synapses are dynamically organized in nanodomains regulated by PSD95. J Neurosci. 2013;33:13204–24.

Nelson AJ, Hess ST. Localization microscopy: mapping cellular dynamics with single molecules. J Microsc. 2014;254:1–8.

Newman ZL, Bakshinskaya D, Schultz R, Kenny SJ, Moon S, Aghi K, Stanley C, Marnani N, Li R, Bleier J, et al. Determinants of synapse diversity revealed by super-resolution quantal transmission and active zone imaging. Nat Commun. 2022;13:229.

Nishimune H, Badawi Y, Mori S, Shigemoto K. Dual-color STED microscopy reveals a sandwich structure of Bassoon and Piccolo in active zones of adult and aged mice. Sci Rep. 2016;6:27935.

Oddone A, Vilanova IV, Tam J, Lakadamyali M. Super-resolution imaging with stochastic single-molecule localization: concepts, technical developments, and biological applications. Microsc Res Tech. 2014;77:502–9.

Osseforth C, Moffitt JR, Schermelleh L, Michaelis J. Simultaneous dual-color 3D STED microscopy. Opt Express. 2014;22:7028–39.

Ottersen OP, Landsend AS. Organization of glutamate receptors at the synapse. Eur J Neurosci. 1997;9:2219–24.

Palay SL. Synapses in the central nervous system. J Biophys Biochem Cytol. 1956;2:193–202.

Park H, Toprak E, Selvin PR. Single-molecule fluorescence to study molecular motors. Q Rev Biophys. 2007;40:87–111.

Park H, Li Y, Tsien RW. Influence of synaptic vesicle position on release probability and exocytotic fusion mode. Science. 2012;335:1362–6.

Park HE, Choi D, Park JS, Sim C, Park S, Kang S, Yim H, Lee M, Kim J, Pac J, et al. Scalable and isotropic expansion of tissues with simply tunable expansion ratio. Adv Sci. 2019;6:1901673.

Park C, Jung S, Park H. Single vesicle tracking for studying synaptic vesicle dynamics in small central synapses. Curr Opin Neurobiol. 2022;76:102596.

Paul TC, Johnson KA, Hagen GM. Super-resolution imaging of neuronal structures with structured illumination microscopy. Bioengineering (Basel). 2023;10

Peng A, Rotman Z, Deng PY, Klyachko VA. Differential motion dynamics of synaptic vesicles undergoing spontaneous and activity-evoked endocytosis. Neuron. 2012;73:1108–15.

Pospisil J, Lukes T, Bendesky J, Fliegel K, Spendier K, Hagen GM. Imaging tissues and cells beyond the diffraction limit with structured illumination microscopy and Bayesian image reconstruction. Gigascience. 2019;8

Rudolph S, Tsai MC, von Gersdorff H, Wadiche JI. The ubiquitous nature of multivesicular release. Trends Neurosci. 2015;38:428–38.

Rust MJ, Bates M, Zhuang X. Sub-diffraction-limit imaging by stochastic optical reconstruction microscopy (STORM). Nat Methods. 2006;3:793–5.

Ryan TA, Reuter H, Smith SJ. Optical detection of a quantal presynaptic membrane turnover. Nature. 1997;388:478–82.

Sakamoto H, Ariyoshi T, Kimpara N, Sugao K, Taiko I, Takikawa K, Asanuma D, Namiki S, Hirose K. Synaptic weight set by Munc13-1 supramolecular assemblies. Nat Neurosci. 2018;21:41–9.

Schermelleh L, Ferrand A, Huser T, Eggeling C, Sauer M, Biehlmaier O, Drummen GPC. Super-resolution microscopy demystified. Nat Cell Biol. 2019;21:72–84.

Schikorski T, Stevens CF. Quantitative ultrastructural analysis of hippocampal excitatory synapses. J Neurosci. 1997;17:5858–67.

Schikorski T, Stevens CF. Quantitative fine-structural analysis of olfactory cortical synapses. PNAS. 1999;96:4107–12.

Schneckenburger H. Total internal reflection fluorescence microscopy: technical innovations and novel applications. Curr Opin Biotechnol. 2005;16:13–8.

Shroff H, White H, Betzig E. Photoactivated localization microscopy (PALM) of adhesion complexes. Curr Protoc Cell Biol. 2008;Chapter 4:Unit 4 21.

Shtrahman M, Yeung C, Nauen DW, Bi GQ, Wu XL. Probing vesicle dynamics in single hippocampal synapses. Biophys J. 2005;89:3615–27.

Sieber JJ, Willig KI, Kutzner C, Gerding-Reimers C, Harke B, Donnert G, Rammner B, Eggeling C, Hell SW, Grubmuller H, et al. Anatomy and dynamics of a supramolecular membrane protein cluster. Science. 2007;317:1072–6.

Sigal YM, Zhou R, Zhuang X. Visualizing and discovering cellular structures with super-resolution microscopy. Science. 2018;361:880–7.

Sigrist SJ, Sabatini BL. Optical super-resolution microscopy in neurobiology. Curr Opin Neurobiol. 2011;22:86–93.

Siksou L, Triller A, Marty S. An emerging view of presynaptic structure from electron microscopic studies. J Neurochem. 2009;108:1336–42.

Sivaguru M, Urban MA, Fried G, Wesseln CJ, Mander L, Punyasena SW. Comparative performance of airyscan and structured illumination superresolution microscopy in the study of the surface texture and 3D shape of pollen. Microsc Res Tech. 2018;81:101–14.

Small A, Stahlheber S. Fluorophore localization algorithms for super-resolution microscopy. Nat Methods. 2014;11:267–79.

Smith CS, Joseph N, Rieger B, Lidke KA. Fast, single-molecule localization that achieves theoretically minimum uncertainty. Nat Methods. 2010;7:373–5.

Sneve MA, Piatkevich KD. Towards a comprehensive optical connectome at single synapse resolution via expansion microscopy. Front Synaptic Neurosci. 2021;13:754814.

Takumi Y, Ramirez-Leon V, Laake P, Rinvik E, Ottersen OP. Different modes of expression of AMPA and NMDA receptors in hippocampal synapses. Nat Neurosci. 1999;2:618–24.

Tang J, Han KY. Extended field-of-view single-molecule imaging by highly inclined swept illumination. Optica. 2018;5:1063–9.

Tang AH, Chen H, Li TP, Metzbower SR, MacGillavry HD, Blanpied TA. A trans-synaptic nanocolumn aligns neurotransmitter release to receptors. Nature. 2016;536:210–4.

Taoufik E, Kouroupi G, Zygogianni O, Matsas R. Synaptic dysfunction in neurodegenerative and neurodevelopmental diseases: an overview of induced pluripotent stem-cell-based disease models. Open Biol. 2018;8

Thompson RE, Larson DR, Webb WW. Precise nanometer localization analysis for individual fluorescent probes. Biophys J. 2002;82:2775–83.

Thomson A, Deuchars J. Temporal and spatial properties of local circuits in neocortex. Trends Neurosci. 1994;17:119–26.

Tiwari A, Myeong J, Hashemiaghdam A, Zhang H, Niu X, Laramie MA, Stunault MI, Sponagel J, Patti G, Shriver L, et al. Mitochondrial pyruvate transport regulates presynaptic metabolism and neurotransmission. bioRxiv : the preprint server for biology. 2024.

Tokunaga M, Imamoto N, Sakata-Sogawa K. Highly inclined thin illumination enables clear single-molecule imaging in cells. Nat Methods. 2008;5:159–61.

Toomre D, Bewersdorf J. A new wave of cellular imaging. Ann Rev Cell Dev Biol. 2010;26:285–314.

Truckenbrodt S, Maidorn M, Crzan D, Wildhagen H, Kabatas S, Rizzoli SO. X10 expansion microscopy enables 25-nm resolution on conventional microscopes. EMBO Rep. 2018;19

Truckenbrodt S, Sommer C, Rizzoli SO, Danzl JG. A practical guide to optimization in X10 expansion microscopy. Nat Protoc. 2019;14:832–63.

Turcotte R, Liang Y, Tanimoto M, Zhang Q, Li Z, Koyama M, Betzig E, Ji N. Dynamic super-resolution structured illumination imaging in the living brain. Proc Natl Acad Sci U S A. 2019;116:9586–91.

Urban BE, Xiao L, Dong B, Chen S, Kozorovitskiy Y, Zhang HF. Imaging neuronal structure dynamics using 2-photon super-resolution patterned excitation reconstruction microscopy. J Biophotonics. 2018;11:e201700171.

Valtschanoff JG, Weinberg RJ. Laminar organization of the NMDA receptor complex within the postsynaptic density. J Neurosci. 2001;21:1211–7.

van den Bogaart G, Meyenberg K, Risselada HJ, Amin H, Willig KI, Hubrich BE, Dier M, Hell SW, Grubmuller H, Diederichsen U, et al. Membrane protein sequestering by ionic protein-lipid interactions. Nature. 2011;479:552–5.

Voie AH, Burns DH, Spelman FA. Orthogonal-plane fluorescence optical sectioning: three dimensional imaging of macroscopic biological specimens. J Microsc. 1993;170:229–36.

von Diezmann L, Shechtman Y, Moerner WE. Three-dimensional localization of single molecules for super-resolution imaging and single-particle tracking. Chem Rev. 2017;117:7244–75.

Wang H, Balice-Gordon R. Editorial: synaptic diseases: from biology to potential therapy. Front Synaptic Neurosci. 2022;14:846099.
Wang B, Yao L, Jing Y, Fei Y, Bai Q, Mi L, Ma J. Multicomposite super-resolution microscopy: enhanced airyscan resolution with radial fluctuation and sample expansions. J Biophotonics. 2020;13:e2419.
Wang B, Ma T, Chen T, Nguyen T, Crouse E, Fleming SJ, Walker AS, Valakh V, Nehme R, Miller EW, et al. Robust self-supervised denoising of voltage imaging data using CellMincer. bioRxiv. 2024;
Westphal V, Rizzoli SO, Lauterbach MA, Kamin D, Jahn R, Hell SW. Video-rate far-field optical nanoscopy dissects synaptic vesicle movement. Science. 2008;320:246–9.
Wildanger D, Medda R, Kastrup L, Hell SW. A compact STED microscope providing 3D nanoscale resolution. J Microsc. 2009;236:35–43.
Willig KI. In vivo super-resolution of the brain - How to visualize the hidden nanoplasticity? iScience. 2022;25:104961.
Willig KI, Rizzoli SO, Westphal V, Jahn R, Hell SW. STED microscopy reveals that synaptotagmin remains clustered after synaptic vesicle exocytosis. Nature. 2006;440:935–9.
Willig KI, Harke B, Medda R, Hell SW. STED microscopy with continuous wave beams. Nat Methods. 2007;4:915–8.
Wolter S, Schuttpelz M, Tscherepanow M, S VDL, Heilemann M, Sauer M. Real-time computation of subdiffraction-resolution fluorescence images. J Microsc. 2010;237:12–22.
Wong MY, Liu C, Wang SSH, Roquas ACF, Fowler SC, Kaeser PS. Liprin-alpha3 controls vesicle docking and exocytosis at the active zone of hippocampal synapses. Proc Natl Acad Sci U S A. 2018;115:2234–9.
Wu X, Hammer JA. ZEISS airyscan: optimizing usage for fast, gentle, super-resolution imaging. Methods Mol Biol. 2021;2304:111–30.
Wu Y, Yeh FL, Mao F, Chapman ER. Biophysical characterization of styryl dye-membrane interactions. Biophys J. 2009;97:101–9.
Wu Y, Wu X, Lu R, Zhang J, Toro L, Stefani E. Resonant scanning with large field of view reduces photobleaching and enhances fluorescence yield in STED microscopy. Sci Rep. 2015;5:14766.
Yeung C, Shtrahman M, Wu XL. Stick-and-diffuse and caged diffusion: a comparison of two models of synaptic vesicle dynamics. Biophys J. 2007;92:2271–80.
York AL, Zheng JQ. Super-resolution microscopy reveals a nanoscale organization of acetylcholine receptors for trans-synaptic alignment at neuromuscular synapses. eNeuro. 2017;4:ENEURO.0232.
Zenisek D. Vesicle association and exocytosis at ribbon and extraribbon sites in retinal bipolar cell presynaptic terminals. Proc Natl Acad Sci U S A. 2008;105:4922–7.
Zenisek D, Steyer JA, Almers W. Transport, capture and exocytosis of single synaptic vesicles at active zones. Nature. 2000;406:849–54.
Zhang Q, Li Y, Tsien RW. The dynamic control of kiss-and-run and vesicular reuse probed with single nanoparticles. Science. 2009;323:1448–53.

Chapter 10
Visualizing Postsynaptic Density in Excitatory Synapses with Electron Tomography

Rong Sun and Qiangjun Zhou

Abstract Electron tomography (ET) has emerged as a critical tool for visualizing the three-dimensional ultrastructure of biological specimens at nanometer resolution. This chapter focuses on the application of ET in studying synaptic ultrastructure, in particular postsynaptic density, providing a detailed overview of the techniques and methodologies used to achieve high-resolution three-dimensional reconstruction of synapses. Beginning with an introduction to electron tomography, this chapter delves into the principles of electron tomography, including sample preparation, data collection, and image processing. Special emphasis is placed on the application of cryogenic electron tomography, which allows for the visualization of biological samples in their near-native state. This chapter also reviews the historical context of postsynaptic ultrastructure studies using conventional electron microscopy and explores the significant insights gained using electron tomography, particularly in understanding the nanoscale organization and structural complexity of excitatory synapses. This chapter concludes by discussing the future potential of electron tomography in advancing our knowledge of synaptic biology and its implications for neuroscience.

Keywords Electron tomography · Transmission electron microscopy · Synapse · Postsynaptic density · Nanoscale organization · Trans-synaptic alignment

R. Sun · Q. Zhou (✉)
Vanderbilt Brain Institute, Vanderbilt University, Nashville, TN, USA

Department of Cell and Developmental Biology, Vanderbilt University, Nashville, TN, USA

Center for Structural Biology, Vanderbilt University, Nashville, TN, USA

Vanderbilt Kennedy Center, Vanderbilt University, Nashville, TN, USA
e-mail: qiangjun.zhou@vanderbilt.edu

E. T. Kavalali (ed.), *Nano-organization of the Synapse*, Advances in Neurobiology 48,
https://doi.org/10.1007/978-3-032-12594-1_10

10.1 Introduction

Synapses are the fundamental units of communication in the nervous system, where the transmission of signals between neurons occurs. Understanding the ultrastructure of synapses is crucial for deciphering the mechanisms underlying synaptic function, plasticity, and overall neural network activity. Traditional methods for studying synaptic structures, such as light microscopy, have been limited by resolution, preventing the detailed visualization of subcellular components at molecular resolution (Dani et al. 2010). Electron microscopy (EM) techniques, particularly transmission electron microscopy (TEM), have played a pivotal role in advancing our understanding of synaptic architecture by providing high-resolution images of synaptic components (Chen et al. 2008; Palay 1956; Sun et al. 2025; Tao et al. 2018a).

However, the two-dimensional (2D) nature of TEM images limits the ability to fully appreciate the three-dimensional (3D) organization of synaptic structures (Palay 1956). Electron tomography (ET) overcomes this limitation by enabling the reconstruction of 3D volumes from a series of 2D projections, offering unprecedented insights into the complex architecture of synapses (Baumeister 2002; Baumeister et al. 1999; Frank 2005; Hurbain and Sachse 2011). Furthermore, the advent of cryogenic electron tomography (cryo-ET) has allowed researchers to visualize synaptic structures in their near-native, vitrified state, preserving the delicate ultrastructure and enabling the study of cellular ultrastructure and macromolecules (e.g., protein complexes) within their cellular context (Fernandez-Busnadiego et al. 2010b; Liu et al. 2020; Sun et al. 2019, 2025; Tao et al. 2018a).

This chapter provides a comprehensive overview of electron tomography, with a focus on its application to synaptic ultrastructure studies. It begins with a discussion of the principles of TEM and its historical development, followed by an in-depth exploration of ET techniques. This chapter then examines the application of ET to synaptic studies, highlighting key discoveries and discussing the future directions for this powerful imaging technique in neuroscience.

10.2 Electron Tomography

10.2.1 Transmission Electron Microscopy

TEM is the foundation upon which electron tomography is built. It has been a cornerstone of biological imaging for decades, offering unparalleled resolution and the ability to visualize the ultrastructure of cells, tissues, and macromolecules (Baumeister 2002; Koster and Klumperman 2003; McIntosh 2001). To fully appreciate the advancements made possible by ET, it is essential to understand the principles and applications of TEM.

10.2.1.1 Historical Development

The development of TEM can be traced back to the early twentieth century, driven by the limitations of optical microscopy. The resolution of optical microscopes is limited by the wavelength of visible light, which restricts the ability to visualize structures smaller than ~200 nm. This limitation prompted researchers to explore alternative imaging techniques that could surpass the diffraction limit of visible light.

In the 1920s, Louis de Broglie proposed the wave-particle duality of electrons, suggesting that electrons could exhibit wave-like behavior with much shorter wavelengths (Scherzer 1949) (e.g., the theoretical wavelength at 200 keV is 0.00251 nm) than visible light (~400–700 nm). This concept laid the theoretical foundation for electron microscopy. In 1931, German physicists Ernst Ruska and Max Knoll constructed the first electron microscope, which used a beam of electrons instead of light for imaging. Ruska's work on the development of TEM earned him the Nobel Prize in Physics in 1986 (Ruska 1987).

The initial designs of TEM were rudimentary, with limited resolution and imaging capabilities. However, advancements in electron optics, vacuum technology, and specimen preparation techniques over subsequent decades led to substantial improvements in TEM resolution and image quality. By the 1950s, TEM had become an essential tool in the biological sciences, enabling researchers to visualize subcellular structures with unprecedented detail and playing a pivotal role in the emergence of cell biology as a scientific field. Through TEM, scientists could examine cellular organelles such as the nucleus, mitochondria, endoplasmic reticulum, and Golgi apparatus, shedding light on their structures and functions (Albert et al. 2020; Bykov et al. 2017; Koster and Klumperman 2003; Mahamid et al. 2016; McIntosh 2001). TEM has also been instrumental in studying key cellular processes, such as cell division, apoptosis, autophagy, and intracellular transport mechanisms (Bieber et al. 2022; Jordan et al. 2018; Navarro et al. 2022; Tamaki et al. 2003). Its high-resolution capabilities have allowed researchers to explore how these processes are organized and regulated at a subcellular or even molecular level. Unlike fluorescence microscopy, which requires labeling specific molecules, TEM provides a comprehensive view of all cellular components, not just those tagged with fluorescent markers. This makes TEM invaluable for observing cellular context, capturing both labeled and unlabeled structures, and providing a holistic perspective on cell morphology and ultrastructure.

In the following decades, further advancements were made in TEM resolution and contrast techniques. Contrast enhancement methods, such as negative staining and heavy metal shadowing, were developed to enhance visualization of biological materials that lacked sufficient electron density (Brenner and Horne 1959; Sasaki et al. 2022). Additionally, improvements in vacuum technology, electron optics, and lens aberration correction contributed to higher image quality and finer resolution (Rose 2006, 2008). By analyzing the unique energy signatures generated when the electron beam interacts with the specimen, researchers could obtain elemental and chemical information alongside high-resolution imaging. Techniques such as energy-dispersive X-ray spectroscopy (EDS), which identifies elements through characteristic X-ray emissions, and electron energy loss spectroscopy (EELS), which provides insights into elemental composition and chemical bonding by analyzing the energy lost by electrons, became integrated with TEM, further expanding its capabilities for elemental and chemical analysis (Haberfehlner et al. 2014).

The 1980s introduced a revolutionary method for biological samples: cryogenic electron microscopy (cryo-EM) (Dubochet et al. 1988; Taylor and Glaeser 1974). This technique involves the rapid freezing of specimens to preserve their native structure without the need for chemical fixation. By the 1990s, cryo-EM had advanced to enable imaging of hydrated biological samples in a near-native state, overcoming many of the previous limitations associated with sample preparation (Dubochet et al. 1988; Erk et al. 1991). Computational methods also began to play a more prominent role, allowing researchers to digitally reconstruct 3D structures from 2D TEM images. These developments paved the way for single-particle analysis (SPA) and cryo-ET.

In the early twenty-first century, TEM entered a "resolution revolution" largely driven by advances in direct electron detectors, powerful image processing algorithms (Punjani et al. 2017; Scheres 2012), and further improvements in aberration correction (Batson et al. 2002). These developments allowed SPA cryo-EM and cryo-ET to reach sub-nanometer resolutions, significantly impacting structural biology by enabling detailed visualization of complex macromolecules (Kühlbrandt 2014). Modern TEMs are now equipped with highly sophisticated electron optics, detectors, and software that allow researchers to determine atomic structures in crystalline materials and biological macromolecules with unprecedented clarity and throughput. The impact of these advances has been recognized with the Nobel Prize in Chemistry in 2017, awarded to scientists Jacques Dubochet, Joachim Frank, and Richard Henderson for their contributions to the development of cryo-EM (Cressey and Callaway, 2017).

10.2.1.2 Basic Principles of TEM

The basic principle of TEM involves transmitting a beam of high-energy electrons through a thin specimen. As the electron beam passes through the specimen, it interacts with the atoms in the sample, resulting in scattering of the electrons. The unscattered electrons, along with those scattered at small angles, are collected and focused by a series of electromagnetic lenses to form an image on a detector, such as a fluorescent screen, a film, a digital camera, or a direct electron detector.

The key components of a TEM include the electron gun, electromagnetic lenses, specimen stage, and imaging system. The electron gun generates the electron beam, typically through thermionic emission or field emission. The electromagnetic lenses focus and direct the electron beam through the specimen, while the specimen stage holds and manipulates the sample. Finally, the imaging system captures the electron beam after it has passed through the specimen, producing a magnified image of the sample's structure. Most electron microscopes operate under high vacuum conditions to minimize electron scattering by air molecules, ensuring stable trajectories within the sample chamber. This vacuum environment enables electrons to travel through the instrument without interference, preserving image quality and optimizing resolution and signal-to-noise ratio. Additionally, maintaining a vacuum is crucial for preventing electrical discharge (arcing) in the electron gun assembly.

The resolution of TEM is fundamentally determined by the wavelength of the electrons and the quality of the electron optics within the microscope. In theory, the wavelength of electrons is inversely proportional to their accelerating voltage, meaning that higher-energy electrons have shorter wavelengths, which enables higher resolution (Scherzer 1949). In practical terms, however, the achievable resolution is also influenced by several other factors, including the stability of the electron source, the precision of the electromagnetic lenses, aberration correction, and the ability to mitigate sample drift and charging effects during imaging (Kisielowski et al. 2008). Currently, advanced TEMs can achieve sub-nanometer and even angstrom-level resolutions (Batson et al. 2002; Kisielowski et al. 2008), making it possible to directly visualize atomic arrangements in crystalline materials, such as metals, semiconductors, and complex oxides. In biological specimens, this high resolution facilitates detailed ultrastructural analysis, enabling researchers to investigate macromolecular complexes, cellular architecture, and even individual protein conformations. For example, SPA cryo-EM has been used to determine the structures of protein complexes at atomic resolution, advancing our understanding of molecular mechanisms and interactions (Nakane et al. 2020; Yip et al. 2020).

10.2.1.3 Limitations in Biological Electron Microscopy

TEM offers several advantages for biological imaging. However, TEM also has limitations when imaging biological samples. One of the main challenges in TEM is the preparation of specimens.

- TEM samples must be extremely thin (typically less than 300 nm) to allow electrons to pass through them. Preparing such thin sections for biological samples is technically challenging and may introduce artifacts.
- Biological samples must be chemically fixed and dehydrated or frozen (cryofixation) before being loaded into high-vacuum sample chamber of the TEM instrument, as living tissue and cells consist of approximately 70–80% water. However, sample fixation limits the ability to study dynamic cellular processes.

Most biological materials are composed of light atoms, such as carbon (C), hydrogen (H), oxygen (O), and nitrogen (N). These atoms generate weak signals and contrasts in TEM due to their low atomic numbers, resulting in a low signal-to-noise ratio (SNR). This makes it difficult to achieve high contrast and resolution for fine structural details. Additionally, the high-energy electron beam required for TEM can cause radiation damage to sensitive biological specimens, potentially altering their structure and leading to artifacts in the images, further complicating the imaging process. Specialized techniques, such as contrast-enhancing stains, are often employed to mitigate these challenges but are not always able to fully preserve the native state of the sample. With advancements in detector technology, particularly the development of direct electron detectors (McMullan et al., 2009; Li et al. 2013), cryo-EM offers an ideal solution to balance preserving the near-native state of samples while minimizing damage and artifacts.

10.2.1.4 3D Reconstruction Techniques

TEM produces 2D images, which are often insufficient for understanding the 3D architecture of complex structures. To overcome these limitations, researchers have developed advanced techniques for reconstructing 3D structures from 2D data. These methods offer detailed insights into the spatial organization and interactions of biological molecules, cells, and tissues.

10.2.1.4.1 3D Registration of Serial-Section Microscopic Images

Serial-section TEM is a method for reconstructing the 3D structure of a specimen by analyzing multiple TEM images of ultrathin sections prepared sequentially from a resin-embedded or vitrified specimen (Fig. 10.1a). The process involves slicing biological specimens into ultrathin sections (typically several tens of nanometers thick) using advanced tools such as an ultramicrotome. These sections are imaged sequentially using TEM (Takemura et al. 2013). The images are then aligned, the target structural features are extracted through a process called segmentation, and the desired 3D structure is computationally reconstructed by stacking the aligned images.

Serial-section TEM is a similar method to automated tape-collecting ultramicrotome scanning electron microscopy (ATUM-SEM) (Kasthuri et al. 2015; Morgan et al. 2016). Alternatively, a diamond knife (serial block-face SEM) (Denk and Horstmann 2004; Hua et al. 2015) or FIB (FIB-SEM) (Knott et al. 2008; Xu et al. 2017) is used to remove thin layers of material from the sample surface, exposing a new cross-section for SEM imaging. By collecting a series of high-resolution 2D images in this "slice-by-slice" manner, a detailed 3D model of the sample can be computationally reconstructed.

This approach has proven particularly valuable in examining the intricate architecture of brain tissue, neuronal circuits, and individual synapses. For instance, by reconstructing 3D models of neural structures, researchers can analyze the organization and spatial distribution of neuronal circuits and synaptic components. This enables a deeper understanding of how neurons communicate and how synaptic structures contribute to overall brain function. For example, Dr. Kristen Harris's pioneering work exemplifies the power of serial-section TEM in neuroscience. Her studies focus on reconstructing synapses in 3D at nanometer resolution, providing critical insights into synaptic ultrastructure and plasticity (Bourne and Harris 2011; Bromer et al. 2018; Cooney et al. 2002). One of her landmark studies used

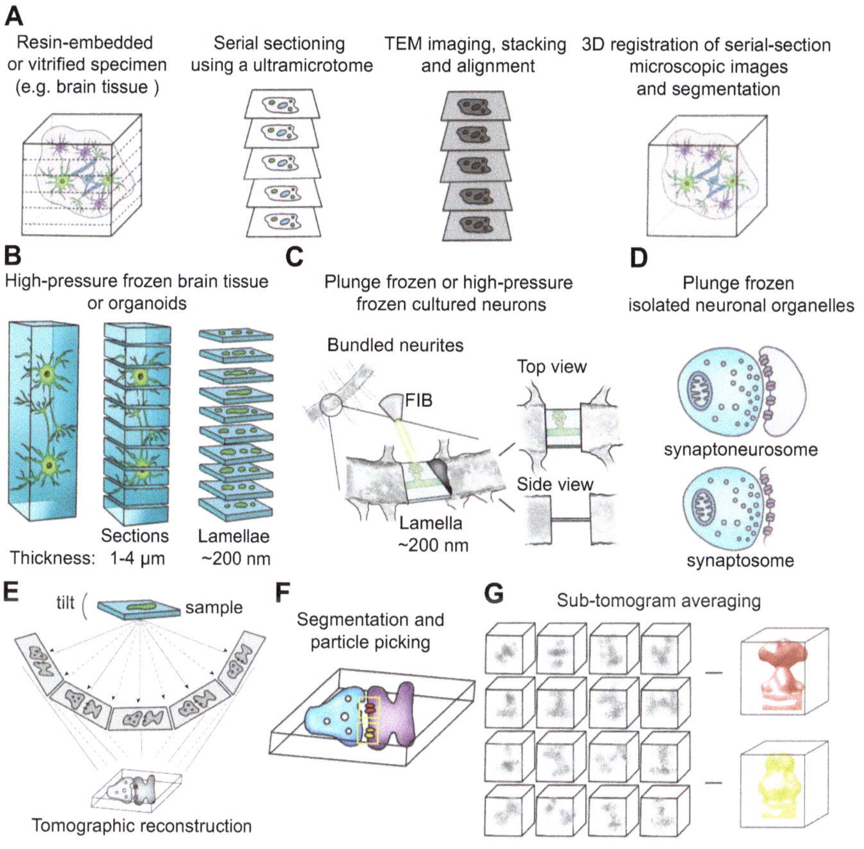

Fig. 10.1 Transmission electron microscopy (TEM) techniques are applied in neuroscience for different sample types. (**a**) Illustration of the procedure for serial-section TEM of resin-embedded or vitrified specimens, such as brain tissue. Ultrathin sections are serially sliced using an ultramicrotome or cryo-ultramicrotome and sequentially arranged on EM grids for imaging. TEM images of each section are captured in the same field of view and precisely aligned. The target structural component (e.g., an organ) is isolated through segmentation, and its 3D structure is reconstructed by stacking the aligned images. (**b**) Illustration of the serial lift-out method, exemplified by its application to brain tissue or organoids embedded in an extracted volume of vitreous ice (blue). (**c**) Illustration of the Cryo-FIB milling method. A schematic representation of bundled neurites arborizing from cell bodies, with the targeted region selected for FIB milling. A cryo-focused ion beam mills the targeted area to create cryo-lamellae (~100–200 nm thick). Scanning electron microscopy provides top-down views for targeting and monitoring progress, while ion beam imaging offers side views to assess lamella thickness. A typical synapse structure is illustrated within an FIB-milled ultrathin section. (**d**) Schematic representation of isolated neuronal terminals, such as synaptosomes and synaptoneurosomes, prepared via plunge-freezing for analysis. (**e**–**g**) Workflow of electron tomography. (**e**) Cryo-ET data collection and 3D tomographic reconstruction. Projection images of samples from panels B–D are collected at various tilt angles using TEM, then computationally reconstructed into a tomogram. A tilt series is acquired over a defined angular range (typically ±60° in 2–3° increments). (**f**) Segmentation and particle picking. Cryo-ET segmentation and particle picking are critical for extracting structural information from tomograms. Segmentation identifies and delineates macromolecular complexes or cellular features within the 3D volume using manual, semi-automated, or AI-driven approaches. Particle picking selects individual particles for sub-tomogram averaging, improving resolution by aligning and averaging multiple instances of the same structure. (**g**) Sub-tomogram averaging (STA). When multiple copies of a macromolecular complex are available, STA enables iterative alignment and averaging of selected particles to produce higher-resolution EM maps

serial-section TEM to investigate the synaptic connectivity and plasticity within hippocampal dendrites, revealing how synapse size and shape correlate with functional states during learning and memory (Harris and Stevens 1989). The ability of this method to capture nanoscale details across entire brain regions or localized neural networks makes it an indispensable tool in neuroscience, advancing our understanding of how structural connectivity underpins neural function (Bourne and Harris 2011; Bromer et al. 2018; Cooney et al. 2002).

10.2.1.4.2 Tilt Series Imaging—Electron Tomography

In this technique (Fig. 10.1e), a sample is tilted at incremental angles under the microscope, and a series of images is captured at each tilt. Advanced computational algorithms are then used to reconstruct a detailed 3D volume. Electron tomography is particularly powerful for studying cellular ultrastructures and macromolecular complexes in their cellular context, providing nanometer-level resolution. More details about this technique will be described in Section 10.2.2.

10.2.1.4.3 Single-Particle Reconstruction

This approach is widely used in TEM to reconstruct the 3D structures of macromolecular complexes, commonly referred to as particles. By collecting, aligning, and computationally averaging thousands to millions of 2D projections of identical particles, researchers can generate high-resolution 3D density maps through single-particle reconstruction (Cheng et al. 2015). These density maps, when resolved to ~4 Å or better, enable the direct fitting of atomic models, with ~4 Å generally considered the practical resolution limit for successful *ab initio* atomic modeling (Kühlbrandt 2014).

A similar strategy is employed in *sub-tomogram averaging*, which focuses on macromolecular complexes imaged in 3D tomograms (Wan and Briggs 2016). Within these tomograms, repeated instances of the same particle are identified, aligned, and averaged computationally, refining the resolution of the resulting structures (Fig. 10.1g). This method is particularly effective for studying macromolecular complexes in their native cellular environments, as it preserves their spatial and contextual integrity.

All three 3D reconstruction techniques—3D registration of serial-section microscopic images, electron tomography, and single-particle reconstruction—can be combined and integrated to achieve a comprehensive understanding of 3D structures at different levels and scales. Serial-section registration enables the reconstruction of large-scale tissue or cellular architectures by aligning consecutive thin sections. Electron tomography provides high-resolution 3D reconstructions of individual cells or organelles, capturing their native ultrastructure. Single-particle reconstruction is particularly powerful for determining the atomic or near-atomic structures of macromolecular complexes. For example, electron tomography can be coupled with serial sectioning, taking tomograms instead of 2D images for each section (Soto et al. 1994). By integrating these approaches, researchers can bridge structural information across multiple scales, from tissue organization down to molecular details, offering a more complete and hierarchical view of biological systems. These complementary techniques have revolutionized structural biology, enabling researchers to visualize intricate biological systems in 3D with unprecedented detail. They continue to evolve with advancements in imaging hardware and computational algorithms.

10.2.2 Electron Tomography

ET is an advanced imaging technique that extends the capabilities of TEM by enabling 3D reconstruction of biological specimens at nanometer-scale resolution or better. Unlike traditional TEM, which provides only 2D images, ET captures a series of 2D projections from different angles and reconstructs them into a 3D volume (Baumeister et al. 1999; Wan and Briggs 2016). Cryo-ET is a powerful variation of ET that visualizes biological specimens in their near-native state. By combining the principles of cryogenic techniques and ET, cryo-ET generates high-resolution 3D reconstructions of vitrified specimens, preserving their native ultrastructure without the need for chemical fixation or staining. Cryo-ET has become an indispensable tool in structural cell biology for studying the ultrastructure of cells, tissues, and macromolecular complexes.

10.2.2.1 Sample Preparation

Sample preparation is a critical step in electron tomography (ET) as the quality of the 3D reconstruction relies on preserving the specimen's ultrastructure. As mentioned in Section 10.2.1.3, biological samples consist of approximately 70–80% water and are composed primarily of light atoms, resulting in low contrast under an electron microscope. Traditionally, biological samples such as cells and tissues are chemically fixed at room temperature using aldehydes (e.g., glutaraldehyde or paraformaldehyde) and stained with heavy metals (e.g., osmium tetroxide or uranyl acetate) (Tizro et al. 2019). Aldehydes primarily cross-link proteins, while osmium tetroxide reacts strongly with membrane lipids and proteins, providing electron density to cellular structures. Following fixation, samples are dehydrated in increasing concentrations of a solvent, such as acetone or ethanol. These solvents extract cellular water and later serve as carriers for embedding resins. However, these methods depend on the diffusion of fixatives and dehydration solvents into the cell, which can lead to issues such as slow infiltration, extraction of cellular contents, and fixation and dehydration artifacts like distorted membranes or organelles. These artifacts can result in a loss of material, making cells appear less dense than in their native state—particularly problematic for specimens with natural diffusion barriers, such as cell walls or cuticles. Despite these limitations, the relative simplicity and speed of traditional fixation protocols make them valuable for routine analysis or initial evaluation of new samples.

Significant advances in sample preparation have come from vitrification, a form of cryofixation in which water is converted into glass-like, amorphous solid without forming daming ice crystals. For thin specimens (e.g., single cells or cell monolayers), vitrification is achieved by extremely rapid cooling such as plunge freezing. For thicker specimens (e.g., tissues or small organisms), high-pressure freezing (HPF) is used to suppress ice crystallization and preserve near-native ultrastructure. These methods offer several advantages, including rapid cessation of cellular activity, superior preservation of ultrastructure, and better retention of antigenicity for subsequent immunolabeling.

Cryofixed samples destined for conventional TEM undergo freeze substitution (FS), in which frozen water within the sample dissolves (rather than melts) in a solvent, typically acetone or methanol, at temperatures around −80 to −90 °C. FS procedures often include fixatives used in traditional methods, and gradual warming allows fixation chemistry to occur efficiently. Embedding follows, involving infiltration of resins that are polymerized into a hard plastic block suitable for ultrathin sectioning. Common resins include epoxy resins like Epon, which are easy to section and provide excellent post-staining but are unsuitable for immunolabeling due to their high polymerization temperatures. Alternatively, methacrylate resins such as Lowicryls can be polymerized at low temperatures using ultraviolet light, making them compatible with HPF/FS and suitable for post-embedding immunolabeling.

Cryo-ET offers distinct advantages over conventional TEM methods that rely on chemical fixation and dehydration. These traditional processes can cause protein aggregation, membrane wrinkling, and loss of native morphology (Dubochet and Sartori Blanc 2001; Korogod et al. 2015). Furthermore, in conventional EM, contrast is generated by heavy metal staining, which limits visibility to cellular structures bound by the stain while obscuring fine features like membrane-bound proteins. In contrast, cryo-ET images are obtained from unstained, fully hydrated specimens. This technique relies on vitrification, where rapid cooling forms amorphous ice instead of crystalline ice, thereby preserving native structures without damage (Dubochet and Sartori Blanc 2001). In this discussion, we focus on sample preparation including vitrification and cryogenic thinning for cryo-ET, with a particular emphasis on different types of neuronal and brain samples.

10.2.2.1.1 Vitrification

Cryo-ET can be applied to a wide range of samples, from purified proteins and isolated organelles to cells, tissues, and even small organisms like *C. elegans*. However, the essential prerequisite for cryo-ET is that samples must be fully vitrified, with no evidence of crystalline ice diffraction. Ice crystals disrupt the specimen's ultrastructure, causing artifacts and a loss of detail in the resulting images. Vitrification preserves the specimen in an amorphous state, maintaining its near-native structure and enabling accurate visualization in the electron microscope.

The most common vitrification method, plunge freezing, involves rapidly plunging the specimen into liquid ethane at approximately −180 °C (Dubochet et al. 1988). Liquid ethane is preferred over other cryogens, such as liquid nitrogen, due to its higher heat capacity, which enables more efficient and uniform cooling. This rapid cooling prevents ice crystal formation, producing a vitrified specimen suitable for imaging in the electron microscope. Samples thinner than a few micrometers (<10 μm) can typically be vitrified directly via plunge freezing (the cooling rate typically exceeds 10^5–10^6 °C/s). However, for thicker specimens (up to 200 μm), high-pressure freezing is required: applying high pressure (~2100 bar) suppresses ice crystallization while rapidly cooling the sample (on the order of ~200 °C within 10 ms), enabling preservation of near-native ultrastructure (Studer et al. 1995; Moor 1987; Sartori et al. 1993; Studer et al. 1989). Once vitrified, specimens can be thinned, localized, and then imaged at cryogenic temperatures. This allows for the collection of high-resolution 3D data without the need for chemical fixation or staining, preserving the ultrastructural integrity of the sample.

10.2.2.1.2 Cryogenic Thinning and Localization

A major challenge in TEM is the need for thin specimens to allow electron transmission. Biological samples, such as neurons and brain tissue, are often too thick for direct imaging, requiring thinning to less than 300 nm. Cellular samples and tissues thicker than ~300 nm cannot typically be imaged to achieve high resolution without thinning. Cryogenic thinning can be achieved through physical cutting with a diamond knife (Fig. 10.1a) or cryo-focused ion beam (cryo-FIB) milling (Fig. 10.1c).

Cryo-electron microscopy of vitreous sections (CEMOVIS) involves using a cryo-ultramicrotomy to cut vitrified samples into ultrathin sections under cryogenic conditions, allowing them to be observed in their vitrified state (Fig. 10.1a) (McDowall et al. 1983). Cryo-ET studies using this method have explored cellular ultrastructure (Cai et al. 2018; Medalia et al. 2002; Ng et al. 2019; Ng and Gan 2020; Nicastro et al. 2000). While cryo-sectioning enables the examination of large cellular volumes through continuous sections, it has a significant drawback: compression along the cutting direction (Dubochet et al. 2007). This deformation can distort cellular components, compromising the accuracy of structural analysis.

Cryo-FIB, adapted from materials science, has become a leading technique for thinning biological samples. FIB-SEM microscopes employ a focused ion beam (e.g., gallium or plasma-based sources) to site-specifically remove material, producing ultrathin (~200 nm) lamellae (Rigort and Plitzko 2015). This method has been widely applied to various samples, including bacteria, yeast, mammalian cells, and cultured primary neurons (Fig. 10.1c) (Gruber et al. 2018; Mahamid et al. 2016; Pi et al. 2023). A key limitation of this technique is the maximum extraction volume, which refers to the largest specimen volume that can be efficiently milled and extracted while maintaining structural integrity. For cryo-FIB, this volume is typically constrained to approximately 20 μm × 20 μm × 10–30 μm, limiting the size of structures that can be analyzed in a single lamella. This constraint necessitates careful selection of the region of interest and, in some cases, the use of multiple lamellae to capture larger biological structures. Additionally, milling-induced artifacts such as curtaining effects, ion beam damage, and redeposition of milled material can impact sample quality and require optimization of milling parameters.

Recent developments in cryo-FIB methodology have enhanced its capabilities, enabling higher throughput, improved automation, and preparation of thicker samples. For example, "DeepFIBing" allows lamella preparation from plunge-frozen samples thicker than ~30 μm (Jentoft et al. 2023). This involves milling perpendicular to the grid to create a clean surface, followed by shallow-angle lamella preparation to minimize surface contamination and improve milling efficiency.

Similarly, the "waffle method" facilitates milling up to 50 μm thick (Kelley et al. 2022). After high-pressure freezing (HPF), trenches are milled at near-perpendicular angles on either side of the region of interest. The grid is then tilted, and a lamella is milled at a shallow angle, incorporating a stress-relief notch to ensure mechanical stability and prevent deformation.

For accessing even thicker biological regions, the lift-out method is employed (Schaffer et al. 2019). This technique involves extracting a tissue block containing the feature of interest, which can then be repositioned and thinned in an optimal orientation. Innovations such as serial lift-out have further increased throughput by enabling the production of multiple sections from a single sample (e.g., *C. elegans*) (Schiøtz et al. 2024). Serial lift-out is especially well-suited for brain tissue studies due to the complex and heterogeneous nature of neural tissue (Fig. 10.1b). Brain structures often contain densely packed and varied cellular components, making it difficult to capture the full anatomy in a single lamella. By utilizing serial lift-out, researchers can extract successive, thin sections from a larger sample, effectively preserving the spatial relationships between cells, synapses, and other features across a series of layers. This approach allows for the detailed analysis of 3D subcellular organization at multiple depths within the brain, overcoming the limitations of traditional FIB-SEM techniques that are constrained by the thickness of the tissue block. Serial lift-out is crucial for capturing large-scale structural data in thick brain samples, such as synaptic connections, subcellular features, and even neural circuits, which are essential for understanding brain function and pathology.

These advancements are driving the next generation of cryo-FIB applications, broadening the scope of biological systems that can be analyzed and facilitating more detailed insights into cellular and subcellular structures.

10.2.2.1.3 Specific Localization

In conventional TEM, one way to localize specific proteins is to use *immuno-electron microscopy* (*immuno-EM*), a technique that labels the protein of interest with antibodies conjugated to electron-dense particles, such as gold nanoparticles

(Chen et al. 2008; Tao-Cheng 2020; Tao-Cheng et al. 2010). This approach provides nanometer-level resolution for visualizing the protein's intracellular location. Typically, a gold-conjugated secondary antibody is used to bind a primary antibody that specifically targets the protein of interest. While highly effective in chemically fixed samples, this method becomes challenging when applied to cryogenic cell or tissue samples. The difficulty arises because the preservation of native ultrastructure limits the use of chemical fixation and permeabilization protocols, and the sparse distribution of macromolecules complicates their identification within the reduced cellular volume of a cryo-lamella. For abundant cellular structures such as ribosomes or microtubules, randomly prepared cryo-lamellae are likely to contain sufficient copies for straightforward identification in cryo-tomograms (Hoffmann et al. 2022; Watanabe et al. 2020). However, for sparsely distributed proteins, locating these macromolecules within a cryo-lamella — which typically represents less than one-tenth of the total cellular volume, and often much less — is far more challenging. Without prior knowledge of their subcellular localization, the process of thinning cells to find a target macromolecule is akin to looking for a needle in a haystack.

Besides immuno-EM, *correlated light and electron microscopy* (*CLEM*) is also an option to localize specific proteins in conventional EM. However, it also suffers from the artifacts caused by conventional EM sample preparation. To overcome these challenges, *cryo-correlated light and electron microscopy* (*cryo-CLEM*) has been developed, enabling the precise localization of fluorescently labeled macromolecules during the cryo-sample preparation process (Sartori et al. 2007; Sun et al. 2019; Tao et al. 2018a). Cryo-CLEM integrates fluorescence imaging, which offers molecular specificity, with electron microscopy, which provides ultrastructural detail. A common cryo-CLEM workflow involves imaging cryo-samples on widefield or confocal microscopes equipped with a cryo-stage, preserving the frozen-hydrated state of the sample (Bieber et al. 2022; Feldmüller et al. 2024; Sartori et al. 2007; Schwartz et al. 2007; Sun et al. 2019; Wozny et al. 2023). Emerging techniques such as *cryo-super-resolution fluorescence microscopy* have the potential to bridge the resolution gap between light and electron microscopy, enabling more precise localization of target molecules (Chang et al. 2014; Dahlberg et al. 2020; Moser et al. 2019; Sexton et al. 2022). However, challenges remain, such as achieving sufficient control over the emissive states of fluorophores at cryogenic temperatures (Dahlberg and Moerner 2021). Furthermore, transferring samples between cryo-light microscopy and cryo-focused ion beam (cryo-FIB) systems for lamella preparation is a significant bottleneck, reducing overall workflow efficiency. Recent advancements in *integrated cryo-CLEM systems* that combine cryo-confocal or widefield fluorescence imaging with cryo-FIB milling have significantly improved efficiency and throughput (Li et al. 2023). Nevertheless, these integrated systems are not yet widely accessible, limiting their application.

To address the challenges of imaging endogenous sparse proteins, the development of novel labeling strategies is critically needed, and some progress has already been made. One approach involves tagging target proteins with multimeric particles that possess distinct structural signatures, enhancing their detectability in tomograms (Fung et al. 2023). Another strategy utilizes nanogold particles, which provide strong electron contrast, making labeled proteins more readily identifiable (Matsui et al. 2024; Young et al. 2024). While these advancements represent important steps forward, there remains significant room for improvement. Future efforts should prioritize enhancing the accuracy of these methods by reducing false-positive and false-negative rates, minimizing potential toxicity, and ensuring that the labeling process does not interfere with protein function or disrupt normal cellular processes.

10.2.2.1.4 Sample Preparation of Neuronal Model Systems

Multiple neuronal model systems have been applied and developed for cryo-ET studies in recent years. An essential step in obtaining high-quality cryo-tomograms is to limit specimen thickness to less than ~300 nm. Currently, sample systems suitable for cryo-ET imaging include synaptosomes (Fernandez-Busnadiego et al. 2010b, 2013; Martinez-Sanchez et al. 2021; Perez de Arce et al. 2015; Sun et al. 2025), cultured neurons (Held et al. 2024; Liu et al. 2020; Lucic et al. 2007; Radhakrishnan et al. 2021; Schrod et al. 2018; Sun et al. 2019, 2025; Tao et al. 2018a), and brain slices (Creekmore et al. 2024; Fernandez-Busnadiego et al. 2010b; Zuber et al. 2005)

Native Nervous Tissues

Isolated tissues and whole organisms, such as *C. elegans* worms, can serve as neuronal model systems for cryo-ET. Recent advancements now allow for cryo-ET imaging of unperturbed *Drosophila* nervous tissue vitrified by plunge-freezing (Bäuerlein et al. 2023). *C. elegans* worms can be preserved using high-pressure freezing, followed by the lift-out technique for cryo-ET studies at molecular resolution (Schaffer et al. 2019; Schiøtz et al. 2024). These model systems are particularly advantageous due to the ease of handling and genetic manipulation of *C. elegans* and *Drosophila*.

Brain slices retain the native organization and cell type diversity of nervous tissue. Thick sections (~100–200 μm) of rodent brains can be vitrified either immediately after preparation (acute slices) or following in vitro culture (organotypic slices) (Fernandez-Busnadiego et al. 2010a; Glynn et al. 2024; Leistner et al. 2023; Matsui et al. 2024). Additionally, integrating advanced high-pressure freezing with light and electrical stimulation enables the precise visualization of highly dynamic processes at the nanometer scale with millisecond resolution (Watanabe et al. 2013, 2014). Human brain tissue is typically only available postmortem, which introduces unavoidable ultrastructural alterations prior to vitrification (Creekmore et al. 2024; Gilbert et al. 2024). Developing methods to minimize tissue damage during preparation will be a critical focus of future research.

Brain Organoids

Brain organoids are 3D self-organizing structures that mimic the early development and function of the human brain. They are generated from pluripotent stem cells by adding small molecules to mimic fetal brain development in vivo. This makes organoids an ideal model for studying neurodevelopmental processes and investigating brain disease. Cryo-ET has been used to examine axon tracts extending outside the organoid body, providing insights into neuronal organization and growth (Hoffmann et al. 2021).

Cultured Neurons

Cultured neurons, derived from dissociated brain tissue, are seeded on a substrate where they spontaneously extend neurites and form functional neuronal networks. These networks allow for electrophysiological stimulation and recording, optical manipulation, and live-cell imaging (Kaech and Banker 2006). For decades, cultured neurons have served as a powerful model system to investigate synaptic formation, transmission, and plasticity (Bi and Poo 1998; Maximov et al. 2007; Tarsa and Goda 2002).

Human-derived neurons, generated from sources such as induced pluripotent stem cells (iPSCs) or direct conversion of somatic cells or stem cells into induced neurons (iNeuron) (Uzay et al. 2023; Vierbuchen et al. 2010), offer a unique opportunity to study neuronal development and function in a physiologically relevant context. These neurons can model human-specific aspects of synaptic architecture, signaling pathways, and disease mechanisms that are not fully recapitulated in animal models. Advances in differentiation protocols now allow for the generation of various neuronal subtypes, including excitatory and inhibitory neurons, as well as specialized populations such as dopaminergic or motor neurons (Bianchi et al. 2018; Sheta et al. 2022). Furthermore, human-derived neuronal cultures provide an essential tool for studying neurological disorders, enabling researchers to investigate disease-related mutations, neurodegenerative processes, and potential therapeutic interventions. The integration of these cultures with advanced imaging techniques, electrophysiological assays, and genetic manipulation further enhances their utility in uncovering the fundamental principles of synaptic function and dysfunction in human neurons.

In cryo-ET experiments, dissociated neurons are seeded onto gold cryo-EM grids and grown under controlled conditions. Following plunge freezing, these cultured neurons can either be directly imaged using cryo-ET or thinned via cryo-FIB milling to produce optimal lamellae for cryo-ET imaging (Fig. 10.1c). This approach has yielded numerous insights, including the structural characterization and spatial organization of synaptic vesicles (Lucic et al. 2007; Radhakrishnan et al. 2021; Schrod et al. 2018; Tao et al. 2018a), postsynaptic densities (Sun et al. 2025), and synaptic receptors (Held et al. 2024; Liu et al. 2020; Sun et al. 2025; Tao et al. 2018a). Human-derived neurons can also be generated for cryo-ET studies by differentiating induced pluripotent stem cells (iPSCs) directly on cryo-EM grids (Woldeyes et al. 2023; Wu et al. 2023).

Dissociated cultures are the most widely used model for cryo-ET due to their accessibility and ease of use. They are straightforward to maintain and can be readily manipulated pharmacologically or genetically to address specific experimental questions. However, cultured neurons may not fully replicate the complexity of synapses found in vivo. While cultured synapses form functional networks, their morphology, composition, and functional properties may differ from those in native nervous tissue. For example, they lack the intricate organizational features and interactions present in the three-dimensional architecture of intact neural circuits.

Synaptosomes
For biological samples thinner than 500 nm, such as thin cells and viruses, thinning prior to imaging may not be necessary. Organelle isolation can reduce the thickness of samples, facilitating easier vitrification and often achieving higher resolution compared to cellular samples, due to the less crowded environments. Synaptosomes are isolated synaptic terminals prepared by fractionating brain homogenates (Fig. 10.1d). They maintain the basic morphological features and molecular contents and retain some functional properties of synapses and can respond to pharmacological treatments (Fernandez-Busnadiego et al. 2010a, 2013; Martinez-Sanchez et al. 2021; Perez de Arce et al. 2015). Using high K^+, hypertonic, or pharmacological treatments, the structural changes during vesicle exocytosis have been well characterized in synaptosome systems (Fernandez-Busnadiego et al. 2010a, 2013; Zuber and Lucic 2019). Synaptosomal suspensions can be vitrified via plunge-freezing and examined directly by cryo-ET without additional thinning. However, while synaptosomes are easily accessible, they are detached from their cellular context, limiting their functionality. In addition, functional interpretations of synaptosomal imaging could be confounded by stimulation through sample preparation or a lack of sufficient context available in an intact neuron network.

10.2.2.2 Cryo-ET Data Collection

Data collection in ET involves acquiring a series of 2D projections of a specimen at different tilt angles (Fig. 10.1e). This process is typically performed in a TEM equipped with a tilting stage, which allows the specimen to be tilted over a range of angles, usually from −60° to +60°. At each tilt angle, the electron beam is transmitted through the specimen, and the resulting projections are recorded by a digital camera or other detectors. Besides stage stability and sample thickness, the quality of the resulting 3D reconstruction also depends on several factors during data collection, including the number of projections, the tilt range, and the signal-to-noise ratio of the images (Baumeister et al. 1999). A higher number of projections and a wider tilt range generally result in more accurate 3D reconstructions. However, these improvements come at the cost of a higher total electron dose, which can lead to radiation damage to the specimen. Optimizing the data collection parameters is therefore critical to achieving a balance between resolution and specimen preservation.

Data collection in cryo-ET follows a procedure similar to conventional ET, but with additional considerations to minimize radiation damage to the vitrified specimen. One benefit of studying samples at cryogenic temperatures is the reduction of artifacts caused by chemicals. However, vitrified specimens in cryo-ET are not chemically fixed, making them more susceptible to radiation damage despite the lower temperatures. Excessive electron dose can alter the structure of the specimen, thereby reducing the quality of the 3D reconstruction. To address this challenge, cryo-ET employs strategies to limit the electron dose while maintaining high-resolution imaging. One key innovation has been the development of low-dose mode, implemented in data collection software packages such as *SerialEM* (Mastronarde 2005). Low-dose mode separates imaging into distinct tasks, each using optimized settings to minimize exposure. Typically, a low-magnification, low-dose setting of search mode is used to locate areas of interest on the grid without damaging the specimen. A high-magnification of focus mode is performed on nearby areas away from the region of interest to avoid exposing the main area to the beam. Then, an exposure mode is taken with the optimized electron dose for data collection. This approach reduces the exposure of sensitive areas of the specimen to the electron beam, preserving the native structure and improving the quality of the 3D reconstruction. The low-dose mode is widely used in SPA cryo-EM and cryo-ET.

Minimizing radiation damage is particularly critical for tilt series acquisition, where the total electron dose must be distributed across multiple projections, unlike in SPA, where a single movie is collected with a total electron dose of 50–60 $e^-/Å^2$. The way the dose is distributed throughout the tilt series significantly affects the alignment of projections and the quality of the resulting 3D reconstruction. Several acquisition schemes have been developed to balance data quality, alignment accuracy, and resolution while mitigating the effects of radiation damage. These schemes, including continuous, bidirectional, and dose-symmetric collection schemes, determine how the electron dose is allocated across the tilt series and influence the quality of information obtained at different tilt angles (Hagen et al. 2017; Turonova et al. 2020). The choice of tilt-series acquisition scheme is critical for achieving high-resolution cryo-ET reconstructions while minimizing radiation damage. The dose-symmetric scheme has emerged as a preferred approach due to its ability to optimize dose allocation, preserve high-resolution information, and produce smooth, high-quality reconstructions (Hagen et al. 2017; Turonova et al. 2020). However, selecting the most appropriate scheme depends on the specific experimental goals, sample characteristics, and imaging conditions.

10.2.2.3 3D Reconstruction

Once the tilt series is acquired, the projections are aligned and used to reconstruct a 3D volume of the specimen (Fig. 10.1e). Alignment is typically achieved by tracking external fiducial markers, such as gold nanoparticles, that are added to the specimen before imaging, or by using internal sample features. These markers or features serve as reference points for aligning the projections, thereby ensuring accurate reconstruction of the 3D volume. The 3D reconstruction of the specimen from the tilt series is a critical step in ET. This process involves computationally combining the 2D projections into a 3D volume, which can then be visualized and analyzed to study the ultrastructure of the specimen.

The most common algorithm used for 3D reconstruction in electron tomography is the weighted back-projection method (Kremer et al. 1996). This algorithm works by back-projecting each 2D projection into the 3D volume and then applying a weighting function to correct for the uneven distribution of data along different angles. The result is a 3D reconstruction that accurately represents the ultrastructure of the specimen. Other reconstruction algorithms, such as iterative reconstruction techniques (Gilbert 1972), have been developed to improve the quality of 3D reconstructions. These techniques use iterative optimization methods to refine the 3D volume by minimizing discrepancies between the projections and the reconstructed volume. Iterative reconstruction methods are particularly useful for reconstructing specimens with high noise levels or limited tilt ranges.

Once the 3D volume is reconstructed, it can be visualized using specialized software that allows researchers to explore the ultrastructure of the specimen in 3D. Features of interest, such as synaptic vesicles, membranes, and protein complexes, can be segmented and analyzed to gain insights into their organization and function. One of the challenges in cryo-ET is the lower contrast of vitrified specimens compared to chemically fixed and stained specimens. To address this issue, advanced image processing techniques, such as contrast enhancement and noise reduction, are used to improve the quality of the 3D reconstruction (Bepler et al. 2019; Costa-Filho et al. 2024).

10.2.2.4 Segmentation

Segmentation is the process of identifying and illustrating specific structures within the 3D volume for further analysis. This step is essential for studying the organization and function of biological structures, as it allows researchers to focus on specific components of interest (Fig. 10.1f).

Segmentation can be performed manually, where researchers use software tools to trace and isolate structures within the 3D volume. Manual segmentation is time-consuming and labor-intensive but offers precise control over the segmentation process (Kremer et al. 1996; Stalling et al. 2005). Automated segmentation methods, which use image processing algorithms including machine learning to identify and segment structures, have also been developed (Chen et al. 2017). These methods can significantly speed up the segmentation process but may require optimization for specific specimens and imaging conditions.

In synaptic studies, segmentation is used to isolate and analyze synaptic vesicles, active zones, and postsynaptic densities, among other structures (Chen et al. 2008; Khosrozadeh et al. 2025; Martinez-Sanchez et al. 2021; Tao et al. 2018a). By segmenting these components, researchers can study their spatial distribution, interactions, and changes in response to different stimuli or conditions (Martinez-Sanchez et al. 2021; Tao et al. 2018a, b). The segmented structures can be visualized in 3D, providing a comprehensive view of the synaptic architecture. Quantitative analysis of the segmented structures, such as measuring the size, shape, and distribution of synaptic vesicles, can provide valuable insights into the functional organization of synapses (Imig et al. 2014; Tao et al. 2018a).

10.2.2.5 Sub-tomogram Averaging

Sub-tomogram averaging is a technique used in cryo-ET to improve the resolution of 3D reconstructions by averaging multiple copies of the same structure within the tomogram (Fig. 10.1g). This technique is particularly useful for studying macromolecular complexes and other structures that are present in multiple copies within the cell.

The basic principle of sub-tomogram averaging involves identifying and extracting multiple copies of the structure of interest from the 3D volume (Fig. 10.1f). These copies are then aligned and averaged to produce a high-resolution 3D reconstruction of the structure. The averaging process reduces noise and enhances the signal, resulting in a more detailed and accurate representation of the structure (Wan and Briggs 2016).

Sub-tomogram averaging has been used to study a wide range of biological structures, including ribosomes (Hoffmann et al. 2022), proteasomes (Guo et al. 2018), and viral particles (Wan et al. 2017). In synaptic studies, sub-tomogram averaging can be used to resolve the structures of protein complexes in synaptic vesicles (Kravčenko et al. 2024; Wang et al. 2024), postsynaptic plasma membrane (Liu et al. 2020; Sun et al. 2025), and other components, providing insights into their organization and function.

10.2.3 Examples of Applications of EM in Neuroscience

TEM plays a critical role in neuroscience, providing unmatched resolution for studying the structure and organization of the nervous system. In particular, ET offers powerful capabilities for visualizing the 3D structure of synapses and organelles within neuronal cells.

10.2.3.1 Discovery of Synaptic Vesicles

The discovery of synaptic vesicles was a pivotal milestone in neuroscience, offering the first direct evidence of the vesicular mechanism underlying neurotransmitter release. Synaptic vesicles are small, membrane-bound organelles that store neurotransmitters and release them into the synaptic cleft in response to an action potential.

In the 1950s, the first electron micrographs of synapses revealed small, round vesicles clustered near the presynaptic membrane (De Robertis and Bennett 1955). These images provided direct support for the quantum hypothesis first proposed by

Bernard Katz and his colleagues (Fatt and Katz 1952; Katz and Miledi 1968), which suggested that neurotransmitters are released in discrete packets. The vesicles observed were later identified as synaptic vesicles, and their role in neurotransmitter release was confirmed through a combination of electrophysiological and biochemical approaches (Ceccarelli et al. 1973; Südhof 2013; Takamori et al. 2006).

The identification of synaptic vesicles was instrumental in elucidating the mechanism of synaptic transmission. It led to the development of the vesicle hypothesis, which posits that neurotransmitters are released via the fusion of synaptic vesicles with the presynaptic membrane. This hypothesis, supported by extensive research, remains a cornerstone of modern neuroscience.

10.2.3.2 Synaptic Vesicle Exocytosis and Endocytosis Visualized by EM

The process of synaptic vesicle exocytosis and endocytosis has been visualized using EM, providing direct evidence of the vesicle hypothesis and revealing the detailed ultrastructure of the fusion event. High-resolution TEM images have shown synaptic vesicles in various stages of exocytosis with the presynaptic membrane, including docking, priming, and fusion pore formation (Heuser and Reese 1973; Imig et al. 2014; Liu et al. 2020; Sun et al. 2018, 2025) and endocytosed synaptic vesicle, providing evidence of distinct synaptic vesicle recycle modes, including clathrin-mediated endocytosis (Heuser and Reese 1973; Tao-Cheng 2020; Watanabe et al. 2014), bulk endocytosis (Wu et al. 2014), and ultrafast endocytosis (Watanabe et al. 2013, 2014).

These studies have provided critical insights into the molecular mechanisms underlying synaptic vesicle endocytosis and exocytosis, including the role of key proteins and other components involved in both the fusion and the endocytosis machinery (Heuser and Reese 1973; Imig et al. 2014; Liu et al. 2020; Shehzad et al. 2020; Sun et al. 2018, 2025; Tao-Cheng 2020; Watanabe et al. 2014; Wu et al. 2014). By visualizing the ultrastructure of the fusion event and endocytosis, researchers have been able to identify key steps in the process and understand how different proteins and lipids contribute to vesicle recycling (Imig et al. 2014; Liu et al. 2020; Shehzad et al. 2020; Sun et al. 2018, 2025; Tao-Cheng 2020; Watanabe et al. 2014; Wu et al. 2014).

ET has further enhanced our understanding of synaptic vesicle fusion and endocytosis by providing 3D reconstructions of the fusion process. These reconstructions have revealed the spatial organization of the fusion and recycling machinery, the arrangement of synaptic vesicles at the active zone, and the vesicle dynamics during neurotransmitter release (Imig et al. 2014; Sun et al. 2025).

10.2.3.3 Electron Tomography for Structure–Function Analysis of Different Synapses

ET has been used to study the structure and function of different types of synapses, providing insights into the diversity of synaptic architectures and their implications for neural circuit function. By reconstructing 3D models of synapses from electron tomograms, researchers have been able to analyze the spatial organization of synaptic components and understand how they contribute to synaptic function (Tao et al. 2018a).

For example, ET has been used to study the ultrastructure of excitatory and inhibitory synapses (Tao et al. 2018a), revealing differences in the organization of synaptic vesicles, active zones, and postsynaptic densities (Tao et al. 2018a). These differences have been linked to the distinct functional properties of excitatory and inhibitory synapses, such as the release probability of neurotransmitters and the kinetics of synaptic transmission (Tao et al. 2018a).

ET has also been used to study the ultrastructure of specialized synapses, such as ribbon synapses in sensory neurons and neuromuscular junctions in muscle cells (Li et al. 2023). These studies have provided insights into the unique adaptations of these synapses for their specific functions, such as rapid and sustained neurotransmitter release in ribbon synapses and precise control of muscle contraction in neuromuscular junctions (Jung et al. 2018).

10.2.3.4 Discovery and Identification of Synaptic Structures at Molecular Resolution

Cryo-ET has unveiled new cellular structures and novel biology (Chaikeeratisak et al. 2017; McCafferty et al. 2024; Nedozralova et al. 2022) and is likely to continue doing so, as it remains the highest-resolution imaging technique in cell biology and is capable of capturing intact molecular networks. Cryo-ET has revealed connectors linking synaptic vesicles to each other and tethers anchoring synaptic vesicles to the presynaptic membrane (Fernandez-Busnadiego et al. 2010b, 2013; Papantoniou et al. 2023). These structures have also been investigated using conventional EM techniques, such as chemical fixation EM and high-pressure freezing with freeze substitution (Cole et al. 2016; Jung et al. 2021). Due to its superior resolution and preservation of native structures, the Lucić group has extensively employed cryo-ET to study these connectors and tethers. Their findings demonstrate that the connectors are dynamic structures essential for vesicle clustering and mobility. Interestingly, synapsin has been implicated in vesicle clustering by other studies (Milovanovic et al. 2018; Park et al. 2021), suggesting it could be a candidate protein component of the connector. However, direct evidence identifying the molecular composition of these connector filaments remains lacking. A similar knowledge gap exists regarding the tethers between synaptic vesicles and the presynaptic membrane. Cryo-ET studies have shown that these tethers play critical roles in synaptic vesicle docking and priming (Fernandez-Busnadiego et al. 2010b,

2013; Papantoniou et al. 2023; Radhakrishnan et al. 2021), yet their precise molecular identities are still unknown. Bridging structural studies with functional experiments will be essential to elucidate the composition and roles of presynaptic connectors and tethers.

10.3 Postsynaptic Density Studies by TEM

Synaptic ultrastructure has been extensively investigated using TEM, providing foundational insights into synaptic organization and function. These studies facilitated the discovery of synaptic vesicles, the visualization of synaptic vesicle fusion, and structural analyses of various synapse types, as discussed above. Additionally, TEM was instrumental in the early characterization of the postsynaptic density (PSD). During the mid-twentieth century, TEM images first revealed the dense, specialized structure of the PSD at the postsynaptic membrane, identifying it as a critical site for synaptic signaling. More recently, studies using cryo-ET have further elucidated the molecular and spatial organization of the PSD.

10.3.1 Conventional EM on PSD Studies

Immuno-EM has significantly advanced our understanding of synaptic components by utilizing antibodies conjugated with gold particles to label specific proteins (Chen et al. 2008; Gonda 1998). This technique offers high spatial precision (~10 nm) and relies on the specificity of antibody-antigen interactions, allowing the precise localization of individual protein molecules. For instance, immuno-EM has been employed to identify specific subtypes of α-amino-3-hydroxy-5-methyl-4-isoxazolepropionic acid receptors (AMPARs) on the postsynaptic membrane (Chen et al. 2008), providing valuable insights into their spatial organization and function. Besides synaptic receptors in the PSD, immuno-EM has been applied to other synaptic proteins like CaMKII and Shank to detect their distribution in synapses during plasticity (Dosemeci et al. 2000; Tao-Cheng et al. 2010).

The introduction of high-pressure freezing (HPF) followed by freeze-substitution (FS) has markedly improved the preservation of synaptic ultrastructure, minimizing morphological distortions caused by chemical fixation. When combined with ET, this approach has enabled the visualization and localization of postsynaptic glutamate receptors and other postsynaptic proteins based on their distinct shapes (Chen et al. 2008, 2011, 2015). The Reese group and others applied the above-mentioned techniques to systematically study the postsynaptic density PSD-95 family of membrane-associated guanylate kinases (MAGUKs), which anchor and organize glutamate receptors in excitatory synapses (Chen et al. 2008, 2011, 2015). These advancements laid the groundwork for deciphering the molecular architecture and functional roles of the PSD in synaptic communication.

10.3.2 Cryo-ET on PSD Studies

The advent of cryo-EM techniques, particularly with the development of direct electron detectors, has ushered in the "resolution revolution," enabling the imaging of biological molecules at unprecedented resolutions (de Oliveira et al. 2020; Kühlbrandt 2014; Li et al. 2013; Moreau and Cochard 2014). This technological breakthrough has also propelled cryo-ET as a more powerful tool for studying molecular assemblies and extracellular structures in situ (within their native cellular environment) at significantly higher resolutions than previously possible (Pyle et al. 2024; Xue et al. 2022; Zens et al. 2024). Early applications of cryo-ET resolved presynaptic vesicular organizations and synaptic clefts in isolated synaptosomes (Fernandez-Busnadiego et al. 2010b; Perez de Arce et al. 2015). More recently, cryo-ET has been used to visualize the 3D ultrastructural features of excitatory and inhibitory synapses (Tao et al. 2018a). It has also enabled the analysis of changes in the distribution of presynaptic dense core vesicles under chronic inactivation in intact cultured hippocampal neurons (Tao et al. 2018b). Using cryo-ET combined with sub-tomogram averaging, GABA receptors in inhibitory synapses have been revealed to form mesoscale assemblies with postsynaptic scaffolding proteins and align with putative presynaptic vesicle release sites (Liu et al. 2020).

The introduction of volta phase plates and electron energy filters has further enhanced the contrast and resolution of cryo-ET reconstructions, facilitating the identification of neurotransmitter receptors (Liu et al. 2020; Tao et al. 2018a). In our recent studies, we applied phase plates to examine the subsynaptic structures of the postsynaptic density (PSD) in isolated synaptosomes and performed cryo-ET without a phase plate in intact cultured hippocampal neurons. These analyses revealed that the PSD consists of membrane-associated nanoblocks of varying sizes, offering new insights into its molecular organization (Sun et al. 2025).

10.4 Structural Complexity of Postsynaptic Density Nanoscale Organization for Synaptic Plasticity

Recent cryo-ET studies have unveiled the nanoscale organization of synaptic components, the precise alignment of trans-synaptic structures, and the in situ architecture of macromolecular complexes involved in neurotransmission (Martinez-Sanchez et al. 2021; Sun et al. 2025). Here we focus on the molecular organization and structural complexity of the postsynaptic compartment in mammalian excitatory synapses, especially neurotransmitter receptors and scaffolding proteins in the postsynaptic densities (PSDs). We also discuss the potential inconsistencies between results obtained using cryo-ET and previous studies such as super-resolution microscopy.

10.4.1 Molecular Basis of Postsynaptic Density

Excitatory glutamatergic synapses are characterized by the clustering of presynaptic vesicles at the axon terminal and the presence of the thick postsynaptic density (PSD), a specialized protein-rich structure localized at the apex of dendritic spines or dendritic shafts beneath the postsynaptic membrane (Palay 1956, 1958; Tao et al. 2018a). The PSD is composed of a dense array of proteins involved in scaffolding, cytoskeletal organization, and signaling (Kaizuka and Takumi 2018). This protein network is readily visualized in conventional EM due to heavy-metal staining, making the PSD a hallmark of excitatory synapses (Palay 1956, 1958). Thin-section EM studies of isolated mammalian PSDs decades ago revealed their disk-like morphology, approximately 200–800 nm in diameter and 30–50 nm thick (Carlin et al. 1980; Cohen et al. 1977). More recent cryo-ET studies have provided a near-native visualization of the PSD, confirming its structural dimensions and densely packed protein organization, all while avoiding chemical treatments that might distort its native architecture.

Biochemical purification and high-resolution SDS-PAGE, coupled with proteomics and amino acid sequencing, have identified key PSD components (Fernandez et al. 2009; Frank et al. 2017; Föcking et al. 2016). These include PSD-95 (DLG4), calcium/calmodulin-dependent protein kinase II (CaMKII), synaptic Ras GTPase-activating protein 1 (SYNGAP1), actin, membrane receptors, cell adhesion molecules, signaling proteins (e.g., kinases and phosphatases), scaffold and adaptor proteins, cytoskeletal regulators, molecular motors, and components of protein synthesis machinery (Boeckers 2006). Mass spectrometry has further characterized PSD interactomes, highlighting the critical role of core scaffolding proteins in synaptic protein networks (Walikonis et al. 2000). A typical PSD contains ~300–400 PSD-95 family MAGUK proteins, ~300–400 Shank family members. N-methyl-d-aspartate receptors (NMDARs) are more or less fixed at ~20 per PSD, whereas the number of AMPARs varies widely between ~5 and 200 (Chen et al. 2005, 2015; Kharazia and Weinberg 1997; Masugi-Tokita et al. 2007; Sugiyama et al. 2005; Takumi et al. 1999). The copy numbers of PSD proteins differ across neuron types and developmental stages, reflecting activity-dependent remodeling that underpins synaptic plasticity. High-resolution ET with antibody labeling has revealed a stratified organization of PSD proteins: (1) membrane receptors and ion channels, including central NMDARs and peripheral AMPARs, form the first layer, enriched with transmembrane cell-adhesion molecules; (2) scaffold proteins such as PSD-95 and other MAGUKs form the second layer, oriented perpendicular to the membrane with lipid-anchored N-terminal domains; and (3) Shank and GKAP-family proteins in the third layer connect MAGUK C-terminal SH3-GK domains to the actin cytoskeleton (Chen et al. 2008, 2011). Together, these proteins create a web-like network that integrates cytoplasmic PSD proteins and enzymes.

Systematic studies using various techniques from multiple research groups have proposed several models to explain the PSD organization. EM and biochemical assays have provided evidence for a meshwork of regularly spaced vertical filaments, which form a structural scaffold within the PSD (Burette et al. 2012; Chen et al. 2008, 2011; DeGiorgis et al. 2006; Tao et al. 2018a). These findings suggest a highly ordered arrangement of proteins that may contribute to the stability and functionality of the synapse. Complementing these structural insights, super-resolution imaging techniques have revealed that the PSD is organized into distinct nanodomains, which may serve as specialized hubs for synaptic signaling and plasticity (Crosby et al. 2019; Hruska et al. 2022; MacGillavry et al. 2013; Nair et al. 2013; Pennacchietti et al. 2017; Tang et al. 2016). This nanodomain organization implies a level of spatial precision in the distribution of synaptic proteins, enabling efficient and localized molecular interactions. Additionally, in vitro studies have proposed that key PSD components can form liquid-like condensates through phase separation, a process driven by weak multivalent interactions among synaptic proteins (Feng et al. 2019; Zeng et al. 2016, 2018). This model suggests that the PSD may exhibit dynamic, fluid-like properties, allowing it to rapidly reorganize in response to synaptic activity. Collectively, these studies provide compelling evidence for the high-level organization of nanoscale supramolecular assemblies within the synapse. Rather than being randomly distributed, synaptic proteins appear to form highly organized, functional nanostructures that are critical for synaptic transmission, plasticity, and overall neuronal communication.

The two major ionotropic glutamate receptors at PSDs, AMPARs, and NMDARs, are central to synaptic signaling (Kharazia and Weinberg 1997; Nusser et al. 1998; Racca et al. 2000; Takumi et al. 1999; Valtschanoff and Weinberg 2001). Detergent-extracted PSD fractions analyzed by biochemistry and mass spectrometry have identified additional proteins associated with AMPAR and NMDAR complexes (Cheng et al. 2006; Lowenthal et al. 2015). The membrane-associated guanylate kinases (MAGUKs), including PSD-95, PSD-93, SAP102, and SAP97, directly interact with NMDARs (Cho et al. 1992; Kim et al. 1996; Müller et al. 1996; Park et al. 2021), binding to conserved motifs in the GluN2 subunits via PDZ domains (Kornau et al. 1995; Niethammer et al. 1996). PSD-95 regulates AMPAR levels in the PSD through interactions with auxiliary proteins such as Stargazin/TARPs (Chen et al. 2000; Nicoll et al. 2006; Schnell et al. 2002). Single-particle tracking shows that AMPAR-Stargazin complexes are stabilized by PSD-95, creating "hot spots" for AMPAR clustering (Bats et al. 2007; MacGillavry et al. 2013; Nair et al. 2013). PSD-95 knockout experiments demonstrate reduced AMPAR-mediated synaptic transmission (Beique et al. 2006), with additional MAGUK knockouts showing broader effects on both AMPAR and NMDAR function (Elias et al. 2006). Conventional EM studies reveal that MAGUKs anchor glutamate receptors and govern PSD organization (Chen et al. 2015).

AMPARs mediate fast excitatory synaptic transmission and are dynamically trafficked in an activity-dependent manner, resulting in long-lasting changes in synaptic strength (Groc and Choquet 2020; Huganir and Nicoll 2013; Malinow and Malenka 2002; Nicoll 2017). NMDARs, in turn, mediate calcium influx essential for triggering these changes. Stargazin-like TARPs and PSD-95 family MAGUKs are key players in AMPAR trafficking and capture. Overexpression of PSD-95 enhances AMPAR clustering (El-Husseini et al. 2000), while knockout of MAGUK family members eliminates most AMPAR-mediated transmission (Chen et al. 2015; Elias et al. 2006; Elias and Nicoll 2007; Levy et al. 2015). Super-resolution imaging supports the idea of PSD-95 organized nanodomains anchoring AMPARs at postsynaptic compartments (Fukata et al. 2013; MacGillavry et al. 2013; Nair et al. 2013). These nanodomains, 100–200 nm wide, align with presynaptic RIM-enriched nanodomains (~80 nm wide), forming trans-synaptic nanocolumns that optimize synaptic transmission (Tang et al. 2016).

Notably, most PSD scaffold proteins contain multiple well-folded domains, such as PDZ, GK, and SH domains, and lack intrinsic enzymatic activity, making them ideal organizers of protein networks (Laursen et al. 2020; Sturgill et al. 2009). These scaffold proteins play a pivotal role in shaping the morphology of the PSD and regulating synaptic plasticity (Laursen et al. 2020; Sturgill et al. 2009). Interestingly, key scaffold proteins often contain repeated domains (e.g., PDZ), and this domain organization is highly conserved across evolution, likely due to selective pressures favoring their function (Sakarya et al. 2010). Studies have shown that tandem PDZ domains exhibit target-binding properties distinct from those of isolated domains or the simple sum of individual domains (Pan et al. 2011). These unique properties arise from direct interactions between PDZ domains, resulting in specific spatial arrangements termed PDZ supramodules (Feng and Zhang 2009; Long et al. 2003). PDZ supramodules, such as those in PSD-95, GRIP1 (glutamate receptor interacting protein 1), X11 (APBA1/MINT), and syntenin, illustrate how tandem domains contribute to molecular complexity in the PSD (Feng and Zhang 2009). Similar supramodules formed by other domain types further enhance the structural and functional diversity of PSD scaffold networks, enabling the assembly of higher-order protein complexes.

The abundance of scaffold proteins with overlapping interacting targets raises intriguing questions about their functional specificity and redundancy. For instance, multiple members of the MAGUK family (e.g., PSD-95/DLG4, PSD-93/DLG2, SAP-102/DLG3, and SAP-97/DLG1) are expressed at excitatory synapses, and each contributes to the trafficking of glutamate receptors during different developmental stages (Zheng et al. 2011). Mouse knockout and short hairpin RNA (shRNA) mediated knockdown experiments have demonstrated that MAGUKs can functionally compensate for one another (Elias et al. 2006), suggesting that this interconnected system ensures proper glutamate receptor expression at synapses.

Collectively, membrane receptors, scaffold proteins, and associated components form a complicated, web-like protein network within the PSD. The presence of multiple domains and supramodules in scaffold proteins adds a layer of molecular complexity, facilitating the formation of higher-order protein assemblies. The overlapping interactions among scaffold proteins further support the dynamic and robust molecular basis of synaptic signaling and plasticity, providing binding sites for additional PSD proteins and enzymes to sustain synaptic function.

10.4.2 PSD Nanoscale Organization and Its Functional Implications

The architecture of the PSD is highly dynamic, undergoing structural and functional remodeling during development and in response to synaptic activity (Araki et al. 2015; Bosch et al. 2014; Compans et al. 2021; Hruska et al. 2018). Evidence suggests that PSD proteins exhibit spatiotemporal regulation, with differential expression patterns across developmental stages and neuronal subcellular compartments (Li et al. 2016, 2017). Advances in super-resolution fluorescence microscopy have revealed nanoscale subsynaptic protein clusters, or nanodomains, within the PSD. Key scaffold proteins, such as PSD-95, are concentrated in nanodomains ranging from 70 to 160 nm in size (Broadhead et al. 2016; Hruska et al. 2018; MacGillavry et al. 2013; Nair et al. 2013; Tang et al. 2016). These nanodomains are dynamically reorganized in response to plasticity-inducing stimuli. For instance, chronic synaptic activity changes during synaptic scaling significantly alter scaffold and receptor organization (Fukata et al. 2013; MacGillavry et al. 2013). Additionally, long-term depression (LTD) leads to the reorganization and reduction of PSD-95 at synapses (Compans et al. 2021; Tang et al. 2016). Environmental enrichment has also been shown to promote PSD nanostructural remodeling, emphasizing the role of experience in reshaping synaptic architecture (Wegner et al. 2022).

Recent cryo-ET studies have provided further insights into the nanoscale organization of PSDs. Our cryo-ET study revealed that the PSD is composed of membrane-associated protein clusters, referred to as nanoblocks, which vary in size (Fig. 10.2) (Sun et al. 2025). Additionally, combining cryo-FIB milling with cryo-ET for imaging synapses demonstrated that PSD scaffolding complexes form nanoclusters, consistent with our findings (Held et al. 2024). Another study used cryo-ET with automated segmentation to reveal that isolated PSDs are composed of modular substructures of varying sizes, which may facilitate PSD remodeling for synaptic function and plasticity (Jung et al. 2023). In our study, most nanoblocks have a median width of ~30 nm in intact synapses and isolated synaptic terminals, which is smaller than PSD-95 nanodomains observed via super-resolution fluorescence microscopy. Larger nanoblocks (>50 nm wide) are comparable in size to PSD clusters, suggesting that nanoblocks and nanodomains may represent the same structures as seen through different imaging modalities. Cryo-ET has also demonstrated that larger nanoblocks are more effective at recruiting receptors. The heterogeneity in the size, assembly, and spatial organization of nanoblocks adds another layer of complexity to PSD architecture. Larger nanoblocks may consist of specific combinations of scaffold proteins forming supramolecular structures that link to synaptic receptors. Smaller nanoblocks may play regulatory roles, organizing into larger assemblies in response to synaptic activity.

Our cryo-ET findings suggest that nanoblocks may consist of defined copies of scaffold proteins that form functional units, linking synaptic receptors to presynaptic release sites via adhesion molecules. Additional proteins between nanoblocks may facilitate the assembly of smaller nanoblocks into larger complexes. The

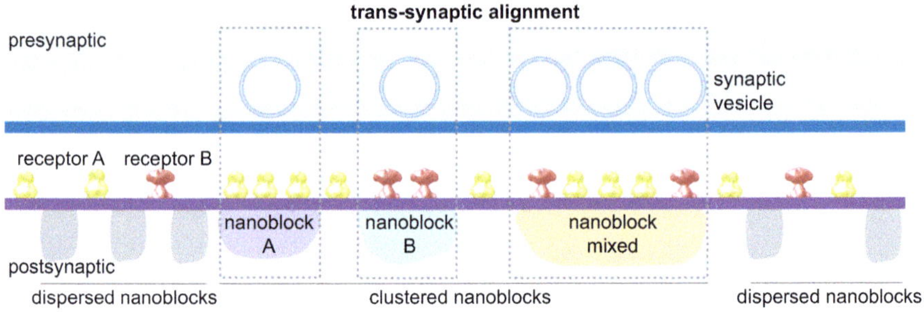

Fig. 10.2 Hypothetical model of trans-synaptic alignment. In our recent study, PSD nanoblocks are defined as subsynaptic, membrane-associated protein clusters in the postsynaptic density, visualized and identified using cryo-ET. Receptor A (yellow) and receptor B (red) represent two distinct receptor types identified through sub-tomogram averaging (Sun et al. 2025). PSD nanoblocks can be either clustered or dispersed, with synaptic receptors assembled or absent. Presynaptic release sites, indicated by tethered, docked, or primed synaptic vesicles, are preferentially aligned with "active" postsynaptic nanoblocks assembled with receptors

variation in the size and assembly of PSD nanoblocks might also reflect diverse protein compositions. It is also possible that nanoblocks of different sizes have distinct compositions and functional roles. Differences in experimental procedures and imaging techniques pose challenges to direct comparisons across studies. However, further investigations into the composition and organization of PSD nanoblocks will help elucidate the relationships between nanoblocks, nanodomains, and nanomodules, providing deeper insights into their distinct roles in synaptic signaling and plasticity.

10.4.3 *Alignment of Trans-synaptic Structures Across the Synaptic Cleft*

Super-resolution studies have provided compelling evidence for the organization of nanoscale supramolecular assemblies spanning the synaptic cleft, forming trans-synaptic nanocolumns that align pre- and postsynaptic compartments in excitatory synapses (Biederer et al. 2017; Chen et al. 2018; Frank and Grant 2017; Gramlich and Klyachko 2019; Guzikowski and Kavalali 2021; Kellermayer et al. 2018; MacGillavry et al. 2013; Tang et al. 2016). At the presynaptic terminal, active-zone scaffold proteins, such as RIM, organize into nanodomains that define synaptic vesicle docking and release sites. These release sites align with postsynaptic receptor clusters, such as AMPA receptors, and scaffolding proteins, like PSD-95, via adhesion molecules spanning the synaptic cleft (Haas et al. 2018; Martinez-Sanchez et al. 2021; Ouni et al. 2022; Ramsey et al. 2021).

While super-resolution microscopy has been instrumental in identifying nanoclusters of labeled synaptic proteins, such as PSD-95, receptors, and RIM, it is inherently limited to visualizing components that can be fluorescently labeled. In contrast, cryo-ET enables the examination of the molecular organization of all cellular components within their native context (Dubochet et al. 1988; Dubochet and Sartori Blanc 2001; Taylor and Glaeser 1974; Turk and Baumeister 2020). Using cryo-ET and sub-tomogram averaging, Martinez-Sanchez et al. demonstrated the alignment between presynaptic vesicles and postsynaptic receptors in synaptoneurosomes (Martinez-Sanchez et al. 2021). However, more recent cryo-ET studies have revealed that postsynaptic scaffolding nanoclusters, such as PSD nanoblocks, do not consistently align with presynaptic vesicle release sites (Held et al. 2024; Sun et al. 2025). Our own findings add further nuance to this complexity. We observed that while synaptic vesicle docking and release sites are trans-synaptically closer to specific "active" PSD nanoblocks associated with receptor-like particles, this alignment is selective and heterogeneous (Fig. 10.2) (Sun et al. 2025). Larger nanoblocks are particularly effective at recruiting receptors and aligning with presynaptic release sites. These results suggest that a simple one-to-one alignment between synaptic vesicles and PSD nanoblocks is insufficient to explain synaptic organization. Instead, nanocolumns should be understood as trans-synaptic structures that selectively modulate evoked neurotransmitter release (Ramsey et al. 2021; Tang et al. 2016). Recent evidence also indicates spatial segregation between regions in the presynapse dedicated to spontaneous and evoked neurotransmitter release (Guzikowski and Kavalali 2021; Wang et al. 2023). Combined with our findings of distinct nanoblocks containing different receptors, this raises the intriguing possibility that spontaneous and evoked release rely on dedicated subsets of nanoblocks. Furthermore, "inactive" nanoblocks that lack receptors may serve as reserve scaffolds or harbor other receptor types.

Future advancements in cryo-ET, particularly in identifying individual protein components within the synapse, will be critical for unraveling the specific alignments between different subgroups of postsynaptic nanoblocks and their corresponding release sites. This line of investigation could help clarify the functional implications of nanoblock diversity in regulating synaptic activity.

10.5 Conclusions

The combination of advanced imaging techniques, such as cryo-ET, has provided unprecedented insights into the ultrastructure of synapses and the molecular mechanisms underlying synaptic transmission and synaptic plasticity. These techniques have allowed researchers to visualize the nanoscale organization of synaptic components, the alignment of trans-synaptic structures, and the in situ structure of macromolecular complexes involved in neurotransmission and synaptic plasticity.

By preserving the native state of biological specimens, cryo-ET has revolutionized our understanding of synaptic function and plasticity, revealing the intricate organization of synaptic vesicles, active zones, and other key structures at nanometer resolution. These insights have provided a deeper understanding of the molecular mechanisms that regulate neurotransmitter release, synaptic plasticity, and other critical processes in the nervous system.

Looking forward, the continued development of imaging techniques and data analysis methods promises to further advance our understanding of synaptic ultrastructure and function. Future studies will likely focus on integrating cryo-ET with other techniques, such as EM-visible labeling methods, live-cell imaging, and super-resolution microscopy, to provide a more comprehensive view of synaptic function in health and disease.

Acknowledgements The authors would like to acknowledge and thank Dr. Minglei Zhao for insightful discussions, Dr. Sarah Croessmann of the Vanderbilt SWERV writing core for assisting in scientific illustrations of this manuscript, and the NIH for support (R01 MH132918 to Q.Z.).

References

Albert S, Wietrzynski W, Lee CW, Schaffer M, Beck F, Schuller JM, Salomé PA, Plitzko JM, Baumeister W, Engel BD. Direct visualization of degradation microcompartments at the ER membrane. Proc Natl Acad Sci U S A. 2020;117:1069–80.

Araki Y, Zeng M, Zhang M, Huganir RL. Rapid dispersion of SynGAP from synaptic spines triggers AMPA receptor insertion and spine enlargement during LTP. Neuron. 2015;85:173–89.

Bats C, Groc L, Choquet D. The interaction between Stargazin and PSD-95 regulates AMPA receptor surface trafficking. Neuron. 2007;53:719–34.

Batson PE, Dellby N, Krivanek OL. Sub-ångstrom resolution using aberration corrected electron optics. Nature. 2002;418:617–20.

Baumeister W. Electron tomography: towards visualizing the molecular organization of the cytoplasm. Curr Opin Struct Biol. 2002;12:679–84.

Baumeister W, Grimm R, Walz J. Electron tomography of molecules and cells. Trends Cell Biol. 1999;9:81–5.

Beique JC, Lin DT, Kang MG, Aizawa H, Takamiya K, Huganir RL. Synapse-specific regulation of AMPA receptor function by PSD-95. Proc Natl Acad Sci U S A. 2006;103:19535–40.

Bepler T, Morin A, Rapp M, Brasch J, Shapiro L, Noble AJ, Berger B. Positive-unlabeled convolutional neural networks for particle picking in cryo-electron micrographs. Nat Methods. 2019;16:1153–60.

Bi GQ, Poo MM. Synaptic modifications in cultured hippocampal neurons: Dependence on spike timing, synaptic strength, and postsynaptic cell type. J Neurosci. 1998;18:10464–72.

Bianchi F, Malboubi M, Li Y, George JH, Jerusalem A, Szele F, Thompson MS, Ye H. Rapid and efficient differentiation of functional motor neurons from human iPSC for neural injury modelling. Stem Cell Res. 2018;32:126–34.

Bieber A, Capitanio C, Erdmann PS, Fiedler F, Beck F, Lee CW, Li D, Hummer G, Schulman BA, Baumeister W, Wilfling F. In situ structural analysis reveals membrane shape transitions during autophagosome formation. Proc Natl Acad Sci U S A. 2022;119:e2209823119.

Biederer T, Kaeser PS, Blanpied TA. Transcellular nanoalignment of synaptic function. Neuron. 2017;96:680–96.

Boeckers TM. The postsynaptic density. Cell Tissue Res. 2006;326:409–22.

Bosch M, Castro J, Saneyoshi T, Matsuno H, Sur M, Hayashi Y. Structural and molecular remodeling of dendritic spine substructures during long-term potentiation. Neuron. 2014;82:444–59.

Bourne JN, Harris KM. Coordination of size and number of excitatory and inhibitory synapses results in a balanced structural plasticity along mature hippocampal CA1 dendrites during LTP. Hippocampus. 2011;21:354–73.

Brenner S, Horne RW. A negative staining method for high resolution electron microscopy of viruses. Biochim Biophys Acta. 1959;34:103–10.

Broadhead MJ, Horrocks MH, Zhu F, Muresan L, Benavides-Piccione R, DeFelipe J, Fricker D, Kopanitsa MV, Duncan RR, Klenerman D, Komiyama NH, Lee SF, Grant SG. PSD95 nanoclusters are postsynaptic building blocks in hippocampus circuits. Sci Rep. 2016;6:24626.

Bromer C, Bartol TM, Bowden JB, Hubbard DD, Hanka DC, Gonzalez PV, Kuwajima M, Mendenhall JM, Parker PH, Abraham WC, Sejnowski TJ, Harris KM. Long-term potentiation expands information content of hippocampal dentate gyrus synapses. Proc Natl Acad Sci U S A. 2018;115:E2410–8.

Burette AC, Lesperance T, Crum J, Martone M, Volkmann N, Ellisman MH, Weinberg RJ. Electron tomographic analysis of synaptic ultrastructure. J Comp Neurol. 2012;520:2697–711.

Bykov YS, Schaffer M, Dodonova SO, Albert S, Plitzko JM, Baumeister W, Engel BD, Briggs JA. The structure of the COPI coat determined within the cell. Elife. 2017;6

Bäuerlein FJB, Renner M, Chami DE, Lehnart SE, Pastor-Pareja JC, Fernández-Busnadiego R. Cryo-electron tomography of large biological specimens vitrified by plunge freezing. bioRxiv. 2023:2021.2004.2014.437159.

Cai S, Chen C, Tan ZY, Huang Y, Shi J, Gan L. Cryo-ET reveals the macromolecular reorganization of. Proc Natl Acad Sci U S A. 2018;115:10977–82.

Carlin RK, Grab DJ, Cohen RS, Siekevitz P. Isolation and characterization of postsynaptic densities from various brain regions: enrichment of different types of postsynaptic densities. J Cell Biol. 1980;86:831–45.

Ceccarelli B, Hurlbut WP, Mauro A. Turnover of transmitter and synaptic vesicles at the frog neuromuscular junction. J Cell Biol. 1973;57:499–524.

Chaikeeratisak V, Nguyen K, Khanna K, Brilot AF, Erb ML, Coker JK, Vavilina A, Newton GL, Buschauer R, Pogliano K, Villa E, Agard DA, Pogliano J. Assembly of a nucleus-like structure during viral replication in bacteria. Science. 2017;355:194–7.

Chang Y-W, Chen S, Tocheva EI, Treuner-Lange A, Loebach S, Sogaard-Andersen L, Jensen GJ. Correlated cryogenic photoactivated localization microscopy and cryo-electron tomography. Nat Methods. 2014;11:737–U179.

Chen H, Tang AH, Blanpied TA. Subsynaptic spatial organization as a regulator of synaptic strength and plasticity. Curr Opin Neurobiol. 2018;51:147–53.

Chen L, Chetkovich DM, Petralia RS, Sweeney NT, Kawasaki Y, Wenthold RJ, Bredt DS, Nicoll RA. Stargazin regulates synaptic targeting of AMPA receptors by two distinct mechanisms. Nature. 2000;408:936–43.

Chen M, Dai W, Sun SY, Jonasch D, He CY, Schmid MF, Chiu W, Ludtke SJ. Convolutional neural networks for automated annotation of cellular cryo-electron tomograms. Nat Methods. 2017;14:983–5.

Chen X, Levy JM, Hou A, Winters C, Azzam R, Sousa AA, Leapman RD, Nicoll RA, Reese TS. PSD-95 family MAGUKs are essential for anchoring AMPA and NMDA receptor complexes at the postsynaptic density. Proc Natl Acad Sci U S A. 2015;112:E6983–92.

Chen X, Nelson CD, Li X, Winters CA, Azzam R, Sousa AA, Leapman RD, Gainer H, Sheng M, Reese TS. PSD-95 is required to sustain the molecular organization of the postsynaptic density. J Neurosci. 2011;31:6329–38.

Chen X, Vinade L, Leapman RD, Petersen JD, Nakagawa T, Phillips TM, Sheng M, Reese TS. Mass of the postsynaptic density and enumeration of three key molecules. Proc Natl Acad Sci U S A. 2005;102:11551–6.

Chen X, Winters C, Azzam R, Li X, Galbraith JA, Leapman RD, Reese TS. Organization of the core structure of the postsynaptic density. Proc Natl Acad Sci U S A. 2008;105:4453–8.

Cheng D, Hoogenraad CC, Rush J, Ramm E, Schlager MA, Duong DM, Xu P, Wijayawardana SR, Hanfelt J, Nakagawa T, Sheng M, Peng J. Relative and absolute quantification of postsynaptic density proteome isolated from rat forebrain and cerebellum. Mol Cell Proteomics. 2006;5:1158–70.

Cheng Y, Grigorieff N, Penczek PA, Walz T. A primer to single-particle cryo-electron microscopy. Cell. 2015;161:438–49.

Cho KO, Hunt CA, Kennedy MB. The rat brain postsynaptic density fraction contains a homolog of the Drosophila discs-large tumor suppressor protein. Neuron. 1992;9:929–42.
Cohen R, Blomberg F, Berzins K, Siekevitz P. The structure of postsynaptic densities isolated from dog cerebral cortex: I. overall morphology and protein composition. J Cell Biol. 1977;74:181–203.
Cole AA, Chen X, Reese TS. A network of three types of filaments organizes synaptic vesicles for storage, mobilization, and docking. J Neurosci. 2016;36:3222–30.
Compans B, Camus C, Kallergi E, Sposini S, Martineau M, Butler C, Kechkar A, Klaassen RV, Retailleau N, Sejnowski TJ, Smit AB, Sibarita JB, Bartol TM, Perrais D, Nikoletopoulou V, Choquet D, Hosy E. NMDAR-dependent long-term depression is associated with increased short term plasticity through autophagy mediated loss of PSD-95. Nat Commun. 2021;12:2849.
Cooney JR, Hurlburt JL, Selig DK, Harris KM, Fiala JC. Endosomal compartments serve multiple hippocampal dendritic spines from a widespread rather than a local store of recycling membrane. J Neurosci. 2002;22:2215–24.
Costa-Filho JI, Theveny L, de Sautu M, Kirchhausen T. CryoSamba: self-supervised deep volumetric denoising for cryo-electron tomography data. J Struct Biol. 2024;217:108163.
Creekmore BC, Kixmoeller K, Black BE, Lee EB, Chang YW. Ultrastructure of human brain tissue vitrified from autopsy revealed by cryo-ET with cryo-plasma FIB milling. Nat Commun. 2024;15:2660.
Cressey D, Callaway E. Cryo-electron microscopy wins chemistry Nobel. Nature. 2017; 550(7675):167.
Crosby KC, Gookin SE, Garcia JD, Hahm KM, Dell'Acqua ML, Smith KR. Nanoscale subsynaptic domains underlie the organization of the inhibitory synapse. Cell Rep. 2019;26:3284–97. e3283
Dahlberg PD, Moerner WE. Cryogenic super-resolution fluorescence and electron microscopy correlated at the nanoscale. Annu Rev Phys Chem. 2021;72:253–78.
Dahlberg PD, Saurabh S, Sartor AM, Wang J, Mitchell PG, Chiu W, Shapiro L, Moerner WE. Cryogenic single-molecule fluorescence annotations for electron tomography reveal in situ organization of key proteins in. Proc Natl Acad Sci U S A. 2020;117:13937–44.
Dani A, Huang B, Bergan J, Dulac C, Zhuang X. Superresolution imaging of chemical synapses in the brain. Neuron. 2010;68:843–56.
de Oliveira TM, van Beek L, Shilliday F, Debreczeni JE, Phillips C. Cryo-EM: the resolution revolution and drug discovery. SLAS Discov. 2020:2472555220960401.
De Robertis ED, Bennett HS. Some features of the submicroscopic morphology of synapses in frog and earthworm. J Biophys Biochem Cytol. 1955;1:47–58.
DeGiorgis JA, Galbraith JA, Dosemeci A, Chen X, Reese TS. Distribution of the scaffolding proteins PSD-95, PSD-93, and SAP97 in isolated PSDs. Brain Cell Biol. 2006;35:239–50.
Denk W, Horstmann H. Serial block-face scanning electron microscopy to reconstruct three-dimensional tissue nanostructure. PLoS Biol. 2004;2:e329.
Dosemeci A, Reese TS, Petersen J, Tao-Cheng JH. A novel particulate form of Ca(2+)/calmodulin-dependent [correction of Ca(2+)/CaMKII-dependent] protein kinase II in neurons. J Neurosci. 2000;20:3076–84.
Dubochet J, Adrian M, Chang JJ, Homo JC, Lepault J, McDowall AW, Schultz P. Cryo-electron microscopy of vitrified specimens. Q Rev Biophys. 1988;21:129–228.
Dubochet J, Sartori Blanc N. The cell in absence of aggregation artifacts. Micron. 2001;32:91–9.
Dubochet J, Zuber B, Eltsov M, Bouchet-Marquis C, Al-Amoudi A, Livolant F. How to "read" a vitreous section. Methods Cell Biol. 2007;79:385–406.
El-Husseini AE, Schnell E, Chetkovich DM, Nicoll RA, Bredt DS. PSD-95 involvement in maturation of excitatory synapses. Science. 2000;290:1364–8.
Elias GM, Funke L, Stein V, Grant SG, Bredt DS, Nicoll RA. Synapse-specific and developmentally regulated targeting of AMPA receptors by a family of MAGUK scaffolding proteins. Neuron. 2006;52:307–20.
Elias GM, Nicoll RA. Synaptic trafficking of glutamate receptors by MAGUK scaffolding proteins. Trends Cell Biol. 2007;17:343–52.

Erk I, Delacroix H, Nicolas G, Ranck JL, Lepault J. Cryo-electron microscopy of vitrified specimens: an approach to the study of bulk specimens. J Electron Microsc Tech. 1991;18:406–10.
Fatt P, Katz B. Spontaneous subthreshold activity at motor nerve endings. J Physiol. 1952;117:109–28.
Feldmüller M, Ericson CF, Afanasyev P, Lien YW, Weiss GL, Wollweber F, Schoof M, Hurst M, Pilhofer M. Stepwise assembly and release of Tc toxins from Yersinia entomophaga. Nat Microbiol. 2024;9:405–20.
Feng W, Zhang M. Organization and dynamics of PDZ-domain-related supramodules in the postsynaptic density. Nat Rev Neurosci. 2009;10:87–99.
Feng Z, Chen X, Zeng M, Zhang M. Phase separation as a mechanism for assembling dynamic postsynaptic density signalling complexes. Curr Opin Neurobiol. 2019;57:1–8.
Fernandez E, Collins MO, Uren RT, Kopanitsa MV, Komiyama NH, Croning MD, Zografos L, Armstrong JD, Choudhary JS, Grant SG. Targeted tandem affinity purification of PSD-95 recovers core postsynaptic complexes and schizophrenia susceptibility proteins. Mol Syst Biol. 2009;5:269.
Fernandez-Busnadiego R, Asano S, Oprisoreanu AM, Sakata E, Doengi M, Kochovski Z, Zumer M, Stein V, Schoch S, Baumeister W, Lucic V. Cryo-electron tomography reveals a critical role of RIM1 alpha in synaptic vesicle tethering. J Cell Biol. 2013;201:725–40.
Fernandez-Busnadiego R, Zuber B, Maurer UE, Cyrklaff M, Baumeister W, Lucic V. Quantitative analysis of the native presynaptic cytomatrix by cryoelectron tomography. J Cell Biol. 2010a;188:145–56.
Fernandez-Busnadiego R, Zuber B, Maurer UE, Cyrklaff M, Baumeister W, Lucic V. Quantitative analysis of the native presynaptic cytomatrix by cryoelectron tomography. J Cell Biol. 2010b;188:145–56.
Frank J. Introduction: principles of electron tomography. In: Frank J, editor. Electron tomography. Springer; 2005.
Frank RA, Grant SG. Supramolecular organization of NMDA receptors and the postsynaptic density. Curr Opin Neurobiol. 2017;45:139–47.
Frank RAW, Zhu F, Komiyama NH, Grant SGN. Hierarchical organization and genetically separable subfamilies of PSD95 postsynaptic supercomplexes. J Neurochem. 2017;142:504–11.
Fukata Y, Dimitrov A, Boncompain G, Vielemeyer O, Perez F, Fukata M. Local palmitoylation cycles define activity-regulated postsynaptic subdomains. J Cell Biol. 2013;202:145–61.
Fung HKH, Hayashi Y, Salo VT, Babenko A, Zagoriy I, Brunner A, Ellenberg J, Müller CW, Cuylen-Haering S, Mahamid J. Genetically encoded multimeric tags for subcellular protein localization in cryo-EM. Nat Methods. 2023;20:1900–8.
Föcking M, Dicker P, Lopez LM, Hryniewiecka M, Wynne K, English JA, Cagney G, Cotter DR. Proteomic analysis of the postsynaptic density implicates synaptic function and energy pathways in bipolar disorder. Transl Psychiatry. 2016;6:e959.
Gilbert MAG, Fatima N, Jenkins J, O'Sullivan TJ, Schertel A, Halfon Y, Wilkinson M, Morrema THJ, Geibel M, Read RJ, Ranson NA, Radford SE, Hoozemans JJM, Frank RAW. CryoET of β-amyloid and tau within postmortem Alzheimer's disease brain. Nature. 2024;631:913–9.
Gilbert P. Iterative methods for the three-dimensional reconstruction of an object from projections. J Theor Biol. 1972;36:105–17.
Glynn C, Smith JLR, Case M, Csöndör R, Katsini A, Sanita ME, Glen TS, Pennington A, Grange M. Charting the molecular landscape of neuronal organisation within the hippocampus using cryo electron tomography. bioRxiv. 2024:2024.2010.2014.617844.
Gonda MA. Electron microscopy, immunological applications. In: Delves PJ, editor. Encyclopedia of immunology. 2nd ed. Oxford: Elsevier; 1998. p. 790–5.
Gramlich MW, Klyachko VA. Nanoscale organization of vesicle release at central synapses. Trends Neurosci. 2019;42:425–37.
Groc L, Choquet D. Linking glutamate receptor movements and synapse function. Science. 2020;368
Gruber A, Hornburg D, Antonin M, Krahmer N, Collado J, Schaffer M, Zubaite G, Lüchtenborg C, Sachsenheimer T, Brügger B, Mann M, Baumeister W, Hartl FU, Hipp MS, Fernández-

Busnadiego R. Molecular and structural architecture of polyQ aggregates in yeast. Proc Natl Acad Sci U S A. 2018;115:E3446–53.
Guo Q, Lehmer C, Martinez-Sanchez A, Rudack T, Beck F, Hartmann H, Perez-Berlanga M, Frottin F, Hipp MS, Hartl FU, Edbauer D, Baumeister W, Fernandez-Busnadiego R. In situ structure of neuronal C9orf72 Poly-GA aggregates reveals proteasome recruitment. Cell. 2018;172(696-705):e612.
Guzikowski NJ, Kavalali ET. Nano-organization at the synapse: segregation of distinct forms of neurotransmission. Front Synaptic Neurosci. 2021;13:796498.
Haas KT, Compans B, Letellier M, Bartol TM, Grillo-Bosch D, Sejnowski TJ, Sainlos M, Choquet D, Thoumine O, Hosy E. Pre-post synaptic alignment through neuroligin-1 tunes synaptic transmission efficiency. Elife. 2018;7
Haberfehlner G, Orthacker A, Albu M, Li J, Kothleitner G. Nanoscale voxel spectroscopy by simultaneous EELS and EDS tomography. Nanoscale. 2014;6:14563–9.
Hagen WJH, Wan W, Briggs JAG. Implementation of a cryo-electron tomography tilt-scheme optimized for high resolution subtomogram averaging. J Struct Biol. 2017;197:191–8.
Harris KM, Stevens JK. Dendritic spines of CA 1 pyramidal cells in the rat hippocampus: serial electron microscopy with reference to their biophysical characteristics. J Neurosci. 1989;9:2982–97.
Held RG, Liang J, Brunger AT. Nanoscale architecture of synaptic vesicles and scaffolding complexes revealed by cryo-electron tomography. Proc Natl Acad Sci U S A. 2024;121:e2403136121.
Heuser JE, Reese TS. Evidence for recycling of synaptic vesicle membrane during transmitter release at the frog neuromuscular junction. J Cell Biol. 1973;57:315–44.
Hoffmann PC, Giandomenico SL, Ganeva I, Wozny MR, Sutcliffe M, Lancaster MA, Kukulski W. Electron cryo-tomography reveals the subcellular architecture of growing axons in human brain organoids. Elife. 2021;10
Hoffmann PC, Kreysing JP, Khusainov I, Tuijtel MW, Welsch S, Beck M. Structures of the eukaryotic ribosome and its translational states in situ. Nat Commun. 2022;13:7435.
Hruska M, Cain RE, Dalva MB. Nanoscale rules governing the organization of glutamate receptors in spine synapses are subunit specific. Nat Commun. 2022;13:920.
Hruska M, Henderson N, Le Marchand SJ, Jafri H, Dalva MB. Synaptic nanomodules underlie the organization and plasticity of spine synapses. Nat Neurosci. 2018;21:671–82.
Hua Y, Laserstein P, Helmstaedter M. Large-volume en-bloc staining for electron microscopy-based connectomics. Nat Commun. 2015;6:7923.
Huganir RL, Nicoll RA. AMPARs and synaptic plasticity: the last 25 years. Neuron. 2013;80:704–17.
Hurbain I, Sachse M. The future is cold: cryo-preparation methods for transmission electron microscopy of cells. Biology of the Cell. 2011;103:405–20.
Imig C, Min SW, Krinner S, Arancillo M, Rosenmund C, Sudhof TC, Rhee J, Brose N, Cooper BH. The morphological and molecular nature of synaptic vesicle priming at presynaptic active zones. Neuron. 2014;84:416–31.
Jentoft IMA, Bäuerlein FJB, Welp LM, Cooper BH, Petrovic A, So C, Penir SM, Politi AZ, Horokhovskyi Y, Takala I, Eckel H, Moltrecht R, Lénárt P, Cavazza T, Liepe J, Brose N, Urlaub H, Fernández-Busnadiego R, Schuh M. Mammalian oocytes store proteins for the early embryo on cytoplasmic lattices. Cell. 2023;186:5308–5327.e5325.
Jordan MA, Diener DR, Stepanek L, Pigino G. The cryo-EM structure of intraflagellar transport trains reveals how dynein is inactivated to ensure unidirectional anterograde movement in cilia. Nat Cell Biol. 2018;20:1250–5.
Jung JH, Chen X, Reese TS. Cryo-EM tomography and automatic segmentation delineate modular structures in the postsynaptic density. Front Synaptic Neurosci. 2023;15:1123564.
Jung JH, Kirk LM, Bourne JN, Harris KM. Shortened tethering filaments stabilize presynaptic vesicles in support of elevated release probability during LTP in rat hippocampus. Proc Natl Acad Sci U S A. 2021;118

Jung JH, Szule JA, Stouder K, Marshall RM, McMahan UJ. Active zone material-directed orientation, docking, and fusion of dense core vesicles alongside synaptic vesicles at neuromuscular junctions. Front Neuroanat. 2018;12:72.

Kaech S, Banker G. Culturing hippocampal neurons. Nat Protoc. 2006;1:2406–15.

Kaizuka T, Takumi T. Postsynaptic density proteins and their involvement in neurodevelopmental disorders. J Biochem. 2018;163:447–55.

Kasthuri N, Hayworth KJ, Berger DR, Schalek RL, Conchello JA, Knowles-Barley S, Lee D, Vazquez-Reina A, Kaynig V, Jones TR, Roberts M, Morgan JL, Tapia JC, Seung HS, Roncal WG, Vogelstein JT, Burns R, Sussman DL, Priebe CE, Pfister H, Lichtman JW. Saturated reconstruction of a volume of neocortex. Cell. 2015;162:648–61.

Katz B, Miledi R. The role of calcium in neuromuscular facilitation. J Physiol. 1968;195:481–92.

Kellermayer B, Ferreira JS, Dupuis J, Levet F, Grillo-Bosch D, Bard L, Linares-Loyez J, Bouchet D, Choquet D, Rusakov DA, Bon P, Sibarita JB, Cognet L, Sainlos M, Carvalho AL, Groc L. Differential nanoscale topography and functional role of GluN2-NMDA receptor subtypes at glutamatergic synapses. Neuron. 2018;100(106-119):e107.

Kelley K, Raczkowski AM, Klykov O, Jaroenlak P, Bobe D, Kopylov M, Eng ET, Bhabha G, Potter CS, Carragher B, Noble AJ. Waffle method: a general and flexible approach for improving throughput in FIB-milling. Nat Commun. 2022;13:1857.

Kharazia VN, Weinberg RJ. Tangential synaptic distribution of NMDA and AMPA receptors in rat neocortex. Neurosci Lett. 1997;238:41–4.

Khosrozadeh A, Seeger R, Witz G, Radecke J, Sørensen JB, Zuber B. CryoVesNet: a dedicated framework for synaptic vesicle segmentation in cryo-electron tomograms. J Cell Biol. 2025;224

Kim E, Cho KO, Rothschild A, Sheng M. Heteromultimerization and NMDA receptor-clustering activity of Chapsyn-110, a member of the PSD-95 family of proteins. Neuron. 1996;17:103–13.

Kisielowski C, Freitag B, Bischoff M, van Lin H, Lazar S, Knippels G, Tiemeijer P, van der Stam M, von Harrach S, Stekelenburg M, Haider M, Uhlemann S, Müller H, Hartel P, Kabius B, Miller D, Petrov I, Olson E, Donchev T, Kenik E, Lupini A, Bentley J, Pennycook S, Anderson I, Minor A, Schmid A, Duden T, Radmilovic V, Ramasse Q, Watanabe M, Erni R, Stach E, Denes P, Dahmen U. Detection of single atoms and buried defects in three dimensions by aberration-corrected electron microscope with 0.5-Å information limit. Microscopy and Microanalysis. 2008;14:469–77.

Knott G, Marchman H, Wall D, Lich B. Serial section scanning electron microscopy of adult brain tissue using focused ion beam milling. J Neurosci. 2008;28:2959–64.

Kornau HC, Schenker LT, Kennedy MB, Seeburg PH. Domain interaction between NMDA receptor subunits and the postsynaptic density protein PSD-95. Science. 1995;269:1737–40.

Korogod N, Petersen CCH, Knott GW. Ultrastructural analysis of adult mouse neocortex comparing aldehyde perfusion with cryo fixation. Elife. 2015;4

Koster AJ, Klumperman J. Electron microscopy in cell biology: integrating structure and function. Nat Rev Mol Cell Biol. 2003; Suppl:SS6–10

Kravčenko U, Ruwolt M, Kroll J, Yushkevich A, Zenkner M, Ruta J, Lotfy R, Wanker EE, Rosenmund C, Liu F, Kudryashev M. Molecular architecture of synaptic vesicles. Proc Natl Acad Sci U S A. 2024;121:e2407375121.

Kremer JR, Mastronarde DN, McIntosh JR. Computer visualization of three-dimensional image data using IMOD. J Struct Biol. 1996;116:71–6.

Kühlbrandt W. Biochemistry. The resolution revolution. Science. 2014;343:1443–4.

Laursen L, Karlsson E, Gianni S, Jemth P. Functional interplay between protein domains in a supramodular structure involving the postsynaptic density protein PSD-95. J Biol Chem. 2020;295:1992–2000.

Leistner C, Wilkinson M, Burgess A, Lovatt M, Goodbody S, Xu Y, Deuchars S, Radford SE, Ranson NA, Frank RAW. The in-tissue molecular architecture of β-amyloid pathology in the mammalian brain. Nat Commun. 2023;14:2833.

Levy JM, Chen X, Reese TS, Nicoll RA. Synaptic consolidation normalizes AMPAR quantal size following MAGUK loss. Neuron. 2015;87:534–48.

Li W, Lu J, Xiao K, Zhou M, Li Y, Zhang X, Li Z, Gu L, Xu X, Guo Q, Xu T, Ji W. Integrated multimodality microscope for accurate and efficient target-guided cryo-lamellae preparation. Nat Methods. 2023;20:268–75.

Li et al. Spatiotemporal profile of postsynaptic interactomes integrates components of complex brain disorders. Nat Neurosci. 2017;20(8):1150–1161. PMID: 28671696 PMCID: PMC564508. https://doi.org/10.1038/nn.4594

Li X, Mooney P, Zheng S, Booth CR, Braunfeld MB, Gubbens S, Agard DA, Cheng Y. Electron counting and beam-induced motion correction enable near-atomic-resolution single-particle cryo-EM. Nat Methods. 2013;10:584–90.

Liu YT, Tao CL, Zhang XK, Xia WJ, Shi DQ, Qi L, Xu C, Sun R, Li XW, Lau PM, Zhou ZH, Bi GQ. Mesophasic organization of GABA(A) receptors in hippocampal inhibitory synapses. Nat Neurosci. 2020;23:1589–U1118.

Long JF, Tochio H, Wang P, Fan JS, Sala C, Niethammer M, Sheng M, Zhang M. Supramodular structure and synergistic target binding of the N-terminal tandem PDZ domains of PSD-95. J Mol Biol. 2003;327:203–14.

Lowenthal MS, Markey SP, Dosemeci A. Quantitative mass spectrometry measurements reveal stoichiometry of principal postsynaptic density proteins. J Proteome Res. 2015;14:2528–38.

Lucic V, Kossel AH, Yang T, Bonhoeffer T, Baumeister W, Sartori A. Multiscale imaging of neurons grown in culture: from light microscopy to cryo-electron tomography. J Struct Biol. 2007;160:146–56.

MacGillavry HD, Song Y, Raghavachari S, Blanpied TA. Nanoscale scaffolding domains within the postsynaptic density concentrate synaptic AMPA receptors. Neuron. 2013;78:615–22.

Mahamid J, Pfeffer S, Schaffer M, Villa E, Danev R, Cuellar LK, Förster F, Hyman AA, Plitzko JM, Baumeister W. Visualizing the molecular sociology at the HeLa cell nuclear periphery. Science. 2016;351:969–72.

Malinow R, Malenka RC. AMPA receptor trafficking and synaptic plasticity. Annu Rev Neurosci. 2002;25:103–26.

Martinez-Sanchez A, Laugks U, Kochovski Z, Papantoniou C, Zinzula L, Baumeister W, Lucic V. Trans-synaptic assemblies link synaptic vesicles and neuroreceptors. Sci Adv. 2021;7

Mastronarde DN. Automated electron microscope tomography using robust prediction of specimen movements. J Struct Biol. 2005;152:36–51.

Masugi-Tokita M, Tarusawa E, Watanabe M, Molnar E, Fujimoto K, Shigemoto R. Number and density of AMPA receptors in individual synapses in the rat cerebellum as revealed by SDS-digested freeze-fracture replica labeling. J Neurosci. 2007;27:2135–44.

Matsui A, Spangler C, Elferich J, Shiozaki M, Jean N, Zhao X, Qin M, Zhong H, Yu Z, Gouaux E. Cryo-electron tomographic investigation of native hippocampal glutamatergic synapses. Elife. 2024;13

Maximov A, Pang ZP, Tervo DG, Südhof TC. Monitoring synaptic transmission in primary neuronal cultures using local extracellular stimulation. J Neurosci Methods. 2007;161:75–87.

McCafferty CL, Klumpe S, Amaro RE, Kukulski W, Collinson L, Engel BD. Integrating cellular electron microscopy with multimodal data to explore biology across space and time. Cell. 2024;187:563–84.

McDowall AW, Chang JJ, Freeman R, Lepault J, Walter CA, Dubochet J. Electron microscopy of frozen hydrated sections of vitreous ice and vitrified biological samples. J Microsc. 1983;131:1–9.

McIntosh JR. Electron microscopy of cells: a new beginning for a new century. J Cell Biol. 2001;153:F25–32.

McMullan G, Chen S, Henderson R, Faruqi AR. Detective quantum efficiency of electron areadetectors in electron microscopy. Ultramicroscopy. 2009.

Medalia O, Weber I, Frangakis AS, Nicastro D, Gerisch G, Baumeister W. Macromolecular architecture in eukaryotic cells visualized by cryoelectron tomography. Science. 2002;298:1209–13.

Milovanovic D, Wu Y, Bian X, De Camilli P. A liquid phase of synapsin and lipid vesicles. Science. 2018;361:604–7.

Moor H. Theory and practice of high pressure freezing. In: Steinbrecht RA, Zierold K, editors. Cryotechniques in biological electron microscopy. Berlin/Heidelberg: Springer; 1987. p. 175–91.

Moreau M, Cochard P. Nobel Prize in Chemistry 2014 – from microscopy to nanoscopy: a revolution in resolution. Med Sci (Paris). 2014;30:1169–76.

Morgan JL, Berger DR, Wetzel AW, Lichtman JW. The fuzzy logic of network connectivity in mouse visual thalamus. Cell. 2016;165:192–206.

Moser F, Pražák V, Mordhorst V, Andrade DM, Baker LA, Hagen C, Grünewald K, Kaufmann R. Cryo-SOFI enabling low-dose super-resolution correlative light and electron cryo-microscopy. Proc Natl Acad Sci U S A. 2019;116:4804–9.

Müller BM, Kistner U, Kindler S, Chung WJ, Kuhlendahl S, Fenster SD, Lau LF, Veh RW, Huganir RL, Gundelfinger ED, Garner CC. SAP102, a novel postsynaptic protein that interacts with NMDA receptor complexes in vivo. Neuron. 1996;17:255–65.

Nair D, Hosy E, Petersen JD, Constals A, Giannone G, Choquet D, Sibarita JB. Super-resolution imaging reveals that AMPA receptors inside synapses are dynamically organized in nanodomains regulated by PSD95. J Neurosci. 2013;33:13204–24.

Nakane T, Kotecha A, Sente A, McMullan G, Masiulis S, Brown PMGE, Grigoras IT, Malinauskaite L, Malinauskas T, Miehling J, Uchański T, Yu L, Karia D, Pechnikova EV, de Jong E, Keizer J, Bischoff M, McCormack J, Tiemeijer P, Hardwick SW, Chirgadze DY, Murshudov G, Aricescu AR, Scheres SHW. Single-particle cryo-EM at atomic resolution. Nature. 2020;587:152–6.

Navarro PP, Vettiger A, Ananda VY, Llopis PM, Allolio C, Bernhardt TG, Chao LH. Cell wall synthesis and remodelling dynamics determine division site architecture and cell shape in Escherichia coli. Nat Microbiol. 2022;7:1621–34.

Nedozralova H, Basnet N, Ibiricu I, Bodakuntla S, Biertümpfel C, Mizuno N. In situ cryo-electron tomography reveals local cellular machineries for axon branch development. J Cell Biol. 2022;221

Ng CT, Deng L, Chen C, Lim HH, Shi J, Surana U, Gan L. Electron cryotomography analysis of Dam1C/DASH at the kinetochore-spindle interface in situ. J Cell Biol. 2019;218:455–73.

Ng CT, Gan L. Investigating eukaryotic cells with cryo-ET. Mol Biol Cell. 2020;31:87–100.

Nicastro D, Frangakis AS, Typke D, Baumeister W. Cryo-electron tomography of neurospora mitochondria. J Struct Biol. 2000;129:48–56.

Nicoll RA. A brief history of long-term potentiation. Neuron. 2017;93:281–90.

Nicoll RA, Tomita S, Bredt DS. Auxiliary subunits assist AMPA-type glutamate receptors. Science. 2006;311:1253–6.

Niethammer M, Kim E, Sheng M. Interaction between the C terminus of NMDA receptor subunits and multiple members of the PSD-95 family of membrane-associated guanylate kinases. J Neurosci. 1996;16:2157–63.

Nusser Z, Lujan R, Laube G, Roberts JD, Molnar E, Somogyi P. Cell type and pathway dependence of synaptic AMPA receptor number and variability in the hippocampus. Neuron. 1998;21:545–59.

Ouni E, Nedbal V, Da Pian M, Cao H, Haas KT, Peaucelle A, Van Kerk O, Herinckx G, Marbaix E, Dolmans MM, Tuuri T, Otala M, Amorim CA, Vertommen D. Proteome-wide and matrisome-specific atlas of the human ovary computes fertility biomarker candidates and open the way for precision oncofertility. Matrix Biol. 2022;109:91–120.

Palay SL. Synapses in the central nervous system. J Biophys Biochem Cytol. 1956;2:193–202.

Palay SL. The morphology of synapses in the central nervous system. Exp Cell Res. 1958;14:275–93.

Pan L, Chen J, Yu J, Yu H, Zhang M. The structure of the PDZ3-SH3-GuK tandem of ZO-1 protein suggests a supramodular organization of the membrane-associated guanylate kinase (MAGUK) family scaffold protein core. J Biol Chem. 2011;286:40069–74.

Papantoniou C, Laugks U, Betzin J, Capitanio C, Ferrero JJ, Sánchez-Prieto J, Schoch S, Brose N, Baumeister W, Cooper BH, Imig C, Lučić V. Munc13- and SNAP25-dependent molecular bridges play a key role in synaptic vesicle priming. Sci Adv. 2023;9:eadf6222.

Park D, Wu Y, Lee SE, Kim G, Jeong S, Milovanovic D, De Camilli P, Chang S. Cooperative function of synaptophysin and synapsin in the generation of synaptic vesicle-like clusters in non-neuronal cells. Nat Commun. 2021;12:263.

Pennacchietti F, Vascon S, Nieus T, Rosillo C, Das S, Tyagarajan SK, Diaspro A, Del Bue A, Petrini EM, Barberis A, Cella Zanacchi F. Nanoscale molecular reorganization of the inhibitory postsynaptic density is a determinant of GABAergic synaptic potentiation. J Neurosci. 2017;37:1747–56.

Perez de Arce K, Schrod N, Metzbower SWR, Allgeyer E, Kong GK, Tang AH, Krupp AJ, Stein V, Liu X, Bewersdorf J, Blanpied TA, Lucic V, Biederer T. Topographic mapping of the synaptic cleft into adhesive nanodomains. Neuron. 2015;88:1165–72.

Pi H, Sun R, McBride JR, Kruse ARS, Gibson-Corley KN, Krystofiak ES, Nicholson MR, Spraggins JM, Zhou Q, Skaar EP. Clostridioides difficile ferrosome organelles combat nutritional immunity. Nature. 2023;623:1009–16.

Punjani A, Rubinstein JL, Fleet DJ, Brubaker MA. cryoSPARC: algorithms for rapid unsupervised cryo-EM structure determination. Nat Methods. 2017;14:290–6.

Pyle E, Miller EA, Zanetti G. Cryo-electron tomography reveals how COPII assembles on cargo-containing membranes. Nat Struct Mol Biol. 2024;

Racca C, Stephenson FA, Streit P, Roberts JD, Somogyi P. NMDA receptor content of synapses in stratum radiatum of the hippocampal CA1 area. J Neurosci. 2000;20:2512–22.

Radhakrishnan A, Li X, Grushin K, Krishnakumar SS, Liu J, Rothman JE. Symmetrical arrangement of proteins under release-ready vesicles in presynaptic terminals. Proc Natl Acad Sci U S A. 2021:118.

Ramsey AM, Tang AH, LeGates TA, Gou XZ, Carbone BE, Thompson SM, Biederer T, Blanpied TA. Subsynaptic positioning of AMPARs by LRRTM2 controls synaptic strength. Sci Adv. 2021;7

Rigort A, Plitzko JM. Cryo-focused-ion-beam applications in structural biology. Arch Biochem Biophys. 2015;581:122–30.

Rose H. Aberration correction in electron microscopy. Int J Mater Res. 2006;97:885–9.

Rose H. Optics of high-performance electron microscopes. Sci Technol Adv Mater. 2008;9

Ruska E. The development of the electron-microscope and of electron-microscopy. Rev Mod Phys. 1987;59:627–38.

Sakarya O, Conaco C, Egecioglu O, Solla SA, Oakley TH, Kosik KS. Evolutionary expansion and specialization of the PDZ domains. Mol Biol Evol. 2010;27:1058–69.

Sartori A, Gatz R, Beck F, Rigort A, Baumeister W, Plitzko JM. Correlative microscopy: bridging the gap between fluorescence light microscopy and cryo-electron tomography. J Struct Biol. 2007;160:135–45.

Sartori N, Richter K, Dubochet J. Vitrification depth can be increased more than 10-fold by high-pressure freezing. J Microsc. 1993;172:55–61.

Sasaki H, Arai H, Kikuchi E, Saito H, Seki K, Matsui T. Novel electron microscopic staining method using traditional dye, hematoxylin. Sci Rep. 2022;12:7756.

Schaffer M, Pfeffer S, Mahamid J, Kleindiek S, Laugks T, Albert S, Engel BD, Rummel A, Smith AJ, Baumeister W, Plitzko JM. A cryo-FIB lift-out technique enables molecular-resolution cryo-ET within native Caenorhabditis elegans tissue. Nat Methods. 2019;16:757–62.

Scheres SH. RELION: implementation of a Bayesian approach to cryo-EM structure determination. J Struct Biol. 2012;180:519–30.

Scherzer O. The theoretical resolution limit of the electron microscope. J Appl Phys. 1949;20:20–9.

Schiøtz OH, Kaiser CJO, Klumpe S, Morado DR, Poege M, Schneider J, Beck F, Klebl DP, Thompson C, Plitzko JM. Serial Lift-Out: sampling the molecular anatomy of whole organisms. Nat Methods. 2024;21:1684–92.

Schnell E, Sizemore M, Karimzadegan S, Chen L, Bredt DS, Nicoll RA. Direct interactions between PSD-95 and stargazin control synaptic AMPA receptor number. Proc Natl Acad Sci U S A. 2002;99:13902–7.

Schrod N, Vanhecke D, Laugks U, Stein V, Fukuda Y, Schaffer M, Baumeister W, Lucic V. Pleomorphic linkers as ubiquitous structural organizers of vesicles in axons. PLoS One. 2018;13:e0197886.

Schwartz CL, Sarbash VI, Ataullakhanov FI, McIntosh JR, Nicastro D. Cryo-fluorescence microscopy facilitates correlations between light and cryo-electron microscopy and reduces the rate of photobleaching. J Microsc. 2007;227:98–109.

Sexton DL, Burgold S, Schertel A, Tocheva EI. Super-resolution confocal cryo-CLEM with cryo-FIB milling for. Curr Res Struct Biol. 2022;4:1–9.

Shehzad A, Brochard P, Matthey R, Kapsalidis F, Shahmohammadi M, Beck M, Hugi A, Jouy P, Faist J, Sudmeyer T, Schilt S. Frequency noise correlation between the offset frequency and the mode spacing in a mid-infrared quantum cascade laser frequency comb. Opt Express. 2020;28:8200–10.

Sheta R, Teixeira M, Idi W, Pierre M, de Rus Jacquet A, Emond V, Zorca CE, Vanderperre B, Durcan TM, Fon EA, Calon F, Chahine M, Oueslati A. Combining NGN2 programming and dopaminergic patterning for a rapid and efficient generation of hiPSC-derived midbrain neurons. Sci Rep. 2022;12:17176.

Soto GE, Young SJ, Martone ME, Deerinck TJ, Lamont S, Carragher BO, Hama K, Ellisman MH. Serial section electron tomography: a method for three-dimensional reconstruction of large structures. Neuroimage. 1994;1:230–43.

Stalling D, Westerhoff M, Hege H-C. 38 – amira: a highly interactive system for visual data analysis. In: Hansen CD, Johnson CR, editors. Visualization handbook. Burlington: Butterworth-Heinemann; 2005. p. 749–67.

Studer D, Michel M, Müller M. High pressure freezing comes of age. Scanning Microsc Suppl. 1989;3:253–68. discussion 268–259

Studer D, Michel M, Wohlwend M, Hunziker EB, Buschmann MD. Vitrification of articular cartilage by high-pressure freezing. J Microsc. 1995;179:321–32.

Sturgill JF, Steiner P, Czervionke BL, Sabatini BL. Distinct domains within PSD-95 mediate synaptic incorporation, stabilization, and activity-dependent trafficking. J Neurosci. 2009;29:12845–54.

Sugiyama Y, Kawabata I, Sobue K, Okabe S. Determination of absolute protein numbers in single synapses by a GFP-based calibration technique. Nat Methods. 2005;2:677–84.

Sun R, Allen J, Mao Z, Wilson L, Haider M, Alten B, Zhou Z, Wang X, Zhou Q. The postsynaptic density in excitatory synapses is composed of clustered, heterogeneous nanoblocks. J Cell Biol. 2025;

Sun R, Chen X, Yin CY, Qi L, Lau PM, Han H, Bi GQ. Correlative light and electron microscopy for complex cellular structures on PDMS substrates with coded micro-patterns. Lab Chip. 2018;18:3840–8.

Sun R, Liu Y-T, Tao C-L, Qi L, Lau P-M, Zhou ZH, Bi G-Q. An efficient protocol of cryo-correlative light and electron microscopy for the study of neuronal synapses. Biophys Rep. 2019;5:111–22.

Südhof TC. Neurotransmitter release: the last millisecond in the life of a synaptic vesicle. Neuron. 2013;80:675–90.

Takamori S, Holt M, Stenius K, Lemke EA, Grønborg M, Riedel D, Urlaub H, Schenck S, Brügger B, Ringler P, Müller SA, Rammner B, Gräter F, Hub JS, De Groot BL, Mieskes G, Moriyama Y, Klingauf J, Grubmüller H, Heuser J, Wieland F, Jahn R. Molecular anatomy of a trafficking organelle. Cell. 2006;127:831–46.

Takemura SY, Bharioke A, Lu Z, Nern A, Vitaladevuni S, Rivlin PK, Katz WT, Olbris DJ, Plaza SM, Winston P, Zhao T, Horne JA, Fetter RD, Takemura S, Blazek K, Chang LA, Ogundeyi O, Saunders MA, Shapiro V, Sigmund C, Rubin GM, Scheffer LK, Meinertzhagen IA, Chklovskii DB. A visual motion detection circuit suggested by Drosophila connectomics. Nature. 2013;500:175–81.

Takumi Y, Ramírez-León V, Laake P, Rinvik E, Ottersen OP. Different modes of expression of AMPA and NMDA receptors in hippocampal synapses. Nat Neurosci. 1999;2:618–24.

Tamaki T, Naomoto Y, Kimura S, Kawashima R, Shirakawa Y, Shigemitsu K, Yamatsuji T, Haisa M, Gunduz M, Tanaka N. Apoptosis in normal tissues induced by anti-cancer drugs. J Int Med Res. 2003;31:6–16.

Tang AH, Chen H, Li TP, Metzbower SR, MacGillavry HD, Blanpied TA. A trans-synaptic nanocolumn aligns neurotransmitter release to receptors. Nature. 2016;536:210–4.
Tao C-L, Liu Y-T, Sun R, Zhang B, Qi L, Shivakoti S, Tian C-L, Zhang P, Lau P-M, Zhou ZH, Bi G-Q. Differentiation and characterization of excitatory and inhibitory synapses by cryo-electron tomography and correlative microscopy. J Neurosci. 2018a;38:1493–510.
Tao CL, Liu YT, Zhou ZH, Lau PM, Bi GQ. Accumulation of dense core vesicles in hippocampal synapses following chronic inactivity. Front Neuroanat. 2018b;12:48.
Tao-Cheng JH. Stimulation-induced differential redistributions of clathrin and clathrin-coated vesicles in axons compared to soma/dendrites. Mol Brain. 2020;13:141.
Tao-Cheng JH, Dosemeci A, Gallant PE, Smith C, Reese T. Activity induced changes in the distribution of Shanks at hippocampal synapses. Neuroscience. 2010;168:11–7.
Tarsa L, Goda Y. Synaptophysin regulates activity-dependent synapse formation in cultured hippocampal neurons. Proc Natl Acad Sci U S A. 2002;99:1012–6.
Taylor KA, Glaeser RM. Electron diffraction of frozen, hydrated protein crystals. Science. 1974;186:1036–7.
Tizro P, Choi C, Khanlou N. Sample preparation for transmission electron microscopy. Methods Mol Biol. 2019;1897:417–24.
Turk M, Baumeister W. The promise and the challenges of cryo-electron tomography. FEBS Lett. 2020;594:3243–61.
Turonova B, Hagen WJH, Obr M, Mosalaganti S, Beugelink JW, Zimmerli CE, Krausslich HG, Beck M. Benchmarking tomographic acquisition schemes for high-resolution structural biology. Nat Commun. 2020;11:876.
Uzay B, Houcek A, Ma ZZ, Konradi C, Monteggia LM, Kavalali ET. Neurotransmitter release progressively desynchronizes in induced human neurons during synapse maturation and aging. Cell Rep. 2023;42:112042.
Valtschanoff JG, Weinberg RJ. Laminar organization of the NMDA receptor complex within the postsynaptic density. J Neurosci. 2001;21:1211–7.
Vierbuchen T, Ostermeier A, Pang ZP, Kokubu Y, Südhof TC, Wernig M. Direct conversion of fibroblasts to functional neurons by defined factors. Nature. 2010;463:1035–41.
Walikonis RS, Jensen ON, Mann M, Provance DW, Mercer JA, Kennedy MB. Identification of proteins in the postsynaptic density fraction by mass spectrometry. J Neurosci. 2000;20:4069–80.
Wan W, Briggs JA. Cryo-electron tomography and subtomogram averaging. Methods Enzymol. 2016;579:329–67.
Wan W, Kolesnikova L, Clarke M, Koehler A, Noda T, Becker S, Briggs JAG. Structure and assembly of the Ebola virus nucleocapsid. Nature. 2017;551:394–7.
Wang C, Jiang W, Leitz J, Yang K, Esquivies L, Wang X, Shen X, Held RG, Adams DJ, Basta T, Hampton L, Jian R, Jiang L, Stowell MHB, Baumeister W, Guo Q, Brunger AT. Structure and topography of the synaptic V-ATPase-synaptophysin complex. Nature. 2024;631:899–904.
Wang CS, Monteggia LM, Kavalali ET. Spatially non-overlapping Ca2+ signals drive distinct forms of neurotransmission. Cell Rep. 2023;42:113201.
Watanabe R, Buschauer R, Böhning J, Audagnotto M, Lasker K, Lu TW, Boassa D, Taylor S, Villa E. The in situ structure of Parkinson's disease-linked LRRK2. Cell. 2020;182:1508–1518.e1516.
Watanabe S, Rost BR, Camacho-Perez M, Davis MW, Sohl-Kielczynski B, Rosenmund C, Jorgensen EM. Ultrafast endocytosis at mouse hippocampal synapses. Nature. 2013;504:242.
Watanabe S, Trimbuch T, Camacho-Perez M, Rost BR, Brokowski B, Sohl-Kielczynski B, Felies A, Davis MW, Rosenmund C, Jorgensen EM. Clathrin regenerates synaptic vesicles from endosomes. Nature. 2014;515:228–33.
Wegner W, Steffens H, Gregor C, Wolf F, Willig KI. Environmental enrichment enhances patterning and remodeling of synaptic nanoarchitecture as revealed by STED nanoscopy. Elife. 2022;11
Woldeyes RA, Nishiga M, Vander Roest AS, Engel L, Giri P, Montenegro GC, Wu AC, Dunn AR, Spudich JA, Bernstein D, Schmid MF, Wu JC, Chiu W. Cryo-electron tomography reveals the structural diversity of cardiac proteins in their cellular context. bioRxiv. 2023;
Wozny MR, Di Luca A, Morado DR, Picco A, Khaddaj R, Campomanes P, Ivanović L, Hoffmann PC, Miller EA, Vanni S, Kukulski W. In situ architecture of the ER-mitochondria encounter structure. Nature. 2023;618:188–92.

Wu GH, Smith-Geater C, Galaz-Montoya JG, Gu Y, Gupte SR, Aviner R, Mitchell PG, Hsu J, Miramontes R, Wang KQ, Geller NR, Hou C, Danita C, Joubert LM, Schmid MF, Yeung S, Frydman J, Mobley W, Wu C, Thompson LM, Chiu W. CryoET reveals organelle phenotypes in huntington disease patient iPSC-derived and mouse primary neurons. Nat Commun. 2023;14:692.

Wu Y, O'Toole ET, Girard M, Ritter B, Messa M, Liu X, McPherson PS, Ferguson SM, De Camilli P. A dynamin 1-, dynamin 3- and clathrin-independent pathway of synaptic vesicle recycling mediated by bulk endocytosis. Elife. 2014;3:e01621.

Xu CS, Hayworth KJ, Lu Z, Grob P, Hassan AM, García-Cerdán JG, Niyogi KK, Nogales E, Weinberg RJ, Hess HF. Enhanced FIB-SEM systems for large-volume 3D imaging. Elife. 2017;6

Xue L, Lenz S, Zimmermann-Kogadeeva M, Tegunov D, Cramer P, Bork P, Rappsilber J, Mahamid J. Visualizing translation dynamics at atomic detail inside a bacterial cell. Nature. 2022;610:205–11.

Yip KM, Fischer N, Paknia E, Chari A, Stark H. Atomic-resolution protein structure determination by cryo-EM. Nature. 2020;587:157–61.

Young LN, Sherrard A, Zhou H, Shaikh F, Hutchings J, Riggi M, Rosen MK, Giraldez AJ, Villa E. ExoSloNano: multi-modal nanogold tags for identification of macromolecules in live cells & cryo-electron tomograms. bioRxiv. 2024:2024.2010.2012.617288.

Zeng M, Chen X, Guan D, Xu J, Wu H, Tong P, Zhang M. Reconstituted postsynaptic density as a molecular platform for understanding synapse formation and plasticity. Cell. 2018;174:1172–87. e1116

Zeng M, Shang Y, Araki Y, Guo T, Huganir RL, Zhang M. Phase transition in postsynaptic densities underlies formation of synaptic complexes and synaptic plasticity. Cell. 2016;166:1163–75. e1112

Zens B, Fäßler F, Hansen JM, Hauschild R, Datler J, Hodirnau VV, Zheden V, Alanko J, Sixt M, Schur FKM. Lift-out cryo-FIBSEM and cryo-ET reveal the ultrastructural landscape of extracellular matrix. J Cell Biol. 2024;223

Zheng CY, Seabold GK, Horak M, Petralia RS. MAGUKs, synaptic development, and synaptic plasticity. Neuroscientist. 2011;17:493–512.

Zuber B, Lucic V. Molecular architecture of the presynaptic terminal. Curr Opin Struct Biol. 2019;54:129–38.

Zuber B, Nikonenko I, Klauser P, Muller D, Dubochet J. The mammalian central nervous synaptic cleft contains a high density of periodically organized complexes. Proc Natl Acad Sci U S A. 2005;102:19192–7.

Chapter 11
Roles of Glia in Synapse Nano-organization

Young-Jin Choi and Won-Suk Chung

Abstract Glia, which are nonneuronal cells in the central nervous system, include astrocytes, microglia, and oligodendrocyte lineage cells. Historically, their passive roles in maintaining central nervous system function—such as supplying substrates for neuronal energy, buffering neurotransmitters, and improving neural conductance—have been primarily highlighted. Importantly, recent research has revealed that glial cells express genes directly related to controlling synapse nano-organization. In particular, astrocytes have the ability to secrete molecules that induce synapse formation and maturation. They also modulate the structure of synapses by expressing proteins that make direct contact with the synaptic membrane. Moreover, astrocytes can actively eliminate synapses through their phagocytic machinery during development and adulthood, thereby establishing circuit homeostasis. Microglia can also assist in the integration of synapses into the neural circuit and regulate synapse formation, maintenance, and elimination. Here, we review key findings on the mechanisms of glial contributions to synapse nano-organization. We will also discuss how different glial cells contribute to the development and homeostasis of synapses through distinct cellular and molecular pathways in both health and disease.

Keywords Astrocytes · Microglia · Oligodendrocyte precursor cells · Synapse formation · Synapse elimination · Synapse nano-organization

Y.-J. Choi · W.-S. Chung (✉)
Center for Vascular Research, Institute for Basic Science, Daejeon, Republic of Korea

Department of Biological Sciences, Korea Advanced Institute of Science and Technology, Daejeon, Republic of Korea
e-mail: wonsuk.chung@kaist.ac.kr

E. T. Kavalali (ed.), *Nano-organization of the Synapse*, Advances in Neurobiology 48,
https://doi.org/10.1007/978-3-032-12594-1_11

11.1 Introduction

For proper signaling to occur, neurons must be intricately interconnected with other neurons through synapses. Their complex organization and structure along neuronal somas and dendrites provide the foundation for the sophisticated computation of neuronal activity, which is essential for our daily learning and behaviors. Interestingly, recent research has established that synapses develop their nano-organization through dynamic bidirectional interactions with glial cells, especially astrocytes, microglia, and oligodendrocyte precursor cells (OPCs). This includes not only the formation of synapses but also their maintenance and elimination, all of which are controlled by these glial cells. In this chapter, we first introduce the different types of glial cells and then delve into their pivotal roles in establishing and maintaining the synaptic nano-organization.

11.1.1 Overview of Different Types of Glial Cells in the Brain and Their Functions

Glial cells in the brain include astrocytes, microglia, oligodendrocytes, and OPCs. These glial cells can support neuronal and synaptic function in various mechanisms. Astrocytes are predominant glial cell types in the central nervous system (CNS), which constitute about 20% of total cells in the rodent brain and 50% in the human neocortex (Sun et al. 2017). Historically, they are primarily known to play supportive roles in maintaining brain homeostasis and supplying energy to neurons. For instance, researchers have revealed that astrocytes express various types of channels and transporters on their plasma membrane to buffer extracellular and synaptic ion concentrations, including K^+ (Chever et al. 2010; D'Ambrosio et al. 2002; Larsen et al. 2014; Seifert et al. 2009) and Cl^-, (Egawa et al. 2013; Kettenmann et al. 1987; Sık et al. 2000) and also H^+ (Chesler 2003; Deitmer and Rose 1996; Grichtchenko and Chesler 1994; Hansen et al. 2015). Additionally, they support neuronal metabolism by uptaking glucose and supplementing lactate after glycolysis (Allen and Messier 2013; Halestrap 2012; Pellerin et al. 2005; Falkowska et al. 2015). Importantly, astrocytes are structurally interconnected with synapses, and this physical association between astrocytic processes and synapses provides strong advantages for astrocytes in responding to changes in synaptic activity. Through this close physical association, astrocytes can modulate the concentration of extracellular neurotransmitters by reabsorbing glutamate (Vandenberg and Ryan 2013; Zhou and Danbolt 2013; Moussawi et al. 2011) and γ-aminobutyric acid (GABA) (Minelli et al. 1996; Scimemi 2014), thus regulating the excitatory-inhibitory (E-I) balance. In the next section, we will further discuss the more recently discovered roles of astrocytes in directly controlling synaptic structures.

Along with astrocytes and microglia, resident macrophages in the CNS (Aguzzi et al. 2013) constantly survey the immune status and play a number of critical roles in modulating the homeostasis of the brain. These functions are largely related to their immune-related functions, such as phagocytosis and inflammatory responses (Brown and Neher 2014; Podleśny-Drabiniok et al. 2020; Janda et al. 2018). Interestingly, gene expression patterns of microglia can undergo dynamic changes upon various stimuli. For example, microglia secrete insulin-like growth factor-1 (IGF-1) after injury, inducing proliferation and differentiation of neural stem cells in the subventricular zone (Thored et al. 2009). They can also produce chemokines that suppress the function of proinflammatory cytokines (IL-4, IL-10, and TGF-β) in chronic neurological diseases, protecting from inflammation-induced damages and controlling neuronal regeneration (Colton 2009; Rasley et al. 2006; Lindholm et al. 1992). On the other hand, they can also be neurotoxic in various neurological disease conditions, causing neuronal dysfunction (Biber et al. 2014). Molecular pathways involving interferon-γ (IFN-γ), toll-like receptor 4 (TLR4), P2X purinoceptor 4 (P2X4), and CX3 motif chemokine receptor 1 (CX3CR1) are thought to cause neuroinflammation and the activation of microglia (Smith 2010; Fan et al. 2010; Rana et al. 2010). In the disease states, microglia can also interact with astrocytes, changing each other's status. For example, during brain injury, an ATP gradient generated by astrocytes can be sensed by microglia through the purinergic receptor P2Y12 (P2RY12) receptor. P2RY12 signaling in microglia subsequently enhances microglial motility to help injury closure (Haynes et al. 2006; Lou et al. 2016). In addition, reactive microglia can secrete factors, such as tumor necrosis factor (TNF), interleukin 1α (IL-1α), and complement component 1q (C1q), inducing the appearance of neurotoxic reactive astrocytes in various disease conditions (Liddelow et al. 2017).

Last but not least, oligodendrocytes and their lineage cells support neuronal signal transmission. Oligodendrocytes are myelinating cells in the CNS, which are produced from OPCs. The myelin sheath, an extended membrane layer from oligodendrocytes, forms a unique structure that enwraps the neural axon, acting as an insulator for efficient electrical conductance. Interestingly, OPCs have gained new attention as an unexpected glial cell type associated with synapses.

Thus, it has been increasingly evident that multiple glial cells can constantly communicate with synapses and influence their formation, maintenance, and elimination. In the following sections, we will primarily discuss the roles of these glial cells—astrocytes, microglia, and OPCs—on regulating synaptic structures.

11.2 Glial Cell Functions in Synapse Nano-organization

In the CNS, astrocytes closely interact with synapses, forming a complex structure known as the "tripartite synapse," which includes astrocytic processes alongside pre- and postsynaptic structures (Araque et al. 1999). A single astrocyte in a mouse brain is estimated to contact up to 100,000 synapses through its numerous fine processes (Halassa et al. 2009). Due to their delicate yet complex structure, astrocytes are considered architects of synaptic structure, supporting their functions in close proximity. Early research involving in vitro coculture experiments with astrocytes and retinal ganglion cells (RGCs) revealed the pivotal roles of astrocytes in synaptogenesis. Intriguingly, RGCs cultured with astrocytes or exposed to an astrocyte-conditioned medium exhibited higher levels of structural synapse formation with increased synaptic activity (Pfrieger and Barres 1997; Ullian et al. 2001). Throughout this research, it has been well established that astrocytes directly induce synapse formation through various secretory factors. Subsequent studies have further identified these factors and their roles in synapse formation and organization.

11.2.1 Role of Astrocytes in Synapse Formation via Secretion

As astrocytes form close contacts with synapses and induce synaptogenesis through secreted factors, further in vitro/in vivo research has been conducted to identify these factors. Thrombospondins (TSPs), the first and most extensively studied protein family secreted by astrocytes, have been shown to induce synaptogenesis (Christopherson et al. 2005). TSPs are extracellular matrix proteins that typically function as trimers and are known to bind various membrane receptors or extracellular proteins, mediating cell-to-cell interactions. During developmental stages, TSPs are highly expressed and secreted by astrocytes, promoting synaptogenesis both in vitro and in vivo (Hennekinne et al. 2013). Interestingly, TSPs induce the formation of synapses that are presynaptically active but postsynaptically silent with the reduced α-amino-3-hydroxy-5-methyl-4-isoxazolepropionic acid receptor (AMPAR) levels, suggesting that developmental synaptogenesis may progress in a regulated and stepwise manner. Further studies have explored how TSPs promote synaptogenesis. The primary binding partner for TSPs was identified as the α2δ1 subunit of calcium channels (Eroglu et al. 2009) and the neuroligin (NLGN)-1 postsynaptic adhesion protein (Xu et al. 2010). The TSPs-α2δ1 interaction controls postsynaptic Rac1 activity, a member of the Rho GTPase family, which modulates the actin cytoskeleton and leads to synaptogenesis (Risher et al. 2018). α2δ1 also influences neuronal protein kinase A (PKA) activity through an unknown pathway, resulting in reduced presynaptic release. Additionally, the interaction between TSPs and NLGN1 promotes synaptogenesis and synaptic plasticity via LIM domain kinase 1 (LIMK1)/cofilin-mediated actin organization (Liu et al. 2016).

As TSPs induce postsynaptically silent synapses, other factors responsible for inducing functional synaptogenesis were investigated (Allen et al. 2012). Glypicans (GPCs), a major family of heparan sulfate proteoglycans, with GPC4 and GPC6 being highly expressed in astrocytes during postnatal development (Farhy-Tselnicker et al. 2021), have been found to be secreted through proteolytic shedding and anchor cleavage (Huang and Park 2021). When released by astrocytes, GPC4/6 increases the number and activity of synapses by recruiting GluA1 AMPAR subunits to the postsynaptic membrane. The mechanism underlying GPCs' role in functional synapse formation involves neuronal pentraxin 1 (NP1), which is released following GPC4 binding to the presynaptic protein tyrosine phosphatase receptor δ (PTPRδ), facilitating AMPAR clustering (Farhy-Tselnicker et al. 2017).

Additionally, chordin-like 1 (CHRDL1) released by astrocytes recruits GluA2 subunits to dendritic spines, fostering postsynaptic maturation and limiting synaptic plasticity, not only during development but also in the adult healthy brain (Blanco-Suarez et al. 2018). While the exact mechanism of GluA2 recruitment remains unclear, it is known to occur via a bone morphogenetic protein (BMP)-independent pathway (Blanco-Suarez et al. 2018), which is typically antagonized by CHRDL1.

Another synaptogenic and synapse maturation factor secreted by astrocytes is high endothelial venule protein (Hevin). Hevin, also known as SPARC-like 1 (SPARCL1), has been found to induce excitatory synaptogenesis (Kucukdereli et al. 2011). Unlike previously described synaptogenic factors, which were mainly participating in early developmental stages, Hevin is involved in the cortical developmental process, which is a relatively later process during CNS development (Risher et al. 2014).

The mechanism of Hevin in synapse formation and maturation is by providing a physical interaction between presynaptic and postsynaptic terminals. Hevin can bind to both presynaptic neurexin (NRXN)-1α and postsynaptic NLGN1β, functioning as a bridging molecule between synaptic junctions. Interestingly, Hevin is antagonized by SPARC (secreted protein acidic and rich in cysteine), which is also expressed and secreted by astrocytes (Kucukdereli et al. 2011). SPARC has no function for directly controlling synapse formation, but it can compete with Hevin by its similar structure, eventually decreasing excitatory synapse numbers. Also in development, SPARC can bind to and inhibit the function of the neuronal β3-integrin complex, which has a critical role in AMPAR stabilization on synapses (Jones et al. 2011). This function of SPARC has been characterized in cholinergic neurons for controlling the maturation of cholinergic presynaptic terminals and cell-autonomous synapse elimination (Albrecht et al. 2012; López-Murcia et al. 2015). This emphasizes that astrocytes can induce synaptogenesis and fine-tune the level of synapse formation and maturation throughout broad ranges of developmental stages.

The discoveries of astrocyte-secreted factors also highlighted their significance in diverse neurological disease conditions. For example, TSPs are known to be expressed at low levels in adult brain astrocytes, but their expression is highly enriched in injury states. Increased expression of *Tsp1* and *Tsp2* after stroke is critical for recovery in synaptic plasticity and function. Also, in the striatum, activation of the astrocytic signaling pathway through the GABAB receptor increased astrocytic *Tsp1* expression, inducing hyperactivity and disrupting attention. Moreover, astrocytes near the lesion site in the spinal cord exhibit highly increased *Tsp4* expression, contributing to hypersensitivity and neuropathic pain (Kim et al. 2012). Additionally, mice lacking *Gpc4* resulted in behavior deficits, such as juvenile hyperactivity and social novelty (Dowling and Allen 2018). CHRDL1 promotes synapse maturation and limits plasticity in the brain, so silencing this signaling pathway can prevent spine loss after ischemic injury and functional recovery in a stroke mouse model (Blanco-Suarez and Allen 2022; Collyer et al. 2023).

Furthermore, hevin secretion from astrocytes plays a critical role in mediating chronic pain in an N-methyl-D-aspartate receptor (NMDAR)-responsive manner (Chen et al. 2022a), and SPARC demonstrated enhanced AMPAR recruitment and synaptic function, which promotes neuronal health in the middle cerebral artery occlusion (MCAO) model (Jones et al. 2018). These overall novel findings suggest that astrocytes play a critical role by secreting various factors that mediate synaptic formation and regulate their organization, which could be a potential therapeutic target for various neurological disorders.

11.2.2 Role of Astrocytes in Synapse Formation via Direct Contact

Besides the aforementioned secreted factors, recent studies have discovered that astrocytes establish close physical association with synapses through adhesion molecules, promoting bidirectional structural complexity and functional communication. NLGNs are cell adhesion molecules that are highly localized in postsynapses and have the ability to contact the presynaptic cell adhesion molecule NRXN, leading to the formation of both excitatory and inhibitory synapses (Südhof 2017). Interestingly, NLGNs are also found to be expressed in astrocytes (Zhang et al. 2014), but their role in astrocytes was unclear. Recent research revealed that NLGN2, expressed during both early and late stages of astrocytic development, plays a crucial role in the formation and function of excitatory synapses, regulating both morphological complexity and the number of excitatory synapses (Stogsdill et al. 2017). Recently, NRXN1 was discovered to be also expressed in astrocytes, which were mainly considered as presynaptic proteins (Trotter et al. 2020). Hippocampal astrocytic NRXN1 is highly modified by heparan sulfate and also displays unique splicing patterns compared with the neuronal one, suggesting that distinct splicing features of NRXN1 lead to differential expression and binding signatures. Astrocytic deletion of *Nrxn1* suppressed excitatory transmission and LTP formation, suggesting that astrocytic NRXN1 has critical roles in controlling synaptic function. These results imply that astrocytes and synapses can communicate through the complex NRXN-NLGN interactions in the formation and functionality of synapses, which deserves further research in the future.

γ-protocadherins (γ-PCDH) are also found in both neurons and astrocytes and interact with each other in a way that is similar to NRXN-NLGN interaction. Homophilic interactions between neuronal and astrocytic γ-PCDH promote synaptogenesis in the spinal cord (Garrett and Weiner 2009) and excitatory synaptogenesis in the brain (Garrett et al. 2012). Furthermore, a recent report suggested that these interactions promote dendritic complexity (Molumby et al. 2016), a crucial aspect of normal neuronal development and function.

More recently, researchers have discovered other signaling pathways that mediate astrocyte-synapse contact. Ephrins, membrane-bound ligands for Eph receptor tyrosine kinase, have been well characterized for their roles in mediating critical bidirectional cell–cell interaction. Interestingly, EphA4 is expressed in neurons and is highly enriched in the dendritic spines in the hippocampus, whereas Ephrin-A3 is expressed in astrocytes. Ephrin-A3 from astrocytes interacts with EphA4 on the dendritic spine, potentially limiting spine morphology, spine maturation, glutamate transportation, and long-term potentiation (LTP) formation (Murai et al. 2003; Nishida and Okabe 2007; Carmona et al. 2009; Filosa et al. 2009). Similarly, Ephrin-B1 from astrocytes binds with EphB localized on excitatory synapses, presumably presynapses, and controls spine density and hippocampus-dependent behavior (Koeppen et al. 2018; Nguyen et al. 2020a). These findings underscore the role of astrocytes in regulating the formation and function of excitatory synapses through direct physical contacts with both pre- and postsynaptic compartments (Fig. 11.1).

Similar to excitatory synapses, a recent study has shown that astrocytes control inhibitory synapse formation and function through contact-dependent mechanisms. Both astrocytes and neurons express neuronal cell adhesion molecule (NRCAM), and its protein is highly localized at astrocytic processes, coupled with gephyrin in

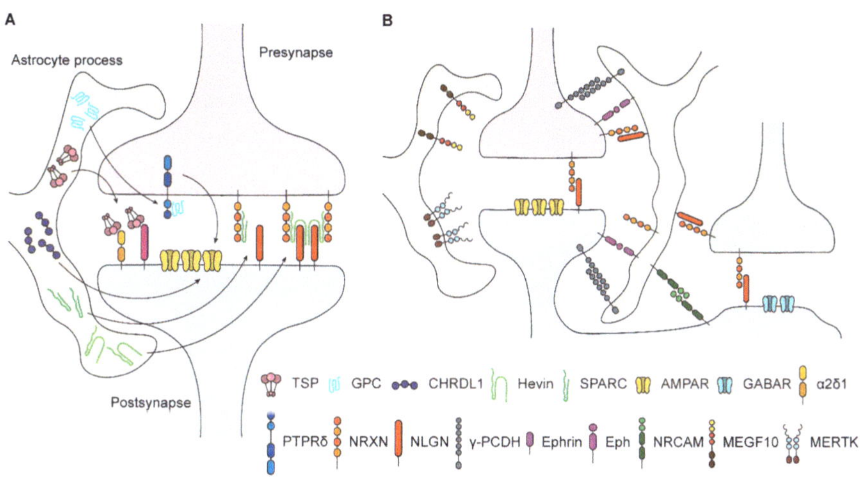

Fig. 11.1 *Role of astrocytes in synapse nano-organization.* (**a**) Overview of astrocyte-secreted factors that modulate synapse nano-organization. TSPs induce synaptogenesis by binding with α2δ1 and NLGN. GPCs bind with PTPRδ and induce functional synapse formation. Hevin binds to presynaptic NRXN and postsynaptic NLGN, leads to the formation of excitatory synapses, which are antagonized by SPARC. CHRDL1 changes postsynaptic properties by recruiting postsynaptic GluA2. (**b**) Overview of contact-mediated modulation of synapse nano-organization. Astrocytic NLGNs form contact with presynaptic NRXNs, trigger both excitatory and inhibitory synaptogenesis. Astrocytic NRXNs also contribute in synapse organization. γ-PCDHs in synapse and astrocyte facilitate excitatory synapse formation through homophilic interactions. Ephrin-Eph signaling between synapse and *astrocyte* restricts synaptic formation and function. NRCAM localized in inhibitory postsynapse binds with astrocytic NRCAM, promoting inhibitory synapse formation. Astrocytes eliminate synapses through phagocytosis via MEGF10 and MERTK, phagocytic receptors

inhibitory postsynapse (Takano et al. 2020). Astrocytic NRCAM interacts with inhibitory postsynaptic NRCAM, which is anchored to gephyrin. In the mouse cortex, this homophilic NRCAM interaction induces inhibitory synaptogenesis and promotes its function (Takano et al. 2020). Similarly, hepatocyte cell adhesion molecule (HepaCAM) is highly localized at astrocytic processes and inhibitory post-synapses. This study has shown that HepaCAM expression in astrocytes regulates gap junction formation between astrocytes and controls E-I balance in the mouse cortex (Baldwin et al. 2021), but its exact role in contacting inhibitory post-synapses remains unclear.

11.2.3 Synapse Elimination by Astrocytes

Throughout one's lifetime, synapses in the CNS are continuously formed and eliminated during the synapse remodeling process. This process is highly prevalent in developmental and adolescent stages, and this developmental synaptic pruning is critical for the maturation and strengthening of the remaining synapses and neural circuits. Despite the previous assumption that the synaptic turnover rate in the adult brain may be substantially low compared to the developmental stages, various brain regions continuously show significant levels of synaptic turnover during learning, addiction, and changes in sensory experiences (Xu et al. 2009; Yang et al. 2009; Russo et al. 2010; Lendvai et al. 2000; Ribic et al. 2019; Tropea et al. 2010). Recent reports suggest that synapse depression could potentially lead to synapse elimination (Stein and Zito 2019; Harris et al. 2024), and optogenetic synapse elimination can reduce synaptic memory in motor learning (Hayashi-Takagi et al. 2015). Those findings suggested that not only synapse formation but also synapse elimination is a critical process for normal synapse organization and brain function. As astrocytes have the ability to secrete molecules and directly contact synapses to regulate their formation, it was plausible to hypothesize that astrocytes can also eliminate synapses, which is a still less appreciated research topic.

Through the initial gene expression analysis of purified astrocytes, it has been shown that astrocytes unexpectedly possess various phagocytosis machineries (Zhang et al. 2014; Cahoy et al. 2008). These include multiple EGF-like domains 10 (MEGF10) and mer proto-oncogene, tyrosine kinase (MERTK) phagocytic receptors as well as their binding partners (Chung et al. 2013). MEGF10 is also known and well characterized by its orthologs, Draper (*D. melanogaster*) and CED-1 (*C. elegans*), for its role in cellular debris clearance during development (MacDonald et al. 2006; Zhou et al. 2001). On the other hand, MERTK, one of the members in the TAM (TYRO3, AXL, MER) receptor tyrosine kinases, binds with growth arrest-specific protein 6 (GAS6) and Protein S to recognize and clear phosphatidylserine (PtdSer)-exposing apoptotic cell membrane debris (Wu et al. 2005; Lemke 2013). Importantly, researchers have found that *Megf10* and *Mertk*-deficient mice display a defect of synapse elimination in the dorsal lateral geniculate nucleus (dLGN), where synapse pruning is highly relevant for eye-specific segregation of synaptic inputs during postnatal development of dLGN. This defect resulted in a failure to refine the neuronal connection from RGCs to dLGN, as well as an increase in the survival of weaker synaptic inputs to dLGN neurons. Moreover, these synaptic eliminations were controlled by neuronal competition between contralateral and ipsilateral RGCs, suggesting that astrocytes can sense the relatively weaker synapses among the synaptic population and preferentially eliminate them through yet-to-be-identified selection mechanisms.

Importantly, a recent study has reported that astrocytes can eliminate synapses not only during developmental stages but also in the adult brain, especially in the hippocampus (Lee et al. 2021). In this study, researchers have applied a fluorescence protein-based pH indicator (mCherry-eGFP system) to detect an acidic influx of synaptic debris in glial cells. By tagging the mCherry-eGFP system onto synapse-specific proteins, it was found that astrocytes can constantly engulf a greater number of synapses than microglia in the hippocampal CA1 region. When *Megf10* is specifically deleted in adult astrocytes, astrocytes fail to phagocytose redundant excitatory synaptic connections in the CA1 region, resulting in increased synaptic basal transmission with conversely decreased LTP and LTD capacity. *Megf10*-deleted mice also showed behavioral defects in learning and memory. These series of findings reveal that astrocytes constantly monitor and reorganize synaptic architectures by eliminating redundant and relatively weaker synapses from the system, both in development and adulthood, and that this phagocytic role of astrocytes is absolutely necessary for establishing synaptic and circuit homeostasis.

Further detailed mechanisms that regulate astrocytic phagocytosis of synapses are still under investigation. By performing chemical screening with in vitro live imaging of astrocytes and synaptosomes, it was recently found that glucocorticoid receptor (GR) activity can enhance astrocytic phagocytosis through transcriptional activation of the *Mertk* phagocytic receptor (Byun et al. 2023). By utilizing early social deprivation as a model of early life stress conditions, the authors showed that increases in the stress hormone evoke excessive developmental pruning of excitatory synapses on interneurons through MERTK, inducing E-I imbalance and various mental disorder-like behaviors in mice.

Astrocytic phagocytosis capacity also appears to be changed in various brain disorders. Recently, it has been reported that astrocytic expression of *Megf10* and *Mertk* is significantly decreased in aged Alzheimer's disease (AD) model mice, but those tendencies were not detected in the normal aging mouse brain (Sanchez-Mico et al. 2021; Boisvert et al. 2018), suggesting that changes in astrocytic phagocytosis capacity may have a correlation with AD progression. Furthermore, in an ischemic stroke mouse model, the authors have found that deletion of *Megf10* and *Mertk* specifically in astrocytes has significantly improved recovery after stroke. These studies suggest that astrocytic phagocytosis of synapses is tightly regulated within the brain, and changes in its regulation can have a direct impact on disease progression.

11.2.4 Role of Microglia in Synapse Elimination

Microglia are the major immune cells residing in the brain, playing primary immune roles in maintaining brain homeostasis. Microglia's cellular properties, similar to those of macrophages, have led to their classification as major phagocytes in the brain. Moreover, microglia, with their higher motility compared to other brain cell types, including astrocytes, provide advantages in rapidly responding to various injuries during neuroinflammatory conditions. One of the initial findings regarding their roles in synapse organization was that microglia can actively participate in synapse elimination through their phagocytic activity. Researchers have discovered that synapses to be eliminated are coated by C1q, and together with its downstream classical complement components, C3, microglia can actively induce synapse phagocytosis through microglial complement receptor 3 (CR3) during LGN development (Stevens et al. 2007). Neuronal activity also regulates this complement pathway-dependent synapse pruning by microglia during development, selectively eliminating weaker synapses (Schafer et al. 2012). Importantly, researchers also found that the microglial complement cascade is hyperactivated in various neurodegenerative diseases, including AD, mediating excessive excitatory synapse loss through microglial phagocytosis (Hong et al. 2016). They have discovered a significant decrease in hippocampal synapses during the early stages of AD accompanied by the accumulation of C1q in the hippocampal synapses. Interestingly, those accumulations of C1q and microglial phagocytosis were accelerated by oligomeric β-amyloid (Aβ), inducing synaptic loss and neuronal dysfunction. This suggests that the complement pathway controlling microglial phagocytosis is critical for normal brain function and disease progression. By using the APP/PS1 AD model, another study has shown that mGluR1 activity in the hippocampus was increased, so that local translation of *C1q* mRNA is condensed in AD excitatory synapses. Consequently, this phenomenon leads to defects in cognitive behavior in APP/PS1 mice, compared with control (Bie et al. 2019). Additionally, in a human postmortem study, patients with epilepsy, multiple sclerosis, and schizophrenia displayed a higher level of complement protein expression and higher microglial phagocytic activity, suggesting that enhanced microglial synapse phagocytosis through the complement pathway might play a key role in various neurological disorders (Wyatt et al. 2017; Andoh et al. 2019; Michailidou et al. 2015; Sellgren et al. 2019).

Along with the complement cascade, other molecular pathways have been discovered that microglia utilize for regulating synapse nano-organization in both health and disease. For example, the microglial CX3CR1 plays a crucial role in normal synapse pruning during hippocampal development (Paolicelli et al. 2011). When *Cx3cr1* is deleted, microglial cells fail to make close contact with synapses, resulting in less engulfment of synaptic materials. C-X3-C motif chemokine ligand 1 (CX3CL1), which is also known as fractalkine, can recruit CX3CR1-expressing microglia, allowing them to interact with synapses (Gunner et al. 2019). In the whisker-lesioned mouse model, microglia appear to use these CX3CL1-CX3CR1 interactions and eliminate impaired thalamocortical synapses in the barrel cortex.

Interestingly, whisker lesions increased *Adam10* expression, which is a metalloprotease known to cleave CX3CL1 and its secretion, in barrel cortex neurons, activating the CX3CL1-CX3CR1 axis. These studies suggest that accurate interaction between synapses and microglial signaling is required for proper synaptic remodeling.

To induce synapse elimination, microglia also use an innate immune receptor, the triggering receptor expressed on myeloid cells 2 (TREM2). *Trem2* deletion caused a decrease in microglial cell number in the hippocampus during early postnatal brain development (Filipello et al. 2018). In addition, *Trem2* deletion causes a defect in microglial phagocytic activity and induces an increase in the number of dendritic spines. *Trem2*-deleted mice displayed defects in the functional connectivity of neurons and their behaviors, highlighting that microglial phagocytosis of synapses through TREM2 is critical for normal brain wiring. Interestingly, it has been found that TREM2 recognizes externalized PtdSer (ePtdSer), a well-known "eat-me" signal exposed to the outer plasma membrane leaflet during apoptosis, causing clearance of cellular debris by nearby phagocytic cells. Interestingly, several studies also reported that PtdSer can be locally exposed to certain synapses during brain development, inducing glia-mediated synapse phagocytosis through different PtdSer-recognizing phagocytic receptors, such as TREM2, MERTK, and G protein-coupled receptor 56 (GPR56) (Rueda-Carrasco et al. 2024; Park et al. 2021; Scott-Hewitt et al. 2020; Li et al. 2020). How certain synapses exhibit ePtdSer on their surfaces for subsequent elimination is currently unclear. In one study, it has been suggested that hyperactive synapses with increased calcium activity tend to expose PtdSer on their synaptic surface and be eliminated by microglial TREM2, indicating that microglial TREM2 plays beneficial roles in maintaining proper neuronal network activity.

In a separate study, it has been shown that ePtdSer can also be recognized by microglial MERTK (Park et al. 2021). PtdSer is known to predominantly reside in inner membrane leaflets due to the constant function of flippases, which can bring ePtdSer back from the outer membrane to the inner membrane. When the authors deleted *Cdc50a*, the coding chaperon protein necessary for proper membrane localization of flippases, they found that microglial MERTK recognizes PtdSer exposed in the neuronal cell soma, leading to the preferential elimination of inhibitory post-synapses, which are highly concentrated in neuronal cell soma compared to dendritic processes. Taken together, these studies suggest that PtdSer can serve as one of the "eat-me" signals in eliminating synapses. It is interesting to note that during developmental synapse pruning, it has been suggested that the weak synaptic connections, rather than strong ones, are preferentially eliminated by glial cells. Thus, further work would be necessary to understand the exact mechanisms of neural activity-dependent PtdSer exposure and subsequent synapse elimination processes.

As the opposite side of the "eat-me" signal, researchers have also discovered that the CD47-SIRPα (signal regulatory protein α) pathway can inhibit the phagocytosis of synapses. SIRPα is a regulatory membrane glycoprotein also expressed in microglia and requires CD47 as its ligand for the "don't eat me" signal. During development, neurons express CD47, which is localized at synapses in dLGN

(Lehrman et al. 2018). CD47 deletion enhanced microglia phagocytic activity, resulting in a decreased number of synapses. Interestingly, more active synapses exhibit a higher enrichment of CD47, indicating that stronger synapses can be preserved based on the CD47-SIRPα axis during developmental pruning. This underscores the dual nature of neuron-to-microglia communication, which can either induce or block the elimination of synapses based on the synaptic activity.

11.2.5 Regulation of Synapse Nano-organization by the Non-phagocytic Function of Microglia

Microglia can also converse with neurons and synapses to induce synapse formation and functional maturation. Dendritic spines are continuously formed and eliminated during motor skill learning. In this dynamic process, microglia have been suggested to secrete brain-derived neurotrophic factor (BDNF) to induce dendritic spine formation (Parkhurst et al. 2013). Microglia depletion and microglia-specific deletion of *Bdnf* led to impaired learning motor skills, lower miniature excitatory postsynaptic Current (mEPSC) frequency, and spine formation. Neuronal tropomyosin receptor kinase receptor B (TrkB) binds to BDNF, causing autophosphorylation and triggering downstream pathways that may trigger synaptic gene expression and spine formation. However, several studies have also pointed out that microglia may not express sufficient levels of BDNF to control synaptic and neuronal function (Zhang et al. 2014; Bennett et al. 2016), suggesting that definitive roles of microglial BDNF in affecting synapses should need further investigation.

Microglia also release another secreted factor, interleukin 10 (IL-10), to induce synapse formation (Lim et al. 2013). IL-10 binds to the neuronal IL-10 receptor, inducing hippocampal dendritic spine formation. Interestingly, IL-10 receptor expression in hippocampal neurons is restricted in postnatal developmental stages, suggesting that IL-10 induces synapse formation during early life, followed by synapse elimination for developmental synapse remodeling and maturation.

Microglial CX3CR1-CX3CL1 signaling can also participate in the synapse maturation process. Researchers have discovered that CX3CR1-CX3CL1 signaling is required for the functional maturation of hippocampal CA3-CA1 synapses (Basilico et al. 2019). The motility of microglial processes was changed by CX3CR1 signaling in microglia, suggesting that direct contact between synapses and microglia plays a role in the maturation of synapses during development. This pathway is also highlighted in olfactory bulb remodeling (Reshef et al. 2017). Microglia-depleted mice showed a decrease in spine density in olfactory bulb adult-born granular cells (abGCs), along with a decrease in synapse formation and elimination. *Cx3cr1*-deleted mice also showed a similar phenotype with lower dendritic spine density and turnover, as well as altered microglia contact with synapses and neurons. This led to an uncontrolled neural activity response in the olfactory bulb, indicating that microglial contact with neurons is crucial to proper synaptic wiring in a brain-wide manner.

As previously discussed, microglia are essential for brain surveillance, largely due to their high motility. Live imaging studies have shown that microglial processes actively extend and retract around synapses (Nimmerjahn et al. 2005), continuously interacting with and sampling the surrounding environment. Recent findings indicate that microglial processes make direct contact with synapses in response to neuronal activity, suggesting that microglia monitor both immune activity and synaptic function (Wake et al. 2009). In the developing somatosensory cortex, microglial interactions with neuronal dendrites have been shown to increase local Ca2+ concentrations in the dendrites, which can lead to filopodia formation and establishment of functional synapses (Miyamoto et al. 2016). This highlights the potential role of microglial-neuronal contact in initiating synapse organization (Fig. 11.2).

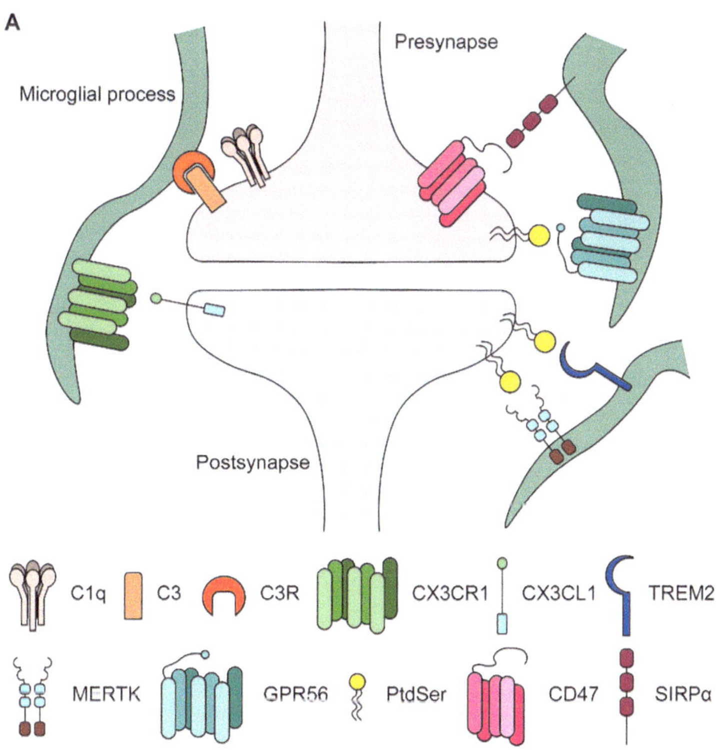

Fig. 11.2 Role of microglia in synapse nano-organization. (**a**) Overview of microglial factors that modulate synapse nano-organization. C1q and C3 localized on synapses induce phagocytosis of synapses through microglial C3R. Microglial CX3CR1 recognize synaptic CX3CL1, inducing microglia-synapse interaction. TREM2, MERTK, and GPR56 recognize PtdSer exposed on synapses, promoting microglial phagocytosis of synapses. CD47 recognition by SIRPα acts as a "don't eat me signal," protecting synapses from elimination

Moreover, recent research suggests that the extracellular matrix (ECM) plays a role in synapse maturation and plasticity (Omar et al. 2017). For example, laminin α5, the major glycoprotein in ECM, binds to postsynaptic Integrin-α3β1 to make ECM. This controls the size and activity pattern of dendritic spines, which in the end leads to normal hippocampal synaptic function. Conversely, further research has confirmed that neural activity also regulates ECM (Padamsey et al. 2017). Increased neural activity induced Ca^{2+} release from the synaptic lysosome, promoting ECM degradation by the secretion of matrix metalloproteinase-9 (MMP-9). This regulation of ECM via neural activity could increase the chance of microglial association with active synapses. Microglia might directly regulate the ECM to control structural synaptic plasticity. Recently, it has been reported that neuronal interleukin 33 (IL-33) induces microglial engulfment of the ECM in the hippocampus (Nguyen et al. 2020b). IL-33 expression in hippocampal neurons is correlated with neuronal activity, and the released IL-33 is locally accumulated near active neurons. IL-33 signaling induces microglial gene expression changes by signaling pathways through the IL-33 receptor, especially upregulating ECM proteases (*Ctsc, Mmp14, Adamts4, Mmp25*), which can lead to degradation and engulfment of ECM by microglia. Reduced dendritic spine density and a defect in hippocampal memory formation in IL-33-depleted mice further confirm the local clearance of ECM and its roles in structural synapse formation. Interestingly, it has also been reported that MMP-9 expression is increased in schizophrenia patients (Dwir et al. 2020), suggesting that ECM regulation by microglia may have gone awry in various neurological disorders.

11.2.6 The Role of Oligodendrocyte Lineage Cells in Synapse Organization

Despite their significant role in myelination, the roles of oligodendrocytes and their precursors (OPCs) in synapse organization remain less studied than those of astrocytes and microglia. One intriguing finding is that BDNF released from oligodendrocytes can regulate glutamate transmission in the brain stem (Jang et al. 2019). Oligodendrocyte-specific deletion of *Bdnf* impairs glutamatergic synapse function, particularly in presynaptic glutamate release. In the *Bdnf* deletion model, administration of TrkB agonists restored presynaptic glutamate release, implying that BDNF-presynaptic TrkB signaling induces neurotransmitter release. Recent intriguing studies have also shown that beyond their canonical role as precursors for myelinating oligodendrocytes, OPCs regulate synaptic function through the secretion of various factors. OPCs in the prefrontal cortex (PFC) secrete fibroblast growth factor 2 (FGF2), which can modulate glutamate signaling of neurons and astrocytes (Birey et al. 2015). Loss of FGF2 signaling from OPCs induced depressive behaviors, suggesting its role in synapse function and impact on behavior.

Along with controlling synaptic function with secreted factors, recent studies have suggested that OPCs can participate in phagocytic elimination of synapses in certain brain regions during postnatal development. In the visual cortex, OPCs eliminate presynapses of the thalamocortical neural circuit during development with unknown phagocytic machineries (Auguste et al. 2022). Phagocytosis of thalamocortical input by OPCs is found to be regulated by the sensory experience of animals, suggesting that thalamocortical neural activity might act as a direct cue for guiding synapse phagocytosis of OPCs.

Furthermore, it has also been increasingly evident that OPCs directly receive both excitatory and inhibitory synaptic inputs from neurons (Bergles et al. 2000; Lin and Bergles 2004; Mount et al. 2019). Although the formation and regulation of synaptic connections between neurons and OPCs need further investigation, it is intriguing to postulate that OPCs may directly recognize and respond to changes in neural activity through the neuron to OPC synaptic connections. Supporting this concept, a distinct population of white-matter OPCs can generate action potential via Na^+ current, which was damaged in the ischemia model (Karadottir et al. 2008). Neuron-to-OPC synapses can also be functionally potentiated, as confirmed with LTP formation through theta-burst stimulation (Ge et al. 2006). These unique characteristics of neuron-to-OPC synapses are presumed to be critical for the maturation and development of not only OPCs but also synapses. Several studies have found that OPCs lose their synaptic input during oligodendrocyte maturation (De Biase et al. 2010). Moreover, optogenetic stimulation of neural activity induced proliferation of OPCs, followed by oligodendrogenesis and myelination (Gibson et al. 2014). In this process, the firing pattern of neural activity can determine the fate of OPCs, indicating that neural activity intricately controls OPC maturation through synaptic communications.

11.3 Conclusions and Future Directions

In the past, neuro-centric views had considered glia cells in the CNS as "glue," mostly emphasizing their structural supportive roles for neurons. However, through the discoveries introduced in this chapter, glia cells are now being recognized as active participants in the development and homeostasis of the brain, particularly in synapse nano-organizations. Glial cells can secrete molecules to induce the formation and maintenance of synaptic structures and functions. Moreover, through phagocytosis and non-phagocytic mechanisms, glial cells actively eliminate synapses for normal circuit function and proper synaptic homeostasis.

However, in spite of these achievements, there are still many unanswered questions. Given that synapse remodeling is very prevalent during developmental periods and neurodegenerative conditions, most of the related studies have been focused on these two cases. Importantly, it has been well suggested that synapse remodeling actually occurs throughout one's lifetime during learning and changes in sensory experiences (Caroni et al. 2014; Magee and Grienberger 2020; Whalley 2014; Wang and Zhang 2008; Whitt et al. 2014), albeit with a relatively reduced ratio compared to developmental stages. Thus, future research will be required to pinpoint the roles of glial cells in mediating constant changes in synapse nano-organization throughout an animal's lifetime including aging stages.

Along with a similar line, roles of glial cells in modulating synapse organization have been studied in the limited brain circuits and regions, such as the visual systems, motor/somatosensory cortex, and hippocampus. Since different types of learning and stimuli have been shown to induce synapse remodeling and plasticity in various brain regions, including the amygdala, striatum, hypothalamus, and cortex (Nabavi et al. 2014; Martin et al. 2000; Zeng et al. 2018; McEwen et al. 2012; Chen et al. 2022b), the current efforts should be expanded to investigate more diverse circuits and brain regions for revealing the novel roles of glial cells in controlling synapse nano-organization.

In this chapter, we introduced the molecular pathways of how glial cells can change the structural aspect of synapses and their organization. However, there is still much to discover regarding the detailed mechanisms of these regulations. For example, how the physical association between glial processes and synapses changes during development, learning and memory formation, and aging processes needs to be worked out more extensively. Glial cells can detect synaptic activity through diverse ligands and receptors and modify their physical association with synapses. Changes in glia-synapse contacting points may occur first, setting up the stage for the subsequent roles of glial cells in either potentiating or de-potentiating target synapses. Recent advances in transcriptomics and proteomics have revealed the cell-type-specific expression of critical molecules between synapses and glial cells (Zhang et al. 2014; Zeisel et al. 2015; Chen et al. 2017; Soto et al. 2023). Investigating how the expression and localization of those molecules change during development, learning and memory formation, and aging would provide broad and comprehensive perspectives in the dynamic regulation of synapse nano-organization by glial cells.

Abnormal changes in synapse numbers and functions have been reported in many brain disorders, such as AD, Parkinson's disease, Huntington's disease, autism spectrum disorder, depression, and schizophrenia (Griffiths and Grant 2023; Rodriguez-Jimenez et al. 2023; Waninger et al. 2020; Nguyen et al. 2019; Ebrahimi-Fakhari and Sahin 2015; Zatkova et al. 2016; Pocklington et al. 2014; Howes and Onwordi 2023). Whether misregulation of glial control of the synapse nano-organization mentioned here can induce the initiation and progression of these various brain disorders would be a fascinating research subject. Previous reports have shown that changes in glial reactiveness and in glial calcium responses can evoke abnormal synaptic function and animal behavior (Sadick et al. 2022; Yun et al. 2018; Olah et al. 2020; Smajic et al. 2022). However, our understanding of the direct mechanisms by which reactive glial cells and changes in calcium responses contribute to the specific disease state requires further investigation.

How synapses are formed, maintained, and eliminated is a fundamental question in understanding the brain. With rapid advances in imaging, sequencing, and data processing technologies, we are now well positioned to gain a comprehensive view of this process. In this quest, we should avoid bias toward any particular cell type, keeping in mind that all cell types in the brain can play critical roles in regulating synaptic structure and function. Over time, these insights may serve as the foundation for developing effective treatments for various neurological disorders.

Acknowledgments This work was supported by the Institute for Basic Science and funded by the Ministry of Science and ICT, Republic of Korea (IBS-R025-A1 (W.-S.C.)), and National Research Foundation of Korea grant funded by the Ministry of Science and ICT, Republic of Korea (2020M3E5D9079912 (W.-S.C.), 2021R1A2C3005704 (W.-S.C.), 2022M3E5E8081188 (W.-S.C.)).

References

Aguzzi A, Barres BA, Bennett ML. Microglia: scapegoat, saboteur, or something else? Science. 2013;339:156–61.

Albrecht D, et al. SPARC prevents maturation of cholinergic presynaptic terminals. Mol Cell Neurosci. 2012;49:364–74.

Allen A, Messier C. Plastic changes in the astrocyte GLUT1 glucose transporter and beta-tubulin microtubule protein following voluntary exercise in mice. Behav Brain Res. 2013;240:95–102.

Allen NJ, et al. Astrocyte glypicans 4 and 6 promote formation of excitatory synapses via GluA1 AMPA receptors. Nature. 2012;486:410–4.

Andoh M, Ikegaya Y, Koyama R. Synaptic pruning by microglia in epilepsy. J Clin Med. 2019;8:2170.

Araque A, Parpura V, Sanzgiri RP, Haydon PG. Tripartite synapses: glia, the unacknowledged partner. Trends Neurosci. 1999;22:208–15. https://doi.org/10.1016/s0166-2236(98)01349-6.

Auguste YS, et al. Oligodendrocyte precursor cells engulf synapses during circuit remodeling in mice. Nat Neurosci. 2022;25:1273–8.

Baldwin KT, et al. HepaCAM controls astrocyte self-organization and coupling. Neuron. 2021;109:2427–2442. e2410.

Basilico B, et al. Microglia shape presynaptic properties at developing glutamatergic synapses. Glia. 2019;67:53–67.

Bennett ML, et al. New tools for studying microglia in the mouse and human CNS. Proc Natl Acad Sci USA. 2016;113:E1738–46. https://doi.org/10.1073/pnas.1525528113.

Bergles DE, Roberts JDB, Somogyi P, Jahr CE. Glutamatergic synapses on oligodendrocyte precursor cells in the hippocampus. Nature. 2000;405:187–91.

Biber K, Owens T, Boddeke E. What is microglia neurotoxicity (Not)? Glia. 2014;62:841–54.

Bie B, Wu J, Foss JF, Naguib M. Activation of mGluR1 mediates C1q-dependent microglial phagocytosis of glutamatergic synapses in Alzheimer's rodent models. Mol Neurobiol. 2019;56:5568–85.

Birey F, et al. Genetic and stress-induced loss of NG2 glia triggers emergence of depressive-like behaviors through reduced secretion of FGF2. Neuron. 2015;88:941–56.

Blanco-Suarez E, Allen NJ. Astrocyte-secreted chordin-like 1 regulates spine density after ischemic injury. Sci Rep. 2022;12:4176.

Blanco-Suarez E, Liu T-F, Kopelevich A, Allen NJ. Astrocyte-secreted chordin-like 1 drives synapse maturation and limits plasticity by increasing synaptic GluA2 AMPA receptors. Neuron. 2018;100:1116–1132. e1113.

Boisvert MM, Erikson GA, Shokhirev MN, Allen NJ. The aging astrocyte transcriptome from multiple regions of the mouse brain. Cell Rep. 2018;22:269–85.

Brown GC, Neher JJ. Microglial phagocytosis of live neurons. Nat Rev Neurosci. 2014;15:209–16.

Byun YG, et al. Stress induces behavioral abnormalities by increasing expression of phagocytic receptor MERTK in astrocytes to promote synapse phagocytosis. Immunity. 2023;56:2105–2120. e2113.

Cahoy JD, et al. A transcriptome database for astrocytes, neurons, and oligodendrocytes: a new resource for understanding brain development and function. J Neurosci. 2008;28:264–78.

Carmona MA, Murai KK, Wang L, Roberts AJ, Pasquale EB. Glial ephrin-A3 regulates hippocampal dendritic spine morphology and glutamate transport. Proc Natl Acad Sci. 2009;106:12524–9.

Caroni P, Chowdhury A, Lahr M. Synapse rearrangements upon learning: from divergent-sparse connectivity to dedicated sub-circuits. Trends Neurosci. 2014;37:604–14. https://doi.org/10.1016/j.tins.2014.08.011.

Chen R, Wu X, Jiang L, Zhang Y. Single-cell RNA-Seq reveals hypothalamic cell diversity. Cell Rep. 2017;18:3227–41. https://doi.org/10.1016/j.celrep.2017.03.004.

Chen G, et al. Hevin/Sparcl1 drives pathological pain through spinal cord astrocyte and NMDA receptor signaling. JCI Insight. 2022a;7:e161028.

Chen M, Qi J, Poo M, Yang Y. Stability and dynamics of dendritic spines in macaque prefrontal cortex. Natl Sci Rev. 2022b;9:nwac125. https://doi.org/10.1093/nsr/nwac125.

Chesler M. Regulation and modulation of pH in the brain. Physiol Rev. 2003;83:1183–221.

Chever O, Djukic B, McCarthy KD, Amzica F. Implication of Kir4. 1 channel in excess potassium clearance: an in vivo study on anesthetized glial-conditional Kir4. 1 knock-out mice. J Neurosci. 2010;30:15769–77.

Christopherson KS, et al. Thrombospondins are astrocyte-secreted proteins that promote CNS synaptogenesis. Cell. 2005;120:421–33.

Chung W-S, et al. Astrocytes mediate synapse elimination through MEGF10 and MERTK pathways. Nature. 2013;504:394–400.

Collyer E, Boyle BR, Gomez-Galvez Y, Iacovitti L, Blanco-Suarez E. Absence of chordin-like 1 aids motor recovery in a mouse model of stroke. Exp Neurol. 2023;370:114548.

Colton CA. Heterogeneity of microglial activation in the innate immune response in the brain. J Neuroimmune Pharmacol. 2009;4:399–418. https://doi.org/10.1007/s11481-009-9164-4.

D'Ambrosio R, Gordon DS, Winn HR. Differential role of KIR channel and Na+/K+-pump in the regulation of extracellular K+ in rat hippocampus. J Neurophysiol. 2002;87:87–102.

De Biase LM, Nishiyama A, Bergles DE. Excitability and synaptic communication within the oligodendrocyte lineage. J Neurosci. 2010;30:3600–11. https://doi.org/10.1523/JNEUROSCI.6000-09.2010.

Deitmer JW, Rose CR. pH regulation and proton signalling by glial cells. Prog Neurobiol. 1996;48:73–103.

Dowling C, Allen NJ. Mice lacking glypican 4 display juvenile hyperactivity and adult social interaction deficits. Brain Plast. 2018;4:197–209.

Dwir D, et al. MMP9/RAGE pathway overactivation mediates redox dysregulation and neuroinflammation, leading to inhibitory/excitatory imbalance: a reverse translation study in schizophrenia patients. Mol Psychiatry. 2020;25:2889–904.

Ebrahimi-Fakhari D, Sahin M. Autism and the synapse: emerging mechanisms and mechanism-based therapies. Curr Opin Neurol. 2015;28:91–102. https://doi.org/10.1097/WCO.0000000000000186.

Egawa K, Yamada J, Furukawa T, Yanagawa Y, Fukuda A. Cl− homeodynamics in gap junction-coupled astrocytic networks on activation of GABAergic synapses. J Physiol. 2013;591:3901–17.

Eroglu C, et al. Gabapentin receptor α2δ-1 is a neuronal thrombospondin receptor responsible for excitatory CNS synaptogenesis. Cell. 2009;139:380–92.

Falkowska A, et al. Energy metabolism of the brain, including the cooperation between astrocytes and neurons, especially in the context of glycogen metabolism. Int J Mol Sci. 2015;16:25959–81.

Fan W, et al. Involvement of microglial activation in the brainstem in experimental dental injury and inflammation. Arch Oral Biol. 2010;55:706–11.

Farhy-Tselnicker I, et al. Astrocyte-secreted glypican 4 regulates release of neuronal pentraxin 1 from axons to induce functional synapse formation. Neuron. 2017;96:428–445. e413.
Farhy-Tselnicker I, et al. Activity-dependent modulation of synapse-regulating genes in astrocytes. elife. 2021;10:e70514.
Filipello F, et al. The microglial innate immune receptor TREM2 is required for synapse elimination and normal brain connectivity. Immunity. 2018;48:979–991. e978.
Filosa A, et al. Neuron-glia communication via EphA4/ephrin-A3 modulates LTP through glial glutamate transport. Nat Neurosci. 2009;12:1285–92.
Garrett AM, Weiner JA. Control of CNS synapse development by γ-protocadherin-mediated astrocyte–neuron contact. J Neurosci. 2009;29:11723–31.
Garrett AM, Schreiner D, Lobas MA, Weiner JA. γ-protocadherins control cortical dendrite arborization by regulating the activity of a FAK/PKC/MARCKS signaling pathway. Neuron. 2012;74:269–76.
Ge WP, et al. Long-term potentiation of neuron-glia synapses mediated by Ca2+-permeable AMPA receptors. Science. 2006;312:1533–7. https://doi.org/10.1126/science.1124669.
Gibson EM, et al. Neuronal activity promotes oligodendrogenesis and adaptive myelination in the mammalian brain. Science. 2014;344:1252304.
Grichtchenko I, Chesler M. Depolarization-induced alkalinization of astrocytes in gliotic hippocampal slices. Neuroscience. 1994;62:1071–8.
Griffiths J, Grant SG. Synapse pathology in Alzheimer's disease. Semin Cell Dev Biol. 2023;139:13–23.
Gunner G, et al. Sensory lesioning induces microglial synapse elimination via ADAM10 and fractalkine signaling. Nat Neurosci. 2019;22:1075–88.
Halassa MM, Fellin T, Haydon PG. Tripartite synapses: roles for astrocytic purines in the control of synaptic physiology and behavior. Neuropharmacology. 2009;57:343–6. https://doi.org/10.1016/j.neuropharm.2009.06.031.
Halestrap AP. The monocarboxylate transporter family—structure and functional characterization. IUBMB Life. 2012;64:1–9.
Hansen DB, Garrido-Comas N, Salter M, Fern R. HCO3−-independent pH regulation in astrocytes in situ is dominated by V-ATPase. J Biol Chem. 2015;290:8039–47.
Harris KM, Kuwajima M, Flores JC, Zito K. Synapse-specific structural plasticity that protects and refines local circuits during LTP and LTD. Philos Trans B. 2024;379:20230224.
Hayashi-Takagi A, et al. Labelling and optical erasure of synaptic memory traces in the motor cortex. Nature. 2015;525:333–8.
Haynes SE, et al. The P2Y12 receptor regulates microglial activation by extracellular nucleotides. Nat Neurosci. 2006;9:1512–9.
Hennekinne L, Colasse S, Triller A, Renner M. Differential control of thrombospondin over synaptic glycine and AMPA receptors in spinal cord neurons. J Neurosci. 2013;33:11432–9.
Hong S, et al. Complement and microglia mediate early synapse loss in Alzheimer mouse models. Science. 2016;352:712–6.
Howes OD, Onwordi EC. The synaptic hypothesis of schizophrenia version III: a master mechanism. Mol Psychiatry. 2023;28:1843–56. https://doi.org/10.1038/s41380-023-02043-w.
Huang K, Park S. Heparan sulfated glypican-4 is released from astrocytes by proteolytic shedding and GPI-anchor cleavage mechanisms. Eneuro. 2021;8:ENEURO.0069.
Janda E, Boi L, Carta AR. Microglial phagocytosis and its regulation: a therapeutic target in Parkinson's disease? Front Mol Neurosci. 2018;11:144.
Jang M, Gould E, Xu J, Kim EJ, Kim JH. Oligodendrocytes regulate presynaptic properties and neurotransmission through BDNF signaling in the mouse brainstem. Elife. 2019;8:e42156.
Jones EV, et al. Astrocytes control glutamate receptor levels at developing synapses through SPARC–β-integrin interactions. J Neurosci. 2011;31:4154–65.
Jones EV, Bernardinelli Y, Zarruk JG, Chierzi S, Murai KK. SPARC and GluA1-containing AMPA receptors promote neuronal health following CNS injury. Front Cell Neurosci. 2018;12:22.

Karadottir R, Hamilton NB, Bakiri Y, Attwell D. Spiking and nonspiking classes of oligodendrocyte precursor glia in CNS white matter. Nat Neurosci. 2008;11:450–6. https://doi.org/10.1038/nn2060.

Kettenmann H, Backus K, Schachner M. γ-Aminobutyric acid opens cl-channels in cultured astrocytes. Brain Res. 1987;404:1–9.

Kim D-S, et al. Thrombospondin-4 contributes to spinal sensitization and neuropathic pain states. J Neurosci. 2012;32:8977–87.

Koeppen J, et al. Functional consequences of synapse remodeling following astrocyte-specific regulation of Ephrin-B1 in the adult hippocampus. J Neurosci. 2018;38:5710–26.

Kucukdereli H, et al. Control of excitatory CNS synaptogenesis by astrocyte-secreted proteins Hevin and SPARC. Proc Natl Acad Sci U S A. 2011;108:E440–9.

Larsen BR, et al. Contributions of the Na+/K+-ATPase, NKCC1, and Kir4. 1 to hippocampal K+ clearance and volume responses. Glia. 2014;62:608–22.

Lee J-H, et al. Astrocytes phagocytose adult hippocampal synapses for circuit homeostasis. Nature. 2021;590:612–7.

Lehrman EK, et al. CD47 protects synapses from excess microglia-mediated pruning during development. Neuron. 2018;100:120–134. e126.

Lemke G. Biology of the TAM receptors. Cold Spring Harb Perspect Biol. 2013;5:a009076.

Lendvai B, Stern EA, Chen B, Svoboda K. Experience-dependent plasticity of dendritic spines in the developing rat barrel cortex in vivo. Nature. 2000;404:876–81.

Li T, et al. A splicing isoform of GPR56 mediates microglial synaptic refinement via phosphatidylserine binding. EMBO J. 2020;39:e104136. https://doi.org/10.15252/embj.2019104136.

Liddelow SA, et al. Neurotoxic reactive astrocytes are induced by activated microglia. Nature. 2017;541:481–7.

Lim S-H, et al. Neuronal synapse formation induced by microglia and interleukin 10. PLoS One. 2013;8:e81218.

Lin SC, Bergles DE. Synaptic signaling between GABAergic interneurons and oligodendrocyte precursor cells in the hippocampus. Nat Neurosci. 2004;7:24–32. https://doi.org/10.1038/nn1162.

Lindholm D, Castren E, Kiefer R, Zafra F, Thoenen H. Transforming growth factor-beta 1 in the rat brain: increase after injury and inhibition of astrocyte proliferation. J Cell Biol. 1992;117:395–400.

Liu A, et al. Neuroligin 1 regulates spines and synaptic plasticity via LIMK1/cofilin-mediated actin reorganization. J Cell Biol. 2016;212:449–63.

López-Murcia FJ, Terni B, Llobet A. SPARC triggers a cell-autonomous program of synapse elimination. Proc Natl Acad Sci. 2015;112:13366–71.

Lou N, et al. Purinergic receptor P2RY12-dependent microglial closure of the injured blood–brain barrier. Proc Natl Acad Sci. 2016;113:1074–9.

MacDonald JM, et al. The drosophila cell corpse engulfment receptor Draper mediates glial clearance of severed axons. Neuron. 2006;50:869–81.

Magee JC, Grienberger C. Synaptic plasticity forms and functions. Annu Rev Neurosci. 2020;43:95–117. https://doi.org/10.1146/annurev-neuro-090919-022842.

Martin SJ, Grimwood PD, Morris RG. Synaptic plasticity and memory: an evaluation of the hypothesis. Annu Rev Neurosci. 2000;23:649–711. https://doi.org/10.1146/annurev.neuro.23.1.649.

McEwen BS, Eiland L, Hunter RG, Miller MM. Stress and anxiety: structural plasticity and epigenetic regulation as a consequence of stress. Neuropharmacology. 2012;62:3–12. https://doi.org/10.1016/j.neuropharm.2011.07.014.

Michailidou I, et al. Complement C 1q-C 3–associated synaptic changes in multiple sclerosis hippocampus. Ann Neurol. 2015;77:1007–26.

Minelli A, DeBiasi S, Brecha NC, Zuccarello LV, Conti F. GAT-3, a high-affinity GABA plasma membrane transporter, is localized to astrocytic processes, and it is not confined to the vicinity of GABAergic synapses in the cerebral cortex. J Neurosci. 1996;16:6255–64.

Miyamoto A, et al. Microglia contact induces synapse formation in developing somatosensory cortex. Nat Commun. 2016;7:12540.

Molumby MJ, Keeler AB, Weiner JA. Homophilic protocadherin cell-cell interactions promote dendrite complexity. Cell Rep. 2016;15:1037–50.

Mount CW, Yalcin B, Cunliffe-Koehler K, Sundaresh S, Monje M. Monosynaptic tracing maps brain-wide afferent oligodendrocyte precursor cell connectivity. elife. 2019;8 https://doi.org/10.7554/eLife.49291.

Moussawi K, Riegel A, Nair S, Kalivas PW. Extracellular glutamate: functional compartments operate in different concentration ranges. Front Syst Neurosci. 2011;5:94.

Murai KK, Nguyen LN, Irie F, Yamaguchi Y, Pasquale EB. Control of hippocampal dendritic spine morphology through ephrin-A3/EphA4 signaling. Nat Neurosci. 2003;6:153–60.

Nabavi S, et al. Engineering a memory with LTD and LTP. Nature. 2014;511:348–52. https://doi.org/10.1038/nature13294.

Nguyen M, Wong YC, Ysselstein D, Severino A, Krainc D. Synaptic, mitochondrial, and lysosomal dysfunction in Parkinson's disease. Trends Neurosci. 2019;42:140–9. https://doi.org/10.1016/j.tins.2018.11.001.

Nguyen AQ, et al. Astrocytic ephrin-B1 controls synapse formation in the hippocampus during learning and memory. Front Synaptic Neurosci. 2020a;12:10.

Nguyen PT, et al. Microglial remodeling of the extracellular matrix promotes synapse plasticity. Cell. 2020b;182:388–403. e315.

Nimmerjahn A, Kirchhoff F, Helmchen F. Resting microglial cells are highly dynamic surveillants of brain parenchyma in vivo. Science. 2005;308:1314–8.

Nishida H, Okabe S. Direct astrocytic contacts regulate local maturation of dendritic spines. J Neurosci. 2007;27:331–40.

Olah M, et al. Single cell RNA sequencing of human microglia uncovers a subset associated with Alzheimer's disease. Nat Commun. 2020;11:6129. https://doi.org/10.1038/s41467-020-19737-2.

Omar MH, et al. CNS neurons deposit laminin α5 to stabilize synapses. Cell Rep. 2017;21:1281–92.

Padamsey Z, et al. Activity-dependent exocytosis of lysosomes regulates the structural plasticity of dendritic spines. Neuron. 2017;93:132–46.

Paolicelli RC, et al. Synaptic pruning by microglia is necessary for normal brain development. Science. 2011;333:1456–8.

Park J, et al. Microglial MERTK eliminates phosphatidylserine-displaying inhibitory postsynapses. EMBO J. 2021;40:e107121.

Parkhurst CN, et al. Microglia promote learning-dependent synapse formation through brain-derived neurotrophic factor. Cell. 2013;155:1596–609.

Pellerin L, Halestrap AP, Pierre K. Cellular and subcellular distribution of monocarboxylate transporters in cultured brain cells and in the adult brain. J Neurosci Res. 2005;79:55–64.

Pfrieger FW, Barres BA. Synaptic efficacy enhanced by glial cells in vitro. Science. 1997;277:1684–7. https://doi.org/10.1126/science.277.5332.1684.

Pocklington AJ, O'Donovan M, Owen MJ. The synapse in schizophrenia. Eur J Neurosci. 2014;39:1059–67. https://doi.org/10.1111/ejn.12489.

Podleśny-Drabiniok A, Marcora E, Goate AM. Microglial phagocytosis: a disease-associated process emerging from Alzheimer's disease genetics. Trends Neurosci. 2020;43:965–79.

Rana I, et al. Microglia activation in the hypothalamic PVN following myocardial infarction. Brain Res. 2010;1326:96–104.

Rasley A, Tranguch SL, Rati DM, Marriott I. Murine glia express the immunosuppressive cytokine, interleukin-10, following exposure to Borrelia burgdorferi or Neisseria meningitidis. Glia. 2006;53:583–92.

Reshef R, et al. The role of microglia and their CX3CR1 signaling in adult neurogenesis in the olfactory bulb. elife. 2017;6:e30809.

Ribic A, Crair MC, Biederer T. Synapse-selective control of cortical maturation and plasticity by parvalbumin-autonomous action of SynCAM 1. Cell Rep. 2019;26:381-393. e386.

Risher WC, et al. Astrocytes refine cortical connectivity at dendritic spines. elife. 2014;3:e04047.

Risher WC, et al. Thrombospondin receptor α2δ-1 promotes synaptogenesis and spinogenesis via postsynaptic Rac1. J Cell Biol. 2018;217:3747–65.

Rodriguez-Jimenez FJ, Urena-Peralta J, Jendelova P, Erceg S. Alzheimer's disease and synapse loss: what can we learn from induced pluripotent stem cells? J Adv Res. 2023;54:105–18. https://doi.org/10.1016/j.jare.2023.01.006.

Rueda-Carrasco J, et al. Author correction: microglia-synapse engulfment via PtdSer-TREM2 ameliorates neuronal hyperactivity in Alzheimer's disease models. EMBO J. 2024;43:3807–11.

Russo SJ, et al. The addicted synapse: mechanisms of synaptic and structural plasticity in nucleus accumbens. Trends Neurosci. 2010;33:267–76.

Sadick JS, et al. Astrocytes and oligodendrocytes undergo subtype-specific transcriptional changes in Alzheimer's disease. Neuron. 2022;110:1788–1805 e1710. https://doi.org/10.1016/j.neuron.2022.03.008.

Sanchez-Mico MV, et al. Amyloid-β impairs the phagocytosis of dystrophic synapses by astrocytes in Alzheimer's disease. Glia. 2021;69:997–1011.

Schafer DP, et al. Microglia sculpt postnatal neural circuits in an activity and complement-dependent manner. Neuron. 2012;74:691–705.

Scimemi A. Structure, function, and plasticity of GABA transporters. Front Cell Neurosci. 2014;8:161.

Scott-Hewitt N, et al. Local externalization of phosphatidylserine mediates developmental synaptic pruning by microglia. EMBO J. 2020;39:e105380.

Seifert G, et al. Analysis of astroglial K+ channel expression in the developing hippocampus reveals a predominant role of the Kir4. 1 subunit. J Neurosci. 2009;29:7474–88.

Sellgren CM, et al. Increased synapse elimination by microglia in schizophrenia patient-derived models of synaptic pruning. Nat Neurosci. 2019;22:374–85.

Sık A, Smith R, Freund T. Distribution of chloride channel-2-immunoreactive neuronal and astrocytic processes in the hippocampus. Neuroscience. 2000;101:51–65.

Smajic S, et al. Single-cell sequencing of human midbrain reveals glial activation and a Parkinson-specific neuronal state. Brain. 2022;145:964–78. https://doi.org/10.1093/brain/awab446.

Smith HS. Activated microglia in nociception. Pain Physician. 2010;13:295.

Soto JS, et al. Astrocyte-neuron subproteomes and obsessive-compulsive disorder mechanisms. Nature. 2023;616:764–73. https://doi.org/10.1038/s41586-023-05927-7.

Stein IS, Zito K. Dendritic spine elimination: molecular mechanisms and implications. Neuroscientist. 2019;25:27–47.

Stevens B, et al. The classical complement cascade mediates CNS synapse elimination. Cell. 2007;131:1164–78.

Stogsdill JA, et al. Astrocytic neuroligins control astrocyte morphogenesis and synaptogenesis. Nature. 2017;551:192–7.

Südhof TC. Synaptic neurexin complexes: a molecular code for the logic of neural circuits. Cell. 2017;171:745–69.

Sun W, et al. SOX9 is an astrocyte-specific nuclear marker in the adult brain outside the neurogenic regions. J Neurosci. 2017;37:4493–507.

Takano T, et al. Chemico-genetic discovery of astrocytic control of inhibition in vivo. Nature. 2020;588:296–302.

Thored P, et al. Long-term accumulation of microglia with proneurogenic phenotype concomitant with persistent neurogenesis in adult subventricular zone after stroke. Glia. 2009;57:835–49.

Tropea D, Majewska AK, Garcia R, Sur M. Structural dynamics of synapses in vivo correlate with functional changes during experience-dependent plasticity in visual cortex. J Neurosci. 2010;30:11086–95.

Trotter JH, et al. Astrocytic neurexin-1 orchestrates functional synapse assembly. bioRxiv. 2020;

Ullian EM, Sapperstein SK, Christopherson KS, Barres BA. Control of synapse number by glia. Science. 2001;291:657–61. https://doi.org/10.1126/science.291.5504.657.

Vandenberg RJ, Ryan RM. Mechanisms of glutamate transport. Physiol Rev. 2013;93:1621–57.

Wake H, Moorhouse AJ, Jinno S, Kohsaka S, Nabekura J. Resting microglia directly monitor the functional state of synapses in vivo and determine the fate of ischemic terminals. J Neurosci. 2009;29:3974–80.

Wang H, Zhang ZW. A critical window for experience-dependent plasticity at whisker sensory relay synapse in the thalamus. J Neurosci. 2008;28:13621–8. https://doi.org/10.1523/JNEUROSCI.4785-08.2008.

Waninger S, et al. Neurophysiological biomarkers of Parkinson's disease. J Parkinsons Dis. 2020;10:471–80. https://doi.org/10.3233/JPD-191844.

Whalley K. Learning and memory: synapse remodelling extinguishes fear. Nat Rev Neurosci. 2014;15:3. https://doi.org/10.1038/nrn3658.

Whitt JL, Petrus E, Lee HK. Experience-dependent homeostatic synaptic plasticity in neocortex. Neuropharmacology. 2014;78:45–54. https://doi.org/10.1016/j.neuropharm.2013.02.016.

Wu Y, Singh S, Georgescu M-M, Birge RB. A role for Mer tyrosine kinase in αvβ5 integrin-mediated phagocytosis of apoptotic cells. J Cell Sci. 2005;118:539–53.

Wyatt SK, Witt T, Barbaro NM, Cohen-Gadol AA, Brewster AL. Enhanced classical complement pathway activation and altered phagocytosis signaling molecules in human epilepsy. Exp Neurol. 2017;295:184–93.

Xu T, et al. Rapid formation and selective stabilization of synapses for enduring motor memories. Nature. 2009;462:915–9.

Xu J, Xiao N, Xia J. Thrombospondin 1 accelerates synaptogenesis in hippocampal neurons through neuroligin 1. Nat Neurosci. 2010;13:22–4.

Yang G, Pan F, Gan W-B. Stably maintained dendritic spines are associated with lifelong memories. Nature. 2009;462:920–4.

Yun SP, et al. Block of A1 astrocyte conversion by microglia is neuroprotective in models of Parkinson's disease. Nat Med. 2018;24:931–8. https://doi.org/10.1038/s41591-018-0051-5.

Zatkova M, Bakos J, Hodosy J, Ostatnikova D. Synapse alterations in autism: review of animal model findings. Biomed Pap Med Fac Univ Palacky Olomouc Czech Repub. 2016;160:201–10. https://doi.org/10.5507/bp.2015.066.

Zeisel A, et al. Brain structure. Cell types in the mouse cortex and hippocampus revealed by single-cell RNA-seq. Science. 2015;347:1138–42. https://doi.org/10.1126/science.aaa1934.

Zeng M, et al. Reconstituted postsynaptic density as a molecular platform for understanding synapse formation and plasticity. Cell. 2018;174:1172–1187. e1116. https://doi.org/10.1016/j.cell.2018.06.047.

Zhang Y, et al. An RNA-sequencing transcriptome and splicing database of glia, neurons, and vascular cells of the cerebral cortex. J Neurosci. 2014;34:11929–47.

Zhou Y, Danbolt NC. GABA and glutamate transporters in brain. Front Endocrinol. 2013;4:165.

Zhou Z, Hartwieg E, Horvitz HR. CED-1 is a transmembrane receptor that mediates cell corpse engulfment in C. Elegans. Cell. 2001;104:43–56.

Correction to: Recent Structural Insights into the Molecular Architecture of Synapses

Axel T. Brunger, Richard G. Held, Yousuf A. Khan, Jeremy Leitz, Jiahao Liang, Chuchu Wang, and K. Ian White

Correction to:
Chapter 2 in: E. T. Kavalali (ed.), *Nano-organization of the Synapse,* Advances in Neurobiology 48, https://doi.org/10.1007/978-3-032-12594-1_2

In Chapter 2, Figure 1 was incorrectly printed. The correct image is provided below:

The updated version of this chapter can be found at
https://doi.org/10.1007/978-3-032-12594-1_2

E. T. Kavalali (ed.), *Nano-organization of the Synapse*, Advances in Neurobiology 48,
https://doi.org/10.1007/978-3-032-12594-1_12

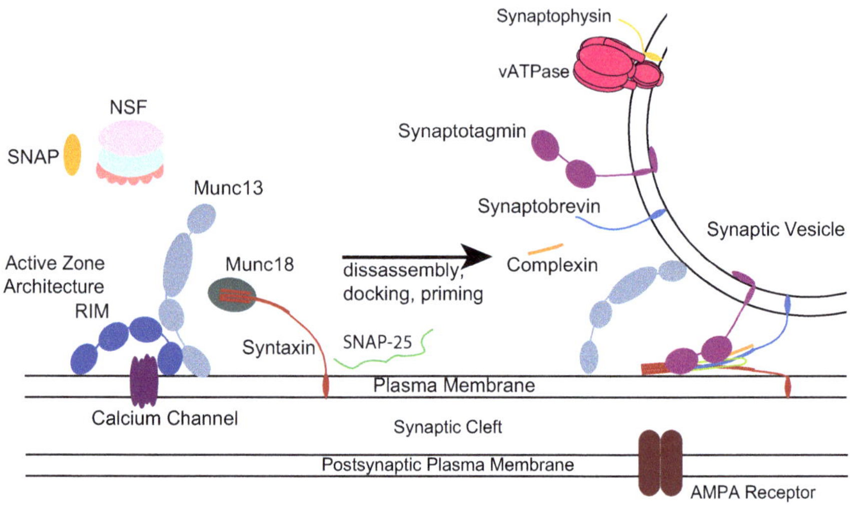

Fig. 2.1 A simplified schema of the presynaptic neurotransmitter release machinery that is discussed in this chapter. Shown are a subset of presynaptic proteins (RIMs, Munc13, Munc18, P/Q- or N-type Ca^{2+}-channels), SNAREs (SNAP-25, syntaxin, and synaptobrevin), complexin, synaptotagmin, NSF, and α-SNAP. Also shown are the synaptic vesicle V-ATPase and synaptophysin, as well as the AMPA receptor in the postsynaptic membrane

Index

A

Adhesion GPCRs (aGPCRs), 111–114, 117, 119–121, 132, 134, 154, 166–169, 172, 173, 180

Alternative splicing, 108, 112, 117, 122, 129, 155, 158, 160, 171, 174, 175, 178–180

AMPA receptors (AMPARs), 12, 29, 30, 124, 126–131, 193, 197, 199, 204–208, 226–229, 232, 234, 235, 268, 310, 312–314, 316, 334–336

Astrocytes, 265, 332–340, 344

Asynchronous neurotransmitter release, 108, 208, 218–220, 222, 223, 225, 233–235, 238–239, 245

C

Cell adhesion molecules (CAMs), 5, 80, 81, 84, 95, 97, 108, 111, 112, 116, 123, 125–129, 131–135, 152–158, 164, 166, 174, 176, 179, 180, 207, 229, 312, 336

Complexins, 12, 16, 17, 22, 26, 46, 80, 95, 220

D

Dense projections (DPs), 83, 88, 92

E

Electron microscopy (EM), 24, 26, 39, 40, 44, 80, 81, 83, 88–92, 95–97, 155, 168, 194–197, 202, 205, 261, 274, 290, 291, 293–295, 298, 301, 307–310, 312, 313

Electron tomography, 81, 87, 152, 179, 290–318

F

Fast neurotransmission, 1–4

I

Ionotropic glutamate receptors, 313

L

Latrophilins, 111, 119, 154, 166–174, 179

Localization microscopy, 96, 266

M

Microglia, 332, 333, 339–344

Multidisciplinary research, 2, 65

Munc13, 12, 14, 16, 17, 19, 20, 23, 24, 30, 44, 46, 82, 84, 88, 91–93, 95, 127, 193, 206, 207, 228, 236, 243, 270

Munc18, 12, 21, 23, 24, 44, 46, 268

N

Nanoclusters, 30, 84, 86, 91, 95–97, 118, 125, 127, 129, 130, 132, 192–209, 227, 243, 267, 268, 315–317

Nanocolumns, 30, 84, 97, 126–130, 132, 134, 193–194, 199–204, 206–208, 228–230, 232, 236, 246, 267, 276, 314, 316, 317

Nanodomains, 20, 21, 46, 59, 125–135, 192–209, 221, 223–225, 227–229, 235, 268, 313–316

Nanoscale organizations, 6, 81, 128, 132, 192, 226, 244, 246, 265, 270, 276, 311–317

N-ethylmaleimide sensitive factor (NSF), 12, 20, 219

E. T. Kavalali (ed.), *Nano-organization of the Synapse*, Advances in Neurobiology 48,
https://doi.org/10.1007/978-3-032-12594-1

Neurotransmitter release, 4, 5, 7, 12, 18, 40, 43, 44, 46–48, 57–63, 97, 108–110, 192, 201, 219–221, 223, 224, 226, 229, 231, 235, 241, 246, 262, 307–309, 317, 318, 344

O

Oligodendrocyte precursor cells (OPCs), 332, 333, 344, 345

P

Phase condensates, 81, 92–97
Photobleaching, 93, 225, 230, 231, 240, 244, 245, 265, 270, 274, 275, 277
Positron emission tomography (PET) imaging, 4–5
Postsynaptic density (PSD), 29, 30, 44, 80, 87, 96, 110, 122, 125, 126, 128, 129, 174, 179, 195–203, 205–207, 209, 219, 226, 228, 230, 234, 236, 242, 279, 290–318
Postsynaptic scaffold proteins, 199, 314, 315
Presynaptic calcium sources, 58
Presynaptic ultrastructure, 310

R

Rapid antidepressant action, 1, 4, 246
Release site organization, 47, 51

S

Signaling, 2–7, 12, 47, 50–52, 58, 62, 94, 108–111, 113, 114, 116, 117, 119–121, 132–135, 152–154, 157, 166, 169, 172, 173, 179, 180, 199, 200, 207–209, 218, 219, 221, 223–226, 229, 232–239, 243, 244, 246, 303, 310, 312–314, 316, 332, 333, 335, 336, 341, 342, 344
Single-particle tracking (SPT), 96, 263, 268, 271–274, 277, 313
Slow neurotransmission, 1
SNAREs, 12, 42, 80, 109, 219, 268
Soluble NSF attachment proteins (SNAPs), 12, 20–22, 44, 220
Spontaneous neurotransmitter release, 64, 225
Stimulated emission depletion microscopy (STED), 82, 83, 85, 86, 89–92, 95, 96, 132, 197, 201, 268–271, 276
Stochastic optical reconstruction microscopy (STORM), 82–84, 86, 91, 92, 96, 129, 131, 132, 230, 263, 266–268, 276, 277
Structured illumination microscopy (SIM), 263–265, 276, 277
Super-resolution imaging, 5–7, 132, 193–195, 197, 208, 245, 268, 276–280, 313, 314
Super-resolution microscopy (SRM), 7, 81, 84, 86, 95, 97, 118, 125, 126, 128, 130–132, 134, 205, 208, 226, 235, 236, 244, 261–269, 311, 316, 318
Synapse, 1, 41, 79, 107, 151, 218, 260, 290, 332
Synapse elimination, 338–342
Synapse formation, 44, 108, 154, 222, 334
Synapse nano-organization, 1, 332–347
Synapse specificity, 116, 133, 134, 232
Synaptic nanostructures, 7, 132, 133, 208, 209, 229, 236, 260, 261, 263, 267, 279
Synaptic transmissions, 2, 4, 5, 7, 12, 40, 43, 44, 60, 61, 109, 110, 129, 134, 193–194, 200, 204, 209, 217, 218, 262, 308, 309, 313, 314, 317
Synaptic vesicle fusion, 12, 16, 25, 30, 40, 44, 46, 49, 61, 62, 64, 80, 219, 226, 229, 268, 308, 310
Synaptic vesicle pools, 5, 7, 39, 50–52, 55, 57, 58, 61–65, 219, 222, 229, 242, 243
Synaptic vesicle recycling, 52, 61
Synaptic vesicles, 5, 39, 80, 109, 154, 193, 217, 260, 303
Synaptotagmins, 12, 16–18, 25, 40, 46, 48, 80, 95, 223, 224
Synchronous neurotransmitter release, 208, 218, 220, 222, 223, 229, 233, 234, 245

T

Transmission electron microscopy (TEM), 290–300, 304, 307, 308, 310–311
Trans-synaptic alignment, 30, 50, 193, 203, 204, 207, 229, 234, 316
Trans-synaptic signaling, 109, 111

MIX
Papier aus verantwortungsvollen Quellen
Paper from responsible sources
FSC® C105338

If you have any concerns about our products,
you can contact us on
ProductSafety@springernature.com

In case Publisher is established outside the EU,
the EU authorized representative is:
Springer Nature Customer Service Center GmbH
Europaplatz 3, 69115 Heidelberg, Germany

Printed by Libri Plureos GmbH
in Hamburg, Germany